CHIRAL SEPARATION METHODS FOR PHARMACEUTICAL AND BIOTECHNOLOGICAL PRODUCTS

CHIRAL SEPARATION METHODS FOR PHARMACEUTICAL AND BIOTECHNOLOGICAL PRODUCTS

Edited by

Satinder Ahuja
Ahuja Consulting
Calabash, North Carolina

WILEY
A JOHN WILEY & SONS, INC., PUBLICATION

Published by John Wiley & Sons, Inc., Hoboken, New Jersey.
Published simultaneously in Canada.

Library of Congress Cataloging-in-Publication Data:

Ahuja, Satinder, 1933–
Chiral separation methods for pharmaceutical and biotechnological products / Satinder Ahuja.
p. cm.
Includes index.
ISBN 978-0-470-40691-5 (cloth)
1. Chiral drugs–Separation. 2. Enantiomers–Separation. I. Title.
RS429.A38 2011
615′.19–dc22

2009049261

Printed in Singapore

10 9 8 7 6 5 4 3 2 1

CONTENTS

CONTRIBUTORS

Satinder Ahuja, Ahuja Consulting, Calabash, North Carolina

Clinton W. Amoss, Chiral Technologies, Inc., West Chester, Pennsylvania

Hasret Ates, Department of Analytical Chemistry and Pharmaceutical Technology, Vrije Universiteit Brussel–VUB, Brussels, Belgium

Mirlinda Biba, Separation and Purification Center of Excellence, Merck Research Laboratories, Rahway, New Jersey

Ricardo E. Borjas, Analytical Development, Vertex Pharmaceuticals, Inc., Cambridge, Massachusetts

William J. Buttner, International Center for Sensor Science and Engineering, Illinois Institute of Technology, Chicago, Illinois; currently at Hydrogen Technologies and Systems Center, National Renewable Energy Laboratory, Golden, Colorado

Cheng-yi Chen, Separation and Purification Center of Excellence, Merck Research Laboratories, Rahway, New Jersey

Robert DePianta, Analytical Chemistry and Sample Logistics, Pfizer Global Research and Development, Groton, Connecticut

Kenneth Douville, Analytical Development, Vertex Pharmaceuticals, Inc., Cambridge, Massachusetts

Fred Fleitz, Separation and Purification Center of Excellence, Merck Research Laboratories, Rahway, New Jersey

Derek H. Henderson, Separation and Purification Center of Excellence, Merck Research Laboratories, Rahway, New Jersey

Amude Kassim, Separation and Purification Center of Excellence, Merck Research Laboratories, Rahway, New Jersey

William R. Leonard, Separation and Purification Center of Excellence, Merck Research Laboratories, Rahway, New Jersey

Norbert M. Maier, Chiral Technologies, Inc., West Chester, Pennsylvania

Debby Mangelings, Department of Analytical Chemistry and Pharmaceutical Technology, Vrije Universiteit Brussel–VUB, Brussels, Belgium

Beverly Nickerson, Analytical Development, Pfizer Global Research and Development, Groton, Connecticut

Jian G. Ning, Schering Plough Research Institute, Union, New Jersey

Peter Sajonz, Separation and Purification Center of Excellence, Merck Research Laboratories, Rahway, New Jersey

Volker Schurig, Institute of Organic Chemistry, University of Tübingen, Tübingen, Germany

Yvan Vander Heyden, Department of Analytical Chemistry and Pharmaceutical Technology, Vrije Universiteit Brussel–VUB, Brussels, Belgium

Ziqiang Wang, TharSFC, A Waters Company, Pittsburgh, Pennsylvania

Gregory K. Webster, Global Analytical Research and Development, Abbott Laboratories, Abbott Park, Illinois

Christopher J. Welch, Separation and Purification Center of Excellence, Merck Research Laboratories, Rahway, New Jersey

Yun K. Ye, Analytical Research and Development, Bristol-Myers Squibb Company, New Brunswick, New Jersey

Mike Zacuto, Separation and Purification Center of Excellence, Merck Research Laboratories, Rahway, New Jersey

Robert L. Zeid, TLI Development, Oak Island, North Carolina

PREFACE

Enantiomers or *chiral* molecules (from the Greek word *cheiro*, meaning "hand"; that is, they are like a pair of hands) relate to each other as an object and its mirror image. This "handedness" of small and large molecules has created a lot of interest in the pharmaceutical and biotechnology industries because they can have different pharmacologic, metabolic, and/or toxicologic activities.

The handedness of the molecules relates to the difference in spatial arrangements of atoms in a molecule. *Stereoisomers* are molecules that are isomeric but have a different spatial arrangement. Symmetry classifies stereoisomers as either enantiomers or *diastereomers*. There are two molecular sources of chirality: molecules that have a stereogenic center and those that have a stereogenic axis. Stereoisomerism is also possible in molecules that have one or more centers of chirality, helicity, planar/axial/torsional chirality, or topologic asymmetry.

The 1960s public health catastrophe brought about by the use of thalidomide reinforced our thinking on the need for regulatory controls, since one isomer can produce a desired effect whereas the other may produce an undesired effect. In 1992, the U.S. Food and Drug Administration issued a policy statement for the development of new stereoisomeric drugs, where the question of stereochemistry was approached directly. To ensure that similar problems are not encountered in the future, the guidelines emphasized the importance of separating and isolating the isomers so that appropriate pharmacologic, metabolic, and/or toxicologic studies could be conducted. Chiral separations entail the most intriguing, and at times difficult, separations of chemical compounds in that the molecules to be separated have the same molecular weight and physical and chemical properties, except for the rotation of polarized light. The molecules with (+) rotation are called *dextrorotatory* and those with (−) rotation are called *levorotatory*. An accurate evaluation of the isomeric purity of active drug substances is critical because the impurities may be carried through the synthesis, preferentially react at one or more steps, and produce an undesirable level of another impurity.

This book provides valuable information on chiral separations of pharmaceuticals and biotechnology products by:

- Covering a variety of modern methods, such as gas chromatography (GC), high-performance liquid chromatography (HPLC), supercritical fluid chromatography (SFC), and capillary electrophoresis (CE).
- Dealing with the impact of chirality on biological activity of small and large molecules.

- Providing detailed information on useful chiral stationary phases (CSPs) for HPLC.
- Including handy information on selection of an appropriate CSP based on mechanistic studies.
- Offering strategies for fast method development with GC, HPLC, SFC, and CE.
- Discussing preparatory methods utilized in the pharmaceutical industry.

The first three chapters in this book provide a broad overview of chiral separations, regulatory considerations in drug product development, and basic considerations in method development.

Chapters 4 to 6 discuss the development of popular polysaccharide columns, various techniques that can be used for separations on them, and mechanistic studies on chiral separations to help us understand how these columns operate so that we can develop better methods with them. Chapters 7 and 8 provide comparisons of chiral columns and chiral separation screens for pharmaceutical analysis and purification.

Separations of chiral compounds by GC, SFC, and CE are discussed at length in Chapters 9 to 11. These methods come in handy in various situations. A strategy of method development for HPLC, SFC, and CE is covered in great detail in Chapter 12. The reader may find significant advantages in this integrated approach to method development with these techniques commonly used in chiral separations today.

Chapter 13 covers preparatory separations in the industrial environment. Chemical sensors provide an interesting promise for chiral separations in the future (Chapter 14). Preliminary studies indicate a vast potential for a variety of chiral applications. The current status of what is being done or not done in terms of chiral separations of biotechnology products is covered in Chapter 15. Some useful suggestions have been made to assist future developments in this field, including the control of biogenerics.

I would like to thank all the authors for their valuable contributions, which make this book a useful resource for laboratory investigators, managers, and regulators who are involved in chiral separations in the pharmaceutical industry.

SATINDER AHUJA

Calabash, North Carolina

CHAPTER 1

Overview of Chiral Separations

SATINDER AHUJA
Ahuja Consulting, Calabash, North Carolina

1 INTRODUCTION

Enantiomers of a molecule relate to each other as an object and its mirror image that are not superimposable. They are also called *chiral* (from the Greek word *cheiro*, meaning "hand"); that is, they are like a pair of hands. The "handedness" of small and large molecules has sparked great interest in pharmaceutical and biotechnology industries [1–10]. This difference in spatial arrangements of atoms in a molecule (i.e., the molecule's stereochemistry) can influence its pharmacologic, metabolic, and toxicologic activity. Molecules that are isomeric but have a different spatial arrangement are called *stereoisomers*. Symmetry classifies stereoisomers as either *enantiomers*, as defined above, or *diastereomers*. Stereoisomerism results from a variety of sources besides the single chiral carbon. There are two simple molecular sources of chirality: molecules that have a stereogenic center and those that have a stereogenic axis. Stereoisomerism is possible in molecules that have one or more centers of chirality, helicity, planar/axial/torsional chirality, or topologic asymmetry.

The amounts of energy necessary to convert given stereoisomers into their isomeric forms may be used for further classification. Stereoisomers with low-energy barriers to this conversion are termed *conformational isomers* (e.g., proteins in the case of biotechnology products), whereas high-energy-barrier conversions are described as configurational isomers (e.g., small molecules). Diastereomers differ in energy content, and thus in every physical and chemical property; however, the differences may be so minute as to be nearly indistinguishable.

Very often, one isomer of a series may produce a desired effect, while another may be inactive or even produce an undesired effect. Chiral separations represent

Chiral Separation Methods for Pharmaceutical and Biotechnological Products, Edited by Satinder Ahuja

the most intriguing and, by some measures, most difficult separations of chemical compounds in that the molecules to be separated have the same molecular weight and physical and chemical properties, except for the rotation of polarized light. As mentioned above, isomeric impurities may have unwanted toxicologic, pharmacologic, or toxocologic effects. Therefore, an accurate assessment of the isomeric purity of substances is essential. Such impurities may be carried through the synthesis, preferentially react at one or more steps, and yield an undesirable level of an additional impurity.

2 REGULATORY CONSIDERATIONS

Regulatory guidance for development of chiral compounds is generally consistent among regulatory bodies in the United States, the European Union, Canada, and Japan (Chapter 2). The focus is to develop specific enantiomeric methods early in the program:

- To determine the relative pharmacological contribution, compared to that of the racemate, of each enantiomer in animals and in humans
- To compare the toxicology profile of the racemate to the individual enantiomers to confirm their relative activity

Based on these data, the sponsor may make a logical choice to proceed with development of the racemate or a single enantiomer.

Although regulatory guidance documents do not specify biologics or biotech-derived products, one can assume that for the generation of a single, purified active pharmaceutical ingredient (API), many of these same concepts apply. A major caveat to the category of products approved under the Public Health Service Act vs. the Food, Drug, and Cosmetic Act is that several approved biologics consist of a pool of heterologous proteins, such as polyclonal antibodies (e.g., intravenous gamma globulin, vaccine antigens, and some isozyme preparations). Given the rigor or orthogonal analytical methods used in biologics development and process validation, it is assumed that issues relating to chiral activity will not be lost in the program.

3 BASIC CONSIDERATIONS IN METHOD DEVELOPMENT FOR CHIRAL COMPOUNDS

Cost considerations, availability of equipment, and know-how play important roles in the selection process for an appropriate method (Chapter 3). Paper chromatography (PC) and thin-layer chromatography (TLC) have been used where cost considerations outweigh other factors. PC is used very rarely these days; however, TLC can be a very useful qualitative technique that entails minimal costs. It can also provide good indications as to which HPLC method would be most suitable for resolving enantiomers. Of course, it can also be used as an independent technique with limitations of resolution and low precision. A significant amount of coverage was provided in earlier texts [3–6] to enable the reader to try TLC; those texts include a number of reference sources for TLC aficionados. Commonly used methods for separation of enantiomers today can be classified broadly into the following four categories:

- Gas chromatography (GC)
- High-performance liquid chromatography (HPLC)
- Supercritical fluid chromatography (SFC)
- Capillary electrophoresis (CE)

Detailed discussion of these methods is provided in this book. Since HPLC methods are generally favored for a variety of reasons, some basic information on selecting a suitable method for HPLC has been included in this chapter. A basic understanding of chiral discrimination by various chiral stationary phases (CSPs) has been provided to help with method development. A strategy for fast method development is also provided in this chapter.

4 SEPARATION OF CHIRAL COMPOUNDS ON POLYSACCHARIDE COLUMNS

The popularity of polysaccharide-based chiral stationary phases has been well documented (Chapter 4). Based on published information, it appears that derivatized polysaccharides are by far the most widely used CSPs in the separation of enantiomers. An incredible number of chiral separations have been and continue to be made with just four commercial chiral stationary phases: Chiralpak AD and AS and Chiralcel OD and OJ. Now these same problems can usually be solved with just three immobilized columns: Chiralpak IA, IB, and IC. In various studies, either of these sets of columns offers resolution for more than 85% of the compounds that have been investigated. Mechanisms of separation and method development are also discussed in this chapter.

5 CHIRAL SEPARATIONS BY VARIOUS TECHNIQUES

Three cases of chiral separations based on phase conversion of a popular Chiralpak AD column are presented in Chapter 5. Examples of successful chiral separation by converting this column from the normal phase to the reversed phase are demonstrated. By phase conversion, some of the compounds changed enantiomeric elution order, whereas others did not. Advantages of phase conversion in chiral separations are also discussed. It should be noted that improper preparation of a normal mobile phase could cause loss of chiral resolution previously observed for various chiral separations; this can result in poor method transference. Finally, a very interesting case of achieving chiral resolution on rotamers with achiral columns is shown that makes one wonder whether the separation is chiral.

6 CHIRAL DISCRIMINATION STUDIES BY NUCLEAR MAGNETIC RESONANCE

Although polysaccharide-based CSPs have been commercialized for more than two decades, the chiral discrimination mechanisms are still unclear at the molecular level (Chapter 6). Chiral recognition exhibited by polysaccharide-based CSPs depends on the higher-ordered structures of the polymers, which makes it difficult to understand the chiral recognition mechanism. Problems often arise with regard to the selection of appropriate systems, with fitting mobile phases, from the polysaccharide-based CSPs available. Unfortunately, no selector–selectand combinations or reliable chiral recognition models have been developed to allow for predictions with respect to separability, magnitude of enantioselectivity, elution order, and suitable chromatographic conditions.

Insight into chiral discrimination at the molecular level for polysaccharide-based CSPs is hindered by the complexities of the polymer, such as the exact stereochemical structure, the geometry of the interaction, the accessible binding sites, and the multiplicity of sites with different affinities for enantiomers. Numerous techniques, such as x-ray crystallography, nuclear magnetic resonance, calorimetric studies, infrared, and computational methods have been used to provide insight into chiral recognition mechanisms for other CSPs. These studies can help improve our understanding of the chiral stationary-phase structures, chiral cavities, and surface properties.

7 COMPARISON OF CHIRAL CHROMATOGRAPHY COLUMNS

Analytical laboratories must be ready continually to address the changing nature of molecules in developments in the pharmaceutical industry (Chapter 7). A majority of compounds screened for chiral method development have been adequately resolved on polysaccharide-based stationary phases, including Chiralpak WH, Chiralpak WM, and Chiralpak WE, AD, OD, AS, and OJ in many laboratories. However, as new phases become available, it is important to characterize their capabilities as well. After optimizing the analysis parameters for several chiral columns produced by different manufacturers, the column series was challenged by chemical entities representative of those developed for commercial use as pharmaceuticals. The chromatographic results were assessed vs. polysaccharide-based phases to gauge how successful various chiral columns are in developing efficient stereoselective methods for resolving chemical entities progressing to market.

8 CHIRAL SEPARATION SCREENS FOR ANALYSIS AND PURIFICATION

The pharmaceutical industry strives to produce effective, safe, and high-quality medicines. Analysts play a critical role in the chiral discovery process because each enantiomer has the potential to produce different therapeutic effects or adverse effects, and may even be metabolized differently (see Chapter 8). Chiral chromatography, analytical and preparative, is now considered an integral part of pharmaceutical analysis and drug discovery. A series of chiral HPLC (normal, polar, and reversed phases), and chiral SFC screens have been developed and implemented. These allow scouting many conditions and columns rapidly and effectively. Parallel chiral HPLC systems and chiral SFC have been found to be very useful. Several examples illustrating the performance of the screens are discussed in detail.

9 SEPARATIONS OF ENANTIOMERS BY GAS CHROMATOGRAPHY

High efficiency, sensitivity, and speed of separation are important advantages of enantioseparation by high-resolution capillary gas chromatography (HRC-GC). Because of the high separation power of HRC-GC (Chapter 9), contaminants and impurities can be separated from the chiral analytes; the simultaneous analysis of multicomponent mixtures of enantiomers (e.g., derivatized proteinogenic α-amino acids). Ancillary techniques such as multidimensional GC (i.e., in series-coupled column operation), interfacing, and coupling methods such as gas chromatography–mass spectrometry (GC-MS) are important tools in chiral analysis. Employing the ion-monitoring mode selected, trace amounts of enantiomers can be detected by GC-MS. The universal flame-ionization detector (FID) is linear over five orders of magnitude, and detection sensitivity can be increased further to the picogram level by electron-capture detection (ECD) and element-specific detection, usually aided by special derivatization strategies. In contrast to liquid chromatograpy or electromigration methods, the delicate choice of solvents (buffers), modifiers, and gradient elution systems is not necessary in GC. However, the prerequisites for the use of GC are volatility, thermal stability, and resolvability of the chiral analyte; these restrict the exclusive use of enantioselective GC.

10 SEPARATIONS OF CHIRAL COMPOUNDS BY SFC

SFC has been used successfully for chiral separations at the analytical, semipreparative, and preparative scales (Chapter 10). Commercial systems have demonstrated excellent performance, robustness, and cost-effectiveness. For industrial purposes, SFC at a simulated moving bed (SMB) on a production scale has been demonstrated on a prototype in the lab. The production capacity can be obtained at the metric tons level. Excellent economic advantages have been demonstrated compared to liquid-based SMB operations.

11 CHIRAL SEPARATIONS BY CAPILLARY ELECTROPHORESIS

Cyclodextrins (CDs) are most frequently used as a selector in chiral CE (Chapter 11). The numerous applications reported over the past several years indicate their potential and popularity. The development of anionic derivatives has boosted their popularity. Some derivatives, such as the highly sulfated CDs, show broad enantioselectivity toward a large number of structurally diverse compounds. They are suitable for developing screening approaches or separation strategies for industries (e.g., in drug development and in quality control). This explains the 18% market share of the applications described from the pharmaceutical industry and the continuous growth predicted in this field. For crown ethers, only small molecules bearing an amino group, such as amino acids, can be separated, although occasionally, separation of a small drug molecule has also been reported. The same applies for ligand-exchange CE, where the analytes must have free-electron pairs and where applications are also limited primarily to amino acids. For macrocyclic antibiotics, the number of applications reported has decreased notably in recent years. This can be attributed to their limited enantioselectivity in CE and the fact that they absorb ultraviolet light at wavelengths below 250 nm. Adsorption onto the capillary wall and limited enantioselectivity may also be reasons that proteins are not used as frequently.

12 HIGH-THROUGHPUT SCREENING AND METHOD-DEVELOPMENT STRATEGIES

Since chiral recognition mechanisms are not fully understood, making the prediction of enantioseparation rather difficult. Some generic screening and method-development strategies have been developed to avoid time-consuming trial-and-error approaches (Chapter 12). These include normal-phase liquid chromatography (NPLC), reversed-phase liquid chromatography (RPLC), polar organic solvent chromatography (POSC), super- and subcritical fluid chromatography, and capillary electrophoresis. When one technique fails to separate certain compounds, it is possible that another technique will succeed in obtaining a baseline resolution. The fact that these techniques complement each other enlarges the spectrum of chiral compounds that can be separated with one of the defined strategies.

13 PREPARATORY SEPARATIONS

Preparatory separations have been employed successfully in a challenging preparation of an enantiopure single diastereomer of a pharmaceutical intermediate from a mixture of four different stereoisomers (Chapter 13). Column screening, modeling, and optimization have led to the identification of an HPLC method employing a step gradient to enhance separation productivity and to reduce solvent consumption. The separation was carried out on a fairly large scale that afforded a substantial amount of the enantiopure single diastereomer.

14 CHIRAL ANALYSIS WITH SENSOR TECHNOLOGY

It is abundantly clear that verification of enantiomeric purity is an important analytical requirement in the pharmaceutical industry. Chiral purity assays are often performed via chromatographic techniques, and performance is controlled by "adsorption" of the analyte onto the coating. Since it is not always known which CSP would provide optimal specificity for a given enantiomeric pair, chromatographic method development can be a time-consuming and expensive process (Chapter 14). Chemical sensors are being investigated to improve the efficiency of column method development. The leading platform for the sensor application is the quartz-crystal microbalance (QCM) because of its ability to make real-time condensed-phase measurements. QCM sensors are coated with stereospecific coatings; the coated sensor readily produces unique responses upon exposure to enantiomeric isomers. Preliminary studies that assess the nature of the analyte–coating interaction indicate vast potential for future chiral applications. Research from various groups is promoting the potential for stereospecific applications for chemical sensors. The research activities are progressing to achieve two important applications: to establish whether sensor technology can be used for direct enantiomeric impurity determinations for pharmaceutical applications, and to determine if sensors make the selection of chiral LC columns more efficient for preparative and analytical needs.

15 CHIRALITY OF BIOMOLECULES AND BIOTECHNOLOGY PRODUCTS

A large number of successful biotechnology products that have been intoduced into our armamentarium of modern medicine are based on proteins, which are complex organic macromolecules whose structures are coded in an organisim's DNA. Each protein has a unique genetically defined amino acid sequence that determines its specific shape and functions. It is well known that proteins are composed of chiral amino acids. Unfortunately, chiral studies are largely ignored on biomacromolecules such as proteins, as they are not monitored to assure that they indeed correspond in terms of all chiral components to the original macromolecules produced biologically. This may stem from the fact that monitoring biological activity is considered adequate in many cases. Alternatively, it is assumed that their unique structure assures appropriate chirality of its components; that is, appropriate folding would not occur if an alternative enantiomer were to be incorporated in the molecule.

Chapter 15 reviews what is being done to monitor biomolecules such as proteins and biotechnology products based on proteins and their building blocks (i.e., amino acids and peptides). With the upcoming advent of biogenerics, it is desirable that all chiral components in new products correspond to the original macromolecules. Furthermore, extensive physicochemical testing needs to be performed to assure that denaturation of proteins has not occurred and that they are refolded properly in case any unfolding occurred during processing.

REFERENCES

1. S. Ahuja. Are chiral separations of drugs really easy? Presented at the American Chemical Society Meeting, Boston, Aug. 19–23, 2007.
2. S. Ahuja. *LCGC*, 25:2, Nov. 2007.
3. S. Ahuja. Overview of preparatory separation science and biotechnology. Presented at the AAPS Biotechnology Conference, Toronto, Ontario, Canada, June 22–25, 2008.
4. W. J. Lough, I. W. Wainer, Eds. *Chirality in Natural and Applied Science*. CRC Press, Boca Raton, FL, 2000.
5. S. Ahuja. *Handbook of Bioseparations*. Academic Press, San Diego, CA, 2000.
6. S. Ahuja. *Chiral Separations by Chromatography*. Oxford University Press, New York, 2000.
7. H. Y. Aboul-Enein, I. W. Wainer. *The Impact of Stereochemistry on Drug Development and Use*. Wiley, New York, 1997.
8. S. Ahuja. *Chiral Separations: Applications and Technology*. American Chemical Society, Washington, DC, 1997.
9. S. Allenmark. *Chromatographic Enantioseparations*. Ellis Horwood, Chichester, UK, 1991.
10. S. Ahuja. *Chiral Separations by Liquid Chromatography*. American Chemical Society, Washington, DC, 1991.

CHAPTER 2

Regulatory and Development Considerations of Chiral Compounds

ROBERT L. ZEID

TLI Development, Oak Island, North Carolina

1 OVERVIEW

The importance of chirality in biological systems has been well recognized for over 50 years, due largely to its observed pervasiveness in nature. Although it may be self-evident that chirality plays a critical role in biological activity, it was not until the 1980s that its importance in drug development was recognized. With the advent of improved assay sensitivity and separation techniques of the day, Ariens et al. correctly identified racemic mixtures as fixed-dose combinations of a *eutomer*, or desired enantiomer, and *distomer*, or biologically inactive and/or toxic enantiomer [1]. Since then, the universe of chiral science has been expanding at an astonishing rate, with significant developments in analytical methods, the discovery of new stereoselective catalyzing agents, improved manufacturing and separation process controls, and in vitro characterization as evidenced by a larger pool of patents and intellectual property litigation in this area.

Although regulatory policies for chiral drug development were put in place nearly 20 years ago, the timing for when to implement these tests in the development plan has not been well defined. Regulatory and development considerations have grown significantly more complicated and layered since then and appear to have resulted in an ad hoc approach instead of a highly integrated program. The focus of this chapter is to provide a more holistic project development plan based on the combined current regulatory guidance and industry experience for the development of racemic drugs and selective enantiomers.

Chiral Separation Methods for Pharmaceutical and Biotechnological Products, Edited by Satinder Ahuja

2 HISTORY OF REGULATORY GUIDANCE

Several regulatory agencies and competent authorities (CAs) have issued guidance documents on chiral development, beginning in the mid-1980s. A summary of the major guidance documents from the U.S. Food and Drug Administration (FDA), European Medicines Evaluation Agency (EMEA), Japanese Ministry of Health, Labor, and Welfare, and Health Canada are summarized below for ease of review and comparison.

2.1 U.S. Food and Drug Administration

In 1987, the FDA published a series of guidelines on the content and format of information to be submitted in new drug applications (NDAs). The guideline on chemistry and manufacturing data noted for the first time that the structural identity of a compound must be verified and all chiral centers identified. For enantiomeric structures that had ratios different from a racemic mixture, those had to be identified, as well as providing a diagram of the chemical structure and all known isomers.

However, it was not until 1992 that the agency issued a policy statement specifically regarding development of new stereoisomeric drugs* as either individual enantiomers or racemates. The rationale for the 1992 policy stemmed from a convergence of two major aspects. First, a growing body of historical data demonstrated different characteristics for enantiomers in biological systems as well as pharmacokinetic differences in absorption, distribution, metabolism, and excretion (ADME); however, these differences in enantiomeric profiles were not studied routinely, due to the difficulty in separation. The second aspect driving the 1992 policy statement was that the technological advances now allowed large-scale commercial separation of chiral compounds and production of single enantiomers. Thus, the agency's perspective shifted to address acceptable manufacturing controls of synthesis, impurities, pharmacological and toxicological assessment, characterization of metabolism and distribution, as well as clinical evaluation of stereoisomeric drugs.

The 1992 policy cited examples of enantiomers in which both had similar qualities, albeit some enantiomers had different concentration–response relationships (or receptor affinity/receptor activation kinetic) profiles compared to each other, such as:

- Dobutamine enantiomers are both positive ionotropes.

*The FDA Policy Statement for the Development of New Stereoisomeric Drugs, issued May 1, 1992. The policy noted that the term *stereoisomers* included enantiomers, geometric isomers (e.g., *cis/trans*), and diastereomers (isomers with more than one chiral center that are not mirror images of one another). Since the agency considered diastereomers and geometric isomers as chemically and pharmacologically distinct, and which can be separated without chiral techniques, these are generally treated as separate drugs, with some very specific exceptions (e.g., in vivo interconversion). There is no reason to develop a mixture of diastereomers or geometric isomers unless they fortuitously present a reasonable fixed-dose combination.

- Ibuprofen enantiomers are both anti-inflammatory agents.
- Warfarin and phenprocoumon enantiomers are both anticoagulants.
- Bupivicaine enantiomers both produce local anesthesia, but the incidence of severe arrhythmias in isolated rabbit heart is much less with the $S(-)$-enantiomer than the $R(+)$- isomer or the racemate.
- Quinolones and β-lactam antibiotic enantiomers are all antibacterial.

Conversely, there are numerous examples of enantiomers with strikingly different qualities, such as:

- The $R(+)$-enantiomer of thalidomide has sedative action but the $S(-)$-enantiomer is a teratogen.*
- The *l*-propranolol enantiomer is a β-adrenergic receptor antagonist (β-blocker), but *d*-propranolol is not.
- The *d*-sotalol enantiomer is a type 3 antiarrhythmic, whereas *l*-sotalol is a β-blocker.
- The *d*-levodopa isomer is associated with granulocytopenia.
- The *d*-levamisole isomer is associated with vomiting.
- The *d*-carnitine isomer is associated with myasthenia gravis symptoms.

In addition to the pharmacological differences, the agency also noted that there were numerous cases of enantiomers having different pharmacokinetic (PK) profiles. Although these differences might not have a clinical impact, they could make nonchiral blood level assays difficult to interpret with accuracy or confuse the interpretation of nonclinical data if the animal disposition of the isomers differed significantly from those in humans.

Despite the differences in enantiomer activity described above, the agency's 1992 policy statement noted that development of racemates historically resulted in only a few recognized adverse profiles. Even in the light of feasibly preparing purified enantiomers, the agency recognized that development of racemates was still appropriate but that the following data should be considered:

- Appropriate manufacturing and control procedures should be used to assure enantiomeric composition with respect to identity, strength, quality, and purity. Manufacturers should notify compendial authorities of these specifications and tests.

*Thalidomide was widely marketed in over 46 countries as a sleep aid. Its use during pregnancy was responsible for over 10,000 children born with profound birth defects. Thalidomide would have been approved in the United States were it not for the vigilance of a medical officer, Frances O. Kelsey. The public health catastrophe prompted stronger regulation (Kefauver–Harris Amendment of 1962) for demonstrating the safety of a product prior to marketing. Although it was commonly thought the toxicity resided solely with the $S(-)$ form of thalidomide, several in vitro studies showed that the drug could racemize quickly in various aqueous media. Metabolic inversion has been observed with other compounds, such as 2-arylpropionic acid (2-APA) and mandelic acid.

- PK evaluations that do not use a chiral assay will be misleading if the disposition of the enantiomers is different. Therefore, techniques to quantify individual stereoisomers should be available. However, if the PK of the enantiomers is demonstrated to be the same or to exist as a fixed ratio in the target population, a nonchiral selective assay may be used.

Based on the synopsis above, the agency's 1992 policy statement was meant as an outline of the particular studies expected during the development of enantiomeric drugs. In general, the enantiomeric composition of a drug should be known for the material used in pharmacology, toxicology, and clinical studies. The finished product specifications should assure identity, strength, quality, and purity with respect to enantiomeric purity.

With respect to PK profiles of enantiomers, quantitative assays should be used for individual enantiomers early in the drug development program, which will allow assessment of any potential for interconversion and the ADME profile. If the drug is a racemate and the PK profiles of the isomers are different, sponsors should monitor the enantiomers individually to determine properties such as dose linearity and the effects of altered metabolic or excretory function as well as drug–drug interactions. If the PK profile is the same for both isomers or a fixed ratio between the plasma levels of enantiomers is demonstrated in the target population, an achiral assay or assay that monitors the most relevant enantiomer is acceptable for later evaluation.

Unless it proves particularly difficult, the main pharmacologic activity of the isomers should be compared in in vitro systems, animals, and/or in humans. A relatively mild toxicology profile of the racemate would usually support further development without separate toxicology assessments of each enantiomer. Conversely, if there are findings other than those associated with a natural extension of the pharmacological activity, unusual toxicity, or toxicity approaching the effective dose in animals or near the planned dose in humans, a toxicology evaluation of the individual enantiomers is warranted.

The 1992 policy statement outlined enantiomer development for chemistry and manufacturing as well as nonclinical and clinical studies. A brief synopsis is provided below.

Chemistry, Manufacturing, and Controls Data

- Analytical methods for racemates and enantiomers should include a stereochemically specific identity test and/or selective assay. This identity and assay method should be in place for drug substance and drug product. With respect to stability testing, a selective method should also be in place, but once it has been demonstrated that stereochemical conversion does not occur, the stereoselective tests might not be required.
- Impurity limits for each isomer should include all isomeric components, impurities, and contaminants. The maximum allowable level of impurity in product used for clinical trials should not exceed the level that was assessed in nonclinical toxicity studies.

- Labeling should include a name with the appropriate stereochemical description.

Nonclinical Studies

- Pharmacologic activity of the individual enantiomers should be characterized for the principal mechanism of action and any other important effects with respect to potency, specificity, maximum effect, and so on.
- Pharmacokinetic profiles should include an in vivo assessment of interconversion and disposition for each enantiomer and be compared later to the clinical PK profile in phase I and phase II.
- Toxicology on the racemate should be sufficient; however, if toxicity other than that predicted by its known pharmacologic properties should occur at low multiples of exposure, toxicity of the individual enantiomers is warranted to ascertain the relative contribution of each for the toxicity observed. If the toxicity of significant concern can be eliminated by development of a single enantiomer, that would be the preferred route of development.
- To study a single stereoisomer from a mixture that has already been studied nonclinically, an abbreviated pharmacology/toxicology evaluation would be required. This program would include a repeated-dose toxicity study up to three months and a reproductive toxicity segment II study in the most sensitive species. The positive control would probably include the racemate. If there is no difference between the single enantiomer and the racemate, no further studies would be required. If the single enantiomer demonstrated a different or more toxic profile than that racemate, further studies would be necessary as well as an evaluation of its implications for human dosing.

Clinical and Biopharmaceutical Studies

- As noted above, a racemate may be considered suitable for development barring no significant differences in the activity and disposition of the enantiomers. Individual enantiomers should be evaluated in case there is toxicity of the racemate at clinical doses that is either unexpected or suggests an unrelated pharmacologic effect. It should be noted that some enantiomers may not have different toxicity profiles, but their metabolites may. Thus, it is also relevant to consider individual programs when the activity of one is clearly superior to that of the other.
- If a racemate is studied, the PK of the two isomers should be evaluated in phase I to confirm a lack of in vivo interconversion. Based on phase I and phase II PK data in the target population, it should then be possible to determine if an achiral assay or monitoring of just one enantiomer is sufficient to evaluate both.
- If a racemate has been the subject of an approved market application and a sponsor wishes to develop the single enantiomer, the FDA requires a new NDA. The evaluation must include a determination of any significant conversion to the other isomer and whether the PK profile of the single

isomer is comparable to that of the racemate. However, bridging with data of a racemate already on the market is not possible.

The combined effect of the FDA policy, refinements in chiral separation, and improved analytical methods over the last 20 years has been a greater effort required from the pharmaceutical industry to define the pharmacology and toxicology of individual enantiomers from racemic drugs. By increasing the scrutiny of a racemate vs. single-enantiomer profiles, the regulatory authorities influenced the industry's decision to develop more single enantiomers in lieu of racemic drug mixtures.

2.2 European Union European Medicines Evaluation Agency

The most specific guidance by either the European Union (EU) or the European Medicines Evaluation Agency (EMEA) was the *Investigation of Chiral Active Substances* issued in October 1993 [2]. This guidance was principally revised from the *Clinical Investigation of Chiral Active Substances/III/3501/91* to include quality and nonclinical development considerations. It mirrored many of the same points as the FDA 1992 *Policy Statement for the Development of New Stereoisomeric Drugs* but provided more specific guidance for chemistry and nonclinical studies.

However, the earliest guidance regarding stereochemistry was published in 1989 by the European Commission (EC) Committee for Proprietary Medicinal Products (CPMP) in Volume II, *Notice to Applicants for Marketing Authorizations for Medicinal Products for Human Use in the Member States of the EC*, which specifically required the Expert Report to include an evaluation of chiral data.

In addition, the CPMP guideline on analytical validation was adopted in July 1989 and published in the July 1990 addendum to Volume III of the *Rules Governing Medicinal Products in the European Community*, which notes that "methods or procedures of analysis proposed must take account of technical and scientific progress and enable the starting material, intermediate, and finished product to be checked by means of generally accepted methods." Although the reference in Volume III does not specifically mention chiral assays or analysis of single enantiomers, it is implied that the most advanced analytical methods available at the time should be used.

With respect to the chemistry aspects, the EU guidance noted that the drug substance (DS) may be a racemate (equimolar), a nonequimolar mixture of enantiomers, a single enantiomer, or a mixture of enantiomers and/or diastereomers. In the case of a single enantiomer, the other enantiomers will be considered an impurity. The analytical methods used [e.g., optical rotation, chiral high-performance liquid chromatography (HPLC), optical rotary dispersion, circular dichroism, nuclear magnetic resonance (NMR)] should be validated to demonstrate the limits of detection and quantitation for both the DS and impurities. Note that for NMR, the method cannot distinguish enantiomers from racemic mixtures without proper shift reagents or latently installed chiral auxiliary agents. Solid-phase methods include polymorph analysis by x-ray particle diffraction (XRPD) and differential scanning calorimetry (DSC).

The EU guidance also provided more detail than the U.S. guidance for nonclinical testing for a single enantiomer vs. a racemate as well as a single enantiomer from an approved racemate. The guidance also noted nonclinical considerations for developing a new racemate from an approved single enantiomer and a nonracemic mixture from an approved racemate or single enantiomer. Similar to what was noted in the U.S. guidance for development of a single enantiomer, the 1993 EU guidance recommended an analysis of animal in vivo ADME data for comparison to human ADME data to ascertain any conversion of one enantiomer into another (e.g., metabolic inversion) or into a racemic metabolic pool. Based on those findings, a chiral bioanalytical assay might be required for continued clinical development.

If the sponsor elected to develop a racemate DS instead of a single enantiomer, justification and support by animal PK/PD and ADME studies were required. Two potential scenarios might occur: rapid conversion of the enantiomer to the racemate in vivo, or no/slow conversion in vivo that allowed for separate enantiomer effects and ADME data pools. The pharmacodynamic (PD) response should be compared for the racemate vs. the single enantiomer. The pharmacokinetic (PK) response should be measured by enantiospecific analytical methods. Both the PK and PD animal data should be compared to human phase I data. With respect to toxicology, it is acceptable to perform toxicology studies using the racemate, especially if sufficient quantities of the preferred enantiomer are unavailable and/or the animal in vivo studies indicate conversion to the racemate. However, if toxicity other than that predicted by an extension of the pharmacological properties occurs at relatively low multiples of the exposure intended for clinical study, relevant toxicity studies should be repeated with the racemate and purified enantiomers to help discern the activity and toxicity of each component.

It is also critical to have proper process controls when dealing with an enantiospecific drug substance. If the API synthetic scheme uses a chiral starting material, regulatory agencies expect strict specification controls on the starting material source and acceptance criteria. The sponsor who procures the chiral starting material or raw material with enantiospecific consistencies should expect, and implement, release testing in-house and batch analysis to establish trending data. On the other hand, if the sponsor chooses to optimize an enantiospecific process by which the desired chirality is induced during a specific step in the process scheme, adequate in-process test controls for enantiospecificity are expected. It is also critical that the enantiospecific process be planned thoroughly and carefully to avoid an early induced chiral center on any intermediates during the process scheme and/or so as not to undergo racemization, due to severe reaction parameters of any step.

In principle, development of a single enantiomer from an approved racemate can utilize existing toxicity data of the racemate to help bridge the toxicity data program of the enantiomer; however, use of the racemate data is decided on a case-by-case basis. Some key considerations include:

- *Pharmacodynamics*. The profile of the enantiomer should be compared to that of the racemate.
- *Pharmacokinetics*. The profile of the enantiomer should be compared to that of the racemate.
- *Toxicology*. A suitable bridging program may be comprised of (1) an acute toxicity study of the enantiomer compared to the racemate; (2) a repeated-dose study up to three months in a single most appropriate species, and (3) a study of pre- and postnatal development, including maternal function with the modification of starting treatment at conception using the racemate as a positive control. Results for the single enantiomer should be compared to those for the racemate.

The EU guidance also outlined clinical considerations, which largely mirrored the U.S. guidance and repeated many of the same points brought out for non-clinical testing (e.g., comparison of PK/PD profiles). In the situation regarding development of a nonracemic mixture from an approved racemate or single enantiomers, the EU guidance noted that this approach was largely similar to a fixed combination product that has been developed to optimize the molar ratio of compounds for therapeutic benefit. Thus, the sponsor developing this type of product should also refer to the EU guidance *Fixed-Combination Medicinal Products*.

In addition to the EU guidance *Investigation of Chiral Active Substances*, points to consider for development of chiral compounds are noted in several other guidances, such as *Investigation of Chiral Active Substances*, Committee for Veterinary Medicinal Products [3].

2.3 Japan (Ministry of Health, Labour, and Welfare)

While the Japanese regulatory authorities have not issued specific guidance on the development of chiral compounds, they appear to be using an approach similar to that adopted by the United States and the EU. A retrospective analysis of drug approvals in Japan from 1988 through 2007 assessed the number of enantiomers, racemates, and achiral drugs that were approved, which showed a definite trend of increased development of single isomers rather than racemates [4].

Shimizawa et al. [4] analyzed 76 drug approvals in Japan from January 2001 through 2003, which were comprised of 29 achiral drugs (39%), 23 single enantiomers with multiple chiral centers (30%), 14 single enantiomers with one chiral center (18%), and 10 racemic drugs (13%). The total number of single enantiomeric drugs was 37. Twenty-nine (29) of the 37 (78%) single enantiomeric drugs were synthesized from an enantiomeric starting material; three products (8%) were derived by asymmetric synthesis, and the remaining five products (14%) were derived by asymmetric resolution (e.g., crystallization and/or chromatographic separation).

Confirmation of stereochemistry was confirmed by XRPD for 22 products (59%); 11 products (30%) did not specify a method; and four products (11%) were confirmed by identification through the authentic samples. Specifications for assuring chirality appeared largely to be those using optical rotation (21 products, or 57%), optical rotation and optical purity by HPLC (11 products, or 30%), optical purity by HPLC only (two products, or 5%), and not set (three products, or 8%). With respect to pharmacokinetic assessment, a PK evaluation was performed for all the products, but only 12 products were associated with any reported assessment of chirality on PK; 10 products were reported with chiral inversion.

An analysis by Shimizawa et al. showed that Japanese government approval of single-enantiomer drug products rose from 39% (of all drugs approved in Japan) in 1988–1990 to 56% in 2000–2002. Conversely, the number of racemic drug products approved in the same period dropped from 32% to 11%. The Japanese approvals were similar to approvals observed worldwide, with single enantiomers accounting for 27% in 1996, up to 39% in 2002 [5,6].

2.4 Canada (Health Canada)

In July 1998, Health Canada published a Therapeutic Products Programme (TPP) guidance document entitled *Stereochemical Issues in Chiral Drug Development*; it became effective May 1, 2000. The overall content followed much of what the U.S. and EMEA guidance documents noted with respect to development considerations for single enantiomers, racemates, and nonracemic mixtures. The key distinction is that since it is one of the most recent guidance documents on this topic, there is more detail in the development plan regarding when certain analytical methods or assessments should be performed.

Of particular note is the preamble, which states that the sponsor should develop and validate an enantiomer-specific assay as early as possible in drug development and use it until the combined data clearly demonstrate that the same results are achieved with a nonselective assay. A list of chiral methods to be considered include NMR, infrared (IR), gas chromatography (GC), chiral HPLC, chiral GC, x-ray crystallography, melting point, circular dichroism (CD), optical rotation, optical rotary dispersion (ORD), and others.

3 CHIRAL SWITCHES AND BRIDGING RACEMIC DATA TO ENANTIOMER PROGRAMS

Although the guidelines detailed in Section 2 touched on chiral switches and bridging data from racemic to enantiomer programs, this was actually an early and intense area of interest by both industry and regulators. Before the FDA issued the 1992 policy statement, the agency formed a stereoisomer committee in 1989 to study and formulate key areas of stereoisomer development. In response, the Pharmaceutical Manufacturers Association (PMA) formed its own committee to articulate industry viewpoints on the development of chiral compounds. The PMA issued a position paper in 1990 which stated that a manufacturer's decision to market an enantiomer or racemate should be case by case after considering all available data and providing the regulatory authorities with data that prove the safety and efficacy of the drug proposed. For economical and practical reasons, it is preferred to make that decision (racemic vs. enantiomer) as early as possible in the development, but the fact remains that the information required is gathered gradually during development. Thus, the industry should not be disposed prematurely against development of racemic drugs and cited numerous examples: antihypertensive agents such as verapamil, atenolol, metoprolol, and captopril, or the bronchodilators salbutamol and antihistamine terfenadine. While the PMA recognized that, more data might be required to justify the selection and development of a racemate over an enantiomer, the agency's requirements should not be interpreted as a discouragement of racemic drug development. The most important point was that if a development team decided to switch from a racemate to a single enantiomer, there should be some conservation of effort.

Thus, a key consideration by the PMA was how and when bridging of racemic development data could be applied to a chiral switch, particularly with respect to avoiding a complete repeat of the racemate nonclinical program with the enantiomer. The industry position was that it might not be necessary to repeat acute and subchronic toxicology studies with the single enantiomer if they have been conducted properly with the racemate; the PMA conceded that chronic toxicity studies should be done with the single enantiomer. The agency concurred (up to a point), and in the FDA 1992 policy, it noted:

> Unless it proves particularly difficult, the main pharmacology activities of the isomers should be compared in in vitro systems, in animals and/or in humans. A relatively benign toxicologic profile using the racemate would ordinarily support further development without separate toxicologic evaluation of the individual enantiomers. If, however, there are toxic findings other than those that are natural extensions of the pharmacologic effects of the drugs, and especially if they are unusual or occur near the effective dose in animals or near the planned human exposures, toxicologic evaluation of the individual isomers in the study where the toxicity was detected should be undertaken.

The 1992 policy specifically noted that an abbreviated pharmacology/toxicology evaluation could be conducted to allow the existing knowledge of the racemate to be applied to the enantiomer. The longest repeated-dose toxicity study (up to three months) and reproductive toxicity segment II study in the most sensitive species would need to be performed with the enantiomer. The positive control group should include the racemate. If there is no difference between the toxicology profiles of the enantiomer and the racemate, no further studies would be needed. However, if the enantiomer were more toxic than the racemate, further studies would be warranted.

The thrust of the agency policy was that toxicity data of the racemate could be used to bridge the nonclinical program of an enantiomer only if they were tested side by side or as controls in these studies. There did not appear to be any allowance for historical comparisons of the racemic toxicology profile to a stand-alone enantiomer profile performed separately. That is an important point to note, especially when a development team may, or may not, know if an enantiomer has potential for development over the racemic mixture. Given the expectation by regulatory authorities to provide these nonclinical data for racemates and enantiomers in the same animal models, it behooves the development team to isolate and purify the enantiomers in sufficient quantity to allow unfettered progress of these studies.

As described previously, the 1992 policy statement defined general guidance with respect to clinical and biopharmaceutical comparisons of racemates vs. the enantiomer. The agency position was that the individual enantiomers of a racemic mixture should be evaluated and compared against the PK of the racemate, preferably in phase I or phase II studies. This would be necessary to discern any distinctions in enantiomer-specific metabolism or toxicity in vivo, especially if both enantiomers possessed pharmacological activity. Based on those phase I or phase II PK data, studies could be performed in the target population to determine if a nonchiral assay can be used. Racemates may be developed in cases where little difference in activity and disposition of the enantiomers is observed.

If a racemate has been marketed and the sponsor wishes to develop the single enantiomer, an evaluation should include the determination of any significant conversion to the other isomer and whether the PK of the single isomer is the same as that of the racemate. The consensus of all the guidance documents is that these changes reflect a new compound that must be studied under a separate investigational new drug (IND) application and hence, a separate NDA or market application.

A review of the FDA and Health Canada guidance documents show that they have a common section that addresses chiral switches (or development of a single enantiomer after the racemate has been developed). They uniformly recommend a comparison of the PK, PD, and acute toxicity of the racemate vs. the individual enantiomers to allow a sufficient body of data to make a determination of the relative activity, toxicity, and in vivo profile. Based on those side-by-side comparisons, the sponsor can scientifically justify why a racemate should be selected in lieu of a specific enantiomer or, conversely, why the acute and subchronic toxicology profile of the racemate can be used to bridge the nonclinical profile of the chronic toxicity for the enantiomer.

Interestingly, in the 1989 publication of Volume II, *Notice to Applicants for Marketing Authorizations for Medicinal Products for Human Use in the Member States of the EC*, bridging studies with data of the already marketed racemate are not allowed. "Where a mixture of stereoisomers has previously been marketed and it is now proposed to market only one enantiomer, full data on this enantiomer should be provided." But in the 1993 guidance *Investigation of Chiral Substances*, this was modified to include language similar to that in the FDA policy statement, which allowed bridging with sufficient PK, PD, and toxicology data comparing the single enantiomer to the racemate.

Thus, there appears to be consensus among the U.S., European, Japanese, and Canadian health authorities that side-by-side comparisons of enantiomers vs. racemic mixtures in animal and human pharmacokinetic, pharmacodynamic, and acute/subacute toxicity studies are requisites to making a sound scientific judgment of the development choice for an enantiomer vs. a racemate and that these data also are a "bridge" to allowing the use of racemate toxicology data in conjunction with the long-term toxicity data for the single enantiomer.

4 ANALYTICAL, PHARMACOKINETIC, AND PHARMACODYNAMIC POINTS TO CONSIDER IN DEVELOPING CHIRAL COMPOUNDS

The guidance documents on chiral compounds are, by their nature, limited to a general or cursory review of the analytical, pharmacokinetic (PK), and pharmacodynamic (PD) aspects that must be considered in chiral drug development. The purpose of this section is to explore more closely the history and key considerations for these major areas as they relate to chiral development.

4.1 Analytical Methods

Louis Pasteur, a self-described researcher of crystallography, laid the foundation for the understanding and separation of enantiomers by his use of optical rotation [7], the ability of chiral compounds to rotate a plane of polarized light while working with tartaric acid crystals that formed during wine fermentation. He observed that the first acid to crystallize was (+)-tartaric acid, which rotated light in a clockwise direction, but in the latter stages of the crystallization process, a second acid, a tartaric acid racemate, was optically inactive. Pasteur noted that the sodium ammonium salt crystals formed from a racemic solution gave rise to two different crystal forms, which were mirror images of each other and could rotate polarized light in opposite directions. From these observations he deduced that the racemic acid was optically inactive, due to equal amounts of both crystalline forms, (+) and (−), that canceled out each other's optical activity. He further postulated that the optical rotation was a characteristic of three- dimensional structures within the molecule. Although optical rotation is still used as a parameter of chiral characterization, it has been superseded by a number of different methods.

Perhaps no other area is so fundamental or has been so pivotal to the development of chiral science as the revolutionary development of analytical methods for the separation of enantiomers. There is a staggering amount of applied methodology on chiral separation for both analytical and preparative methods in manufacturing, so anything in this section is cursory at best. Analytical methods are discussed in several chapters of the book. However, there is some value in providing even a brief overview (Table 1) to lend the reader an understanding of the scope and limits of commonly used analytical methods, with cross-references to more detailed reviews [8–10].

4.2 Pharmacokinetics and Pharmacodynamics

The influence of chirality on pharmacokinetics (PK) and pharmacodynamic (PD) profiles has been reviewed extensively elsewhere [11], so it is summarized only

TABLE 1 Overview of Analytical Methods for Chiral Separation

Method	Description of Principle and Application
Chiral HPLC	Chiral HPLC may be used to separate mixtures of enantiomers directly through the use of chiral stationary phases (CSPs) or chiral mobile-phase additives in conjunction with regular (achiral) columns. A key component of many CSPs is a polysaccharide-based material (e.g., derivatized amylose or cellulose polymers), although other materials have been used, such as proteins or vinyl polymers (e.g., Pirkle-type columns that use π-electron donor or π-electron acceptors).
Chiral GC	CSPs are also used and often modified with chiral agents for the separation of enantiomers. The first CSPs were derivatized amino acids, but they were not thermally stable. The CSPs used most commonly now are modified cyclodextrins.
Differential scanning calorimetry (DSC)	DSC is used routinely for polymorph assessment. The melting points may be used in distinguishing individual enantiomers from the racemate.
Optical rotation	This method can be used to distinguish between enantiomers because they rotate the plane of polarized light in opposite directions but in equal amounts. It is widely used but is not considered a very specific method for quantitative purposes.
Nuclear magnetic resonance (NMR)	NMR is a useful tool for the determination of enantiomeric purity or enantiomeric composition. This is accomplished by making the NMR signals for the protons of the enantiomers nonequivalent by the use of chiral lanthanide shift reagents, chiral solvating agents, or chiral derivatizing agents.
Optical rotary dispersion (ORD)	ORD measures the change of specific rotation of an optically active compound with the wavelength of the light used.
Circular dichroism (CD)	CD measures the differential absorption of left and right circularly polarized light by an optically active compound. These chiroptical methods can be used to identify and/or quantitate enantiomers.
Supercritical fluid chromatography (SFC)	A supercritical fluid is one that is above a critical temperature and pressure and exists in a supercritical state where its viscosity approaches that of a gas whereas the solvent strength is closer to that of a liquid. The lower-diffusion mobile phase can be pumped through the column at a higher rate, and diffusion is faster, both of which aid column efficiency.
Capillary electrophoresis (CE)	Although this method was developed in the 1980s and has been used for chiral separations. The advantage of CE is the minimal amount of sample preparation needed. However, the coupling of CE with mass spectroscopy has greatly enhanced the method utility.

(*Continued overleaf*)

TABLE 1 ***(Continued)***

Method	Description of Principle and Application
Nonaqueous CE (NACE)	Nonaqueous CE (NACE) offers the advantage of analysis of compounds poorly soluble in water and avoids the solvents required for HPLC or GC.
Chiral CE	Chiral CE is based largely on separation by complex formation with cyclodextrins.
X-ray crystallography or x-ray powder diffraction (XRPD)	X-ray crystallography in the solid state could be used to determine the absolute configuration of molecules and to distinguish conglomerates from racemic compounds.

briefly here. Drug absorption from the gastrointestinal (GI) tract is largely passive and mediated by drug solubility and permeability; blood levels [e.g., the area under the curve (AUC)] are also generally a reflection of the dose administered or concentration released over time. Numerous studies have demonstrated differences in the dissolution and absorption of crystalline forms of the racemate vs. the individual enantiomers (e.g., ketoprofen [12], disopyramide [13], lansoprazole [14], verapamil [15], tramadol [16], anesthetic drugs [17], and ibuprofen [18]).

Active transport across the GI epithelium occurs with a select number of drugs, such as L-dopa and L-methotrexate, owing largely to the receptor interactions at endogenous binding sites on the membrane surface. Thus, it is not surprising that stereoselectivity is revealed, with the L-enantiomers being more rapidly absorbed than the D-enantiomers. The bioavailability of D-methotrexate in one study was 2.5% compared to that of the L-enantiomer [19]. However, stereoselective absorption is a factor in only a small subset of drugs; most are absorbed without any regard to chirality.

Although drug distribution is enormously complicated and beyond the scope of being fully described in this chapter, it is largely mediated by key factors such as permeability across epithelial cell membranes, the concentration of free drug available to cross cell membranes and/or the blood–brain barrier, and the relative binding affinity to plasma proteins vs. tissue proteins. Diffusion across these membranes can be by passive diffusion, facilitated diffusion through channels or carriers, and active transport [20].

Unlike absorption, which is largely mediated without influence by chirality, it has been shown that plasma protein binding can be stereoselective. Thus, the level of free active drug can be significantly different for enantiomers of highly protein-bound agents such as ibuprofen and propranolol. For example, the *S*(+)-enantiomer of ibuprofen is 50% less protein bound than is the inactive *R*(−)-enantiomer [21]. The opposite is true with the anticoagulant warfarin, where the active *S*(+)-enantiomer is more than 50% bound [22,23]. Since the degree of protein binding may not be comparable for one enantiomer vs. the other, there is the potential for the inactive enantiomer (either when given as a racemate or when formed by metabolic inversion) to bind to the protein of interest and effectively compete with the pharmacological effect or alter the PK profile [24,25]. This phenomenon is especially important when developing an enantiomer from a previously approved racemate (e.g., chiral switch), and the goal is to bridge the PK and distribution data of the racemate. It is important to evaluate the PK of the individual enantiomers vs. the racemate to determine if there is significant displacement of one enantiomer by the other in vivo. Additionally, the relative pharmacological effect of this displacement should be assessed. An additional consideration of stereoselective binding is that genetic disorders (e.g., impaired CYP2D6 metabolism), disease states, and/or concomitant drugs can alter the plasma protein/tissue protein-binding sites ratio and affect safety or efficacy, necessitating an examination of PK in select patient populations or subsets of patients with unique conditions that would warrant warnings or contraindications in the labeling.

Although there is ample evidence of stereoselective binding to plasma proteins, the potential for stereoselective distribution to various target tissues is only theoretical and has not been shown in vivo. The only evidence for stereoselective distribution stems from in vitro studies with β-blockers such as atenolol which show that the storage and release from cardiac nerve endings (in cultured PC12 cells) were stereoselective for the *S*(+)-enantiomer [26]. But even these observations may be a reflection of the stereoselectivity shown to influence pharmacological activity and the pharmacodynamic profile, not selective distribution per se. For example, the enantiomers of mepivacaine and bupivacaine exert different effects on local blood flow vasoconstriction, which may be responsible for the prolonged differences in rates of absorption and duration of action; the *S*(+)-enantiomer was favored significantly for the duration of nerve block and toxicity compared to the *R*(−)-enantiomer [27–29].

Numerous studies have shown differences in metabolism by the P450 hepatic cytochromes of one enantiomer over another, such as the following:

- Clearance of the inactive *R*(−)-enantiomer of warfarin is 42% greater than for the *S*(+)-enantiomer [30], due largely to differences in metabolic enzyme systems. The *R*(−)-enantiomer is metabolized primarily by CYP1A2 to 6- and 8-hydroxywarfarin, by CYP3A4 to 10-hydroxywarfarin, and by carbonyl reductases to diastereoisomeric alcohols. The *S*(+)-enantiomer is metabolized by CYP2C9 to 7-hydroxywarfarin. The efficacy of warfarin is affected by changes in the metabolism of the *S*(+)-enantiomer [31].
- The *R*(−)-enantiomer of the β-blocker propranolol shows 40% greater clearance than that of the *S*(+)-enantiomer [32,33].
- The calcium channel antagonist verapamil demonstrates selective elimination but has the additional complication that this difference changes depending on the route of administration (e.g., oral vs. intravenous). The *S*(+)-enantiomer is preferentially eliminated 70% more than the *R*(−)-enantiomer when given by intravenous route [34], but this ratio is only fourfold higher when given orally [35]. As the *S*(+)-enantiomer is the pharmacologically active form, this means that administration by the oral route has greatly diminished efficacy.

With respect to renal clearance, three processes govern the rate of excretion: glomerular filtration, tubular reabsorption, and excretion. The glomerular filtration rate (GFR) is a reflection of the plasma filtration rate and unbound plasma concentration, and where there are large differences in protein binding between the two enantiomers, the GFR is similarly affected to show selectivity. But overall, clinically significant differences of enantiomer excretion by GFR are lacking. For tubular reabsorption and tubular secretion, a small fraction of drugs can also be affected stereoselectively, such as with disopyramide [36] and terbutaline [37], where the enantiomers can differ as much as twofold but there are many instances of no difference. The stereoselectivity may be a reflection of the enantiomers for the epithelial transporters involved in renal tubular secretion. However, for the majority of compounds filtered in the glomerulus and not secreted actively by epithelial transport mechanisms, there is not likely to be any clinically significant difference in renal metabolism. Renal insufficiency would be expected to result in proportionally diminished excretion of both enantiomers.

4.3 Enantioselective Drug–Drug Interactions

Although numerous instances of enantioselective drug–drug interactions are detailed elsewhere [38], perhaps the example most often studied is the interaction of racemic warfarin and phenylbutazone, which was observed to cause a life-threatening potentiation of the hypoprothrombinemic response [39] due to inhibited clearance of the *S*(+)-enantiomer by increased clearance of the inactive enantiomer [40,41]. Other instances of drug–drug interactions with an impact on metabolic pathways include rifampin, a strong inducer of hepatic cytochrome P450 enzymes, which was shown to affect CYP isoforms to varying degrees that resulted in stereoselective metabolism of mephenytoin [42]. In another study, one enantiomer of propafenone, an antiarrhythmic agent, was stereoselective in affinity for the CYP2D6 enzyme and inhibited the 5-hydroxylation inactivation of the active enantiomer, thus potentiating its effect [38,43].

A powerful in vitro tool used to discern potential drug–drug interactions is evaluation by drug-metabolizing enzymes (DMEs) that employ suspensions of primary hepatocytes, precision-cut liver slices, or various hepatic subcellular fractions [44]. The method not only allows the potential to reveal drug metabolism issues but also pharmacologically active metabolites, subject variability, and human-specific metabolites, to name a few. Most of these DME systems isolated from tissue contain several families of enzymes with various isoforms: cytochrome P450, UDP-glucuronosyltransferases, *N*-acetyltransferases, sulfotransferases, and NADPH-dependent flavin-containing monooxygenases. Because of this complex interplay and the confounding variables, the evolution of recombinant DMEs have evolved that allow the select expression of a particular enzyme system in a microbial host for a more discerning evaluation. As a result, researchers can now obtain commercially prepared recombinant DMEs with a variety of enzyme systems. These recombinant DME systems cannot replace the necessity of evaluating drug interactions and metabolic fate in healthy adults and certain patient populations, but they can add to the breadth of data used to flag a potential risk, warning, or contraindication as clinical development proceeds.

5 PROJECT MANAGEMENT AND INTEGRATED DEVELOPMENT: POINTS TO CONSIDER

Perhaps one of the most difficult aspects of product development is coordination of the multiple development "engines": analytical, nonclinical, manufacturing, and clinical—such that no particular single discipline proceeds too far ahead without matching milestones met by the other disciplines (e.g., NDA registration lots put on stability before a stability-indicating assay is fully validated). A development program that lacks holistic project management with a focus on integrated development may suffer delays by having to repeat critical studies in order to bridge the gaps from one island of data to another. The basic aspects of drug development and project management are not repeated here, but key considerations specific to enantiomeric development are reviewed below. Figure 1 shows the major components of the program: analytical, nonclinical, manufacturing, and clinical.

A decision on whether to develop a single enantiomer, racemate, or nonracemic mixture should be based on scientific data relating to quality, safety, and efficacy, with the ultimate consideration the risk/benefit ratio of the product under the conditions of use proposed. The development of a racemate or a nonracemic mixture may be justified under the following conditions:

- The enantiomers are unstable in vitro or undergo racemization in vivo.
- The enantiomers have similar pharmacokinetic, pharmacodynamic, and toxicological properties.
- It is not technically feasible to separate the enantiomers in sufficient quantity and/or with sufficient quality.
- A nonracemic mixture may be justified where a specific ratio of enantiomers is expected to improve the therapeutic profile.

Sometimes the data for that decision may not be immediately apparent at the outset of the program, so the development team must be alert to detect and decide on enantiomer vs. racemate development as early as possible since that decision will affect the remainder of the program. Perhaps the best tool is to place milestones along the development plan that query the team in terms of this effort.

5.1 Analytical Methods

The first and foremost step in any development program is to establish an orthogonal battery of analytical methods to confirm the structural identity and purity of the enantiomer desired. Typically, the methods fall into two major categories:

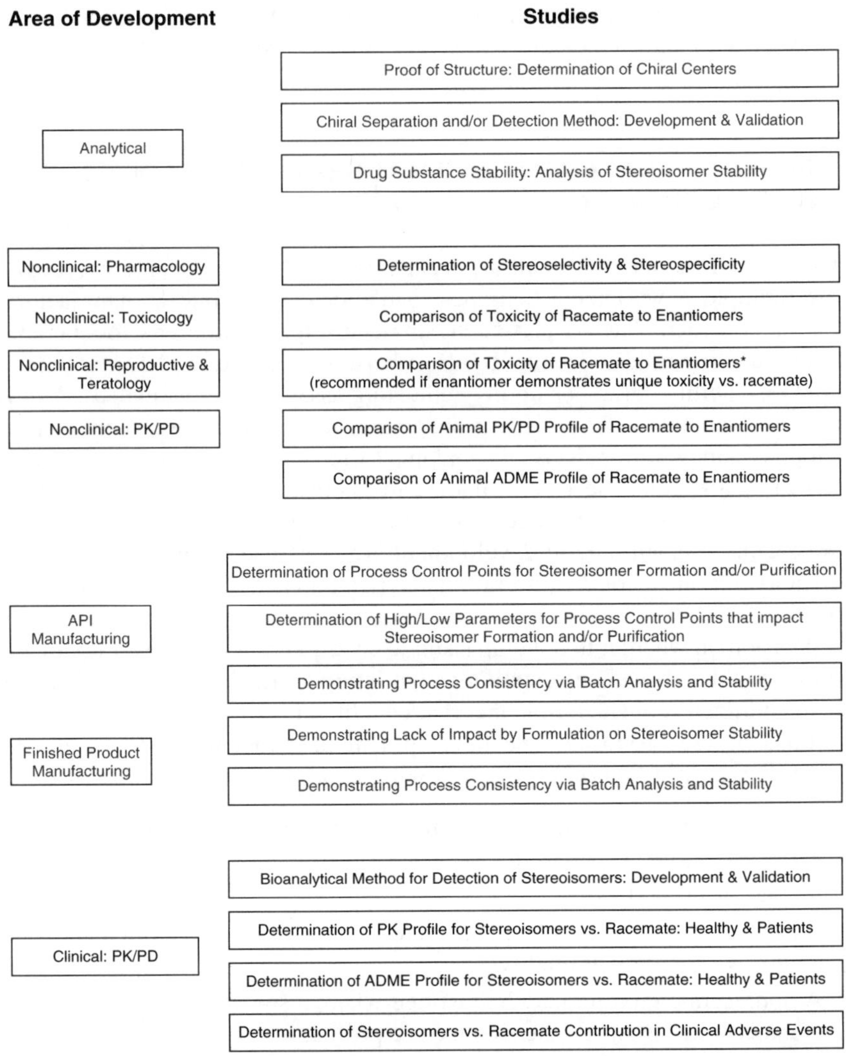

Figure 1 Development plan: outline of chiral studies.

qualitative and quantitative. The qualitative methods (e.g., NMR, circular dichroism) are extremely valuable to establish the purity and structural identity. These methods are generally used earliest in the program and usually are not rate limiting to any development plans. The quantitative methods (e.g., chiral HPLC, ORD, GC, CE) are much more valuable for ongoing assessment of drug substance purity and stability. A key part of the quantitative method validation is to establish the limit of detection (LOD) and limit of quantitation (LOQ) for the enantiomeric impurity.

The timing and extent of quantitative method development is more involved and may affect the timing of other program milestones. A key consideration in the development of quantitative methods is having adequate amounts of reference standard and individual enantiomers available, but this can be hampered by manufacturing delays in scale-up or purification. At times, synthesis of the unnatural enantiomer as a reference standard may be tedious and time consuming. However, it is still useful to proceed with early development of these methods, even with impure reference standards, to assess the linearity, LOD, and LOQ. Another aspect that is often overlooked is that these quantitative methods are important tools for in-process controls and gaining windows on manufacturing process consistency.

Another consideration in integrated analytical method development is that as new methods are validated, adequate bridging studies should be performed on drug substance and finished product retains for comparison. This is especially important when comparing drug substance used in clinical studies (e.g., PK/PD or ADME) and/or nonclinical pharmacology and toxicology studies.

The development of enantiomer-specific methods should be linked to milestones in other programs, such as assessing the stability of the scaled-up active pharmacentical ingredient (API) used in nonclinical toxicology or clinical trial formulation stability. The development of an enantiomeric specific assay for bioanalytical samples should be performed before starting animal PK/PD studies to have a better window on potential metabolic inversion or other aspects of enantiomer-specific ADME. Basically, an enantiomer-specific method should be developed as early as possible in the development program and used extensively throughout API characterization, manufacturing process controls, and bioanalytical method development. Perhaps the only exception to this approach is in the circumstance where a chiral molecule may be demonstrated to be extremely stable both in vitro and in vivo and without any potential for metabolic inversion (e.g., camptothecin derivatives).

5.2 Nonclinical Program

The nonclinical program will vary tremendously depending on the inherent toxicity of the compound and the proposed clinical indication and duration of dosing, as well as a multitude of other features unique to each program. Although it is impossible to address every permutation under the rubric of nonclinical studies, it is possible to identify salient features applicable to every type of program. For the purposes of this review, the nonclinical program is divided into pharmacology, toxicology, reproductive and teratology studies, and animal PK/PD.

Early screening tests conducted in vitro and in vivo are typically meant to identify mechanisms of action, selectivity, and some aspects of physiologic toxicity. These experiments are often conducted either as non-GLP or "GLP-like" to quickly ascertain the potential of the candidate for further development. So it is not expected that the drug substance analytical methods at this stage would have much rigor, since it would be an extraordinary expense for projects that may not proceed beyond this nascent stage. These structure–activity relationship (SAR) studies usually allow only limited evaluation, since there may be milligram-to-gram quantities of material. Even though the program is still early, it is prudent to have a sound understanding of whether or not the physiological effects are enantiomer specific. If it is possible to have selective receptor antagonists, one can study receptor binding kinetics (e.g., dissociation constants, K_d) to ascertain receptor affinity and activation by the racemate. If the program proceeds to the next level, it is practical to isolate and test the separate enantiomers against the racemate in the same battery of tests.

When programs are formally elevated to the next level of funding and/or development, the project team should flag a series of commensurate tasks in the analytical and stability areas that will support nonclinical testing. As noted below, a quantitative method for the drug substance or API should be in place to confirm the stability and purity of material used in nonclinical testing. But it is understood that some acute toxicity studies or in vitro studies (e.g., cell-based assays) may be accomplished without this in place if the material is prepared fresh from frozen stock. Enantiomeric purity may be determined by qualitative methods (e.g., optical rotation, CD) at this point.

Since the project team may not have sufficient data to decide on enantiomer vs. racemate development until after key nonclinical studies are completed, the development of an enantiomeric specific assay for bioanalytical samples can usually wait until later. However, it is critical to have this assay in place prior to animal PK/PD studies to have a better window on potential metabolic inversion or other aspects of enantiomer-specific ADME. As stated in several regulatory policies on chiral drug development, the use of an enantiomer-specific assay should be developed and validated at an early stage and used until it has been demonstrated that a nonselective assay provides results equal to those obtained with the specific assay. Once in place, the assay may be used to compare the PK/PD and ADME of the racemate to the enantiomer selected. The use of an enantiomeric-specific assay may also be essential in longer-term repeat-dose toxicity, reproductive, and teratology studies, particularly if an enantiomer is being compared to the toxicity of the racemate.

5.3 Active Pharmaceutical Ingredient: Manufacture and Characterization

The manufacture and characterization of an API are such highly integrated development programs that they are combined for the purposes of this review. Developing process controls and optimizing process parameters [e.g., quality by design (QbD), critical quality attributes] cannot occur without adequate methods as "windows" to the process, but it takes pure material to help develop and validate the methods. Thus, it is often a stepped and complementary program that uses early, less pure enantiomer to develop the analytical methods that enable process improvements, which in turn, produce higher-purity enantiomer. It is particularly important for enantiomer synthesis that chiral assays be in place to optimize the high and low parameters for enantiomer specific formation and purification.

Drug substance or API is typically characterized early in development (e.g., a lead compound program) to confirm its structure by a battery of qualitative methods. This battery of testing may suffice for the monitoring of chirality required through its early in vitro and in vivo screening in pharmacology for activity and assessment of SARs. Since the scale of synthesis at this early stage is probably bench scale, generating milligram-to-gram quantities, these qualitative tests also serve as an early indicator of process control. For example, some extractions may use multiple phases, with each pool tested for quantity and purity before combining and proceeding to the next steps.

It is not unusual to see some of these same qualitative tests used as part of the release testing of early bench-scale or pilot-scale lots until they can be replaced by quantitative methods. However, when an API is made at a scale sufficient to support GLP (good laboratory practice) nonclinical studies, there should be a commensurate quantitative assay in place to assure the overall stability of the API during the toxicologic or pharmacologic studies. This effort can be hampered by the fact that early process development may generate only scant quantities that are preciously guarded, and thus API stability is relegated to a relatively low priority. It is understood that companies do not have the resources to make every substance at the GMP (good manufacturing practice) level, but the team should impress on senior management that this method development is also vitally important to the improved process control of manufacturing. Thus, it should receive the highest priority.

It is also important to have API stability data to assure the enantiomeric purity of material stored at recommended conditions and forced degradation (e.g., heat, light, pH, oxygen, light); this can be demonstrated through some of the quantitative methods noted above. API polymorph analysis may also be meaningful at an early stage to show the stability of crystalline structure over time, which can be assessed by XRPD and DSC. Should any changes in polymorph be detected, this should be a flag to the development team to implement the development of a quantitative enantiomeric assay as a corollary to the nonclinical studies. That is, the quantitative enantiomeric assay can be used to determine enantiomeric purity of API used in the nonclinical studies to confirm the stability of material. The combined process control data, API characterization data, stability data, and QbD studies of critical parameters will allow a determination of the relative risk (i.e., low, moderate, high) for process control of enantiomers and, in turn, will drive the relative frequency of enantiomer-specific testing needed for each batch of API.

Although the regulatory guidance documents do not specify biologics or biotech-derived products, one would assume that for the generation of a single, purified API, many of these same concepts apply. A major caveat to the category of products approved under the Public Health Service Act vs. the Food, Drug, and Cosmetic Act is that several approved biologics are comprised of a pool of heterologous proteins, such as polyclonal antibodies (e.g., immune globulin intravenous, vaccine antigens, and some isozyme preparations). Given the rigor of orthogonal analytical methods used in biologics development and process validation, it is not likely that issues relating to chiral activity will be lost or missing in the program.

5.4 Finished Product: Manufacture and Characterization

The development of a prototypic finished product formulation includes API stability assessments under accelerated stability and forced degradation conditions. These tests should include an enantiomer-specific assay. Since formulation stability studies are generally short (e.g., three to six months) and under accelerated conditions, it is useful to include the same testing profile for early batches of finished product and for major manufacturing changes. Based on the combined data throughout the program, if the enantiomer stability is demonstrated in finished product, it may not be necessary to include this testing routinely for commercial product.

If changes in enantiomer profile and/or polymorph are observed during finished product manufacture, it would be prudent to incorporate enantiomer-specific testing as an in-process control to ascertain where these changes are occurring. Some processes may require significant revision if they are altering the enantiomeric purity of a finished product. For example, a wet granulation step may result in altered polymorph structure and be replaced by dry blending.

If the enantiomer-specific assay is implemented at the proper time in the development program, it can add value to characterization of the finished product release testing, stability, and optimize manufacturing process controls. This is especially important since the process development data will be used to support the process validation parameters. If the development team shows a wide range of operating parameters with no impact on enantiomer purity or polymorph, postapproval manufacturing changes will be more facile.

5.5 Clinical PK/PD

By far the most comprehensive aspect of enantiomer-selective assays and assessments will be in the clinical realm. As noted in the guidance documents and published literature, there are numerous examples of enantiomers with diverse pharmacology, toxicology, PK/PD, and metabolic profiles. Thus, the clinical program represents the most laborious and calibrated exercise to divine the true potential impact of enantiomer specific activity on safety and efficacy.

The foundation for any clinical endeavor must begin with development and validation of a bioanalytical method that is enantiomer specific. Once that is in place, the clinical group can identify the routine PK/PD parameters associated with the enantiomer as well as detect any metabolic inversion or in vivo conversion to a racemic form. This is typically done in healthy adult volunteers in late phase I or early phase II. Where it is demonstrated that the enantiomer does not undergo racemization or metabolic inversion, it is also useful to include a comparison of the PK/PD to the racemate in healthy adults.

A second tier of ADME studies should be conducted in healthy volunteers to identify metabolic routes and rates of excretion, which will aid in fleshing out what potential intrinsic (e.g., genetic enzyme deficiencies) and extrinsic (e.g., renal or hepatic impaired, certain disease states) patient populations need to be tested for comparison. The data from these patient populations will be vital for ascertaining any potential drug–drug interactions or comorbid conditions to include in the product labeling under warnings or contraindications.

The combined data from the clinical program should allow comparison of ADME of a racemate with the selected enantiomer and demonstrate the stability of the enantiomer and its metabolic fate in healthy adults as well as the proposed patient population. These data should be compared to animal PK/PD data to ascertain similar profiles vs. divergent data sets. Where a racemate is administered to patients, the primary and secondary PK/PD of each enantiomer in the nonclinical studies should be compared to the clinical PK data.

6 CONCLUSIONS

The regulatory guidance for development of chiral compounds is very consistent among regulatory bodies in the United States, the European Union, Canada, and Japan. The focus is (1) to develop enantiomer-specific methods early in the program; (2) to determine the relative pharmacological contribution of each enantiomer in vitro, in animals, and in humans compared to the racemate; and (3) to compare the toxicology profile of the racemate to the separate enantiomers to confirm their relative activity. Based on those data, the sponsor may make a scientifically based decision about whether or not to proceed with developing the racemate or a single enantiomer. An additional consideration is the use of quality by design and high–low parameter studies to determine critical process steps and their impact on enantiomer formation and purity.

Given the complexity of these programs, there have been several reviews of marketed products to ascertain if the chiral development requirements were influencing sponsors to develop enantiomers over racemic mixtures. Although the data indicate that a larger percentage of approved products are enantiomers, increasing from 27% in 1996 to 39% in 2002, it is not clear that regulatory requirements alone are driving this shift. Indeed, some of this may be a reflection of life-cycle management for some of these products as they encounter generic competition; and some of this shift reflects the cost considerations for expedited development of an enantiomer vs. a racemate. As the number of racemate drug developments and approvals declines compared to enantiomer development, it is likely that the projected growth from chiral switches will plateau. This is supported by the United States Adopted Names (USAN) experience, which showed that almost half (47%) of all drugs receiving USANs in 2004 were chiral and that the number of USANs for chiral switches were fewer.

However, one of the most pressing areas relating to chiral development is a better definition of timing and project management, which does not allow one discipline to proceed out of synch with the other parts of the program. It would behoove any chiral development team to specify these milestones, gleaned from guidance documents and industry experience, to allow a more facile determination of pros and cons of developing a racemate vs. a single enantiomer. Finally, it is critical to be vigilant throughout a chiral development program to ascertain quickly when development of a racemic mixture should cede to a specific enantiomer and what appropriate bridging studies will be required.

REFERENCES

1. E. J. Ariens. Stereochemistry, a basis for sophisticated nonsense in pharmacokinetics and clinical pharmacology. *Eur. J. Clin. Pharmacol*., 26:663–668, 1984.
2. *Investigation of Chiral Active Substances* (Directive 75/318/EEC) was first adopted in October 1993 and put into effect in April 1994. The previous title was *Clinical Investigation of Chiral Active Substances*, (II/3501/91).
3. *Investigation of Chiral Active Substances* (EMEA/CVMP/128-95-Final) was issued in June 1997, effective January 1, 1998.
4. R. Shimazawa, N. Nagai, S. Toyoshima, H. Okuda. Present state of new chiral drug development and review in Japan *J. Health Sci*., 54(1):23–29, 2008.
5. S. C. Stinson. Counting on chirality. *Chem. Eng. News*, 76(38):83–104, 2000.
6. A. M. Rouhi. Chiral business. *Chem. Eng. News*, 81(18):45–55, 2003.
7. (1913)/Louis Pasteur. In *Catholic Encyclopedia*. en.wikisource.org/wiki/Catholic_Encyclopedia_(1913)/Louis_Pasteur.
8. S. Ahuja. *Chiral Separations: Applications and Technology*. American Chemical Society, Washington, DC, 1996.
9. G. Gübitz, and M. G. Schmid. *Chiral Separations: Methods and Protocols*. Springer-Verlag, New York, 2004.
10. G. Subramanian. *Chiral Separations Techniques: A Practical Approach*. Wiley-VCH Weinheim, Germany, 2007.
11. P. J. Van Ess, J. Fleishaker. Development of chiral compounds in *Pharmacokinetics in Drug Development: Regulatory and Developmental Paradigms*, P. Bonate, D. Howard, Eds., Springer-Verlag, New York, 2004.
12. C. Álvarez, J. J. Torrado, R. Cadórniga. Stereoselective drug release from ketoprofen and ricobendazole matrix tablets. *Chirality*, 11(8):611–615, Aug. 1999.
13. H. Takahashi, H. Ogata, M. Shimizu, K. Hashimoto, K. Masuhara, K. Kashiwada, K. Someya. Relative bioavailability of two disopyramide capsules in humans based on total, unbound, and unbound enantiomer concentrations. *Biopharm. Drug Dispos*., 14(5):409–418, Feb. 8, 2006.
14. M. Miura, H. Tada, N. Yasui-Furukori, T. Uno, K. Sugawara, T. Tateishi, T. Suzuki. Effect of clarithromycin on the enantioselective disposition of lansoprazole in relation to CYP2C19 genotypes. *Chirality*, 17(6):338–344, Apr. 26, 2005.

15. Y.-W. Kwon, D. J. Triggle. Chiral aspects of drug action at ion channels: a commentary on the stereoselectivity of drug actions at voltage-gated ion channels with particular reference to verapamil actions at the Ca^{2+} channel. *Chirality*, 3(5):393–404, 1991.

16. L. Hui-chen, Y. Yang, W. Na, D. Ming, L. Jian-fang, X. Hong-yuan. Pharmacokinetics of the enantiomers of *trans*-tramadol and its active metabolite, *trans*-*O*-demethyltramadol, in healthy male and female Chinese volunteers. *Chirality*, 16(2):112–118, 2004.

17. T. N. Calvey. Isomerism and anesthetic drugs. *Acta Anaesthesiol. Scand*., 39(s106):83–90, Aug. 1995.

18. A. García-Arieta, F. Abad-Santos, M. A. Rodríguez-Martínez, Y. Varas-Polo, J. Novalbos, N. Laparidis, S. Gallego-Sandín, K. Orfanidis, J. Torrado. An eutomer/distomer ratio near unity does not justify non-enantiospecific assay methods in bioequivalence studies. *Chirality*, 17(8):470–475, 2005.

19. J. Hendel, H. Brodthagen. Entero-hepatic cycling of methotrexate estimated by use of the D-isomer as a reference marker. *Eur. J. Clin. Pharmacol*., 26(1):103–7, 1984.

20. R. J. Ott, K. M. Giacomini. Stereoselective transport of drugs across epithelia. In *Drug Stereochemistry*, 2nd ed., I. Wainer, Ed. CRC Press, Boca Raton, FL, 1993.

21. A. M. Evans, R. L. Nation, L. N. Sansom, F. Bochner, A. A. Somogyi. Stereoselective plasma protein binding of ibuprofen enantiomers. *Eur. J. Clin. Pharmacol*., 36(3):283–290, May 1989.

22. S. Toon, L. K. Low, M. Gibaldi, W. F. Trager, R. A. O'Reilly, C. H. Motley, D. A. Goulart. The warfarin–sulphinpyrazone interaction: stereochemical considerations. *Clin. Pharmacol. Ther*., 39:15–24, 1986.

23. E. Chan, A. J. McLachlan, M. Pegg, A. D. MacKay, R. B. Cole, M. Rowland. Disposition of warfarin enantiomers and metabolites in patients during multiple dosing with racemic warfarin. *Br. J. Clin. Pharmacol*., 37:563–569, 1994.

24. F. K. Główka, J. Caldwell. Protein binding of indobufen enantiomers: pharmacokinetics of free fraction—studies after single or multiple doses of rac-indobufen. *Chirality*, 14(9):736–741, Sept. 16, 2002.

25. W. Lin, T. Hayakawa, H. Yanaguimoto, M. Kuzuba, T. Obara, G. Ding, F. Cui, N. Inotsume. Pharmacokinetic interaction of ibuprofen enantiomers in rabbits. *J. Pharm. Pharmacol*., 56(3):317–321, Mar. 1, 2004.

26. J. G. Webb, J. A. Street, E. E. Bagwell, T. Walle, T. E. Gaffney. Stereoselective secretion of atenolol from PC12 cells. *J. Pharmacol. Exp. Ther*., 247(3):958–964, Dec. 1, 1988.

27. F. P. Luduena, E. F. Bogado, B. F. Tullar. Optical isomers of mepivacaine and bupivacaine. *Arch. Int. Pharmacodyn. Ther*., 200(2):359–369, Dec. 1972.

28. G. Aberg. Toxicological and local anesthetic effects of optically active isomers of two local anesthetic compounds. *Acta Pharmacol. Toxicol*. (*Copenhagen*), 31:273–286, 1972.

29. C. Aps, F. Reynolds. An intradermal study of the local anesthetic and vascular effects of the isomers of bupivacaine. *Br. J. Clin. Pharmacol*., 6:63–68, 1978.

30. S. Toon, K. J. Hopkins, F. M. Garstang, L. Aarons, A. Sedman, M. Rowland. Enoxacin–warfarin interaction: pharmacokinetic and stereochemical aspects. *Clin. Pharmacol. Ther*., 42:33–41, 1987.

31. L. S. Kaminsky, S.-Y. Zhang. Human P450 metabolism of warfarin. *Pharmacol. Ther.*, 73(1):67–74, 1997.

32. S. A. Ward, T. Walle, K. Walle, G. R. Wilkinson, R. A. Branch. Propranolol's metabolism is determined by both mephenytoin and debrisoquine hydroxylase activities. *Clin. Pharmacol. Ther.*, 45:72–79, 1989.

33. T. Walle, U. K. Walle, T. D. Cowart, E. C. Conradi, T. E. Gaffney. Selective induction of propranolol metabolism by smoking: additional effects on renal clearance of metabolites. *J. Pharmacol. Exp. Ther.*, 241:928–933, 1987.

34. B. Vogelgesang, H. Echizen, E. Schmidt, M. Eichelbaum. Stereoselective first-pass metabolism of highly cleared drugs: studies of the bioavailability of L- and D-verapamil examined with a stable isotope technique. *Br. J. Clin. Pharmacol.*, 18(5):733–740, Nov. 1984.

35. M. Eichelbaum, G. Mikus, B. Vogelgesang. Pharmacokinetics of (+)-, (−)-, and (±)-verapamil after intravenous administration. *Br. J. Clin. Pharmacol.*, 17:453–458, Nov. 1984.

36. J. J. Lima, H. Boudoulas, B. J. Shields. Stereoselective pharmacokinetics of disopyramide enantiomers in man. *Am. Soc. Pharmacol. Exp. Ther.*, 13(5):572–577, Sept. 1, 1985.

37. L. Borgström, L. Nyberg, S. Jönsson, C. Lindberg, J. Paulson. Pharmacokinetic evaluation in man of terbutaline given as separate enantiomers and as the racemate. *Br. J. Clin. Pharmacol.*, 27(1):49–56, Jan. 1989.

38. M. Gibaldi. Stereoselective and isozyme-selective drug interactions. *Chirality*, 5:407–413, 1993.

39. R. A. O'Reilly, P. M. Aggeler. Studies on coumarin anticoagulant drugs: initiation of warfarin with a loading dose. *Circulation*. 38:169–177, 1968.

40. J. M. Eble, B. D. West, K. P. Link. A comparison of the isomers of warfarin. *Biochem. Pharmacol.*, 15:1003–1006, 1966.

41. C. Banfield, R. O'Reilly, E. Chan, M. Rowland. Phenylbutazone–warfarin interaction in man: further stereochemical and metabolic considerations. *Br. J. Clin. Pharmacol.*, 16(6):669–675, Dec. 1983.

42. H. H. Zhou, L. B. Anthony, A. J. Wood, G. R. Wilkinson. Induction of polymorphic 4′-hydroxylation of *S*-mephenytoin by rifampicin. *Br. J. Clin. Pharmacol.*, 30(3):471–475, Sept. 1990.

43. H. K. Kroemer, C. Fischer, C. O. Meese, M. Eichelbaum. Enantiomer/enantiomer interaction of (*S*)- and (*R*)-propafonone for cytochrome P450IID6 catalyzed 5-hydroxylation: in vitro evaluation of the mechanism. *Mol. Pharmacol.*, 40:135–142, 1991.

44. M. Yao, M. Zhu, S.-Y. Chang, D. Zang, A. D. Rodrigues. Applications of recombinant and purified human drug metabolizing enzymes: an industrial perspective. In *Handbook of Drug Metabolism*, 2nd ed., P. G. Pearson, L. C. Wienkers. Eds.; Informa Healthcare, New York, 2009.

CHAPTER 3

Basic Considerations in HPLC Method Development of Chiral Compounds

SATINDER AHUJA
Ahuja Consulting, Calabash, North Carolina

1 INTRODUCTION

Chiral molecules (from the Greek word *cheiro*, meaning "hand") relate to each other as an object and its mirror image that are not superimposable (i.e., they are like a pair of hands). These molecules, also called *enantiomers*, can have significant differences in biological activity. The differences in pharmacological, metabolic, or toxicologic activity relate to the arrangements of atoms in a molecule (i.e., its stereochemistry). This explains why regulatory requirements in the pharmaceutical industry demand detailed investigations of chiral molecules [1,2]. Small molecules are covered in this chapter; large biomolecules are discussed in Chapter 15.

The reader may find of some interest earlier publications and presentations that deal at length with this subject [1–3]. Before initiating method development, it is desirable to develop a basic knowledge of stereochemistry [1–7]. Some basic information on stereochemistry is provided in Section 2.1 to help the reader develop a better understanding of the separation mechanisms that come into play in various separation methods used for chiral compounds. This knowledge will help the reader select a suitable chiral separation method, based on the molecular structure of the chiral compound of interest. Logical reasons for the selection process are discussed later in the chapter.

Chiral Separation Methods for Pharmaceutical and Biotechnological Products, Edited by Satinder Ahuja

2 DETERMINING THE CHIRALITY OF A MOLECULE

Look at the structure of the molecule that has to be resolved, and then answer the following question: Is there a stereogenic center? The simplest example of a stereogenic center is an asymmetric carbon with four different substituents. This process appears to be more cumbersome when you are looking at many carbon atoms in a molecular structure. The simplest way to solve this problem is to number all carbon atoms in the structure and look at each in turn to see whether they are asymmetric.

1. If one asymmetric carbon or stereogenic center is found, you can expect two enantiomers. For two stereogenic centers, the number of enantiomers is four. It should be clear that this number grows rapidly as the number of asymmetric centers increases; the 2^n rule applies, except for fused rings where n is the number of asymmetric centers.
2. If no asymmetric carbon is found, look at the plane of symmetry of the entire molecule and/or other atoms, such as sulfur and nitrogen, which can also confer chirality.

2.1 Stereoisomerism

Molecules that are isomeric but have a different spatial arrangement are called *stereoisomers*. Molecular symmetry classifies stereoisomers as either *enantiomers*, as defined above, or *diastereomers* (not related as mirror images). Stereoisomerism can result from a variety of sources besides the single chiral carbon mentioned above. These topics are discussed in detail in several books and review articles [1,3–10]; a short summary is provided here. There are two simple molecular sources of chirality: molecules having a stereogenic center and those having a stereogenic axis. Stereoisomerism is possible in molecules that have one or more centers of chirality; helicity; planar, axial, or torsional chirality; or topological asymmetry.

The amounts of energy necessary to convert given stereoisomers into their isomeric forms can be used for their classification. Stereoisomers with low-energy barriers to this conversion are termed *conformational isomers*; high-energy-barrier conversions are described as *configurational isomers*. Diastereomers differ in energy content, and thus in every physical and chemical property; however, the differences may be so minute as to be nearly indistinguishable.

2.2 Stereochemistry and Biological Activity

The importance of determining the stereoisomeric composition of chemical compounds, especially those of pharmaceutical importance, has been well recognized for some time [11–14]. However, a fairly large number of drugs are still used in racemic or diastereomeric forms. It is important to remember that some enantiomers may exhibit potentially different pharmacologic activities, and the patient may be taking a useless or even undesirable enantiomer when ingesting a

TABLE 1 Activities of Some Chiral Compounds

Compound	Activity
Amphetamine	The D-isomer is a potent central nervous system stimulant, whereas the L-isomer has little effect.
Propoxyphene	The α-L is antitussive; α-D is analgesic.
Epinephrine	The L-isomer is 10 times more active as a vasoconstrictor than is D-isomer.
Synephrine	The L-isomer has 60 times the pressor activity of the D-isomer.
Propanolol	The *S*(−) isomer has only β-adrenergic blocking activity.
Warfarin	The *S*(−) isomer is a five times more potent anticoagulant than the *R*(+) isomer.
Ascorbic acid	The D-isomer is antiascorbutic, whereas the L-isomer is not active.

racemic mixture. To ensure the safety and effect of both drugs used currently and those newly developing, it is necessary to isolate and examine both enantiomers separately. At least three situations demand that we measure and control the stereochemical composition of drugs. Each situation presents a specific technical problem during (1) manufacture, where problems of preparative-scale separations may be involved; (2) quality control (or regulatory analysis), where analytical questions of purity and stability predominate; and (3) metabolic and pharmacologic studies of plasma disposition and drug efficacy, where ultratrace methods may be required [3].

Frequently, one isomer of a series may produce a desired effect while another may be inactive or even produce some undesired effect. Furthermore, accurate assessment of the steroisomeric purity of substances is critical since stereoisomeric impurities may have unwanted toxicologic, pharmacologic, or metabolic effects. Such impurities may be carried through the synthesis, preferentially react at one or more steps, and yield an undesirable level of another impurity. Some examples of activity differences are given in Table 1. Large differences in activity between stereoisomers point out the need to accurately assess isomeric purity of pharmaceuticals.

3 REGULATORY REQUIREMENTS

The U.S. Food and Drug Administration (FDA) issued a set of initial guidelines in 1987 on the submission of new drug applications (NDAs), where questions relating to stereochemistry were approached directly in the guidelines on the manufacture of drug substances [15,16]. The finalized guidelines require a full description of the methods used in the manufacture of the drug, including testing to demonstrate its identity, strength, quality, and purity. Submissions to the FDA should show the applicant's knowledge of the molecular structure of the drug substance. For chiral compounds, this includes identification of all chiral centers. The enantiomer ratio, although 50 : 50 by the definition for a racemate, should be defined for any other admixture of stereoisomers. The proof of structure should consider stereochemistry and provide appropriate descriptions of the molecular structure. An enantiomeric form can be considered an impurity, and therefore it is desirable to explore potential in vivo differences between these forms. For further information on regulatory requirements, see Chapter 2.

4 SEPARATION METHODS

Even though many experts claim that chiral separations are easy [2], some investigators in chiral separations have a difficult time determining where to start. The first step is to determine which method to use for a given separation. Cost considerations, availability of equipment, and know-how play important roles in the selection process [1–7]. Paper chromatography (PC) and thin-layer chromatography (TLC) have been used where cost considerations are overwhelming factors. PC provided a basis for the three-point interaction model proposed by Dalgliesh for chiral separations (see additional information later in the chapter). It is used very rarely these days; however, TLC can be a very useful qualitative technique that entails minimal costs. It can provide good indications as to which HPLC method would be best suited for resolving enantiomers. Of course, it can also be used as an independent technique even with its limitations of resolution and low precision. A significant amount of coverage was provided in an earlier text [3] to enable the reader to try TLC; it also included a number of reference sources for TLC aficionados. Commonly used methods for the separation of enantiomers today can be broadly classified into the following four categories:

1. Gas chromatography (GC)
2. High-performance liquid chromatography (HPLC)
3. Supercritical fluid chromatography (SFC)
4. Capillary electrophoresis (CE)

It is readily apparent that all of these methods are based on chromatography, except for capillary electrophoresis, in which two phases are not involved in the separation process. It may be recalled that the two phases in chromatography are designated the stationary phase and the mobile phase, based on their role in the separation process. Technically, there is no stationary phase in capillary electrophoresis unless the capillary walls are assigned that role. Some chromatographers promote this concept, but it is not entirely correct. A modification of this technique called *capillary electrochromatography* (CEC) has found numerous applications. It is worthwhile noting here that CE is a very useful technique for enantiomeric separations (see Chapter 11).

Chromatographic methods are generally favored over nonchromatographic methods for separation of chiral compounds, based on the information available in scientific literature. All of these methods except PC and TLC are discussed at length in later chapters. However, it should be noted that of the various methods mentioned above, HPLC methods are likely to be the most useful (exceptions may be found in the literature [3,7–9]). The next important step is to select an appropriate column, based on various considerations (see the discussion below). The primary concern for any investigator in this area, notwithstanding the high cost of the columns, is which column will work best in a particular situation. In this chapter we answer these questions and provide examples that will help the reader make intelligent decisions in this complex field.

Over 40 years ago, systematic research was initiated for the design of chiral stationary phases functioning to separate enantiomers by gas chromatography. This led to molecular design and preparation of chiral-phase systems for liquid chromatography. Lately, these efforts have been directed toward finding new types of chiral stationary and mobile phases on the basis of the stereochemical viewpoint. Many factors may be responsible for the extent of interactions of stereoisomeric molecules in any environment, such as dipole–dipole interactions, electrostatic forces, hydrogen bonding, hydrophobic bonding, inductive effects, ion–dipole interactions, ligand formation, partition coefficient differences, pK differences, resonance interactions and stabilization, solubilities, steric interference (size, orientation, and spacing of groups), structural rigidity and conformational flexibility, temperature, and van der Waals forces. The nature and effects of some of these factors can influence the chromatography of stereoisomers; therefore, they should be reviewed carefully before developing a separation method (see Sections 5.1 and 6).

5 SEPARATION OF CHIRAL COMPOUNDS BY HPLC

It is generally recognized now that HPLC methods offer distinct advantages over classic techniques in the separation and analysis of stereoisomers, especially for enantiomers that are generally much more difficult to separate. HPLC methods show promise for moderate-scale separations of synthetic intermediates as well as for final products [17–21]. For large-scale separations and in consideration of the cost of plant-scale resolution processes (see [7], Chapter 2), the separation methods offer substantial increases in efficiency over recrystallization techniques.

There are basically two approaches to the separation of an enantiomer pair by HPLC: indirect and direct. In the indirect approach, used rarely today, the enantiomers may be converted into covalent, diastereomeric compounds by a reaction with a chiral reagent, and these diastereomers are typically separated on a routine achiral stationary phase. In the direct approach, which is often called *chiral HPLC*, the enantiomers or their derivatives are passed through a column containing a chiral stationary phase (CSP). Chiral HPLC and SFC (see Chapter 13) are ideally suited for large-scale preparation of optical isomers.

As mentioned above, the chromatographic separation of enantiomers can be achieved by various methods; however, it is generally desirable to use some type of chiral discriminator or selector [22,23]. Two different types of selectors can be distinguished: a chiral additive in the mobile phase or a chiral stationary phase.

Of these two selectors, chiral stationary phases are in more common use for separations of enantiomers, and they are discussed at some length below.

5.1 Basic Information on the Mechanisms of Chiral Separations by HPLC

To develop an optimum method of chiral separation, it is important to understand the mechanisms. Our understanding of chiral separations with some of the systems is quite good, but it remains poor for proteins. The mechanism of separations on polysaccharide stationary phases is discussed in later chapters. The separation basis with various chiral stationary phases in their respective group is discussed below; only some general comments are included here. A number of chiral recognition models have been proposed to account for optical resolutions by HPLC; these are often based on the three-point interaction rule advanced by Dalgliesh [24] in 1952. He arrived at this conclusion from paper chromatographic studies of certain aromatic amino acids. He assumed that the hydroxyl groups of the cellulose were hydrogen-bonded to the amino carboxyl groups of the amino acid. A third interaction was caused, according to these views, by the aromatic ring substituents. It led to the postulation that three interactions operating simultaneously between an enantiomer and the stationary phase are needed for chiral discrimination. However, this is not always necessary, as steric discrimination could also result from steric interactions.

Chiral separations are also possible through reversible diastereomeric association between an enantiomeric solute and a chiral environment that is introduced into the column. Because chromatographic resolutions are possible under a variety of conditions, it may be concluded that the necessary difference in association can be obtained by many types of molecular interactions. The association, which may be expressed quantitatively as an equilibrium constant, is a function of the magnitudes of the binding as well as the repulsive interactions involved. The latter are usually steric, although dipole–dipole repulsions could also occur, whereas various kinds of binding interactions may operate. These include hydrogen bonding, electrostatic and dipole–dipole attractions, charge-transfer interaction, and hydrophobic interaction (in aqueous systems).

With CSPs, where steric fit is of primary importance, including those based on inclusion phenomena, such as cyclodextrin and crown ether phases, it is possible to construct chiral cavities for the preferential inclusion of only one enantiomer. Molecular imprinting techniques are very interesting in this regard [3,25]. They create rigid chiral cavities in a polymer network in such a way that only one of two enantiomers will find the environment acceptable, and as a result, separation occurs.

6 A PRIMER FOR SELECTING A SUITABLE CSP

As mentioned earlier, it is important to study an analyte to determine what types of interactions it can bring about. This information is very helpful in selection

TABLE 2 Conventional Classification of Chiral Stationary Phases

Type	Description	Examples	Usual Mode
1	Brush type (donor–acceptor)	DNB-glycine, DNB-leucine, naphthylalanine, Whelk-O 1, etc.	Normal phase[a] (polar modifier)
2	Polysaccharides	Chiralcel AD, OD, AS, OJ, etc.	Normal phase[a] (polar modifier)
3	Inclusion CSPs (e.g., cyclodextrins, chirobiotics)	Cyclobond 1–3, Chiralpak OP, OT, etc., Chirobiotic V, T, etc.	Reversed phase (e.g., aqueous MeCN, aqueous MeOH)
4	Ligand exchangers	Proline, OH-proline, etc.	Reversed phase (e.g., aqueous buffers)
5	Proteins	Albumin, glycoproteins, human serum albumin, etc.	Reversed phase (e.g., aqueous buffers)

[a]Reversed-phase separations are also possible.

of the desired CSP. A conventional classification of common chiral stationary phases is provided in Table 2. The table also provides a quick review of various interactions on some CSPs. The chromatographic analysis of entantiomeric compounds can be based on the separation mode: normal-phase or reversed-phase mode, depending on the chromatographer's preference [3]. A conventional flow diagram (Figure 1) is provided to help the reader select an appropriate CSP. It should be noted that the preferred approach is to select a CSP based on various considerations discussed below.

6.1 Brush-Type Columns (Type 1)

The chiral stationary phase in brush-type, or Pirkle, columns is composed of various selectors capable of ionic or covalent bonding. These CSPs are generally composed of an optically pure amino acid bonded to γ-aminopropyl–silanized silica gel. An amide or urea linking of a π-electron group to the asymmetric center of amino acid provides π-electron interactions and one point of chiral recognition. These CSPs were designed on the premise that at least three points of interaction are necessary between the chiral molecule and the CSP. One of these interactions must be stereochemically dependent. When three points of interaction occur, a transient diastereomeric complex is formed between the solute and the CSP. The π acids, π bases, hydrogen-bond donors and/or acceptors, amide dipoles, and other functional groups linked to a chiral selector of the CSP provide the required three-point interaction and confer overall selectivity to the column. Other points of attractive and/or repulsive interactions also contribute to enantioselectivity. The compounds that can be separated with these CSPs should:

- Form hydrogen bonds

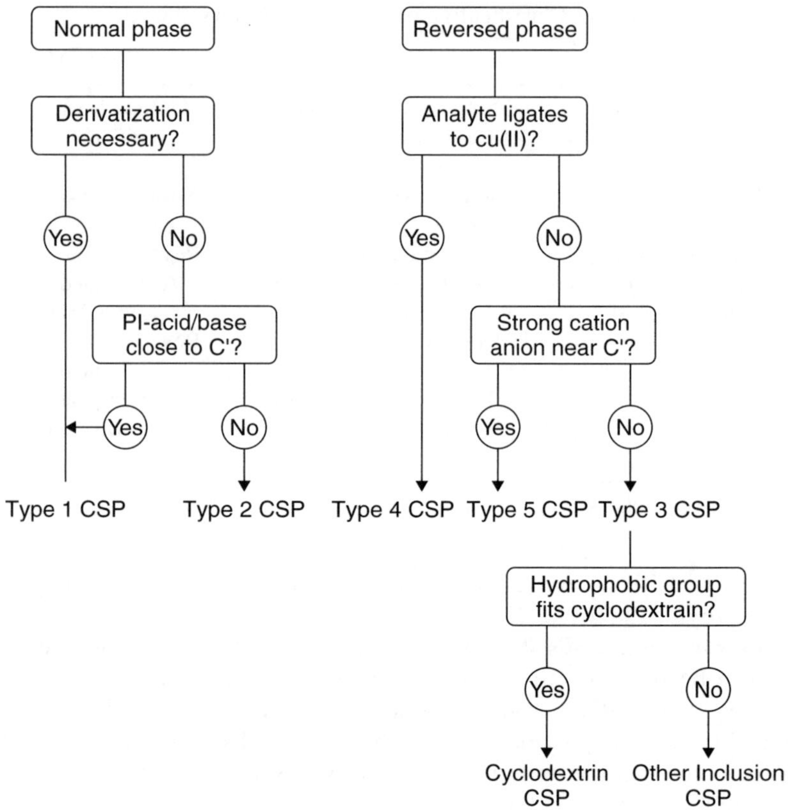

Figure 1 Flow diagram to select an appropriate CSP. (From [3].)

- Have $\pi - \pi$ interactions
- Have dipole stacking
- Have other attractive interactions

If the sample of interest does not contain the necessary sites of recognition, the sites have to be added by forming appropriate derivatives, a process not too popular with separation scientists. Enantiomeric compounds that can be separated on these columns are as follows: amines, amino acids, carboxylic acids, esters, ethers, hydroxy acids, ketones, and lactones.

Two types of columns can be used. For the older brush-type columns, derivatization was frequently required. This was cumbersome, and it complicated separations unnecessarily. Columns such as α-Burke and Whelk-O that obviate the need for derivatization have evolved. Examples of underivatized separations on these columns are given below.

Underivatized Separations An α-Burke II column is useful for resolution of metoprolol and related compounds. A mobile phase composed of methylene chloride/ethanol/methanol (85 : 10 : 5) containing 10-mm ammonium acetate is used at a flow rate of 1 mL/min for a 4.6-mm ID × 25-cm column. The α value of the enantiomers is 1.28. Related compounds, such as alprenolol, atenolol, betaxolol, bufralol, bupranolol, oxprenolol, practolol, pindolol, pronethalol, and propanolol, can be resolved with slight modifications of mobile phases.

The Whelk-O 1 column was originally developed for separating the enantiomers of naproxen (Figure 2). Naproxen can be resolved by using a normal-phase method with hexane/IPA/HOAc (80 : 20 : 0.5) on Whelk-O 1 (4.6 mm ID × 25 cm) at a flow rate of 1 mL/min in 16 min [7]. Alternatively, methanol/0.1% phosphate (60 : 40) can be used at the same flow rate with the same run time. The α value is 2.1 in the normal-phase mode and 1.7 in the reversed-phase mode. Other arylpropionic acids that have been resolved in the normal-phase mode are ibuprofen, ketoprofen, and flurbiprofen. Various pharmaceuticals that have been resolved on this column are cyclothiazide, bendroflumethiazide, oxazepam, mephenytoin, bupivacaine, and *p*-chlorowarfarin.

An improved version of this column, Whelk-O 2, has recently been designed to improve the resistance of stationary phase to hydrolysis while using strong organic modifiers such as trifluoroacetic acid.

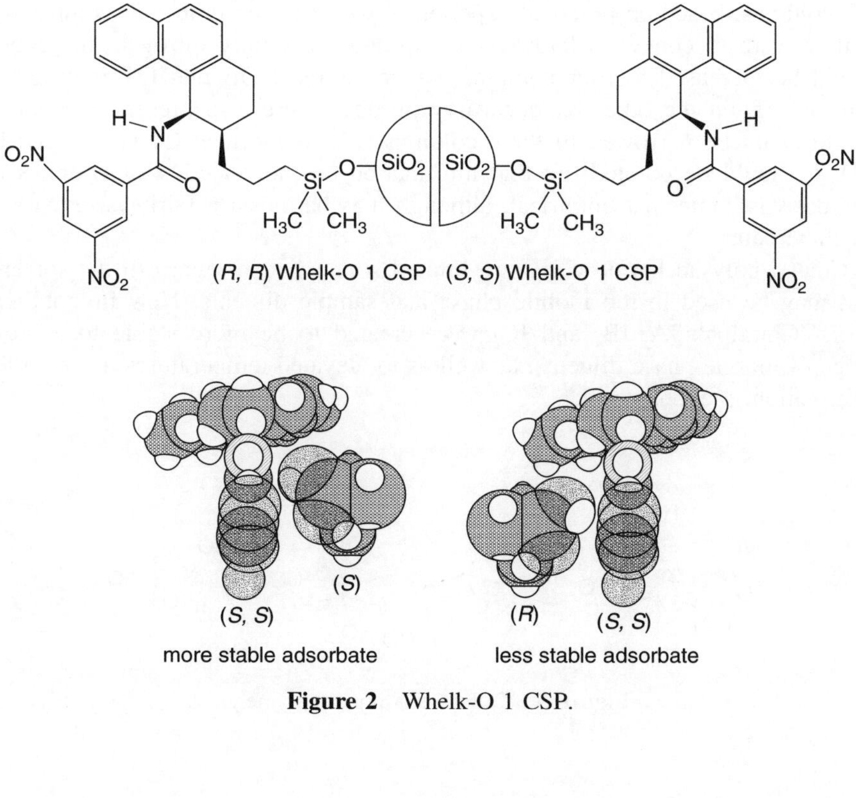

Figure 2 Whelk-O 1 CSP.

6.2 Polysaccharide Columns (Type 2)

As mentioned above, Dalgliesh was the first to realize the resolving capacity of polysaccharides, such as cellulose, by his observation that a racemic amino acid could occasionally give two spots in paper chromatography. It is also possible to achieve optical resolution of amino acids on cellulose by using TLC. This has led to the use of cellulose and cellulose derivatives, as well as amylose, for chiral separations by liquid chromatography.

Cellulose is a linear polymer that has the chemical constitution of poly-β-D-1,4-glucoside (see Figure 3). It forms long chains of at least 1500 (+)-D-glucose units per molecule. The molecular weight of cellulose ranges from 2.5×10^5 to 1×10^6 or more. Each of the (+)-D-glucose repetitive units contains five chiral centers and three hydroxyl groups. All the ring substituents are equatorial.

Broadly speaking, the substituted cellulose columns can be divided into two major categories: cellulose esters and cellulose carbamates. The popular phases (Chiralpak AD and AS and Chiralcel OD, OJ, OB, and OC) are now available on 5-μm silica. These columns have found a wide range of applicability. A handy guide to selection of columns based on functionalities of the solute is provided in Figure 4.

The compounds to be resolved can be divided into four major categories, based on whether they contain a given functionality (e.g., aromatic carbonyl, tertiary nitrogen, or hydroxyl). Aromatic compounds containing carbonyl groups and no hydroxyls or tertiary nitrogen are more readily separated on Chiracel OJ, OD, and OC columns, based on practical experience with a large number of compounds. If they were to contain a hydroxyl group and no tertiary nitrogen, the choice would be increased to four columns, where Chiracel OB and OT replace OD and OC. If, on the other hand, tertiary nitrogen were also present, the choice would again be narrowed to three columns, where Chiracel OJ is replaced by AD. It should be pointed out that information on the selection of columns on this basis is somewhat empirical, although it is based on a fairly large base of practical data.

Coated polysaccharide CSPs are limited somewhat in regard to the solvents that may be used in the mobile phase and sample diluents. New immobilized CSPs (Chiralpak IA, IB, and IC) were created to be more stable to a broad range of mobile-phase diluents, as well as to elevated temperatures. (For further information, see Chapter 4.)

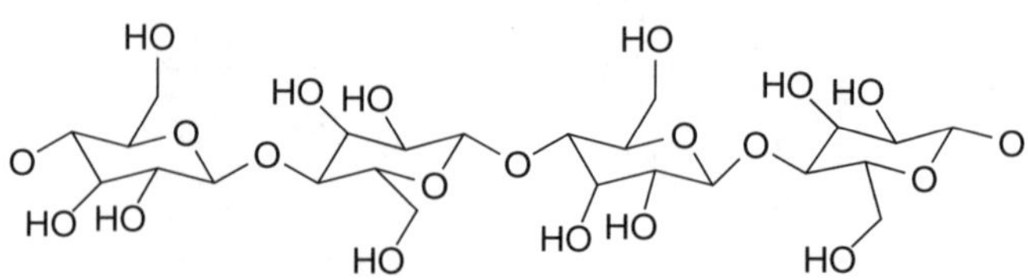

Figure 3 Cellulose, a linear polymer.

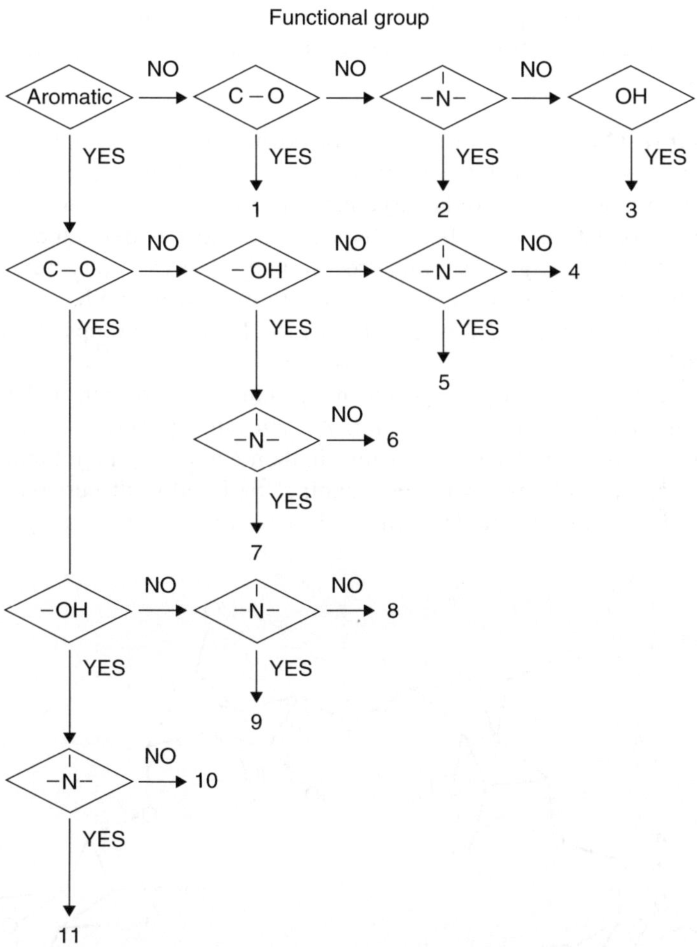

1. Chiralcel OB, OD, AD, OJ, MA, WH
2. Chiralcel CR, WF, WH
3. Chiralcel AD, OB, OD, OJ
4. Chiralcel OD, OJ, OC
5. Chiralcel OC, OD, OJ
6. Chiralcel AD, OD, OJ, OB
7. Chiralcel CR, OJ, OD
8. Chiralcel OB, OJ, OD, OT
9. Chiralcel OB, OJ, OD, OC, OT, WH
10. Chiralcel OD, OJ
11. Chiralcel OC, OD, AD

Figure 4 Guide to selection of columns, based on functionalities of the solute. (From [3].)

6.3 Inclusion CSPs (Type 3)

Chiral discrimination on a stationary phase can be achieved by creation of chiral cavities, in which stereoselective guest–host interactions influence the resolution. Included in this group are cyclodextrins, crown ethers, polyacrylates, and polyacrylamides. The applicability of this approach has been expanded by utilizing macrocyclic antibiotics as chiral stationary phases by Armstrong and colleagues.

Cyclodextrin Columns Various cyclodextrins are produced by the partial degradation of starch and the enzymatic coupling of cleaved glucose units into crystalline, homogeneous toroidal structures of different molecular sizes. Alpha-, β-, and γ-cyclodextrins have been characterized most widely; they are chiral and contain 6, 7, and 8 glucose units, respectively. For example, β-cyclodextrin (Figure 5) has 35 stereogenic centers, and the toroidal structure has a hydrophilic surface resulting from the 2-, 3-, and 6-position hydroxyl groups, making them water soluble.

A cyclodextrin cavity is composed of the glucoside oxygen and methylene hydrogen, giving it an apolar character. As a result, cyclodextrins can include polar molecules of appropriate dimensions in their cavities and bind them through dipole–dipole interactions, hydrogen bonding, or London dispersion forces. In general, cyclodextrins are stable from pH 3 to 14. Listed below is a general plan

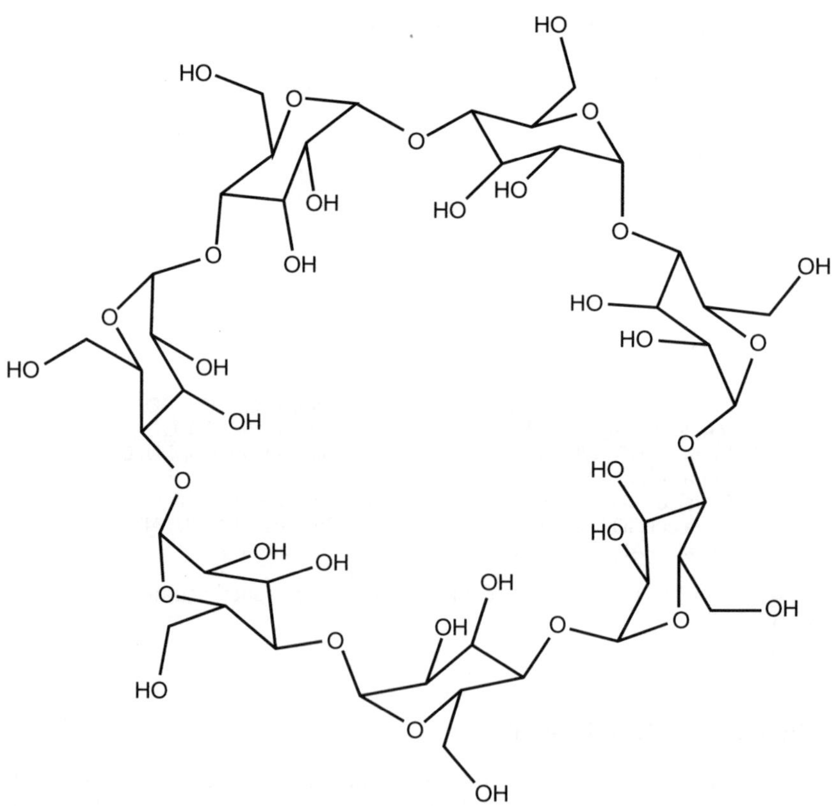

Figure 5 β-Cyclodextrin.

that can be utilized for making initial choices. β-Cyclodextrin has been used in the largest number of applications because it has been found useful for low-molecular-weight analytes in the pharmaceutical and environmental areas. These columns can be operated in reversed-phase or normal-phase mode.

To achieve a chiral separation in the reversed-phase mode, it is essential that the analyte have at least one aromatic ring. Exceptions are heterocyclic analytes and *t*-boc amino acids. For cyclodextrin inclusion, the molecular weight of the polyaromatic ring structure is not as critical as its bulk. The most important consideration is proper fit of the molecule in the cyclodextrin cavity. This fit is a function of the size and shape of the analyte relative to the cavity. For example, an analyte such as norgestrel, a five-ring steroid structure, is better separated on a γ-cyclodextrin column, whereas the enantiomer of a naphthalene-like structure or single substituted aromatic ring would fit better on a β-cyclodextrin column. For chiral recognition, solvent strength is independent in most cases because it affects only displacement of the analyte from the cavity [3].

Normal-phase separations on cyclodextrin columns have generally been carried out with mobile phases such as hexane/isopropyl alcohol, acetonitrile/methanol, methanol, or ethanol. π–π Hydrogen bonding forces influence primarily enantiomeric separations. It is possible to override inclusion complexation in favor of interacting directly with secondary hydroxyl groups across the larger opening of a cyclodextrin toroid or the appendant carbamate, acetate, or hydroxypropyl functional groups. β-Blockers such as propanolol, timolol, and atenolol, and compounds such as warfarin can be separated on these columns.

Bonded Cyclodextrins A number of bonded cyclodextrins are currently available. The carbamate coupling of the π bases, 1-naphthylethyl to a bonded cyclodextrin, creates a complex environment that has demonstrated diverse chiral separations. It has been labeled a multimodal chiral stationary phase because it can be used in either a normal-or reversed-phase mode with appropriate modifiers. Proper selection of the mobile phase is dictated by analyte structure, solubility, and stability. For example, if the analyte is π acidic, normal-phase solvents can be used. If the analyte is not π acidic but contains two hydrogen-bonding groups, one on or near the stereogenic center, polar modifiers must be used. Of the three carbamates available, the (*S*)-naphthylethylcarbamate has shown the greatest selectivity and versatility. Because the naphthylethylcarbamate plays a role in enantioselectivity, the (*R*)-form can be useful if the separation does not occur on the (*S*)-form. A variety of compounds, such as catelol, labetolol, nadolol, metoprolol, pindolol, propanolol, timolol, oxazepam, suprofen, comuachlor, and warfarin, have been resolved on cyclodextrin columns [3].

Macrocyclic Antibiotics The chiral selectors in this group include vancomycin, ristocetin A, teicoplanin, avoparcin, rifamycin B, and thiostrepon. Chirobiotic V is based on covalently bonded amphoteric glycopeptide vancomycin to 5-μm silica gel. These ligands are linked to assure their stability while retaining essential components for chiral interactions. For example,

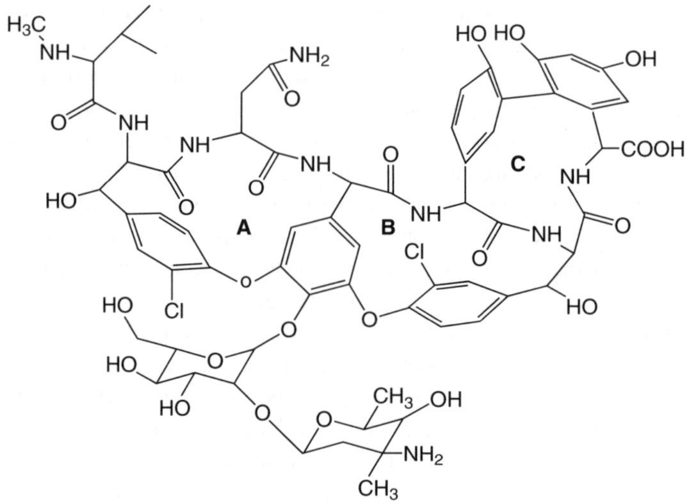

General Physical Data:
Molecular Weight - 1449
pK's - 2.9, 7.2, 8.6, 9.6, 10.4, 11.7
Isoelectric Point - 7.2
Chiral Centers - 18
Sugar Moieties - 2
Inclusion Cavities - A,B,C

Figure 6 Vancomycin.

vancomycin (Figure 6) contains 18 chiral centers surrounding three pockets or cavities. Five aromatic ring structures bridge these cavities. Hydrogen donor and acceptor sites are readily available close to the ring structures. It has been claimed that the selectivity of this column is similar to that of glycoprotein AGP (see Section 6.5), and it is stable when 0 to 100% organic modifier is used.

Vancomycin has exhibited broad selectivity in both reversed- and normal-phase solvents, and limited selectivity with polar organic mobile phases. Since vancomycin contains peptide, carbohydrate, and other ionizable groups, it would be expected to offer different selectivity in these modes. The structure of vancomycin indicates that various typical interactions outlined for protein phases and other cellulosic polymer-type phases are possible with this phase. The potential interactions and their relative strengths are as follows: $\pi-\pi$ interactions, very strong; inclusion, weak because of shallow pockets; dipole stacking, medium to strong; steric interactions, weak; and anionic or cationic binding, strong.

Compared to cyclodextrins, the shallow pockets for inclusion yield weaker energies. This leads to faster kinetics, which can in turn lead to faster separations. Reversed-phase conditions favor inclusion and hydrogen bonding. Under these conditions, changes in pH can produce cationic or anionic interactions. Dipole stacking and $\pi-\pi$ complexation are favored in normal-phase solvents. Polar organic mobile phases enhance the potential for all the interactions noted above. Analytes such as acids, amides, esters, and neutral compounds can be resolved.

Chirobiotic T is based on bonding the amphoteric glycoside teicoplanin to a 5-μm silica gel through covalent linkage. Teicoplanin contains 20 chiral centers surrounding four pockets or cavities. Hydrogen donor and acceptor sites are readily available close to seven aromatic rings.

A wide range of amino acid derivatives has been resolved on these columns. Analytes such as neutral molecules, amides, acids, esters, and cyclic amines show considerable enantioselectivity. Other amines have been separated with varying degrees of success.

Selection of Mobile Phases The reversed- or normal-phase mobile phases function equally because of the complex structure of the macrolide and ionizable groups. In the reversed-phase mode, optimization is accomplished by controlling the type and amount of organic modifier, type of buffer, and pH level. Efficiency and selectivity are affected by ionic strength, buffer type, flow rate, and temperature. Of the various organic modifiers, tetrahydrofuran (THF) gives the greatest selectivity and efficiency. Typical composition of organic modifier/buffer (pH 4.0 to 7.0) is 10 : 90. Alcohols as modifiers generally require higher concentrations, e.g., 20% for comparable retention to acetonitrile or THF. Ammonium nitrate and triethylamine acetate buffers have been found useful. In general, analytes act more favorably at a pH where they are not ionized. Lower column temperatures are favored because of the enhancement of weaker bonding forces.

In normal-phase separations, peak efficiency and resolution can be improved with ethanol as the polar modifier of hexane. A good starting point may be 20% ethanol. In most cases, ethanol works better than isopropanol.

Applications The comparability of separations on Chirobiotic T and V for warfarin is shown in Figure 7. Benoxaprofen, ibuprofen, fenoterol, mephobarbital, naproxen, warfarin, albuterol, citrulline, dopa, and phenylalanine have been resolved on chirobiotic columns [3]. Recently, Chirobiotic V2 and T2 have been created to produce higher selectivity and capacity.

6.4 Ligand Exchange (Type 4)

Davankov and Kurganov were the first to indicate that cross-linked resins with fixed ligands, (R)-N, N'-dibenzyl-1,2-propanediamine in the form of copper(II) complexes, display high enantioselectivity for alanine, serine, and leucine [18]. Various amino acids, including baclofen [3], can be resolved on a reversed-phase C_{18} column with a chiral mobile phase of aqueous cupric acetate and N, N-di-n-propyl-L-alanine (DPA) containing 15% acetonitrile [19,20]. Cation-exchange chromatography can then be used to break the Cu–DPA–baclofen complex on a Dowex-50 W column to yield small quantities of the optical isomers for mechanistic studies. A number of ligand exchangers are sold by Daicel, including Chiralpak WH, WM, and WE. These columns are useful for resolution

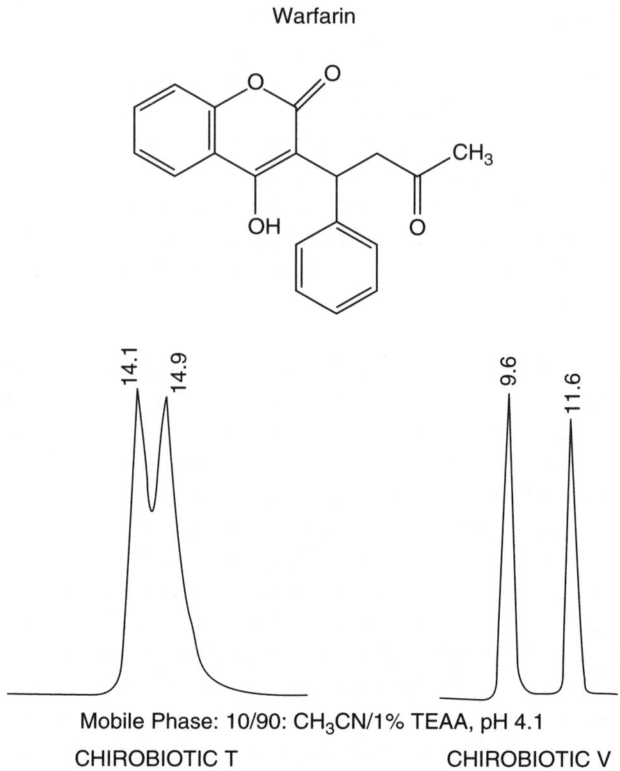

Figure 7 Comparison of separations of warfarin on Chirobiotic T and V. (From [3].)

of amino acids and their derivatives [17]. The mobile phase is simply aqueous copper sulfate. Chiralpak Ma(+) is another ligand-exchange column that is useful for hydroxycarboxylic acids.

Investigations have revealed that a ligand-exchange column may be a simpler and more useful approach for separating enantiomers of baclofen [3], a drug similar to phenylalanine. The selection of a Chiralpak WH column, a column designed to serve as a ligand exchanger [21], considerably shortened method development time. It is fairly well known that the separation of amino acids is significantly affected by the molarity of $CuSO_4$ used for the separation. However, the effect of organic solvents such as methanol on the retention of free amino acids is not well known, except for valine, which shows a decrease in retention time with an increase in methanol concentration up to 20%. Temperature has a significant effect on retention, e.g., the retention time of phenylalanine decreases significantly with increasing temperature. These considerations have led to an optimal mobile phase (containing 0.25 mM copper sulfate, run at a flow rate of 1.5 mL/min at 50°C) for the resolution of D- and L-forms of baclofen from the racemate [3].

6.5 Protein Phases (Type 5)

Type 5 CSPs are based on albumin or glycoproteins. Chiral AGP, based on the α_1-acid glycoproteins, is a second-generation chiral selector that is used as the chiral stationary phase. The process of immobilizing AGP on porous spherical 5-μm silica has been patented. This CSP has been found useful for resolving a broad range of compounds, such as racemic amines, acids, and nonprotolytic compounds, without derivatization. A number of drugs, such as alprenolol, atenolol, bupivicane, chlorthalidone, disopyramide, ephedrine, ethotoin, felodipine, fenoprofen, hexobarbital, metoprolol, pheniramine, and verapamil, have been resolved on chiral AGP [3]. The resolution ability of this column is due to the unique nature of the chiral stationary phase and the resulting enantioselectivity. It is generally recognized that the greatest selectivity can be induced by choosing the proper mobile-phase composition. For bioanalytical work, this CSP has been highly recommended. The typical mobile phase for this column is phosphate buffer with an organic modifier. Enantioselectivity and retention can be regulated by changing the mobile-phase composition with respect to any of the following parameters: pH, organic modifier, and modifier concentration. For the types of various modifiers that have been used and their respective concentrations (see [3], Table 10.6). Typical operating conditions entail the use of 10 mM buffer at pH 7.0, with or without an organic modifier. Modifier concentrations as high as 15% isopropanol or 10% acetonitrile have been used. The pH can affect the resolution of different analytes. A number of other protein-based columns are also available for certain applications.

7 QUICK APPROACH TO CSP SELECTION

Of the various CSPs listed above, criteria such as resolution, efficiency, capacity, column stability, mobile-phase compatibility, and analysis time can offer some help in selection of a suitable CSP (Table 3). However, this information should be updated by each investigator on the basis of his or her experience.

A significant saving in time can be made by the use of a searchable chiral database that is based on molecular structures. Christian Roussel and others [26] have provided Chirbase, a molecular-oriented factual database that includes tens of thousands of entries, and the entire system can be searched by Isis software. The list provides comprehensive structural, experimental, and bibliographic information on successful and unsuccessful chiral separations [27]. Also, assistance can be sought from the major column manufacturers (e.g., Astec/Supelco, Chiral Technologies, Regis Technologies, and ChromTech AB), which offer practical background information on their columns as well as technical assistance in developing analytical and/or preparative enantiomeric separations.

Even though almost 200 CSPs are commercially available, reliable and rapid enantioselectivity prediction of a new chiral molecule with any CSP remains elusive. Of the commercially available CSPs, the polysaccharides and macrocyclic glycopeptides have been favored by various groups for screening strategies in

TABLE 3 Comparison of Various CSPs[a]

	Brush Type	Polysaccharide	Protein	Cyclodextrin	Ligand Exchange	Imprinted Polymer	Crown Ether
Resolves enantiomers	++	++	++	+	+	+	+
High efficiency	++	+	+	—	+	—	+
Column stability	++	—	—	+	+	+	+
Mobile-phase compatibility	++	—	+	+	—	+	+
Inversion of elution order	Yes	No	No	No	No	Yes	Yes
High capacity	++	+	—	+	+	—	—
Low analysis time	+	—	—	—	+	—	—

Source: American Chemical Society Short Course by S. Ahuja, W. Pirkle, and C. Welch; and [3].

[a]Chirobiotic columns are an inclusion-type CSP that have been ascribed selectivity similar to that of protein CSPs.

HPLC. Two screening strategies have been proposed by Perrin and colleagues [28,29] for the enantiomeric separation of drugs using polysaccharide columns in the isocratic normal- and reversed-phase modes. In both modes, three complementary columns are used that have broad enantiorecognition abilities for a wide range of pharmaceutical compounds. These columns are Chiralcel OD-H, Chiralpak AD, and Chiralcel OJ for the normal phase. The compounds are screened on each column using two different mobile phases containing 90 : 10 (v/v) of *n*-hexane/2-propanol or *n*-hexane/ethanol at a flow rate of 1 mL/min. DEA [0.1% (v/v)] is added to the mobile phase for the analysis of basic compounds; however, for the acidic compounds 0.1% (v/v), TFA is used instead. This strategy was applied to a set of 36 different drugs. The study has also shown that for basic compounds, screening on Chiralcel OD-H and the Chiralpak AD columns is usually sufficient to achieve separation. Resolution of the enantiomers was observed for 32 compounds on at least one column. Short analysis times (i.e., 20 min or less) were usually achieved.

In the reversed-phase strategy, the three columns (Chiralcel OD-RH, Chiralpak AD-R, and Chiralcel OJ-R) are used with two mobile phases. The compounds are first analyzed with a mobile phase consisting of an aqueous phosphate buffer, pH 2.0, containing 100 mM KPF_6, mixed with acetonitrile [60 : 40 (v/v)]. The chaotropic agent PF_6^- is added to the mobile phase to achieve the separation of basic analytes at a low pH. According to Perrin [29], most enantiomers can be separated with this approach. If very little or no enantioselectivity is achieved for some components, all three columns are investigated with a basic mobile phase, consisting of an aqueous 20-mM borate buffer, pH 9.0, mixed with acetonitrile [60 : 40 (v/v)] at a flow rate of 0.5 mL/min. Enantioselective separation is achieved on at least one of the columns for 89% of the drugs (33 of 37) analyzed. Analysis times are usually less than 30 min.

Another screening strategy using the same types of columns but with normal-phase gradient elution has been proposed [30]. Each compound is analyzed on four columns: Chiralcel OD-H, Chiralpak AD, Chiralpak AS, and Chiralcel OJ. *n*-Hexane/2-propanol and an *n*-hexane/ethanol gradient elution system are used to screen the compounds except for the Chiralpak AD column, where only the former system is used. To speed up the analysis, a column-switching device is used. The gradient is run from 20 to 70% alcohol in 20 min with a flow rate of 0.75 mL/min. Baseline resolution is achieved with this strategy for 85% of the substances tested (more than 800 compounds have been investigated). When the baseline separation is not achieved, isocratic optimization of the separations is necessary.

Four polysaccharide-based CSPs and three macrocyclic glycopeptide-based CSPs have been evaluated for rapid screening for over 55 chiral compounds of pharmaceutical interest [31]. The polysaccharide columns are employed in the normal-phase and polar organic modes and showed overall enantioselectivity for 87% of the compounds tested. The macrocyclic glycopeptide columns are employed in the reversed-phase and polar organic modes and showed enantioselectivity for 65% of the analytes. When the results from the polysaccharide and macrocyclic glycopeptide screen are combined, they show enantioselectivity for 53 of 55 enantiomeric pairs (96%). This shows that the two screens are complementary, i.e., both types of columns should be included to achieve the highest probability of success. The screens can be automated using a column switcher that allows for different combinations of CSP and mobile phase to be tested overnight. It is claimed that the method development can be completed within 24 hours for a given compound. The column coupling approach has been applied on macrocyclic glycopeptide CSPs. It allows evaluation of this entire class of chiral selectors with a single coupled column for the ability to separate a molecule. Even if a partial separation is obtained on the coupled column, a baseline separation is possible with one of the columns in this class [32].

A parallel multicolumn screening approach has been reported by Zhang et al. [33]. The modified HPLC system allows simultaneous screening of five CSPs in parallel using a regular HPLC autosampler and a pump with 5 ultraviolet detectors.

A future chiral screening program will focus on various separation techniques besides HPLC [34]. It should be noted that a unified strategy for chiral method development screening should include gas GC, SFC, supercritical fluid chromatography, and capillary electrophoresis. All of these methods are discussed at length in this book.

8 SUMMARY AND CONCLUSIONS

The trial-and-error approach generally used in chiral separations can be extremely time consuming. The material in this chapter will help in selecting an appropriate CSP for HPLC method development of chiral compounds. In this selection process, selectivity, mode of operation, compatibility, robustness, efficiency, loadability, and reproducibility should be considered carefully. The literature cited here can be helpful in column selection. The chiral databases as well as the column screening approach discussed above can shorten the selection process. Most chiral separations can be achieved on two types of CSPs: polysaccharide-based and macrocyclic glycopeptides. These columns can be operated in the reversed-phase, normal-phase, and polar organic modes with complementary enantioselectivities. Improved columns still need to be designed based on careful evaluation of the molecular architecture of CSPs and its potential interactions with the analyte.

REFERENCES

1. S. Ahuja. *LCGC*, 25:2, Nov. 2007.
2. S. Ahuja. Presentation at the American Chemical Society Meeting, Boston, Aug. 19–23, 2007.
3. S. Ahuja. *Chiral Separations by Chromatography*. Oxford University Press, New York, 2000.

4. E. Eliel, S. Wilen. *Stereochemistry of Organic Compounds*. Wiley, New York, 1994.
5. S. Allenmark. *Chromatographic Enantioseparation*. Ellis Horwood, Chichester, UK, 1991.
6. H. Y. Aboul-Enein, I. W. Wainer. *The Impact of Stereochemistry on Drug Development and Use*. Wiley, New York, 1997.
7. S. Ahuja. *Chiral Separations: Applications and Technology*. American Chemical Society, Washington, DC, 1997.
8. J. Gal. *LCGC*, 5:106, 1987.
9. S. Hara, J. Cazes. *J. Liq. Chromatogr*., 9(2–3), 1986.
10. R. W. Souter. *Chromatographic Separations of Stereoisomers*. CRC Press, Boca Raton, FL, 1985.
11. S. Ahuja. Presentation at the First International Symposium on Separation of Chiral Molecules, Paris, May 31–June 2, 1988.
12. G. Blaschke, H. P. Kraft, K. Fickentscher, F. Koehler. *Arzneim.-Forsch*., 29:1640, 1979.
13. Top 200 Drugs in 1982. *Pharm. Times*, p. 25, Apr. 1982.
14. Y. Okamoto. *Chemtech*, p. 176, Mar. 1987.
15. W. H. DeCamp. *Chirality*, 1:2, 1989.
16. U.S. FDA. *Guidelines for Submitting Supporting Documentation in Drug Applications for the Manufacture of Drug Substances*. Office of Drug Evaluation and Research (HFD-l0O), U.S. Food and Drug Administration, Rockville, MD, 1987; Policy Statement on Stereoisomeric Drugs, www.fda.gov/cder/guidance/stereo.htm, 1997.
17. S. Ahuja. *Chiral Separations by Liquid Chromatography*. ACS Symposium Series, Vol. 471. American Chemical Society, Washington, DC, 1991.
18. M. D. Palamareva, L. R. Snyder. *Chromatographia*, 19:352, 1984.
19. J. Bitterova, L. Soltes, T. R. Fanovee. *Pharmazie*, 45:H.6, 1990.
20. E. Gil-Av. In *Chiral Separations by Liquid Chromatography*, S. Ahuja, Ed. ACS Symposium Series, Vol. 471. American Chemical Society, Washington, DC, 1991.
21. V. Schurig, M. Jung, D. Schmalzing, M. Schliemer, J. Duvekot, J. C. Buyten, J. A. Peene, P. Musschee. *J. H. R. C. C*., 13:470, 1990.
22. I. Wainer, D. E. Drayer, Eds. *Drug Stereochemistry: Analytical Methods and Pharmacology*. Marcel Dekker, New York, 1988.
23. R. Dappen, H. Arm, V. R. Mayer. *J. Chromatogr*., 373:1 1986.
24. C. Dalgleish. *J. Chem. Soc*., 137, 1952.
25. G. Wulff. In *Polymeric Reagents and Catalysts*, W. T. Ford, Ed. American Chemical Society, Washington, DC, 1986, p. 186.
26. C. Roussel, P. Piras. *Pure Appl. Chem*., 65:235, 1993.
27. P. Piras, C. Roussel, J. Pierrot-Sanders. *J. Chromatogr. A*, 906:443–458, 2001.
28. C. Perrin, V. A Vu, N. Matthijs, M. Maftouh, D. L. Massart, Y. Vander Heyden. *J. Chromatogr. A*, 947:69, 2002.
29. C. Perrin, N. Matthijs, D. Mangelings, M. Maftouh, C. Granier-Loyaux, D. L. Massart, Y. Vander Heyden. *J. Chromatogr. A*, 966:119, 2002.
30. M. L. De La Puente, C. T. White, A. Rivera-Sagredo, J. Reilly, K. Burton, G. Harvey. *J. Chromatogr. A*, 983:101, 2003.

31. M. A. Andersson, D. Aslan, A. Clarke, J. Roeraade, G. Hagman. *J. Chromatogr. A*, 583:100–101, 2003.
32. X. Wang, J. T. Lee, T. E. Beesley. *LCGC*, 18:626, 2000.
33. Y. Zhang, W. Watts, l. Nogle, O. McConnell. *J. Chromatogr. A*, 104:975–984, 2004.
34. A. Akin, F. J. Antosz, J. L. Ausec, K. F. Greve, R. L. Johnson, L. E. Magnusson, T. Ramstad, S. L. Secreast, D. S. Seibert. *Curr. Pharm. Anal.*, 3:53, 2007.

CHAPTER 4

Separation of Chiral Compounds on Polysaccharide Columns

CLINTON W. AMOSS and NORBERT M. MAIER
Chiral Technologies, Inc., West Chester, Pennsylvania

1 INTRODUCTION TO POLYSACCHARIDE-BASED CHIRAL STATIONARY PHASES

Biochemical interactions vital to life depend on chirality, and chirality must be taken into account when developing new pharmaceutical compounds having specific and beneficial activity profiles. Driven by this need, a multitude of new tools has been established over the past three decades to separate chiral compounds, allowing quantification and isolation of the individual enantiomers at preparative and industrial scale. Liquid chromatographic enantiomer separation using enantioselective adsorbents, called *chiral stationary phases* (CSPs), is currently the dominating methodology used to achieve these goals. The design, synthesis, and evaluation of new CSPs remains an active field of research, and more than 1500 CSPs have been reported in the scientific literature. However, it is a remarkable fact that only a small number of these CSPs offer the crucial advantages of ease of preparation, broad applicability, and multimodal operation conditions.

CSPs derived from easily accessible and naturally abundant polysaccharides—cellulose and amylose—have taken the dominating position among the current commercially available enantioselective media for liquid chromatographic applications. These CSPs are well known to any practitioner of the art of HPLC-based enantiomer separation under the product names Chiralcel and Chiralpak. These phases were developed originally by Professor Yoshio Okamoto and co-workers at Osaka University and commercialized by Daicel Chemical Industries. Various extensive studies performed in academic and industrial settings have demonstrated that Daicel's polysaccharide-type CSPs are capable of resolving up to 95% of all enantiomer separation challenges relevant to

Chiral Separation Methods for Pharmaceutical and Biotechnological Products, Edited by Satinder Ahuja

drug development efforts focusing on low- and medium-molecular-weight chiral molecules [1–4]. A similar situation is encountered when analyzing the relative contribution of these CSPs to preparative- and industrial-scale enantiomer separations. Most of the preparative-scale enantiomer separation projects reported, performed in normal phase, polar organic, and supercritical fluid mode, utilize one or another of these polysaccharide-type CSPs. Also, production processes of commercial chiral drugs based on simulated moving-bed (SMB) technology, a highly productive and continuously operating chromatographic process, use polysaccharide-based phases exclusively [5].

Major progress in the field of enantioselective chromatography has been achieved recently by the development of robust immobilized polysaccharide-based CSPs. With the advent of the new generation of commercial enantioselective adsorbents, chromatographers are in a position to explore solvent systems that are entirely incompatible with coated CSPs. These immobilized CSPs show broad chiral capability similar to that of the coated versions, but have additional practical advantages, such as unique selectivities, broader temperature operation range, the possibility of solubility guided method development, and the possibility to be regenerated using harsh solvent systems. For this reason, in this chapter we provide a combined treatment using the established coated polysaccharide-based and the advanced generation of immobilized CSPs. The intent of this account is to provide novices in the field of polysaccharide-type CSPs-based enantiomer separation with guidance concerning the materials available, their properties, the mechanistic nature of solute–CSP interactions, operation modes, and approaches to analytical method and preparative development. Although the focus of the chapter is an introduction to the field, we are confident that its contents will provide the experienced practitioner with some useful information, and the expert with entertaining reading.

2 ORIGINS OF POLYSACCHARIDE-BASED CHIRAL STATIONARY PHASES

Early success in using cellulose for chromatographic enantiomer separation was reported in 1951 by Kotake et al., [6] and shortly after by Dalgliesh [7], both achieving resolution of amino acid derivatives with paper chromatography. Later expansion of this work demonstrated that amino acids could also be separated by this technique. However, the first practical chromatographic use of polysaccharides came in 1973, when Hesse and Hagel prepared cellulose triacetate from microcrystalline cellulose and used it as a liquid chromatographic stationary phase [8–15]. These authors postulated that the microcrystalline structure of cellulose is essential for chiral recognition; when microcrystalline cellulose was dissolved in solvent and precipitated, the authors obtained a much less effective phase which sometimes exhibited a reversal in elution order [15–18]. An important breakthrough that would lead to a dramatic increase in the effectiveness and variety of polysaccharide-based chiral stationary phases was made in the early

1980s. Taking advantage of previously developed technology [19–22], in 1984, Yoshio Okamoto, then a professor at Osaka University, coated cellulose triacetate and cellulose tribenzoate on macroporous silica. The resulting phases proved far superior to unsupported acetylated microcrystalline cellulose (CTA-I) for high-performance liquid chromatography (HPLC) applications [22]. In addition, the new materials exhibited enantiomer resolution completely different than CTA-I, and for some analytes (e.g., Tröger's base) the elution order was reversed on the new coated cellulose triacetate column. As an additional advantage, the mechanical strength of the coated materials was much higher than the strength of the parent unsupported microcrystalline material, allowing higher pressure drops and higher efficiency [10]. Essentially, with the breaking of the "spell" of crystallinity and the recognition of the multimodal operation characteristics of these new enantioselective adsorbents [18], the way was paved for development of the most useful generation of chiral stationary phases. Independent research efforts at Daicel carried out at about the same time led to the development of a variety of coated cellulose ester derivatives that would see commercialization [23]. Further research in Okamoto's group led to the development of silica-coated phenylcarbamate derivatives of polysaccharides for HPLC [24], which would soon emerge as the most generally effective of all classes of known CSPs [23].

2.1 Development of Optimal Polysaccharide-Based CSPs

Extensive investigations into the structure–enantiomer relationships of polysaccharide-type CSPs have been conducted by Okamoto and co-workers [12,25]. In general, cellulose derivatives having electron-donating substituents were more selective than derivatives having electron-withdrawing groups [25]. However, when these studies were extended to ester derivatives of amylose, such as the tribenzoates, little chiral recognition potential was observed.

The preparative potential of cellulose esters has been investigated by several groups [17,26]. As an alternative to coated cellulose tribenzoate, Rimbock et al. [27] and later Francotte et al. [28–30] prepared benzoylcellulose beads of 10- to 30-μm diameter from the pure polymers. These cellulose beads exhibited a very high loading capacity, good mechanical stability, and good column efficiency. A large number of derivatives of cellulose and amylose phenylcarbamates were also investigated [12,31,32]. As found with cellulose benzoate derivatives, electron-donating groups improved chiral recognition properties. Also, electron-withdrawing groups such as halogens also contributed to enhanced chiral recognition. However, derivatives with nitro or methoxy groups showed relatively poor chiral recognition properties [31]. Disubstituted cellulose phenylcarbamates having both methyl and chloro or fluoro groups on the phenyl ring were also found to exhibit high enantioselectivity [32–34]. Especially effective cellulose phenylcarbamate phases include 3-chloro-4-methyl-, 4-chloro-3-methyl-, and 3-fluoro-4-methyl-phenyl substituted. Recently, cellulose tris(3-chloro-4-methylphenylcarbamate) has been marketed as a new chiral stationary phase [35].

Of all derivatives of cellulose, the 3,5-dimethylphenylcarbamate (CDMPC) and 3,5-dichlorophenylcarbamate appear to have the most general chiral selectivity. This high degree of selectivity is also exhibited by the 3,5-dimethylphenylcarbamate of amylose (ADMPC). In this context it is interesting to note that in studies [31,36] involving 510 racemic compounds, 315 showed either full or partial resolution on the CDMPC column. Similarly, of 384 racemates examined on an ADMPC column, 107 were resolved completely, and another 102 were resolved partially. In a combined screening experiment with 510 compounds, 129 were resolved only on CDMPC, 85 were resolved only on ADMPC, and 129 were resolved on both columns, with a combined success rate of about 78% (400 of 510 compounds) [12]. Within the course of the pioneering studies, it was recognized that cellulose tris(3,5-dichlorophenylcarbamate) also shows high selectivity to a broad selection of chiral compounds. Unfortunetely, in coated format the utility of this phase was rather limited, due to its high solubility in alkane and alcohols. However, as discussed below, this limitation has been overcome successfully via immobilization.

Two other derivatives of amylose show remarkable chiral recognition capabilities. The chloromethyl (5-chloro-2-methyl)phenylcarbamate of amylose exhibits a rather broad chiral recognition profile [37,38], in structural preference similar to that of ADMPC. This column was introduced into the market recently [35]. Out of a considerable number of benzyl-substituted amylose derivatives, Okamoto's group identified the corresponding tris[(*S*)-α-methylbenzylcarbamate] derivative as the most effective member (Chiralpak AS) [39–42]. In this particular case, the carbamate group provides an additional stereogenic center, endowing this polymer with a unique chiral recognition profile. It should be noted that alkyl carbamates are not readily used as CSPs in coated format, due to enhanced solubility in a broad range of solvents. However, amylose tris(cyclohexylcarbamate) has been shown to exhibit excellent chiral recognition properties for HPLC and thin-layer chromatography (TLC) applications [43,57]. All of this development work led to the polysaccharide CSPs that are commercially available today. Figure 1 shows the general structure of cellulose- and amylose-based chiral stationary phases. Table 1 shows the most important phases that are currently commercially available.

3 STRUCTURE AND MECHANISMS OF OPERATION

Cellulose and amylose are biopolymers found widely in nature; cellulose is, in fact, the most abundant biomacromolecule on Earth and is found in a variety of sources; amylose is a major component of starch. Both types of polymers are made up of repeating D(+)-glucose units; cellulose has β-(1,4) linkages and amylose has α-(1,4) linkages. For both polymers, the glucose units are in a chair conformation, with three alcohol groups in equatorial positions. Each of the five carbons in the glucose ring is asymmetric. As these materials are derived from natural sources, the molecular weights may vary considerably, with either cellulose or amylose spanning a general range of several hundred to several thousand

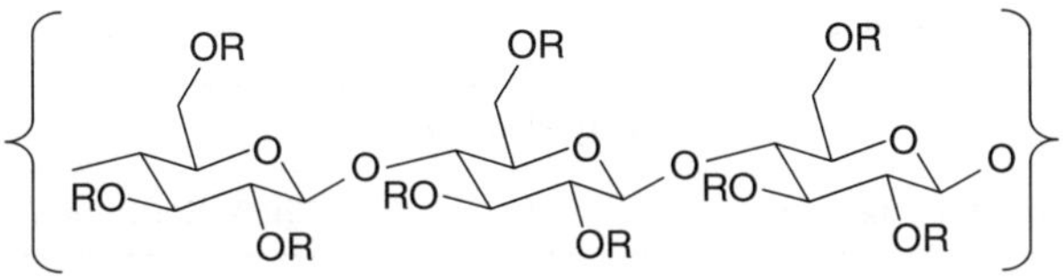

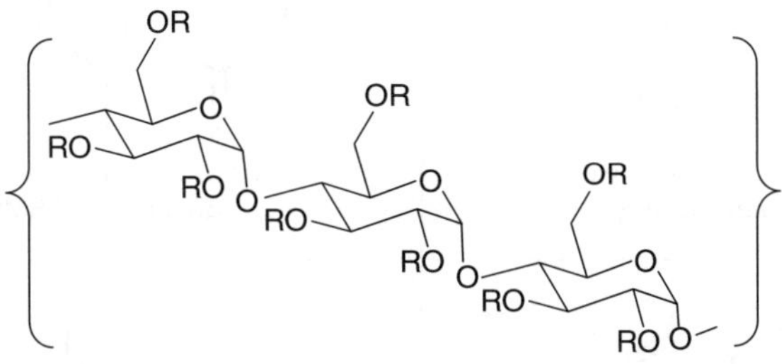

Figure 1 Repeating units of derivatized amylose and cellulose chiral stationary phases. R = ester, phenylcarbamate, or benzylcarbamate.

repeat glucose units; however, individual sources of either polysaccharide tend to have a much narrower molecular-weight distribution.

Studies by Vogt and Zugenmaier determined that amylose tris(phenylcarbamate) (amylose tricarbanilate) has a 4/1 helical structure and that cellulose tris(phenylcarbamate) (cellulose tricarbanilate) has a left-handed threefold (3/2) helical structure [44,45]. Subsequent studies by Steinmeier and Zugenmaier using x-ray diffraction data showed that cellulose tris benzoate (CTB) also has a threefold (3/2) helical structure, whether it is prepared heterogeneously (CTB I) or homogeneously (CTB II) [46]. It has been stated widely that the mechanisms of interaction between derivatized polysaccharides and enantiomers are not well understood, and for specific interactions at the molecular level this is still true. However, systematic studies employing supplemental techniques to chromatography have been used to improve our understanding of how these phases work. A number of reviews have covered various aspects of these studies [2–4,10–14,47–57].

Any model of enantiomeric selectivity for polysaccharide CSPs must take into account spatial effects of the cavities within the collective matrix of the polysaccharide chains, as well as intermolecular interactions such as hydrogen bonding, π–π interactions, dipole–dipole interactions, and steric effects. Obviously, the more strongly retained isomer will have strong interactions and a better fit with the CSP, leading to stronger adsorption, and the weaker retained isomer will either be rejected or have a poor fit, leading to early elution.

Early mechanistic studies were conducted on cellulose triacetate, cellulose tribenzoate, and derivatives of cellulose tribenzoate. Francotte and co-workers [58] advanced the theory, proposed originally by Hesse and Hagel [15,16], and

TABLE 1 Structures of Commercially Available Polysaccharide CSPs

Amylose-O-R	
Chiral Stationary Phase	**Chiral Stationary Phase**
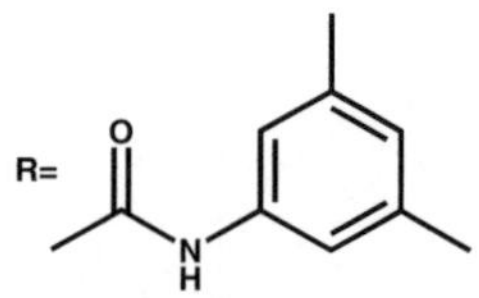 Amylose tris (3,5-dimethylphenyl-carbamate) [ADMPC]**	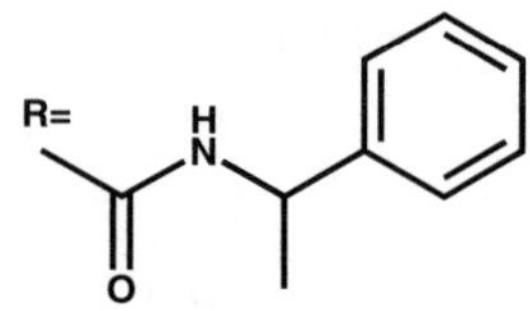 Amylose tris [(S)-α-methylbenzyl-carbamate] [ASMBC]
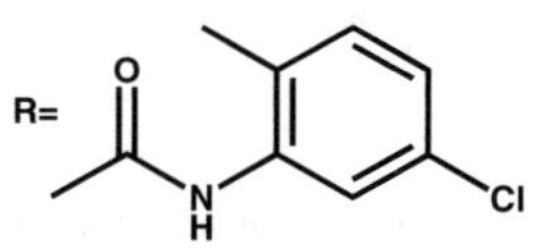 Amylose tris (5-chloro, 2-methyl-phenyl carbamate)	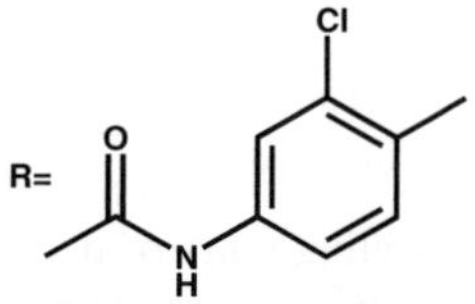 Amylose tris (3-chloro, 4-methyl phenyl carbamate)
Cellulose-O-R	
Chiral Stationary Phase	**Chiral Stationary Phase**
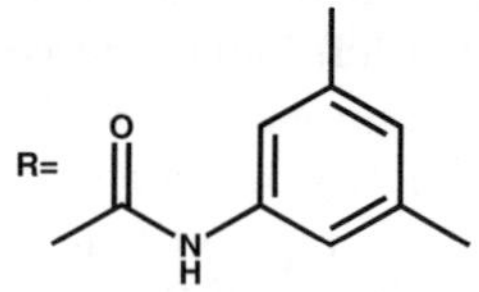 Cellulose tris(3,5-dimethyl-phenyl carbamate) [CDMPC]**	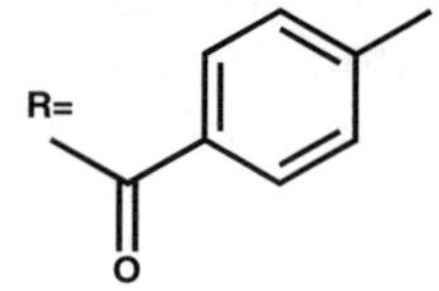 Cellulose tris(4-methylbenzoate) [CTMB]
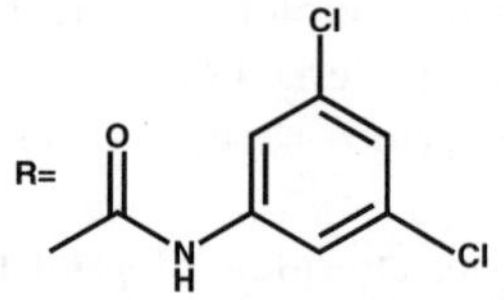 Cellulose tris(3,5-dichloro-phenyl carbamate)*	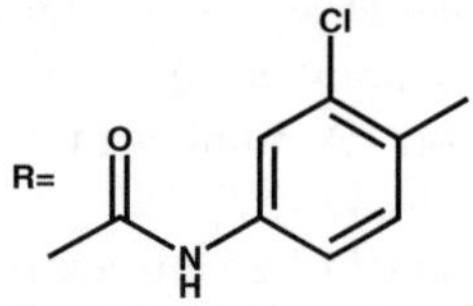 Cellulose tris (3-chloro, 4-methyl phenyl carbamate)

TABLE 1 (*Continued*)

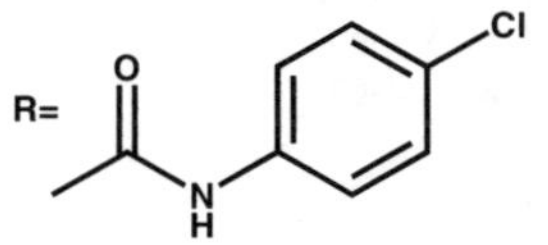 Cellulose tris (4-chlorophenyl-carbamate)	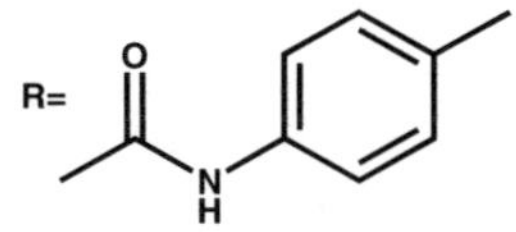 Cellulose tris(4-methylphenyl-carbamate)
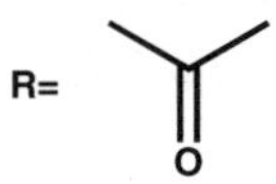 Cellulose triacetate [CTA]	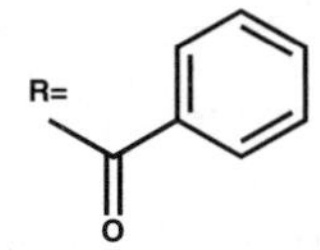 Cellulose tribenzoate [CTB]

*Immobilized on silica.
**Coated and immobilized on silica.

based on work by Staudinger and Döhle [59] that the mechanism of chiral discrimination on CTA I involved an inclusion-complex formation, in which the more or less rigid molecules can enter (totally or partially) the space between the polysaccharide chains and some of its sidechains [58]. CTA-I can be a very useful CSP for preparative applications, due to its good selectivity and high loading capacity. At that time, it was thought that the crystallinity of cellulose triacetate was vital to its chiral recognition properties, especially as CTA II showed greatly reduced resolving power compared to CTA I, due to its loss in crystallinity. However, Shibata and co-workers showed that amorphous CTA II coated on silica had better resolution properties than crystalline CTA II coated on silica, particularly if the mobile phase was changed from the conventional ethanol to 9 : 1 hexane/isopropanol [10]. Consequently, they concluded that crystallinity was not required for chiral recognition on CTA II.

In the ensuing years, many studies were performed to understand how the coated polysaccharide esters and carbamates actually worked [60–73]. Most of the these studies were concerned with the effects of substituents on the benzoate ester or phenylcarbamate rings, the effect of size and shape of various alcohol mobile-phase modifiers, and the systematic variation of solute structure to determine the effects that various structural elements had on the type of interactions between chiral stationary phase and solvent. Many of these studies relied on chromatographic results only, but spectroscopic and modeling techniques were added gradually to corroborate the chromatographic results. Much useful information was obtained, even though initially this did not lead to true mechanistic models that could generally be applied.

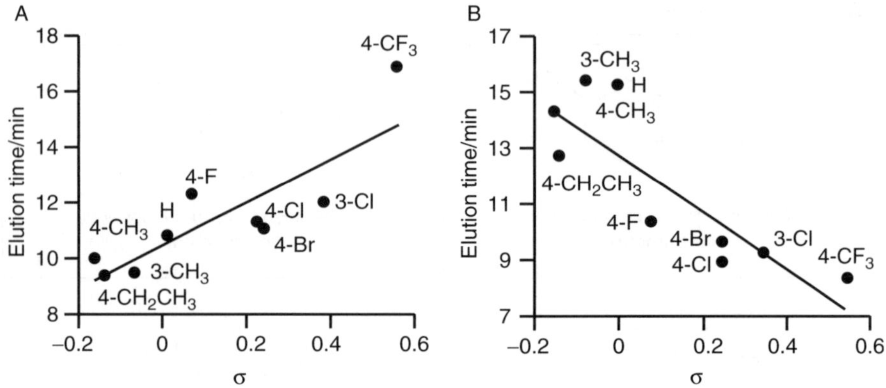

Figure 2 Retention vs. Hammett's σ on cellulose tris(phenylcarbamate) phases. (A) Retention times of acetone and (B) the first eluting enantiomer of 2,2,2-trifluoro-1-(9-anthryl)ethanol on cellulose tris(phenylcarbamate) derivatives against the Hammett σ values. (From [31], with permission. Copyright © 1986 Elsevier Science B.V.)

A study by Okamoto et al. [25] looked at sustituent effects for mono- and disubstituted cellulose tris(phenylcarbamates). The chromatographic behavior of 10 model racemic compounds was determined for 19 cellulose tris(phenylcarbamates). Of these CSPs, the ones with the strongly electron-withdrawing groups showed poor selectivity. However, the most effective phases did have electron-donating groups, particularly at the 4-position. In this study, the retention time of acetone, an achiral molecule, was also studied. When the retention time of acetone was plotted against the electron-withdrawing power of the substituents, as measured by Hammett σ values, good correlation was obtained (Figure 2). A second plot was made of the retention time of the first eluting isomer of trifluoroanthrylethanol vs. Hammett σ values; in this case the retention time of the first peak (as well as the second peak) decreased roughly with the increasing electron-withdrawing power of the phenyl substituent. However, taken in reverse, it can be seen that the retention time increased with increased electron-donating power of substituents that would increase the electron density of the carbonyl group. Thus, a stronger hydrogen bond between the alcohol proton of the solute and the carbonyl oxygen of the CSP could take place.

Further, when the ^{1}H-NMR spectra of these CSPs was taken in [$^{2}H_6$]acetone, the NH proton of the cellulose triphenylcarbamate was shifted downfield in an amount proportional to the electron-withdrawing ability of the substituent. This NH resonance shifted greatly when [$^{2}H_5$]pyridine was added, and in fact at 9% pyridine, three distinct NH peaks are observed, indicating that the three carbamate functions from different glucose positions have different adsorbing strengths [25]. From this observation it was inferred that the acidity of the NH proton is increased as the electron-withdrawing power of the substituents on the phenyl group increases [25]. They surmise further that the acetone is adsorbed on the stationary phase via hydrogen bonding with the NH proton and that this binding is stronger when the proton is more acidic [25]. Conversely, electron-donating groups such as methyl will increase the electron density of the carbonyl [11]. A mode of interaction is thus proposed in which the main chiral adsorbing sites are the NH and C=O carbamate groups, which can hydrogen bond, and C=O groups, which can also participate in dipole–dipole interactions. The π–π interactions between the aromatic groups of solutes and the phenyl groups of the stationary phases may be less important [25]. However, these π–π interactions may be important under reversed-phase conditions, as evidenced by the separation of some β-blockers on CDMPC [11,60,61]. These π–π interactions also offer a possible explanation for the separation of some nonpolar aromatic [2.2]paracyclophanes on ADMPC and CDMPC [11,62].

Okamoto et al. found a good correlation between the ability of a substituted cellulose tris(phenylcarbamate) to form a lyotropic liquid-crystalline stationary phase in a concentrated solution and the chiral recognition of such a phase. A few cellulose phenyl- and alkylcarbamate derivatives that do not form such lyotropic phases demonstrated low chiral recognition [27]. Later, Okamoto et al. [25] also studied substituent effects on 10 cellulose tribenzoate CSPs, in which the phenyl ring was substituted with various groups. As with the substituted phenylcarbamates, inductive effects were quite important. Phases with electron-donating groups were the most effective, although the strongest electron-withdrawing group, methoxy, was not a suitable substituent, due to the strength of the group itself [25]. Again, acetone adsorption was compared to the electron-withdrawing power of the substituent, and although the same correlation can be made as for the phenylcarbamates, the reason is less clear. Still, for many compounds it appears that the main adsorbing site is the ester group. The group can interact with a solute through dipole–dipole interaction and/or hydrogen bonding [25]. The effect of the substituents on chiral selectivity was less pronounced than on the tris(phenylcarbamate) CSPs, and this may be because other factors, such as the morphology of CTB, may play an important role. It is known that the chiral recognition abilities of these derivatives can depend on the coating process. For example, 1,3-diacetoxybutane is not resolved on cellulose tribenzoate when it is coated on silica from dichloromethane solution, but it is resolved when it is coated from a 10 : 1 dichloromethane/nitrobenzene solution [11,63].

Wainer et al. made several studies of cellulose tris(benzoate) phases coated on silica [64–66]. Studies were made of the retention of a series of aliphatic and aromatic amides on the CTB phase, and a preliminary chiral recognition model was proposed. Among the key features of this model is the formation of a diasteromeric solute–CSP complex via attractive interactions, which include hydrogen bonding and π–π and dipole interactions between the amide bond of the solute and the ester bond of the CSP. In addition, the model calls for positioning of the solute and CSP such that at least two possible interactions are involved. Further, the improvements in the steric fit in the chiral cavity will increase the magnitude of the chiral resolution [64]. This proposed mechanism for the Chiralcel OB CSP is an attractive binding–steric fit formulation which differs from the inclusion mechanism previously proposed for CTA I [64]. Further studies looked at the effect of various alcohol modifiers in the mobile phase on this system [65]. One important observation of these studies was that the mobile-phase modifier not only competes for binding sites with the chiral solute, but can also alter the steric nature of the chiral cavities by binding to achiral sites at or near the chiral cavity [65]. Finally, Wainer et al. refined this model after studying the effects of a series of alcohols adsorbed on the CTB-coated silica-based phase [66]. In this case, the solute's alcoholic hydrogen is involved in a hydrogen bond with the ester carbonyl of the solute to form a diasteromeric complex. This complex is stabilized by the insertion of the aromatic portion of the solute into a chiral cavity or ravine of the CSP. Chiral discrimination between enantiomeric solutes is then determined by differences in steric fit in the chiral cavity [66].

In a later study, Francotte and Wolf prepared cellulose tribenzoate beads in pure polymeric form [28]. The beads showed virtually the same x-ray powder diffractograms as CTB coated on silica [18] and CTB prepared by heterogeneous conditions [27]. Using these beads, they were able to effect good resolution of a series of alcohol and diol acetates that could not form hydrogen bonds and which did not have aromatic groups. This partially contradicts the mechanism proposed by Wainer et al. [66] and suggests that the interaction mechanism is dependent on a multitude of factors, some of which will be determined by the type of solute [17]. It should be noted here that the discrepancies between interaction mechanisms postulated for derivatized polysaccharide phases is not necessarily a cause for concern. Many solutes do fit certain mechanism models quite well. The power of polysaccharide phases is that they show chiral discrimination for so many different types of solutes, even when the mechanisms proposed would not predict successful enantiomeric selectivity.

To really understand the interactions of phenylcarbamate chiral stationary phases at a molecular level, it is helpful to employ nuclear magnetic resonance (NMR). Unfortunately, many of the phenylcarbamates with really high resolving power are soluble only in polar solvents such as pyridine and tetrahydrofuran (THF). These polar solvents can interact strongly with the carbamate residues of the phase, strongly masking the chiral interactions that might be detected by NMR [12]. To get around this problem, Yashima and Okamoto were able to use cellulose tris(4-trimethylsilylphenylcarbamate) and cellulose tris(5-fluoro-2-methylphenylcarbamate) chiral stationary phases, which were found to have good chiral discrimination [30,67]. When the NMR spectra of *trans*-stilbene oxide was taken alone and in the presence of tris(4-trimethylsilylcarbamate), the methine proton of the (−)-enantiomer shifted downfield, but the same proton for the (+)-enantiomer hardly shifted. Thus, the (−)-enantiomer must be more strongly bound to the CSP. The elution order for these two isomers on this CSP is (+), then (−). Thus, the elution order and the downfield shift of the NMR spectra are well correlated.

Since some important phenyl carbamates are not soluble in useful NMR solvents, Yashima et al. performed force-field calculations using Quanta/CHARMM and Molecular Interaction software to determine the interaction of cellulose trisphenylcarbamate (CTPC) with *trans*-stilbene oxide (TSO) and *trans*-1,2-diphenylcyclopropane [68]. Whereas the first compound is very well separated on CTPC, the second compound is not separated at all. The lowest interaction energies calculated for the CTPC–solute simulation show a difference of 2.4 kcal/mol for the two TSO isomers, and almost no difference for the two *trans*-1,2-diphenylcyclopropane isomers. The interaction for *trans*-stilbene oxide shows $\pi-\pi$ interactions between the TSO phenyl rings and the phenyl groups of the CSP, while the ether oxygen of TSO is located near the NH proton of CTPC.

Oguni et al. discussed an improved cellulose triacetate phase (CTA II) though the choice of coating solvent [13]. For example, CSP prepared from dichloromethane/phenol or dichloromethane/trifluoroacetic acid had improved chiral recognition compared to the phase coated from dichloromethane only [61]. If the cellulose acetates were allowed to solidify from these various solvents, and these solids were then subjected to crossed polarization/magic-angle spinning (CP/MAS) ^{13}C-NMR, there was a narrowing of the half-widths of the C_1 signal in going from CH_2Cl_2 (315 Hz) to CH_2Cl_2/TFA (243 Hz) to CH_2Cl_2/phenol 218 Hz), but no change in the chemical shift. The narrow line width could be the result of more regular conformation or of greater mobility. As there was no difference in the spin-lattice relaxation time (T_1) in the solid polymers, which would be related to mobility, the authors concluded that the shift narrowing was cause by higher regularity of the conformation of cellulose triacetate about the β-(1,4)-glycoside linkage. This appears to be the earliest application of CP/MAS to the study of these stationary phases.

Yashima et al. used circular dichroism to study the interaction between bilirubin and a series of substituted phenylcarbamates [69]. Phases such as cellulose tris(4-bromo-, 4-chloro-, and 4-methylphenylcarbamate) were able to bind preferentially bilirubin having an (*S*)-helix conformation. In both chloroform–ethanol solution and solid film, these phases in the presence of bilirubin exhibited induced circular dichroism (CD) spectra. The intensity of these absorbances depended on the type and position of the substituents on the phenyl groups of CTPC, and in particular, the *para*-halogenated substituents exhibited the most intense induced CD [69].

In 1996, Yashima et al. reported an extensive study using several NMR techniques in order to better understand the chiral recognition mechanism of these phases [70]. Expanding on earlier work with chloroform-soluble CTPC phases [67], they studied the interaction of cellulose tris(5-fluoro-2-methylphenylcarbamate) with enantiomers of bi-2-naphthol and 2,2′-dihydroxy-6,6′-dimethyldiphenyl. These CSP–solute interactions exhibited strong chromatographic selectivity ($\alpha > 3$) as well as strong ^{1}H- and ^{13}C-NMR chemical shifts. If either of the alcohol protons on bi-2-naphthol was replaced with a methyl group, the retention and chromatographic recognition were greatly reduced. Evidently, hydrogen bonding by both hydroxyl groups of bi-2-naphthol is essential for chromatographic recognition [70]. Using COSY and NOESY (two-dimensional NMR) experiments, the ^{1}H chemical shift was determined for the hydroxy proton, and the ^{1}H and ^{13}C chemical shifts were determined for each of the 10 aromatic positions on both (*R*)- and (*S*)-bi-2-naphthol. From these measurements it was determined that the (*S*)-OH protons were more strongly shifted downfield with line broadening than were the (*R*)-OH protons, and that two of the (*S*)-aromatic protons were more strongly shifted upfield with line broadening [70]. These effects were ascribed to hydrogen-bonding effects for the hydroxy protons and to π-stacked or shielding effects by a neighboring aromatic group of the CSP [70]. The larger chemical shifts and line broadening for the (*S*)-bi-2-naphthol are in agreement with the stronger retention of (*S*)-bi-2-naphthol on cellulose tris(5-fluoro-2-methylphenylcarbamate) when racemic bi-2-naphthol is eluted with 90 : 10 hexane/isopropanol.

By studying the spin-lattice relaxation of the bi-2-naphthol enantiomers in the presence of the cellulose tris(5-fluoro,2-methylphenylcarbamate) stationary phase, it was possible to make conclusions about the relative binding ratios of this solute–CSP complex [70]. ^{1}H-NMR titrations were carried out in $CDCl_3$, in which the concentration of the (*S*)-bi-2-naphthol was kept constant at 1.94 mM and the concentration of cellulose tris(5-fluoro,2-methylphenylcarbamate) stationary phase was varied from 0 to 63.2 mM. The chemical shifts of the OH and H_4 protons of bi-2-naphthol were plotted as a function of CSP concentration. A strong downfield shift of the OH resonance is indicative of hydrogen bonding, and a strong upfield shift of the H_4 and H_6 aromatic protons is indicative of probable π-stacking. A determination of the binding constants from these data supports the formation of a 1 : 1 complex between the (S)-bi-2-naphthol and each unit of the CSP. This conclusion was further supported by the study of a continuous variation plot (Job plot) in which the OH proton resonance was measured against the mole ratio of (*S*)-bi-2-naphthol/CSP in which the total concentration of (*S*)-bi-2-naphthol and CSP was held constant at 25 mM. The amplitude of this plot reached a maximum at a mole fraction of the CSP of approximately 0.5. This implies that all glucose units of the CSP have the same binding affinity to the (*S*)-bi-2-naphthol. This could happen only if the CSP has a regular structure, even in a solution. If the CSP were a random coil, different energy-binding sites would appear, and the Job plot maximum would occur at a different mole ratio than 1 : 1 [70].

For this same system, calculations were performed to determine the selectivity (α) and the difference in free energy ($\Delta\Delta G^\circ$) of the bi-2-napthol/CSP diasteromeric complex as determined by NMR and HPLC. The NMR values of 10.6 and 1.39 kcal/mol were approximately 2.5 times the HPLC values of 4.23 and 0.84 kcal/mol. This relative difference is probably due to the difference in solvent employed: $CDCl_3$ for NMR and hexane/2-propanol for HPLC. There is some evidence that the 2-propanol weakens the hydrogen-bond interaction between bi-2-naphthol and this CSP.

Finally, for this same system, molecular modeling was applied to determine the most probable fit for the bi-2-naphthol/CSP complex. Cellulose tris(phenylcarbamate) has a left-handed 3/2 helical conformation, and the polar carbamate residues can be found along the chiral axis in a kind of helical groove. These carbamate groups are located inside, and the hydrophobic aromatic groups are placed outside the polymer chains. This arrangement allows for access of the polar enantiomers into the chiral groove, where they can interact with the carbamate residues via hydrogen-bond formation. The model predicts that both of the hydroxyl groups of the (*S*)-bi-2-naphthol participate in hydrogen-bonding interactions, whereas for the (*R*)-bi-2-naphthol only one hydroxyl group can participate in a hydrogen bond. This model would thus corroborate the relative results obtained by chromatography and NMR, which show that the (*S*)-bi-2-naphthol/CSP complex is stronger than the (*R*)-bi-2-naphthol/CSP complex [70].

In a series of papers, Booth et al. developed a series of relationships with the goal of being able to predict the enantioselective behavior of many types of compounds on amylosic chiral stationary phases [71–73]. A goal of these studies was to develop a quantitiative structure–enantioselective retention relationship (QSERR) which takes into account hydrogen bonding and aromaticity of the solutes [71]. In the first of these studies they used a series of 28 α-alkyl arylcarboxylic acids, some of which were also esterified, to determine chromatographic behavior on ADMPC. Using molecular modeling and QSERR calculations coupled with the result of van't Hoff plots, they came to the conclusion that a molecule such as benoxaprofen can dock in the chiral ravine of ADMPC via a two-step process. In the first step, the enantiomers form a diastereomeric complex with the CSP, for which the (*R*)- and (*S*)-enantiomers have equal stability. This complex involves the formation of two hydrogen bonds between each enantiomer and the CSP. Then in the second step, a third hydrogen bond is formed after the solutes and the CSP adjust conformationally to each other. Since the conformational changes will be different for the two enantiomers in this step, chiral recognition will take place. The authors then postulate that the mechanism process is "conformationally driven" rather than proceeding through the traditional "three-point" interaction model [71]. Later these studies were expanded to examine the retention of 17 amide compounds on ADMPC, ASMBC, and ARMBC, in which ARMBC is the amylose tris[(*R*)-α-methylbenzylcarbamate] [72]. It was concluded that this chirality of the side chain has almost no effect on the chiral selectivity of the CSP. Thus, the difference in ARMBC and ASMBC is the conformation of the chiral center at the α-methyl position of the side chain. Once again, structure–retention relationships were developed to describe the results. For all compounds, the elution order was determined by the chirality of the amylosic backbone, while the magnitude of the enantiomeric separations was affected by the chirality of the carbamate side chain [72].

Later, an additional predictive model was applied to these results [73]. In this case, chemometric predictions were applied to QSERRs developed in the earlier studies. This process can best be described as a multivariate regression/neural network approach. Predictions were made for the separation and retention of six compounds on the ADMPC, ARMBC, and ASMBC chiral stationary phases. When these predictions were compared to experimental chromatographic results, 11 of 16 predictions were in close agreement with the experimental results. The predictions were generally in agreement when no separation was observed, and in only two cases was the experimental elution order opposite that predicted. Innumerable additional studies have been carried out in which groups of related compounds have been separated on cellulosic or amylosic chiral stationary phases, and mechanisms of interaction have been postulated based primarily on chromatographic results. It is, unfortunately, not possible to cover all of these studies in depth. Many of these studies stress the importance of supramolecular structure and the shape of chiral cavities or chiral ravines in determining chiral selectivity. There are a few studies of particular interest [74–78].

Later, Yamamoto et al. applied NMR and molecular modeling to the determination of the ADMPC structure and mechanism of interaction [79]. First it was necessary to make a relatively low-molecular-weight (DP $\sim$ 100) chloroform-soluble ADMPC prepared by the enzymatic polymerization of α-D-glucose 1-phosphate dipotassium catalyzed by a phospholylase isolated from potato using maltopentaose as a primer [79,80]. In this study, the coupling between ADMPC and (*S*)-1-(9-anthryl)-2,2,2-trifluoroethanol was studied using ^{1}H-NMR titration and molecular modeling. Also, the structure of ADMPC was studied by two-dimensional NMR (NOESY) and molecular modeling, and the NMR and chromatographic recognition of ADMPC was determined in the same solvent (chloroform). A problem with the structural determination of ADMPC is that x-ray data are not available, so the structure determination depends much more on NMR. Based on modeling of an ADMPC dimer and the determination of various interproton distances from the NMR data, the authors concluded that ADMPC has a left-handed 4/3 helical structure [79]. Information on the interproton distances was obtained by measuring the peak volumes of the cross and diagonal peaks at different mixing times in the NOESY spectra [57]. When this information was combined with the energy profile of the dimer of ADMPC, two dihedral angles in the glycoside bond were determined and the polymer was then constructed using three-dimensional periodic boundary conditions [57]. As with CDMPC, the polar carbamate groups are located inside, and the hydrophobic aromatic groups are placed outside, the chain's main chain. Chiral interactions then take place in the chiral grooves along the main chain, where the carbamate groups can participate in hydrogen bonding. Yamamoto et al. and Wenslow also examined the structure of ADMPC by solid-state NMR and reported that ADMPC has a helical structure with fewer than six folds [79,81].

The HPLC separation of 1-(9-anthryl)-2,2,2-trifluoroethanol (TFAE) was determined on coated ADMPC with 90 : 10 hexane/2-propanol eluent and on chemically bonded ADMPC with chloroform eluent. Both ADMPC phases showed good resolution of this compound, but with reversal of elution order. The chromatographic interaction of chemically bonded ADMPC in chloroform was correlated with ^{1}H titration data for the ADMPC/(*S*)-TFAE complex. Several glucose protons were shifted significantly upfield, indicating that the anthryl ring of the alcohol solute may be closely above these protons on the same side of the glucose ring, so that it can significantly affect its ring current [79]. These same protons were barely shifted in the presence of (*R*)-TFAE. Unfortunately, it was not possible to observe clear NOEs for the ADMPC solute complex, so a precise model for the complex could not be proposed. Nevertheless, a model was proposed in which the TFAE molecule fits into the chiral groove of the polymer in such a way that the alcohol proton from TFAE is hydrogen bonded to the carbonyl oxygen of the carbamate at the 2-position on the glucose ring [79].

Using a system that had been studied previously, Matsumoto et al. looked at the complexation of bi-2-naphthol with cellulose tris(5-fluoro-2-methylphenylcarbamate), this time using mass spectrometry [52,82,83]. In these experiments, racemic mixtures of binaphthol were prepared in which one of the two enantiomers was partially deuterated at the 3- and 3′-positions. These samples were then complexed with the carbamate polymer in chloroform and inserted directly into the ion source of the mass spectrometer. Frequent mass spectral scans were taken as the sample temperature was raised from 25°C to 400°C. These scans showed two dominant m/z peaks, one at $m/z = 286$ (from binaphthol) and one at $m/z = 288$ (from deuterated binaphthol). In one typical experiment, the (*S*)-enantiomer was deuterated, whereas the (*R*)-enantiomer was not. As the temperature increased, successive scans showed an increase in the ratio of 288 to 286 (m/z). This change was attributed to the slower desorption of the more strongly complexed (*S*)-enantiomer. If the racemic mixture was switched to the opposite distribution of deuterated (*R*) and nondeuterated (*S*), an equivalent decline in the 288/286 m/z ratio was observed. In experiments in which achiral polystyrene was substituted for the carbamate polymer, no change in the 288/286 ratio was observed. These results were again corroborated by the chromatographic separation of bi-2-naphthol on the carbamate polymer, in which no isotope effect was observed. In a related experiment using negative-ion fast-atom bombardment (FAB)/mass spectrometry (MS), similar chiral recognition was also observed.

Yamamoto et al. further applied computational studies to the chiral discrimination mechanism of cellulose phenylcarbamates [52,84]. A problem with CTPC and CDMPC is that they are chloroform insoluble and hence not easily studied by NMR in solution. Two approaches were taken to calculate the interaction energy between these polymers and the solutes *trans*-stilbene oxide and benzoin. *Trans*-stilbene oxide is resolved on both polymers with good selectivity but with reversal of elution order; benzoin is resolved on CDMPC but not CTPC. In one molecular modeling approach, enantiomers were generated around the NH proton and the C=O oxygen of the carbamoyl group of CTPC and CDMPC, which are considered the most important binding sites, and the interaction energy was then calculated. In another approach, enantiomers were randomly generated by the Monte Carlo method on the surface of CTPC and CDMPC, and the interaction energy was calculated. In both approaches, an octamer or nonamer of the CTPC or CDMPC was constructed based on x-ray data for CTPC, followed by optimization with molecular mechanics or molecular dynamics calculations. The calculations suggest that lowest or averaged interaction between CTPC and (*S*,*S*)-*trans*-stilbene oxide is lower than that between CTPC and (*R*,*R*)-*trans*-stilbene oxide, whereas the reverse is true for the interaction between CDMPC and these enantiomers. When similar calculations were applied to the interaction of CTPC and benzoin, almost no interaction energy differences could be found for the two enantiomers, which is in agreement with the chromatographic experiment in which they coelute [14].

Polar hydrogen-bonding interactions do not always dominate, as in the case of nonpolar aromatic compounds that have been resolved [62]. In this case, $\pi-\pi$ interactions between the phenyl group of the CTPC and the aromatic groups of the solvent may dominate in effecting the chiral recognition [14]. Also, for reversed-phase conditions with aqueous eluents, it is proposed that hydrophobic chiral cavities between the CTPC chains must play the dominant role for effective chiral recognition [14].

We mentioned earlier that Wenslow used solid-state NMR to examine the structure of ADMPC [81]. In a pair of papers, Wang et al. studied the effects of mobile and stationary phases on the enantiomer resolution of some structurally similar compounds [85,86]. They noted that some compounds had reversals of elution order when they were eluted from Chiralpak AD and Chiralcel OD columns. They also noted several instances of reversal of elution order when the mobile phase was switched between hexane/ethanol and hexane/isopropanol. They reported that some compounds had higher k' values with hexane/ethanol mobile phases than with hexane/isopropanol, even though lower retention might be expected for the more polar ethanol. They concluded that the polarity was not the dominant factor in determining k' on the Chiralpak AD column [85]. However, the retention of a solute does depend on its bulkiness and polar functional groups such as NH and C=O. Since the Chiralcel OD and Chiralpak AD columns have the same 3,5,-dimethylphenylcarbamate derivative, differences in retention on the two columns must be due to the higher-order structure of the CSP. Also, reversals in elution order on the Chiralpak AD column when the mobile phase was switched from isopropanol to ethanol might be due to an alteration of the steric environment of the chiral cavities on the CSP by ethanol [86].

In their $^{1}H/^{13}C$ CPMAS NMR studies, Wenslow and Wang first flushed a Chiralpak AD column with various ratios of hexane/isopropanol or hexane/ethanol and then pumped a short section of packing out of an opened end of the column. When the CPMAS NMR spectra of dry packing were compared to those of hexane-wetted packing, it was clear that hexane had been incorporated into the wetted packing. However, this solvent incorporation was weak, as evidenced by the narrow line widths of the hexane peaks, which are associated with high mobility of the solvent. Flushing the packing with hexane/isopropanol or hexane/ethanol resulted in displacement of the hexane by the alcohol. More ethanol than isopropanol was required for this displacement. This incorporation was accompanied by a narrowing of the ^{13}C-NMR line widths for the Chiralpak AD backbone and side-chain carbons. This narrowing could have been caused by increased mobility and/or increased crystallinity of the CSP. It was concluded that the line narrowing was caused by the increased crystalline character of the Chiralpak AD [86], and that this was increased for isopropanol as compared to ethanol. Of further interest was the chemical shift value of the C_1 carbon of the amylose backbone. Previously, Gidley and Bociek [87] had studied the glycosidic linkage conformation of amylose by CPMAS NMR and reported that for butanol-complexed amylose, which had a left-handed sixfold helical structure (with a repeat of six glucose units), the C_1 shift was 104.1 ppm. Based on the upfield shift of ADMPC C_1 signal to about 99 ppm, which would be associated with an increased degree of twisting or helical folding of the glucose units, a helix with a repeat of less than six glucose units is more likely. This would be consistent with the fourfold helical structure reported for amylose tris(phenylcarbamate) reported by Vogt and Zugenmayer based on x-ray analysis [44].

Later, Wang and Wenslow extended this work by studying the effects of 1-propanol, 1-butanol, and *t*-butanol on the CPMAS spectra of Chiralpak AD [88]. As in the previous study, the solid-state NMR spectra indicated that these additional alcohols were incorporated into the Chiralpak AD polymer. For hexane/1-propanol, hexane/2-propanol, and hexane/1-butanol no changes were seen in the NMR spectra, even when the alcohol concentration was increased from 10% to 30%. For these alcohols, this probably means that the CSP structure is unchanged if the alcohol concentration exceeds 10%. However, for *t*-butanol, some definite changes were observed. As the *t*-butanol concentration increases, C_6, C_1, and the C=O carbon peaks all shifted upfield. At 20% *t*-butanol there appeared to be a transition of the CSP structure in which the C_4. C_3, and $C_{2,5}$ peaks all broadened and merged together, which would indicate that the CSP was becoming less ordered. However, at about 35% *t*-butanol, this trend started to reverse, and the CSP again became more ordered. The insertion of alcohols of different size and shape is likely to cause different degrees of twisting of the Chiralpak AD polymer at the glucosidic linkages. Definite changes were thus observed in the NMR spectra and in the chromatographic selectivity when these various alcohols were incorporated into the Chiralpak AD polymer. Compared to the linear alcohols (ethanol, 1-propanol, and 1-butanol), the branched alcohols (2-propanol and t-butanol) caused more twisting of the glucose units on the helical structure of the CSP [88]. The authors believe that alcohols of different bulk caused structural changes in the CSP. These structural changes affect the specific stereo environment of the chiral cavities. Chiral selectivity is then controlled not only by the attractive interactions between CSP and solute, but also by the steric fit of the solute in the chiral cavities [88]. Another group of mechanistic studies has been carried out by Grinberg and co-workers [53–55]. In one study, the mechanism of interaction for cellulose tris(4-methylbenzoate) (Chiralcel OJ) was studied with a combination of solvent modification, temperature studies, infrared spectroscopy, differential scanning calorimetry, and molecular modeling. Their basic model for the retention of a large racemic diol molecule was that retention was governed by a combination of hydrogen bonding and inclusion in the chiral cavities of the CSP. It was postulated that increasing the length of linear alcohol mobile-phase modifiers would lead to an opening of the chiral cavities, causing the second eluted isomer to lose its inclusion interactions, with a consequent loss in k'. However, as both enantiomers are less soluble in the longer alcohols, an increase in achiral elution would occur for both enantiomers. For branched alcohols such as 2-propanol, 2-butanol, and 2-methyl-2-propanol, increased steric bulk was associated with a decrease in enantioselectivity.

A series of interesting observations was made regarding temperature effects. Theory would predict that the free energy associated with the separation of an enantiomeric pair is determined by a combination of enthalpy and entropy contributions:

$$\ln \alpha = -\frac{\Delta\Delta G^\circ}{RT} = -\frac{\Delta\Delta H^\circ}{RT} + \frac{\Delta\Delta S^\circ}{R} \tag{1}$$

Therefore, a plot of ln α vs. $1/T$ (van't Hoff plot; Figure 3) should yield a straight line with a slope of $-\Delta\Delta H^\circ/RT$ and an intercept of $\Delta\Delta S^\circ/R$, provided that the enantioselective interactions do not change over the temperature range studied [53]. When plots were made over a range of 5 to 50°C, a discontinuity in the van't Hoff plot was observed at approximately 18°C. Differential scanning calorimetry in three different typical mobile phases also showed exothermic transitions near this temperature range.

Although a detailed discussion of the various ramifications of these observations is beyond the scope of this review, some summarized conclusions follow [53]. Chromatographic experiments indicated that the leading interaction between the diol and chiral stationary phase is hydrogen bonding. At temperatures below 18°C, the Chiralcel OJ phase separates enantiomers primarily through an inclusion mechanism, which is entropy driven. In this temperature region the CSP is swollen due to strong alcohol adsorption on the cellulose phase. Above 18 to 19°C, the phase undergoes a transition, and the alcohol molecules are released. Also, the chiral cavities open up, which allows for faster solute transfer between the stationary and mobile phases. In this higher temperature range, separation occurs by a selective adsorption process that is enthalpy driven. Molecular modeling results also correlated well with the chromatographic results concerning the elution order of the diol enantiomers and a related pair of hydroxy ester enantiomers.

In a subsequent study, Bereznitski et al. made similar mechanistic studies of ADMPC (Chiralpak AD) [54]. In this case, the primary probe was a complex triazole molecule with three chiral centers, also containing two fluorinated or triflorinated aromatic groups and a teriary amine. The (*S*,*S*,*R*)-enantiomer and (*R*,*R*,*S*)-enantiomer were well separated on the Chiralpak AD phase. To mimic the effect of the 3,5-dimethylphenylcarbamate derivative of the CSP, the triazole compound was complexed with 3,5-dimethyphenylacetamide (DPMA). Fourier transform attenuated total reflectance infrared (FTATR-IR) spectra were taken of the triazole and DPMA separately and as a 1 : 1 complex, all in dichloromethane. Based on these spectra, they concluded that the primary interaction between the triazole and the Chiralpak AD stationary phase was a hydrogen bond between the NH group of the triazole moiety and the C=O of the carbamate group of the CSP. However, this proton is involved in a tautomerism, so it is likely that the hydrogen bond may be weaker than normal [54]. As in the case of the cellulosic phase, van't Hoff plots of the traiazole in several hexane/alcohol mobile phases showed a change in slope at about 16 to 20°C. Also, as before, differential

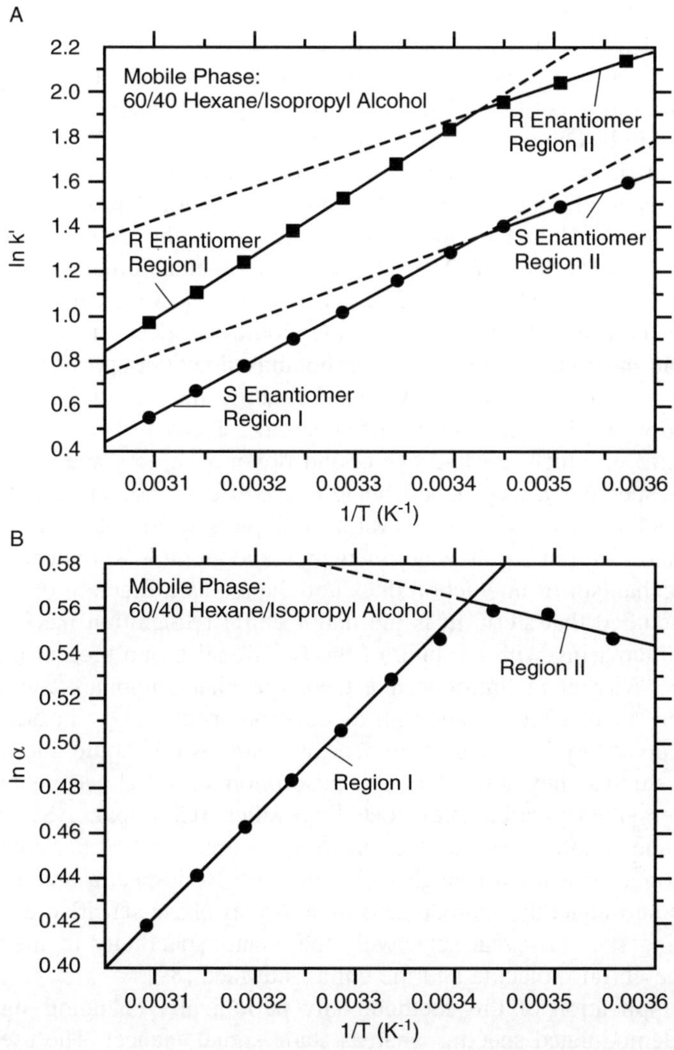

Figure 3 van't Hoff plots for (A) k' and (B) α of diol enantiomers with hexane/2-propanol 60:40 (v/v). Column: Chiralcel OJ 4.6 × 250 mm. (From [53], with permission.)

scanning colorimetry (DSC) traces showed an exothermic transition in this same temperature range. It is suggested that this behavior may indicate a breakage of the hydrogen bonds between the amylose carbamate chains, which would cause a conformational change [54]. An interesting observation emerged from the molecule modeling of the triazole/Chiralpak AD interactions. The more strongly retained (*R*,*R*,*S*)-enantiomer forms a hydrogen bond with the C=O group of the CSP, and the remainder of the molecule can fit nicely in a cleft of the amylose carbamate. However, the less retained (*S*,*S*,*R*)-enantiomer is predicted to have a bifurcated hydrogen bond with two of the C=O groups of the stationary phase, which would lead to a weaker interaction.

Ding et al. also looked at the retention mechanism for a reversed-phase cellulosic ester, Chiralcel OJ-RH [55]. In this study, indandiol, a neutral molecule with two chiral centers, was used as the primary probe. Comparative studies were also made with nine other structurally related neutral molecules. Mobile phases consisted of water mixed with methanol, ethanol, 1-propanol, 2-propanol, or acetonitrile. At 10 to 15% modifier in water, all of the alcohols gave α values of 1.15 to 1.19 for the (1*R*,2*R*)–(1*S*,2*S*) pair of enantiomers, and of 1.27 to 1.28 for the (1*R*,2*S*)–(1*S*,2*R*) pair of enantiomers. However, no selectivity was observed when acetonitrile was used as the modifier. Selectivity was attributed to hydrophobic interactions, and hydrogen bonding does not appear to be involved directly in retention. However, hydrogen bonding may contribute indirectly to the separation mechanism, as the difference in selectivity between the alcohols and acetonitrile, which are hydrogen-bond donor/acceptors and hydrogen-bond acceptors respectively, may be attributed to the effect these solvents have on the tertiary structure of the CSP [55]. Van't Hoff plots were obtained for both *cis*- and *trans*-indandiol in 70 : 30 water/methanol, and all plots were linear, indicating that the mechanism of interaction does not change as a function of temperature. It was concluded that steric fit is the major chiral recognition mechanism, with selectivity increasing with the bulk of the functional group present [55].

In 2003, Wirz et al. introduced a technique that combines attenuated total reflection infrared spectroscopy with modulation spectroscopy in order to probe the different adsorption of enantiomers at chiral solid–liquid interfaces [89]. For these studies they looked at the adsorption of ethyl lactate on amylose tris(*S*-α-methylbenzylcarbamate) coated on silica (Chiralpak AS). To employ this technique, solutions of the two enantiomers are admitted periodically to the chiral interface in a flow-through cell. Time-resolved spectra are recorded and the signals subsequently demodulated by a digital phase-sensitive data analysis. The resulting spectra reveal selectively the enantiospecificity of the interaction between the chiral molecule and the chiral interface [89].

In the application of this method, only periodically changing signals show up in the demodulated spectra, whereas static signals cancel. The use of phase-sensitive data analysis results in high-quality spectra; there is a large signal-to-noise ratio compared to that from conventional difference spectroscopy [89]. In this technique, only signals that result from different diasteromeric interaction with the CSP are detected; other signals, such as those from those of dissolved species or those that result from nonspecific interaction with the CSP, are filtered out. To perform these experiments, a slurry of Chiralpak AS coated on 20-μm silica is dissolved in dichloromethane, then deposited on a Ge ATR-IR prism, and the solvent is allowed to evaporate. After several repetitions of this procedure, a film of CSP is deposited on the prism. As a result of these experiments, the authors were able to conclude that D-ethyl lactate is more strongly adsorbed than L-ethyl lactate, with an estimated relative $\Delta\Delta G$ value of about 7 kJ/mol. In complexing with ethyl lactate, the NH group of the CSP served as a hydrogen donor, and the C=O group served as a hydrogen acceptor. Also, the ester group of the ethyl lactate served as a hydrogen acceptor, and the O–H group of ethyl lactate served as a hydrogen donor. The spectral shift of the NH bending mode of the CSP (amide II) and the C=O stretching mode of the lactate were larger for D-ethyl lactate than for L-ethyl lactate. This indicates that the N–H–O=C hydrogen bonding is stronger in the case of D-ethyl lactate [48]. Other conclusions are that one of the three side chains of Chiralpak AS is more involved than the other two in interacting with ethyl lactate. However, the data also indicate that more than one interaction is populated at room temperature and that interaction with the ethyl lactate may induce a conformational change of the amide group of the chiral stationary phase [89].

In a recent series of papers, Kasat et al. undertook systematic studies to better understand the mechanism of interaction on several polysaccharide phases [90–94]. In their approach, they studied sorbent–solute and sorbent–solvent interactions separately before bringing them together to develop chiral recognition models based on the combined sorbent–solute–solvent interactions. Kasat et al. first looked at the interaction of ADMPC (Chiralpak AD) with 10 simple solutes, using the techniques of ATR-IR spectroscopy, density functional theory (DFT) modeling, and HPLC [90]. The 10 solutes were 1-propanol, heptane, heptanol, benzene, propylbenzene, benzyl alcohol, pyridine, tetrahydrofuran, diethylamine, and aniline. The functional groups of these compounds could interact with the CSP through hydrogen-bond, hydrophobic (aromatic–aromatic or $\pi-\pi$ dispersive), other dipole–dipole, and dipole–induced dipole interactions.

In general, mostly qualitative conclusions were reached using the ATR-IR spectra obtained from these solutes in the presence of the Chiralpak AD polymer. More quantitative results were obtained from the DFT modeling, which can provide energies of interaction for hydrogen bonds and other types of interaction, as well as providing estimates of the shifts in amide I and amide II bands. In agreement with other studies, the combined results from ATR-IR and DFT indicate that the C=O and NH groups of ADMPC are the key interaction sites. Good correlation was also obtained for the k' values of these solutes determined in 95 : 5 hexane/isopropanol, with the IR and DFT results.

Later, Kasat et al. studied sorbent–solvent interaction between ADMPC and hexane, methanol, ethanol, 2-propanol, and acetonitrile [91]. In addition to ATR-IR, HPLC, and DFT, these studies included x-ray diffraction (XRD) of thin films and ^{13}C cross-polarization/magic-angle spinning (CP/MAS) and MAS solid-state NMR of polymer-coated silica beads. This appears to be the first paper in which all of the combined techniques have been used collectively to study the interaction of a polysaccharide chiral stationary phase with common mobile phases. These studies begin with a model of an ADMPC 8-mer rod, with a fourfold helix, in which about half of the C=O and NH groups are expected to form intrarod hydrogen bonds. Further, this rod structure shows the presence of various cavities of dimension 0.4 to 0.7 nm, in which solvents and solutes may penetrate and interact with the sorbent [91]. From ATR-IR measurements it was concluded that all solvents, polar and nonpolar, are incorporated into the polymer structure. The alcohols cause the breakage of some intra- and interrod hydrogen bonds of the NH and C=O groups of the polymer, changing the bulk structure of the polymer. However, these hydrogen bonds are replaced by stronger hydrogen bonds between the alcohols and the chiral polymer. Incorporation of acetonitrile also changes the structure of the polymer; however, acetonitrile is a weaker hydrogen-bond donor and acceptor than the alcohols, and the shift assigned to the amide I band of the polymer is caused by dipole–dipole interactions of the amide I group with the acetonitrile.

The 8-mer model would predict a rod diameter of about 20 Å, while the *d*-spacing obtained from the XRD analyses indicates a repeat rod spacing of 15 Å for the dry polymer. Thus, some interrod overlap is postulated, and this would account for the interrod hydrogen bonds predicted by the ATR-IR data. This *d*-spacing does not increase with the incorporation of hexane; however, it does increase by 1.5 to 3 Å with the incorporation of polar solvents, with the *d*-spacing increasing in proportion to the bulk of the solvent. On the other hand, the width of the x-ray peak narrows considerably with incorporation of polar solvent, which would indicate that the polymer is becoming more ordered. In the case of methanol, it is hypothesized that methanol forms hydrogen-donor H-bonds with the polymer C=O groups and hydrogen-acceptor H-bonds with the NH groups, allowing for a more ordered arrangement of the polymer rods [91]. When CP/MAS solid-state NMR is applied to solvated polymer, narrowing of various ^{13}C peaks is observed. The presence of these peaks again corroborates the IR and XRD conclusions that solvent is incorporated into the polymer. The narrow peaks could be the result of increased crystallinity of the polymer and/or greater mobility of the carbamate derivatives on the polymer backbone. Since the XRD data indicate an increase in polymer crystallinity, this is likely to be the reason for the peak narrowing, which further supports the inferences of Wenslow and Wang [40,47] of the same effect [91]. However, from observation of the ^{13}C-MAS spectra, which are quite different from the ^{13}C-CP/MAS spectra, additional conclusions about the mobility of the side chains can be reached. By studying four aromatic bands in the ^{13}C-MAS spectra, it was concluded that polar solvents cause an increase in the polymer side-chain mobility, whereas the polymer backbone mobility remains unchanged [91].

Next, Kasat et al. compared the effects of backbone and side chain on the molecular environments of chiral cavities in three related polysaccharide stationary phases [92]. The three phases chosen were ADMPC (Chiralpak AD), CDMPC (Chiralcel OD), and cellulose tris[(*S*)-α-methylbenzylcarbamate] (ASMBC or Chiralpak AS). ADMPC and CDMPC have the same derivative (3,5-dimethylphenylcarbamate) but different backbones (amylose vs. cellulose), whereas ADMPC and ASMBC have the same amylose backbone but different derivatives [3,5-dimethylphenylcarbamate vs. (*S*)-α-methylbenzylcarbamate]. As before, the techniques employed included ATR-IR, XRD, CP/MAS and MAS NMR, and DFT modeling. By measuring chemical shifts in the amide region of the IR spectra, a number of relative conclusions about the three phases were obtained [92].

It was concluded that the strengths of the hydrogen bonds of the NH and C=O groups in the three different polymers are significantly different, since the respective chiral cavities are different. ADMPC has smaller chiral cavities than CDMPC, with the result that the intramolecular hydrogen bonds are stronger in ADMPC. The relative strengths of intrapolymer hydrogen bonds for the NH group are in the order ADMPC > CDMPC > ASMBC, whereas the relative strengths of the intrapolymer C=O bonds are in the order ASMBC > ADMPC > CDMPC. Also, the coupling of the CH_3 and NH groups with the phenyl ring results in much higher intensities for the phenyl peaks in ADMPC and CDMPC than in ASMBC. This coupling results in different π-acidities and basicities of the phenyl rings, yielding different face-to-face and face-to-edge phenyl–phenyl interaction with the phenyl groups of the solutes.

XRD measurements for the three polymer chains found some differences in the *d*-spacings. The smaller *d*-spacing for the ASMBC vs. ADMPC is attributed to the slight kink in the ASMBC side chain. The larger *d*-spacing for CDMPC vs. ADMPC could indicate that CDMPC can have weaker hydrogen bonds. Evaluation of the CP/MAS and MAS NMR data leads to the conclusion that the glycosidic bond conformations and helicities are similar for ASMBC and ADMPC, but are different from ADMPC vs. CDMPC. On the other hand, DFT calculations predict that both ADMPC and CDMPC side chains will have a planar conformation, while the ASMBC chain has nonplanar and possibly multiple conformations. Collectively, all of this information points to significant differences in the molecular environments of the C=O, phenyl, and NH groups in each chiral cavity, which may be major factors in the chromatographic recognition of solutes by the three chiral stationary phases [92].

Having laid the groundwork for the understanding the structure and interactions of the ADMPC, CDMPC, and ASMBC chiral stationary phases, Kasat et al. then took a further step by exploring the interactions of these phases with the compound 2-amino-1-phenyl-1-propanol [93]. This compound has two chiral centers; the isomer with the configuration (1*R*,2*S*) is −PPA, and the isomer with the configuration (1*S*,2*R*) is +PPA. This compound has amino and alcohol groups capable of participating in hydrogen-bond interactions, as well as a phenyl group capable of participating in π–π interactions. By HPLC in 90 : 10 hexane/isopropanol, the selectivities (−PPA/+PPA) were 2.4, 1.0, and 1/1.2, respectively, on ADMPC, ASMBC, and CDMPC, with the elution order reversed on ADMPC and CDMPC. As before, analysis of the amide I and amide II bands by ATR-IR show significant changes following the uptake of both +PPA and −PPA by all three polymers. The main adsorption sites appear to be NH, C=O, and phenyl, and the changes in intensities and wavenumbers of the amide bands are different for all three polymers.

Preliminary evaluation of wavenumber shifts based on DFT calculated interactions reveal that −PPA and +PPA interactions with a single ADMPC side chain could not account for enantioselectivity, because the side chains can be approached from either side. However, more information is available from the XRD studies. For each of the three polymers, adsorption of either −PPA or +PPA caused a splitting and narrowing of the principal *d*-spacing bands observed in the XRD spectra. It is postulated that PPA causes the rods to have a more ordered arrangement as the interrod interactions are altered. As an example, the *d*-spacing for ADMPC is a broad band at 15.5 Å. Upon adsorption of either −PPA or +PPA, two narrower bands appear at 14.6 Å and 18.9 Å. Similar observations are also made for ASMBC with an initial *d*-spacing of 14.1 Å and CDMPC with an initial *d*-spacing of 16.3Å. Unfortunately, these measurements do not account directly for the enantioselectivity of the three phases. However, information about the helical pitch and interrod spacing could then be incorporated into the molecular dynamics modeling.

Molecular dynamics (MD) calculations for all CSP-PPA interactions are consistent with the HPLC observations; however, the MD calculations would predict stronger interactions than those observed. This is presumably because the MD calculations consider only enthalpic contributions and do not take into account entropic and achiral contributions, such as those arising from solvation or uncoated surface effects. Table 2 lists the various interactions predicted using molecular modeling. It can be noted that only between ADMPC and −PPA and CDMPC and +PPA are there three probable interactions of complexation; all of the other CSP–PPA complexations lead to two interactions. In the case of ADMPC/−PPA, two of the interactions are stronger hydrogen bonds, and thus it is not surprising that the enantioselectivity for PPA by ADMPC is high, with +PPA (two interactions) eluted first and −PPA (three interactions) eluted second. These interactions are illustrated in Figure 4. On the other hand, in the case of CDMPC/+PPA, although there are three interactions, two of them are the weaker π–π interactions, so the two PPA enantiomers are separated on CDMPC, but the enantioselectivity is only 1.2. This model has great potential for the understanding of many other chiral discrimination mechanisms with these polysaccharide chiral stationary phases.

In an additional study conducted by Kasat et al., the enantioresolution of 13 structurally similar solutes were examined on CDMPC chiral stationary phase [94]. In this case, the principal probe molecule was methylephedrine, which is structurally similar to PPA but with two methyl groups substituted on the amine portion of the molecule. Also, some nonchiral solutes were included in the study to evaluate the effects of some structural elements. Once again, the ATR-IR results focused on the amide portions of the infrared spectra. As expected, *n*-propanol interacts with CDMPC primarily through hydrogen bonds with the carbamate NH and C=O groups. When benzene is adsorbed on the polymer, weak hydrogen bonds can form with the NH groups of the polymer. It is inferred that the alcohol protons of benzyl alcohol can form hydrogen bonds with the NH groups of the polymer, and that it may also be possible that benzyl alcohol may have π–π interactions with the phenyl groups of the polymer. When polymer interactions with methylephedrine are studied by ATR-IR, it is inferred that (+)-methylephedrine is incorporated into the polymer and that the OH group of this compound can form hydrogen bonds with either or both polymer NH and C=O groups. Similar observations are noted for (−)-methylephedrine. In addition, (−)-methylephedrine may

TABLE 2 Interactions Predicted between PPA and Polysaccharide CSPs

Polymer	Complex with +PPA		Complex with −PPA		PPA (−/+) α
	Type	Location	Type	Location	
ADMPC	H-bond	C= O(polymer)–HO(+PPA)	H-bond	C=O(polymer)–H_2N(−PPA)}	
		or NH(polymer)–HO(+PPA)	H-bond	NH(polymer)–HO(−PPA)	2.4
	π–π	Phenyl–Phenyl	π–π	Phenyl–Phenyl	
ASMBC	H-bond	C= O(polymer)–HO(+PPA)	H-bond	NH(polymer)–HO(−PPA)}	
	π–π	Phenyl–Phenyl	π–π	Phenyl–Phenyl	1
CDMPC	H-bond	C= O(polymer)–HO(+PPA)	H-bond	C=O(polymer)–HO(−PPA)}	
	π–π	Phenyl–Phenyl	π–π	Phenyl–Phenyl	0.8
	π–π	Phenyl–Phenyl			

Source: Data from [93].

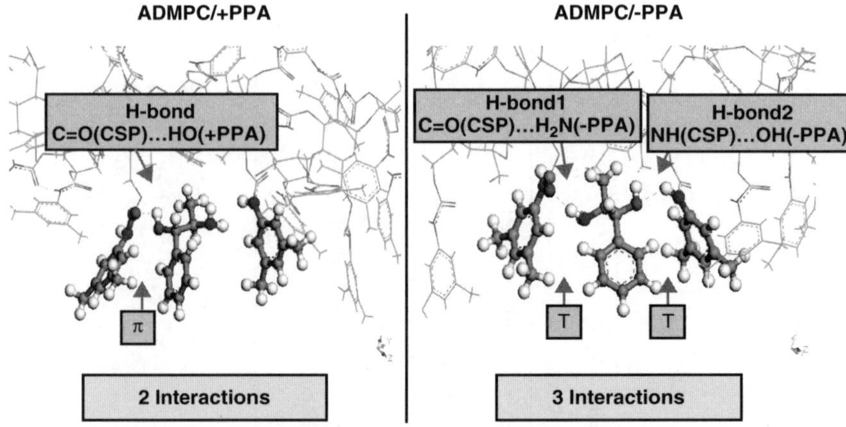

Figure 4 Interactions between ADMPC (Chiralpak AD) and PPA enantiomers. −PPA: (1*R*,2*S*)-2-amino-1-phenyl-1-propanol; +PPA: (1*S*,2*R*)-2-amino-1-phenyl-1-propanol; hydrogen bonds as noted; [π]; π−π interactions; T, phenyl–phenyl interaction. (From [93], with permission.)

be retained by π−π interactions between solute and polymer. From the collective observations of both chiral and nonchiral solutes, it is concluded that the NH and C=O groups of the polymer are the principal binding sites. However, some NH groups may reside deep in the chiral cavity, where they can only be accessed by relatively small solutes. Thus, a small solute such as *n*-propanol might access these sites, but a larger molecule, such as methylephedrine, might not. For these studies, the DFT model was constructed from a 9-mer of CDMPC units, in which the polymer has a threefold helix. The size of a triangular chiral cavity is determined by the distance between the most closely approaching C_2 and C_3 side chains of monomer 4 and the C_6 side chain from monomer 5. The three side chains are mostly parallel to each other, with C_3 shifted slightly compared to C_2 and C_6. The π−π interactions can be categorized into four conformations: parallel staggered, parallel displaced, herringbone, and T-shaped [95].

The model for interaction of CDMPC with (−)-methylephedrine includes a NH(C-3)–HO(−)-methylephedrine hydrogen bond and three π−π interactions, one of which is T-shaped, and the others are herringbone. This interaction is weakened by the steric hindrance from the fairly large $CH_3CHN(CH_3)_2$ group. In contrast, the interactions for the stronger retained (+)-methyl ephedrine include the same NH(C-3)–HO(+)-methylephedrine hydrogen bond but two displaced parallel π−π interactions. These interactions are not affected by steric hindrance. The authors thus concluded that for this pair, significant enantiorecognition can occur from a combination of steric hindrance, one hydrogen bond, and π−π interactions. Weaker enantiorecognition can occur from either a combination of weak steric hindrance, hydrogen bond and π−π interactions, or one strong hydrogen bond, one weak hydrogen bond and π−π interactions without any steric hindrance. Evidence from this study and previous studies by this group

[92,93] strongly support the three-point interaction model as the mechanism of enantiorecognition for the CDMPC chiral stationary phase [94]. Recently, Wirz et al. reported on an improved method that used modulation excitation spectroscopy in conjunction with ATR-IR to study the interactions between ASMBC and pantolactone [96]. In the newer method, enantiomers are separated on a chiral chromatography column, which is connected directly to a flow-through ATR-IR cell coated with the same polymer as the chiral stationary phase in the column. Here the enantiomers are detected, and the retention time of each enantiomer is used to select a frequency in which pulses of the racemate concentration are varied in such a way that the individual enantiomer pulses have a 90° phase lag as they elute from the column. Through demodulation of the resulting signals, it is possible to gain information about the CSP–selectand interaction. An advantage of this procedure is that it can utilize racemic mixtures rather than the individual enantiomers that were required before. Thus, the method was verified by applying the technique first to a racemic mixture of pantolactone and then by performing similar experiments with the individual enantiomers. To test the method further, they followed the change in retention of both enantiomers of pantolactone over the temperature range 12 to 36°C. The chromatograms showed that the more strongly retained (*R*)-pantolactone was more affected than the (*S*)-pantolactone by this temperature change. It was determined that the longer retention of the (*R*)-pantolactone was due to strong hydrogen bonding between the OH and C=O groups of the pantolactone, and the OC=O and NH groups of the CSP, while the retention of (*S*)-pantolactone was determined by the weaker interaction of its OH group bonded to the C=O of the CSP.

Also recently, Grinberg and co-workers have reported on the use of vibrational circular dichroism (VCD) to study conformational changes of ADMPC as the result of polar solvent uptake [56]. In these studies the VCD spectra were taken of dry films of ADMPC, as well as of ADMPC in contact with solutions of hexane/ethanol, hexane/*n*-propanol, and hexane/isopropanol. Of particular interest is the portion of the spectra from 1800 to 1500 cm^{-1} where amide I and amide II bands can be found. Uptake of *n*-alcohols by the ADMPC leads to enhanced intensity of some VCD bands, which indicates a more ordered structure of the polymer. In the dry film, NH groups can hydrogen-bond with neighboring C=O groups. However, as alcohol is added, these hydrogen bonds become weaker in favor of new hydrogen bonds between the type I C=O and the alcohol groups. In fact, in the presence of *n*-alcohols, the amide I band splits into two additional negative bands, indicating that not all of the C=O groups have equivalent interactions with the alcohols [56]. In the amide II region, two types of NH groups are found. In the presence of *n*-alcohols, a dry ADMPC VCD band at 1554 cm^{-1} first decreases in intensity and then switches direction from positive to negative. This change is even more drastic in the presence of isopropanol. Another VCD band at 1531 cm^{-1} of dry film becomes a shoulder as the *n*-alcohol concentration increases, initially undergoing a red shift, and then it gets broader with the addition of alcohol. These changes are attributed to a conformational change of the ADMPC [56], which had previously not been verified by other techniques.

4 IMMOBILIZATION STRATEGIES AND CHEMISTRIES

Although coated polysaccharide chiral stationary phases have been and continue to be extremely successful, they suffer from a single serious drawback. The polymer coatings of most useful polysaccharide chiral stationary phases dissolve in a variety of commonly used medium-polarity organic solvents. These prohibited solvents comprise, among others, dichlormethane, chloroform, ethyl acetate, tetahydrofuran, acetone, toluene, and methyl *tert*-butyl ether. Consequently, these solvents cannot be used as either mobile phases or solvent diluents, since contact with these agents will irreversibly damage the chiral stationary phase by uncontrolled swelling or even stripping the polymer from the silica support. To address these limitations, a great deal of research has been devoted to identifying strategies to immobilize polymers to a solid support, rendering them insoluble toward the full repertory of application-relevant solvent systems. This task is complicated by the fact that an ideal protocol must achieve effective immobilization while largely retaining the productive molecular and supramolecular organization and thermodynamic interaction characteristics between chiral stationary phase, solvent, and solute. Published strategies for immobilizing polysaccharide chiral stationary phases generally fall into two strategies: chemically bonding the CSP to a solid support, and coating the chiral stationary phase onto the support followed by (photo)chemical cross-linking procedures to yield insoluble networks. A number of reviews covering specific details of these immobilization chemistries are available in the literature [4,47,97–100].

4.1 Fixation via Bifunctional Grafting Agents

An early paper by Okamoto reports a quite involved approach based on regioselective functionalized cellulose 3,5-dimethyl- or 3,5-dichlorophenycarbamate derivatives [101]. In a multiple-step process, 6-*O*-tritycellulose was coated on γ-aminopropylsilica gel, followed by cleavage of the trityl group. The unmasked hydroxyl group at the cellulose was then linked to γ-aminopropylsilica gel via a diisocyanate spacer, and the corresponding materials were exhaustively capped with 3,5-dimethyl- or 3,5-dichlorophenylisocyanate to form the corresponding isocyanates. Several related protocols have been used by this or other groups [102–104] with some experimental modifications, one of them pursuing the immobilization of cellulose derivatives on polymer beads [105].

4.2 Mixed Cellulose Vinyl–Derived Carbamates + Polymerization

Kimata and Tsuboi [106,107] and Kubota et al. [108] reported a three-step immobilization process in which γ-aminopropylsilica was first transformed into acrylamidopropylsilica. Cellulose derivatized with 4-vinylbenzoyl chloride was then coated on the modified silica. The tris(4-vinylbenzoyl) cellulose was then radically copolymerized with the acrylamidopropylsilica gel to yield a covalently bonded CSP. A related but more sophisticated protocol has been reported by Okamoto's group [108]. In this technique a 3,5-dimethylphenylcarbamate is formed selectively at positions 2 and 3 of glucose, whereas the 6-positions are only partially derivatized (ca. 90%). The remaining 10% of these 6-hydroxy groups are converted into 4-vinylphenylcarbamate. This mixed cellulose phenylcarbamate is then coated on γ-aminopropylsilica along with a solution of styrene, then heated in the presence of a radical initiator to effect copolymerization with surface-anchored functionalities. More recently, this technique has also been employed with aliphatic vinyl substituents [109,110]. Chen et al. have utilized an analogous approach with methacrylate groups rather than styrene, to make immobilized CSPs for HPLC and capillary electrochromatography [111–113].

4.3 Mixed Polysaccharide 10-Undecenoyl/Carbamates or Benzoates

Oliveros et al. reported a new type of immobilization [114,115] based on cellulose derivatives bearing both 10-undecenoyl groups and phenylcarbamate groups. After coating, these mixed cellulose derivatives could then be thermally polymerized to form an immobilized phase on a variety of supports, such as modified or unmodified silica, alumina, or graphite. With allyl-modified silicas, heterogeneous coupling was observed, in addition to the reticulation caused by the reaction of the 10-undecenoyl groups with each other [116]. Additional studies investigated the characteristics of these phases [117–120].

4.4 Enzymatic Immobilization at Amylose Reducing Terminal Residue

A highly elegant but rather elaborate immobilization approach was reported involving enzymatic solid-phase synthesis of an amylose oligomer followed by postderivatization with the corresponding isocyanate [80,121]. This approach started with an enzymatic polymerization to form amylose chains with the desired chain length and narrow molecular-weight distribution. Two routes are reported to fix the amylose to silica; one involves introduction of a terminal triethoxysilane group to the amylose polymer, which is then reacted with silica gel. The other involves the use of γ-aminopropylsilica gel to react with the terminal reducing group. By either approach, once the amylose is fixed to the surface, it is derivatized to form a tris(carbamate) that makes up the final CSP. The resulting CSP is reported to have chromatographic properties similar to the properties of the corresponding coated CSPs; unfortunately, due to the highly substrate-specific enzyme used for the polymerization, this technique cannot be applied to cellulose.

4.5 Cross-linking of the Polymer by Photochemical or Free-Radical Reactions

Francotte et al. patented several methods for the insolubilization of polysaccharide derivatives [122–127]. In these methods, the polysaccharide derivative is first coated on silica and subsequently exposed to ultraviolet light to render insoluble polysaccharide derivatives. As an alternative to these methods, protocols utilizing a radical initiator combined with thermal treatment [127] for immobilization have been described.

4.6 Cross-linking Through the Use of Silyl Derivatives

Duval has patented the preparation of novel silane derivatives of polysaccharides or oligosaccharides containing chlorosilane, hydroxysilane, or alkoxysilane functions. These silane derivatives are easily polymerized and cross-linked into a three-dimensional network [128,129]. Later, Chen et al. described the preparation of polysaccharides derivatized with 3-(triethoxysilyl)propyl residues that could be polymerized [130]. More recently, Okamoto's group has described the immobilization of polysaccharide derivatives bearing a triethoxysilyl group via intermolecular polycondensation [100,131]. In this process, amylose or cellulose can be derivatized sequentially with an isocyanate such as 3,5-dimethylphenylcarbamate, followed by 3-(triethoxysilyl)propyl isocyanate. After coating these coderivatized polysaccharides on unmodified silica gel, the cross-linking is achieved via treatment of the triethoxysilyl groups of the polymer with trimethylchlorosilane at 110°C for 10 mins. It is reported that this immobilization strategy combines the benefits of simple processing, high immobilization efficiency, high chiral recognition, and wide applicability to various polysaccharide derivatives [100].

4.7 Immobilization of Polysaccharides in Beads

The preparation of cellulose benzoate beads by Francotte and Wolf was described previously [28,30], but these beads had solvent restrictions similar to the coated phases. To overcome this limitation, a joint effort was undertaken between Francotte and the research group of Minguillón. In this case, a thermal radical reaction was used to insolubilize cellulose that had been derivatized with a mixture of 10-undecenoyl- and 3,5-dimethylphenylcarbamates of cellulose [132]. Later, Okamoto et al. utilized diisocyanate cross-linking to prepare insoluble cellulose tris(3,5,-dimethylphenylcarbamate) beads [131,133]. Although these beads can have high loading capacity, they generally lack mechanical stability. Thus, Okamoto et al. sought to improve mechanical strength by preparing organic–inorganic hybrid beads from a cellulose derivative [4,134]. In this process a polysaccharide containing mixed 3,5-dimethylphenyl- and 3-(triethoxysilylpropyl)carbamate was combined with tetraethylorthosilicate and cross-linked in a sol–gel reaction. The resulting hybrid beads were composed of 69% organic and 31% inorganic material. The beads could be packed at 400 kg/cm^2 without mechanical deformation, and showed significantly higher loading and throughput than cellulose tris(3,5-dimethylphenylcarbamate) immobilized on silica (Chiralpak IB) [4].

4.8 Miscellaneous Immobilized Polysaccharides

Breitlinger has described the preparation of immobilized amylosic CSPs made from oligosaccharide primers coated on silica. The underivatized materials are then lengthened by solid-phase enzymatic reaction [135]. Chankvetadze et al. prepared silica monolith columns that were covalently modified with cellulose tris(3,5-dimethylphenylcarbamate) via an epoxide group [136]. The resulting phases exhibited excellent solvent compatibility and high enantioselectivity. Kasuya et al. have described a process to attach the terminal reducing residue of low-molecular-weight cellulose to silica via a chemical reaction [137]. Phenyl- and 3,5-dichlorophenylcarbamates were then prepared from the attached cellulose, and the resulting materials were compared to silica-coated CSPs with the same derivatives.

5 HPLC METHOD DEVELOPMENT ON COATED POLYSACCHARIDE COLUMNS

5.1 Introduction

The high complexity of chiral recognition processes governing enantioselectivity on polysaccharide-type CSPs render elusive a rational approach to method development. Whereas commercial databases compiling information on successful enantiomer separations (e.g., ChirBase [138]) may offer assistance in case of known molecules, method development for new chiral compounds involves rather low sophistication trial-and-error experimentation. This screening approach essentially involves the systematic evaluation of a number of columns under various mobile-phase conditions. A comprehensive screening program covers the full repertory of commercially available polysaccharide-type CSPs (six coated columns: Chiralpak AD, Chiralpak AS, Chiralpak AY, Chiralcel OD, Chiralcel OJ, Chiralcel OZ; and three immobilized columns: Chiralpak IA, IB, and IC) and various multimodal mobile-phase systems with normal phase, polar organic, supercritical fluid, extended range and reversed-phase characteristics. Intelligently designed screening platforms using instrumentation equipped with fully automated column and solvent selection features allow the user to take full advantage of the broad chiral recognition complementarity of the highly versatile polysaccharide-type CSPs, identifying in many cases a successful enantiomer separation within a single working day. Subsequently, however, a considerable amount of method optimization may be required to adapt the separation achieved to the analytical or preparative applications projected. Unfortunately, method optimization requires iterative steps of informed decision making from the side of qualified personnel and is therefore difficult to automate. As a result, the optimization-stage step may well account for the major part of the entire chiral method development effort in terms of time and resources. In the sub-chapters that follow, the state of the art of chiral screening methodology is discussed briefly and some general recommendations are given for method development in different mobile-phase modes.

5.2 HPLC Normal-Phase Method Development

Historically, normal-phase mobile phases composed of mixtures of alkanes containing various amounts of polar modifiers have been the most frequently employed solvent system for method development with polysaccharide-type CSPs. The relatively apolar nature of these mixtures supports effective enantioselective solute–CSPs interactions, such as the formation of intermolecular hydrogen bonding and dipole stacking. Retention and, to some extent, selectivity can be tuned readily by adjusting the relative content of a polar modifier, most often an alcohol, such as ethanol or isopropanol. At low percentages of polar modifiers, normal-phase mobile phases feature media of low viscosity, a property that supports fast mass transfer chromatographic operation at high flow rates and fast mass transfer. In addition, normal-phase mobile phases tend to be inert toward sensitive solutes and have high volatility, making them a good choice for preparative applications.

Various schemes have been proposed to develop normal-phase separations with coated polysaccharide phases [2,47,139–152]. Most reported normal-phase screening schemes utilize the four most successful polysaccharide CSPs—ADMPC, CDMPC, ASMBC, and CTMB—in combination with isocratic mobile phases composed of hexane containing 10 to 25% of an alcohol. However, screening approaches employing normal-phase solvent gradients rather than isocratic mobile phases have also been reported. Thus, method development with a normal-phase gradient has been implemented as a routine operation in several industrial research laboratories, using rapid solvent gradients spanning the range of 20 to 70% alcohol modifier in hexane as a means of obtaining high-throughput screening [140–142]. However, several inherent issues associated with normal-phase gradients on polysaccharide-type CSPs need to be considered. Potential problems with the reproducibility of gradient experiments may arise from polar components in mixed mobile phases due to preferential adsorption onto the stationary phase. Particularly problematic might be traces of water found even in commercial-grade HPLC solvents, as they may adsorb cumulatively on the stationary phase, deactivating the stationary phase [139]. To address these issues it is recommended that very dry solvent or solvent with a known water content be used and that column temperature be controlled carefully to obtain reproducible results. Another issue that needs to be considered in context with gradient-based method development schemes is the relatively slow column equilibration of polysaccharide-type CSPs upon changing mobile-phase conditions. The rapid change in solvent composition associated with step solvent gradients may in many cases be too fast to allow the full establishment of interphase equilibrium. Consequently, the separation observed may suffer from poor reproducibility, and it may be difficult to transfer into the isocratic mobile-phase mode.

An alternative method of obtaining higher throughput is to perform multiple experiments simultaneously rather than sequentially. Fully automated parallel chromatographic screening schemes using two [143], four [144], and eight [144–147] polysaccharide-type CSPs have been reported, allowing for a major enhancement in sample throughput. Certainly, such systems offer great advantages to laboratories required to screen high volumes of compounds in short periods of time, such as R&D units concerned with synthetic method development and optimization, and screening of catalysts and enzymes. Unfortunately, there are currently only a few companies offering dedicated and sufficiently robust parallel chromatography equipment.

5.3 HPLC Polar Organic Method Development

Most of the coated polysaccharide-based columns are fully compatible with mobile phases consisting of individual or mixtures of polar organic solvents such as methanol, ethanol, isopropanol, and acetonitrile, generally referred to in the literature as the polar organic mode. Chiral compounds tend to show relatively weak retention on polysaccharide-type CSPs operated with polar organic mobile phases as a consequence of the highly competing interactions of these highly polar media at specific binding sites. Nevertheless, polar organic mobile-phase systems often prove to be very efficient mobile phases for multifunctional chiral molecules, which often suffer from poor solubility and/or excessive retention in alternative mobile-phase mixtures. Various polar organic method development strategies have been reviewed [2,35,47,80,153–164] on polysaccharide CSPs. Interestingly, in systematic screening study of a considerable range of chiral analytes on coated polysaccharide-type CSPs with methanol, ethanol, and acetonitrile, solvent-triggered reversal of elution orders was observed [153]. In addition, it was demonstrated that β-lactams sensitive to alcohols can be effectively separated with acetonitrile, acting as an essentially highly inert aprotic mobile phase. In a related study focusing on the chiral separation of eight β-blockers [154], it was demonstrated that amylose- and cellulose-based CSPs have distinct chiral recognition profiles, depending on the polar organic mobile-phase systems employed. The cellulose-based Chiralcel OD-RH CSP produced the best selectivities in pure acetonitrile, whereas with the amylose-based Chiralpak AD-RH CSP, the best selectivity values were achieved with acetonitrile modified with 5 to 10% isopropanol or ethanol. For both columns, ethanolamine (0.1%) proved to be the most favorable basic additive. Another study focused on the effectiveness of methanol and acetonitrile mobile phases on the separation of 23 chiral drugs and chiral drug analogs on four polysaccharide phases [155]: Chiralcel OD and OJ, Chiralpak AD, and cellulose tris(3,5-dichlorophenylcarbamate). The latter phase, which has poor compatibility with alkane/alcohol mobile phases due to high solubility in these media, was found to have higher chiral recognition ability than that of Chiralpak AD and was able to resolve all 23 compounds when used with acetonitrile mobile phase.

A recent investigation into the utility of polar organic solvent [156] experimental design was undertaken to develop a simple-to-use and broadly applicable generic screening protocol for a set of four coated polysaccharide–type CSPs (Chiralpak AD-RH and AS-RH and Chiralcel OD-RH, and OJ-RH) and two solvents (methanol and ethanol), with two levels of 2-propanol concentration

(0 and 5%). In addition, the issues of potential additive-induced memory effects were addressed using a mixture of 0.1% each of diethylamine (DEA) and trifluoroacetic acid (TFA), with the claim that this mixture leads to the easiest restoration of the original performance characteristics of the additive-exposed CSPs. As an initial screening step, an evaluation of the chiral compounds of interest on the CSPs selected, using acetonitrile/DEA/TFA and methanol/DEA/TFA, was recommended. Compounds not resolved in these screens were subjected to a secondary screening round employing the same mobile phases containing 5% of various alcohols (for CH_3CN, the order of modifier preference is 2-propanol (IPA) > EtOH > MeOH; for MeOH the order of modifier preference is IPA > butanol > EtOH). This basic screening strategy was subsequently expanded to include optimization steps to achieve or improve resolution (via variation of mobile-phase organic modifier content and temperature), peak shape (via variation of additive type and level), and analysis time (via adjustment of temperature and flow rate) [157]. Recently, a refined version of this screening concept was published, including two new CSPs based on cellulose tris(3-chloro-4-methylphenylcarbamate) and amylose tris(5-chloro-2-methylphenylcarbamate) [35].

A more comprehensive study [2,158] of the chiral recognition behavior of 82 structurally diverse compounds on coated Chiralpak AD, Chiralcel OD, Chiralpak AS, and Chiralcel OJ chiral stationary phases in polar organic, 50 : 50 methanol/ethanol, and acetonitrile mobile phases has been conducted to assess the relative success rates of polar organic mode screening. Out of this set of analytes, 25 compounds (30%) were resolved with selectivity values of 1.5 or greater, while some additional compounds were separated with selectivity values of less than 1.5. In line with the findings of others [159], it was found that the addition of water to the alcoholic mobile phases increased retention in most cases, but generally diminishes retention when added to acetonitrile. Addition of alcohols to acetonitrile also caused a decrease in retention; however, this addition also resulted in an increase in selectivity, with k_1' values ≥ 0.5 with neat acetonitrile. Extensive studies with acetonitrile modified with 10% methanol and 0 to 25% isopropanol frequently showed improvement in selectivity relative to neat acetonitrile.

5.4 Use of Additives in Method Development

Successful enantiomer separation of basic and acidic chiral compounds on polysaccharide-based CSPs requires appropriate basic and acidic additives to control the ionization status of the analytes. Omission or failure to choose an appropriate additive results in poor peak shapes, lack of or excessively long retention, extensive nonspecific adsorption, and loss of resolution. The additives most generally employed are DEA and TFA used at concentrations of around 0.1% in the mobile phase when analyzing acidic and basic compounds, respectively [165,166]. In general, the use of acidic additives is an essential requirement for acidic compounds, whereas it may be optional for basic compounds (weakly basic analytes, such as heterocyclic aromatic amines or bulky substituted tertiary amines), which may show satisfactory band shapes even in the absence of basic additives. Enantioselective interactions leading to chiral separations on polysaccharide-based CSPs are generally nonionic in nature. Consequently, the most favorable chromatographic performance is achieved if the ionization of solutes is controlled. The essential function of additives is to suppress ionization of strongly acidic or basic groups in a molecule. An alternative strategy that is sometimes effective is to use an oppositely charged additive to promote the formation of more or less stable ion pairs. For example, an acidic additive can be coupled with a basic solute to form an ion pair that elutes through the column as an overall neutral species.

Various groups have proposed somewhat different approaches to the use of additives in screening. Occasionally, screens for neutral, basic, and acidic compounds are carried out exclusively using only acidic additives, such as TFA [143,145]. The rationale behind this strategy is to suppress the relatively strong background ultraviolet adsorption inherent in amine-type additives, such as dimethylethylamine, which prove particularly problematic in gradient screen applications. Another approach consists of the simultaneous use of basic and acidic additives in a single mobile phase, such as a mixture of DEA and TFA (0.1% each) in polar organic mobile phases when screening. It is claimed that this mixture is less prone to producing persistent memory effects, compromising method development efforts employing nondedicated CSPs [156]. A possible disadvantage of this approach is that the two additives will "cancel each other out," leaving the mobile phase less effective than if only one of the two additives was included. However, this approach is sometimes of value during the optimization stage. An example is the separation of nicotine enantiomers on a Chiralcel OJ column with 90 : 10 hexane/ethanol, in which the most reproducible separation is achieved with a mixture of DEA and TFA in the mobile phase [167].

Although much screening is performed with DEA and TFA, many other basic or acidic additives of lesser or greater strength may be employed. For example, in approximate increasing order of strength, diethylmethylamine, cyclohexylamine, butylamine, triethylamine, ethylenediamine, and ethanolamine can all be used in place of DEA as additives at up to 0.5% in nonpolar and polar organic mobile phases on most polysaccharide-based CSPs. Similarly, acetic acid, formic acid, and various alkane sulfonic acids (particularly ethanesulfonic acid and methanesulfonic acid) can all be used up to 0.5% in these same mobile phases. In many cases stronger additives yield better resolution. However, other factors to be considered are ultraviolet transparency, ease of removal in preparative applications, and the ability of the additive to catalyze chemical change in labile molecules. Another possible consideration is the removal of adsorbed strong additives from the column after the run is completed. Methods have been found to eliminate any lingering "memory effects" that could alter the results when developing methods for future compounds [156,168,169].

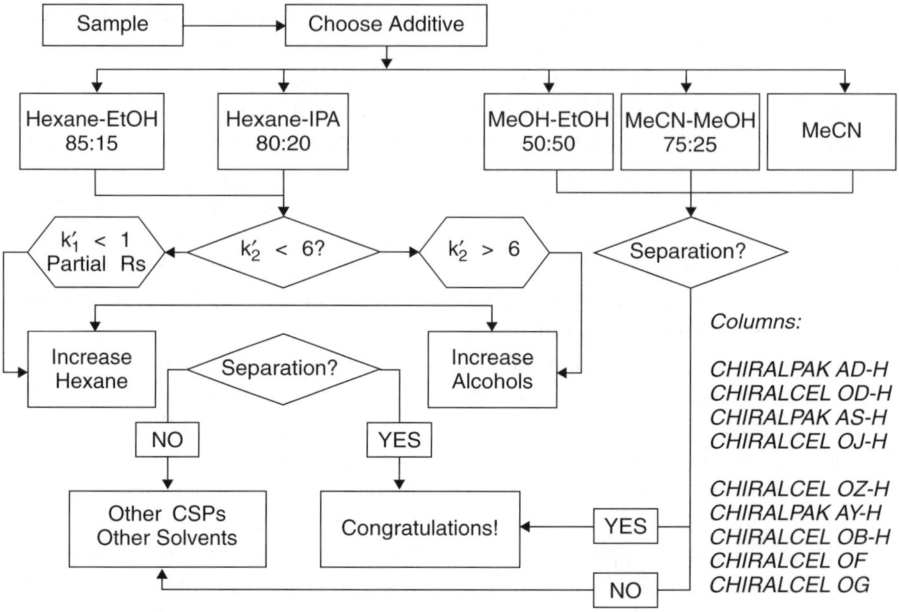

Figure 5 Method development strategy for coated columns using normal phase and polar organic modes.

5.5 HPLC Screening Strategy for Coated Polysaccharide–Based CSPs

A flow diagram for a general screening protocol of coated polysaccharide–based CSPs is shown in Figure 5. This screening process may also be applied as a portion of the global screening process for immobilized columns as well. Ideally, the structure of the enantiomers to be separated will be known. Generally, if the structure contains basic functional groups, an amine additive such as DEA can be chosen for incorporation into the mobile phase, and TFA can be chosen if the compound has acidic functional groups. At a minimum, the four standard CSPs, Chiralpak AD-H and AS-H and Chiralcel OD-H and OJ-H, should be chosen for primary normal-phase screening. This set of CSPs can be supplemented with two other valuable CSPs: Chiralpak AY-H and Chiralcel OZ-H. A more complete screening set might also include Chiralcel OB-H, OF, and OG. Care should be taken with the Chiralcel OF and OG phases, as these phases have limited tolerance to some alcohols. A separate group of the same columns, less Chiralcel OF and OG, is recommended for polar organic screening. In normal phase, the recommended starting mobile phases are 80 : 20 hexane/isopropanol and 85 : 15 hexane/ethanol. Depending on the results of the initial runs, more alcohol or more hexane may be added to the mobile phase to keep the retention factors of both enantiomers ideally between 1 and 6.

Some of the recently introduced chloro,methylphenylcarbamate CSPs offer complementarity to the standard four screening columns. These chloro- and methylphenylcarbamate derivatized polysaccharides are among the most effective to emerge out of an entire library of dozens of CSPs that were synthesized and studied by Okamoto's group. Thus, a new phase such as Chiralpak AY-H may offer a different elution order than Chiralpak AD-H for complex elution profiles. Two examples of this are illustrated in Figure 6, in which impurities in the samples must be separated from the enantiomers. In such cases, in which the amount of minor peaks might need to be quantitated, elution order could be quite important. Other HPLC-grade alkanes can be used in place of hexane, such as cyclohexane, isooctane, and *n*-heptane. Generally, at room temperature or around 20 to 25°C, changing the ratio of hexane to alcohol will cause the retention factors to change significantly, but the selectivity factors may not change much. On occasion, a separation will be found for which the two enantiomers are well resolved only with a low fraction of alcohol (0.1 to 5%). Such separations are difficult to reproduce, and it is better, if possible, to find a separation that uses a higher proportion of alcohol. For some phases, especially Chiralcel OD or OJ, one can use hexane/methyl *tert*-butyl ether (0 to 10% methyl *tert*-butyl ether) for such separations. Another alternative is to investigate the use of extended-range

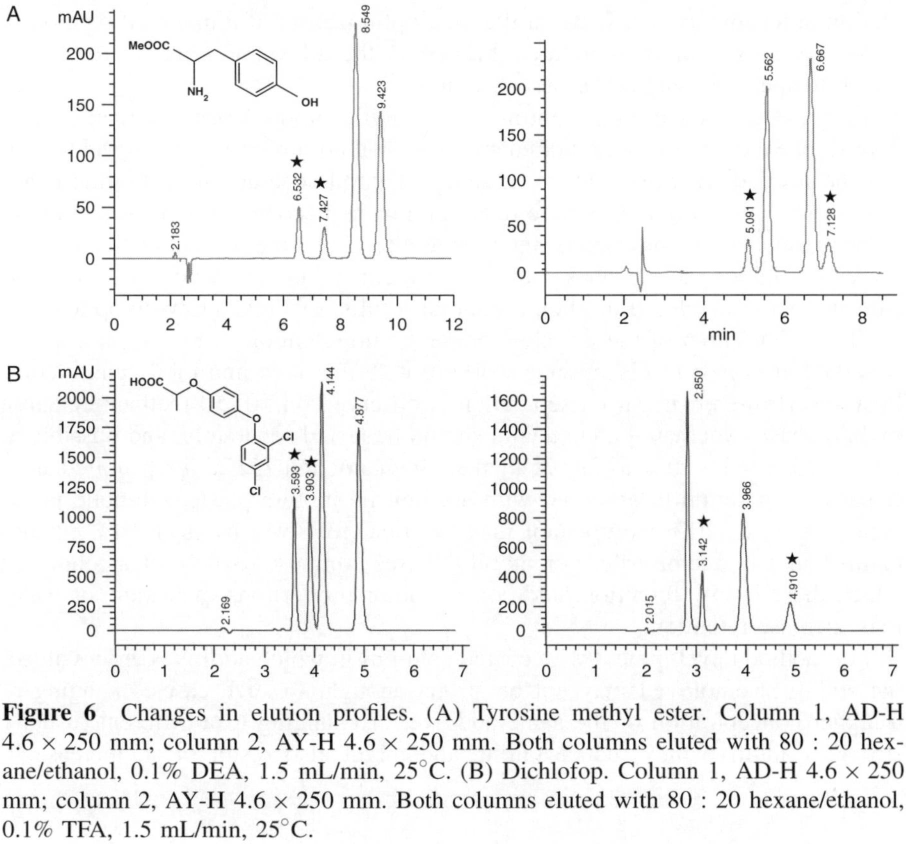

Figure 6 Changes in elution profiles. (A) Tyrosine methyl ester. Column 1, AD-H 4.6 × 250 mm; column 2, AY-H 4.6 × 250 mm. Both columns eluted with 80 : 20 hexane/ethanol, 0.1% DEA, 1.5 mL/min, 25°C. (B) Dichlofop. Column 1, AD-H 4.6 × 250 mm; column 2, AY-H 4.6 × 250 mm. Both columns eluted with 80 : 20 hexane/ethanol, 0.1% TFA, 1.5 mL/min, 25°C.

solvents on an immobilized column in which a higher fraction of a less polar modifier may be used in the mobile phase.

All of the strategies discussed above focus on achieving a separation to optimize selectivity by identifying optimal CSP/mobile phase combinations. Another way to improve separation, especially in the case of partial resolution, is to enhance the efficiency of the separation system. Improvements in the efficiency of a given separation system may be achieved by using a longer column, by using columns packed with identical CSPs with smaller particle sizes (3 μm instead of 5 or 10 μm), or by optimization of the flow rate. For higher speed in screening, flow rates of 1.5 mL/min for 4.6-mm-diameter columns and 0.3 mL/min for 2.1-mm-diameter columns are recommended. However, for 5-μm columns used in screening, flow rates of 0.5 mL/min for 4.6-mm-diameter columns and 0.1 mL/min for 2.1-mm columns offer better efficiency and, consequently, better resolution. Changing temperature can have mixed effects. In general, higher temperatures, up to 40°C, yield higher-efficiency separations, but changing temperatures can have mixed effects on selectivity. Lower temperatures, down to 5°C, may result in higher selectivity, and higher temperatures may result in lower selectivity, but not always. Cases are known in which selectivity can change dramatically over this temperature range; in some cases a reversal in elution order may occur. Thus, in the final optimization of a marginal method, it is best to try several temperatures throughout the allowable range to determine which temperature yields the best resolution.

Frequently, a separation identified in an initial normal phase screening will deliver an excellent resolution of enantiomers and no further method development will be needed. However, in the absence of good separations in normal-phase mode, or in cases where too-large retention factors become an issue, screening in polar organic mode may return interesting alternative methods. For this type of screening, initial mobile phases of 50 : 50 methanol/ethanol and 100% acetonitrile are recommended. Based on the initial results, several follow-up options are available. For either of these mobile phases, if no retention and no separation are observed, it is best to discontinue working with that column/mobile phase combination. However, if partial separation is observed on 50 : 50 methanol/ethanol mobile phases, methanol and ethanol should be tested separately, and possibly at intermediate ratios such as 25 : 75 methanol/ethanol and 75 : 25 methanol/ethanol. If partial separation is observed with acetonitrile mobile phase, a mobile phase containing 5 or 10% isopropanol may be tried, followed by 5 or 10% ethanol or methanol in acetonitrile. For mobile phases composed solely of alcohols, in which there is low retention, addition of minor proportions of hexane or water may increase retention.

For method development with coated columns in which additives are involved, the additives employed may not be strong enough. In such cases, changing to a higher concentration of the same additive, or changing to a different (usually stronger) additive may reduce tailing and might also result in an increase in selectivity. For those cases in which only one set of coated columns is available, switching from normal phase to polar organic mode, or switching from one polar organic solvent to another, should be done with care to avoid damaging the column. When changing back and forth between normal-phase mode and polar organic mode, columns should be flushed with isopropanol between modes. Due to the high viscosity of isopropanol, the flow rate must be kept low (ca. 0.3 mL/min for a 4.6 × 250 mm 5-μm column, and ca. 0.06 mL/min for a 2.1 × 150 mm 5-μm column). These low flow rates should be maintained when the next mobile phase is introduced, and kept low until all of the isopropanol is flushed from the column, as evidenced by a steady low-pressure drop across the column. A similar isopropanol flush is recommended when switching from one polar organic mobile phase to another. This safe course of action is recommended to avoid having a mixture of three polar organic solvents on the coated column simultaneously, a combination that can sometimes, inexplicably, cause irreversible catastrophic damage to the coated column.

5.6 HPLC Reversed-Phase Method Development

Reversed-phase mobile phases can be used in many cases to resolve enantiomers successfully on polysaccharide-based CSPs. A frequent misconception is that polysaccharide-based CSPs will be damaged by water. Some main motivations for the use of reversed-phase chiral chromatography are the desire to simplify the analysis of samples derived from aqueous matrices and/or the need to operate in aqueous mobile phases in order to make the separation compatible with LC-MS (particularly electrospray) detection. Work with reversed-phase HPLC on polysaccharide-based CSPs has been the subject of several excellent papers [60,61,170] and has been discussed in depth in several reviews [2,47,164,172]. Reversed-phase separations on neutral compounds can be performed with simple solvent mixtures such as water/acetonitrile, or water/alcohol, where the alcohol is typically methanol, ethanol, or isopropanol. For acidic or basic compounds, control of mobile-phase pH is needed to control the ionization of solutes [164]. Selectivity for acidic compounds generally increased as the pH was reduced from 4 to 2, while the same is true for basic compounds increases in the pH range 8 to 10. Strategies for method development then depend on whether or not the reversed-phase method will be used in conjunction with mass spectrometric detection. The normal allowable pH range for polysaccharide-based CSPs is 2 to 7, although the upper limit can be extended to pH 9 if special buffers are used. For separations that do not require volatile components, phosphate buffers are recommended for preparing acidic mobile phases, and borate buffers can be used up to pH 9. An alternative strategy for basic compounds is to use a chaotropic ion [171] in the mobile phase, which forms a neutral ion pair that is also more hydrophilic toward the aqueous mobile phase. Two of the most effective chaotropic ions for this purpose are hexafluorophosphate (PF_6^-) and perchlorate (ClO_4^-) [164]. However, concentrated solutions of perchloric acid and perchlorate salts in organic solvents are easily detonated; hence, due to the danger of concentrating any perchlorate solution, preparative applications with perchlorate salts should never be attempted. Instead, it is better to use PF_6^- buffers prepared from potassium hexafluorophosphate (the potassium salt has solubility superior to that of sodium salt) at a concentration of 50 to 100 mM. For chiral separation applications utilizing LC-MS, volatile buffers such as formic acid, ammonium formate, and ammonium bicarbonate are recommended. A 20 mM NH_4CO_3 + 0.1% (v/v) DEA (pH 9) buffer has been reported as the aqueous component of aqueous/organic gradient used for screening on the four standard polysaccharide-based CSPs [14]. Recent investigation suggest that this buffer system is also an excellent choice for isocratic mobile phases for the screening of three immobilized phases [173].

Figure 7 shows a recommended isocratic reversed-phase method development scheme that can be applied to immobilized columns, coated columns, or a blend of both types. For reversed-phase screening, 150-mm or shorter columns are recommended. For coated columns, dedicated reversed-phase columns (e.g., the Daicel RH series columns) are recommended for operation in the reversed-phase mode. This scheme can be applied to method development either with or without LC/MS. In all cases, the starting ratio of aqueous to organic is 60 : 40 aqueous/acetonitrile or 40 : 60 aqueous/methanol. Adjustments in these ratios can be made according to k' ratios obtained for the first and second eluting peaks; if the first peak is not retained strongly enough, a higher ratio of aqueous to organic may be used; and conversely, if the k' value for the second peak is too high, a lower ratio of aqueous to organic can be used.

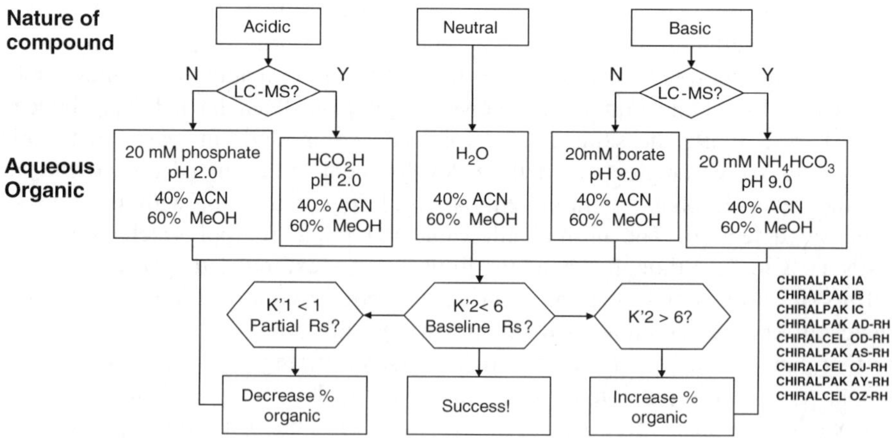

Figure 7 Method development strategy for coated and immobilized columns in reversed-phase mode.

6 HPLC METHOD DEVELOPMENT WITH IMMOBILIZED COLUMNS

6.1 Advantages

Beginning in 2004, Daicel has introduced to the market a series of immobilized columns. These new immobilized polysaccharide-based CSPs exhibit compatibility with all organic mobile phases. Due to this extended solvent compatibility, the immobilized phases have a number of significant advantages over coated columns [99]:

1. Immobilized CSPs can be used with any organic solvent; consequently, the immobilized columns are very rugged and behave like any other HPLC column.
2. Any organic solvent can be used to dissolve a sample into solution. This feature can be especially useful for preparative separations, in which productivity is usually linked to sample solubility.
3. A broader range of solvent polarities, including a large group of medium-polarity organic solvents, is now available to be used as mobile phases.
4. A greater variety of solvents is available for method development. Numerous separations have been developed on immobilized columns with extended-range solvents, for cases in which conventional solvents and coated solvents were unsuccessful.
5. More appropriate solvents can be used to flush undesired adsorbed material from the columns.
6. Solvent flushing can also be used to regenerate column selectivity.
7. Mobile phases can be chosen that are least likely to cause racemization of samples.
8. Immobilized columns can be used over a higher temperature range.

6.2 Comparison of Coated and Immobilized Polysaccharide-Based CSPs

Several articles have compared the performance of coated and immobilized polysaccharide-based CSPs [174–183]. Typically, such comparisons are made with hexane/ethanol or hexane/isopropanol mobile phases, although a few studies have been made with polar organic solvents. The consensus of these various comparison studies indicates that the immobilization process does somewhat affect the selectivity of phases having the same chiral selector. As a general rule, when the selectivity toward a group of selected compounds by both types of columns eluted with the same solvents is compared, the coated columns exhibit a little higher selectivity. However, the immobilized columns exhibit a distinct overall advantage because they can be used with many solvents that are prohibited for use with coated columns. Chief among these solvents are dichloromethane, methyl *tert*-butyl ether (MTBE), chloroform, tetrahydrofuran, and ethyl acetate. Other, less frequently employed solvents include toluene, dioxane, acetone, methylal, and ethylal. Many unique separations can be achieved with these extended-range solvents in combination with immobilized columns that cannot be achieved with coated columns. Also, these extended-range solvents may be used as sample diluents that can be safely injected onto the immobilized columns [178–180,184]. Occasionally, upon switching from the coated to the corresponding immobilized columns, a reversal is observed [181]. Other solvents, such as dimethyl sulfoxide (DMSO), dimethylformamide, pyridine, and dioxane, have been used safely as an injection solvent on a Chiralpak IA column [99,185].

6.3 Normal-Phase Method Development on Immobilized Columns

Certainly, with virtually no restriction in the choice of mobile phases, an efficient strategy providing guidance in the selection of promising solvents becomes desirable. After an extensive research program evaluating numerous possible mobile-phase systems, Franco, Zhang, et al. identified five promising solvent mixtures that are likely to be effective on consolidated Chiralpak IA, IB, and IC columns [186,187]. Four of these mixtures can be seen in Figure 8, which shows a currently recommended screening strategy when using immobilized columns. It can be seen from this strategy that two of the four screening mixtures utilize traditional alkane/ethanol or alkane/isopropanol. The other two extended-range mixtures incorporate midpolarity extended-range solvents, in the order of lowest to highest eluting strength: dichloromethane (DCM), then ethyl acetate (EtOAc). Typical starting conditions are listed in the table. Depending on the retention factors established from the initial conditions, the elution strength of

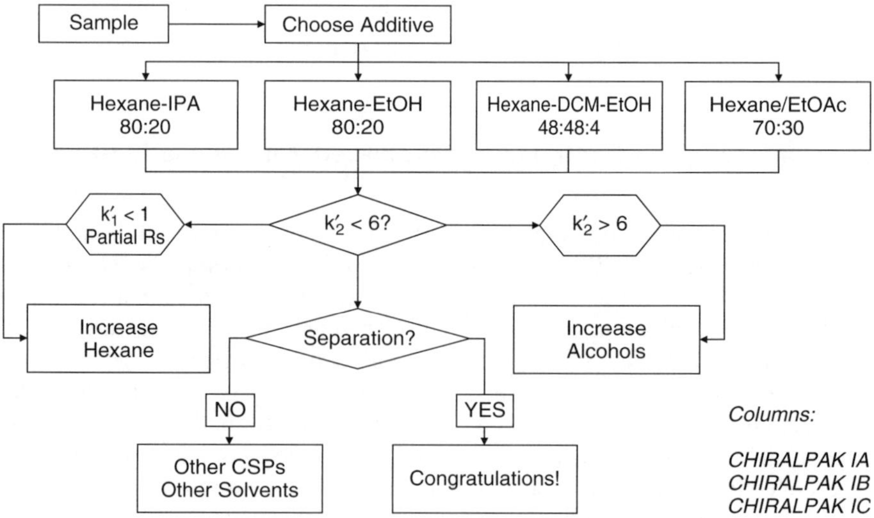

Figure 8 Method development strategy using immobilized columns employing conventional and extended-range mobile phases.

extended-range solvents can be attenuated through the addition of more hexane, or enhanced by increasing the amount of extended-range modifier, alcohol modifier, or both. Beyond the optimization strategies proposed in the table other secondary strategies might be chosen for particular applications. For example, chloroform might be substituted for dichloromethane, and it appears that chloroform may be more effective than dichloromethane on the Chiralpak IB column [99,177]. Also, ethyl acetate and tetrahydrofuran can be used in effective mobile phases for numerous applications on the immobilized columns [99,175,177]. As the strongest extended-range solvent, tetrahydrofuran does not typically require any alcohol solvent as comodifier, although there are some cases in which tetrahydrofuran modified with ethanol or methanol might prove useful. A possible concern with some of the extended-range solvents is UV adsorption. All of the conventional solvents, such as hexane, methanol, ethanol, isopropanol, and acetonitrile, have UV cutoffs at 205 nm or lower, so there is generally not a problem employing them with an UV detector. However, solvents such as dichloromethane, chloroform, ethyl acetate, acetone, and toluene have higher UV cutoffs, and the latter two solvents in particular would be precluded from use with an UV detector in nearly all cases. To overcome this problem, it is best to use a refractive index detector or a detector employing a solvent evaporation principle. In particular, evaporative light-scattering detectors (ELSDs) have proven to be an excellent alternative to UV detectors with extended-range solvents on immobilized columns [188]. Using this detector, applications employing such solvents as ethyl acetate, chloroform, and even toluene were demonstrated. Unlike a refractive index detector, an ELSD can be used with solvent gradients, and the sensitivity of an ELSD is generally much greater. However, the response of an ELSD is proportional to mass rather than concentration. The practical effect of this ELSD response is that the second eluted enantiomer produces smaller peak area than the first eluted enantiomer. As a result of this unequal response, the peak areas cannot be used directly to determine enantiomeric ratio. Therefore, a calibration curve specific to the particular ELSD and its operating conditions must be constructed. Fortunately, a plot of log peak area vs. log component mass yields a linear relationship, and enantiomeric ratios determined from the equations describing this linear calibration very closely match those obtained using an ultraviolet detector for the same peaks [188].

As an example of the versatility of immobilized columns with extended-range solvents, Figure 9 illustrates the separation of two compounds with five different extended-range mobile phases. Note that for the praziquantel separation, neat tetrahydrofuran was used, whereas the other separations employed more typical alkane/extended range solvent mixtures.

6.4 Method Development in Polar Organic Mode on Immobilized Columns

Many useful solvent mixtures can be obtained from combining different extended-range solvents. For example, although not as nonpolar as hexane, MTBE is still a

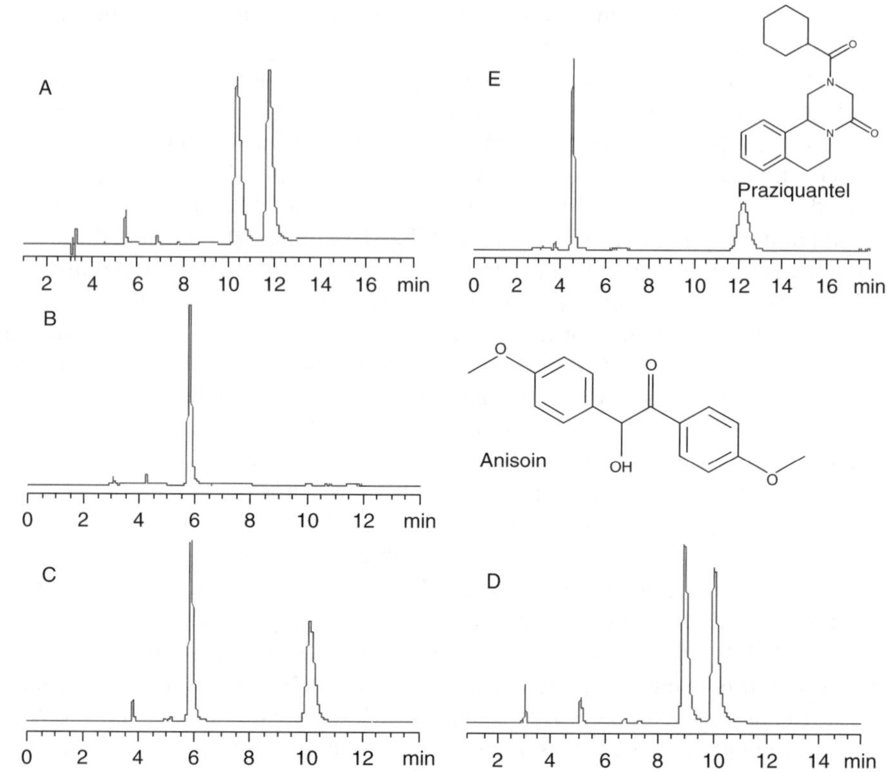

Figure 9 Extended-range solvent applications on Chiralpak IC column. Column dimensions 4.6 × 250 mm ID (A) *n*-Hexane/CH_2Cl_2 (20 : 80 v/v), $k_1' = 2.39$, $\alpha = 1.19$, Rs = 2.75; (B) *n*-hexane/EtOAc (60 : 40 v/v), $k_1' = 0.89$, $\alpha =$ 1.00, Rs = 0.00; (C) MTBE–MeOH (98 : 2 v/v); $k_1' = 1.11$, $\alpha = 2.16$, Rs = 10.67; (D) *n*-hexane/THF (80:20 v/v) $k_1' = 2.02$; $\alpha = 1.18$, Rs = 2.75; (E) 100% THF.

fairly nonpolar solvent. Like hexane, it can be mixed with ethanol, isopropanol, THF, chloroform, dichloromethane, ethyl acetate, and acetone. However, unlike hexane, MTBE can be mixed with much higher proportions of acetonitrile and methanol (only about 5% methanol is soluble in hexane, and the fraction of acetonitrile that is soluble in hexane is even less).

Immobilized columns can be screened with polar solvents, just as coated polysaccharide-based CSPs, using the same scheme as that shown in Figure 1. With coated columns, if a component is not retained strongly with acetonitrile or methanol, little can be done to increase retention other than to add water, as in the case of methanol. However, with immobilized columns, acetonitrile can be mixed with MTBE or ethyl acetate to increase retention as well as resolution. Such mixtures were used successfully to resolve supiride and aminoglutethimide enantiomers on Chiralpak IA columns [99]. On the other hand, 20% tetrahydrofuran was added to dichloromethane to reduce retention for a Ca-sensitizing drug by a factor of 5 on a Chiralpak IA column [176]. For this preparative application, THF provided additional advantages of higher solubility and lower viscosity while maintaining high resolution.

6.5 Use of Additives with Immobilized Columns

Acidic and basic mobile-phase additives are necessary when using immobilized columns just as they are for coated polysaccharide-based CSPs. However, alternative types of basic modifiers appear to be especially favorable in combination with immobilized CSPs. A study assessing the variety of effects of diethylamine, *n*-butylamine, ethanolamine, and ethylenediamine on the chromatographic performance characteristics on a Chiralpak IB column revealed that ethanolamine and ethylenediamine gave much improved peak shapes and efficiency [177]. However, for general screening of all immobilized columns, it may be preferable to screen initially with DEA as the standard additive, and optimize later with alternative amine additive on the Chiralpak IB column, if necessary.

6.6 Method Development in Reversed-Phase Mode on Immobilized Columns

Method development in the reversed-phase mode on immobilized columns may be conducted in the same manner as for coated columns. All of the strategies shown in Figure 6 can also be applied to immobilized columns. Unlike coated columns, for which dedicated RH-series CSPs should be employed for reversed-phase separations, the Chiralpak IA, IB, and IC columns can be used in all HPLC modes, including normal-phase, polar organic, and reversed-phase modes. Sometimes the use of reversed-phase conditions can enhance enantioselectivity compared to alternative chromatographic modes. For example, reversed-phase conditions can be applied to the separation of 2,3-dibenzoyl-D,L-tartaric acid or FMOC-D,L-leucine. Whereas methanol containing 0.1% TFA failed to produce any selectivity on a Chiralpak IC column for these two compounds, excellent resolutions were obtained for both compounds under reversed-phase conditions [189]. Other reversed-phase examples, employing unbuffered water, and water buffered with 0.1 M KPF_6, 0.02 M borate (pH 9), and 0.05 M phosphate (pH 2), were also demonstrated for the Chiralpak IC column [189].

It also has been demonstrated that immobilized polysaccharide-based CSPs can be operated safely at higher temperatures than can coated polysaccharide-based CSPs, which traditionally have an upper temperature limit of 40°C. Thus, when exposed to 80°C, a Chiralpak IA column produced with MTBE mobile-phase separation with enhanced efficiency for five chiral compounds [190]. Importantly, no change in performance was noted when the column was returned to 25°C. Operation at higher temperatures thus becomes an additional advantage for immobilized columns compared to coated columns.

7 SFC APPLICATIONS OF POLYSACCHARIDE-BASED CSPs

7.1 General Considerations

Chromatography exploiting mobile phases based on supercritical (or subcritical) carbon dioxide (SFC) is increasingly appreciated in academic and industrial laboratories as an invaluable separation technology, complementary to conventional HPLC. Compared to conventional solvents, supercritical carbon dioxide offers the advantages of lower viscosity and enhanced diffusivity, which translates chromatographically into rapid column equilibration and highly efficient interphase mass transfer, providing the practical benefits of superior separation performance at high flow rates. From a preparative viewpoint, SFC is a particularly attractive option because of the ease of product recovery (mobile phase is a gas at ambient pressure) and the benign nature of carbon dioxide. It has been demonstrated that polysaccharide-based CSPs are fully compatible with super- and subcritical fluid mobile-phase media [2,47,191–196], and currently, chiral separation might well be the main field of applications for SFC chromatography. The mobile phases generally employed for chiral SFC, carbon dioxide with small amounts of various polar modifiers, are believed to be similar in polarity and chemical nature to those used to normal-phase chromatography, producing comparable and sometimes even superior enantioselectivities with polysaccharide-based CSPs. Among the few disadvantages of SFC applied to polysaccharide-based CSPs is that enantiomeric polar compounds that are best analyzed in the reversed-phase HPLC mode are not as likely to be separated successfully by SFC. Another potential disadvantage is that SFC does require specific equipment, which although not necessarily more difficult to operate, is currently not nearly so prevalent as the ubiquitous HPLC equipment that is likely to be available in nearly every laboratory.

7.2 SFC Mobile Phases and Additives

In terms of polarity, CO_2 behaves like an alkane but with the added characteristic that it is somewhat acidic. Therefore, it would be expected that CO_2-containing mobile phases without acidic additives are more likely to elute easily and separate acidic enantiomers, and less likely to elute easily and separate basic enantiomers. This is exactly the opposite of HPLC mobile phases not containing acidic or basic additives. Ye et al. have studied the effect of amine additives on polysaccharide-based CSPs [197]. Amine additives may reduce nonchiral silanol interactions from the silica support. In doing so, retention may be reduced, but selectivity could also be enhanced. Also, amine additives may compete with amine samples for specific binding sites on the CSP, and this effect would also reduce retention but could have either a positive or a negative effect on selectivity. When amine additives are used in SFC mode, acidic samples are generally unable to elute; the amine additives may reduce the acidity of carbon dioxide by forming complexes with the sample amine groups [197–202]. This has been demonstrated by comparing NMR spectra for the compound propranolol before and after it was exposed

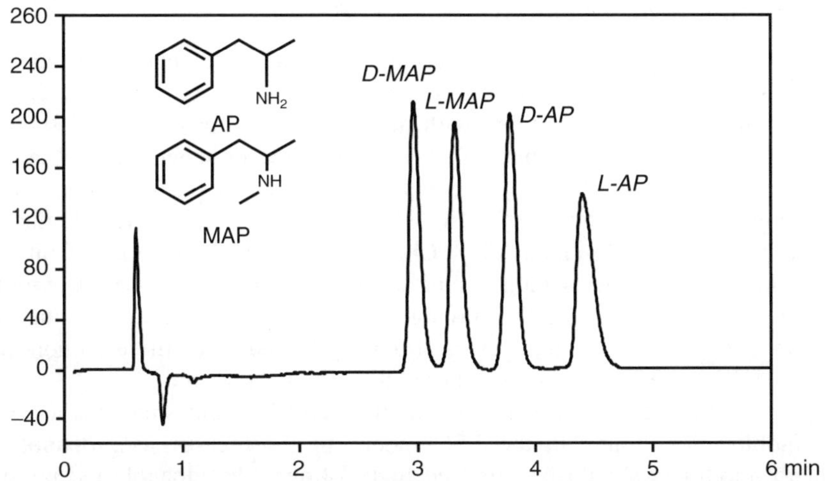

Figure 10 Rapid SFC separation of methamphetamine and amphetamine enantiomers. Conditions: column, Chiralpak AD-H (250 × 4.6 mm). Mobile phase: 10% IPA (0.5% cyclohexylamine) 5 mL/min; 150 bar backpressure. Detection: UV, 210 nm. Peak **1**, D-methamphetamine, 2.94 min; peak **2**, L-methamphetamine, 3.28 min; peak **3**, D-amphetamine, 3.75 min; peak **4**, L-amphetamine, 4.37 min. (From [197], with permission.)

to carbon dioxide [199]. The group of Ye et al. studied the effect of linear alkyl and cycloalkyl amines and found that cyclic amines larger than cyclobutylamine gave increased retention times and enhanced selectivity. Using a Chiralpak AD-H column and a mobile phase of 90 : 10 : 0.5 CO_2/isopropanol/cyclohexylamine, Ye et al. were able to develop a 5-minute separation of the two enantiomers of amphetamine and the two enantiomers of methamphetamine in a single run [197]. The separation of underivatized methamphetamine enantiomers has not been demonstrated in the HPLC mode using polysaccharide-based CSPs. The separation of methamphetamine/amphetamine enantiomers is illustrated in Figure 10.

In a later study, Stringham investigated the use of acidic additives for the separation of basic compounds by subcritical fluid chromatography (SFC) [203]. Using a mobile phase of 80 : 20 : 0.1 CO_2/ethanol/ethanesulfonic acid and a Chiralpak AD-H column, he was able to resolve 36 of 45 basic compounds that had not previously been separated by SFC. The mechanism appears to involve the separation of a tight salt pair formed between the basic compound and ethanesulfonic acid [203]. Geiser et al. have also demonstrated that amine hydrochloride salts can be separated directly by SFC using CO_2/methanol mobile phases [204,205]. Such separations are particularly advantageous in the preparative mode, since many hydrochloride salts have excellent solubility in methanol. For normal-phase preparative chromatography, the salts typically have low solubility, and hence such compounds would need to be separated in the free-base form.

7.3 Method Development Strategies in SFC

Various automated SFC method development schemes have been proposed using the four standard polysaccharide-based CSPs [206–208]. At minimum, each of these schemes uses CO_2 modified with methanol and isopropanol. All schemes use acidic and basic additives but with differences between schemes. For example, 0.1% triethylamine [206], 0.2% isopropylamine [207], or 0.5% isopropylamine [208] might be used as the basic additive. Maftouh et al. screen the columns and solvents in the order Chiralpak AD > Chiralcel OD > Chiralcel OJ > Chiralpak AS and methanol > isopropanol, and proceed only as far is necessary to produce the first good separation [208]. Having compared the separation of hundreds of compounds by normal- and reversed-phase LC and SFC, these authors now consider SFC to be the method of choice for primary screening.

Figure 11 illustrates a typical basic method development strategy for coated or immobilized columns in the SFC mode. In most cases, isopropanol and methanol modifiers should be screened first, followed by ethanol, and possibly acetonitrile. A primary screen can be conducted on the four standard coated columns, the three immobilized columns, or a combination of both types. Beyond the primary screen, additional columns such as Chiralcel OB-H, Chiralpak AY-H, and Chiralcel OZ-H could also be screened. Columns such as Chiralcel OF can also be used in SFC mode if care is taken with modifiers. Concentrated isopropanol can damage a Chiralcel OF column, but the Chiralcel OF column

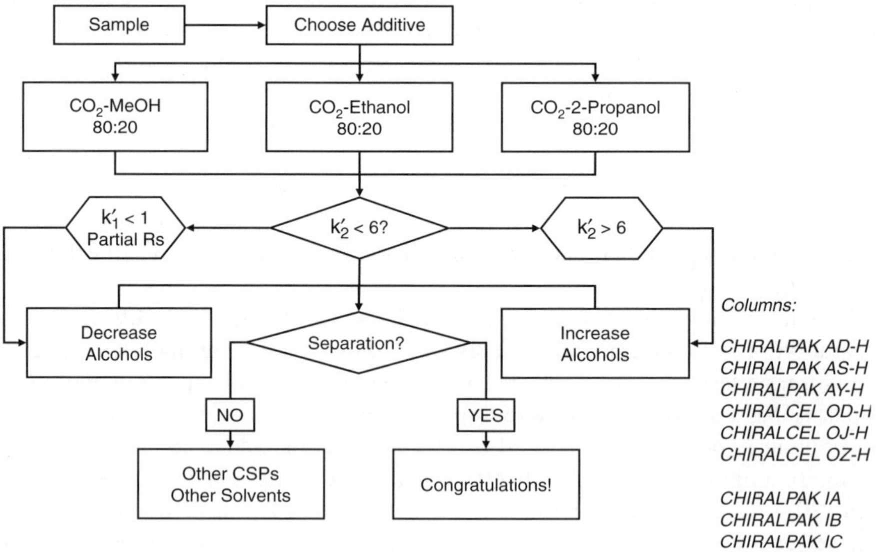

Figure 11 SFC method development strategy for coated and immobilized polysaccharide-based columns. Additional screening recommendations: immobilized columns: (1) 25% THF in CO_2; (2) 25% (80 : 20 MTBE/MeOH) in CO_2. Do not use these conditions with coated columns.

appears to be somewhat more robust toward solvent exposure in the SFC mode. In general, most analytical-sized columns can be used interchangeably between the HPLC and SFC modes. However, because of the poor solubility of hexane in CO_2, normal-phase HPLC columns should be flushed with isopropanol before introduction of the SFC mobile phase.

7.4 SFC Mobile Phases for Immobilized Columns

As is the case with HPLC, extended-range modifiers can be used quite effectively in SFC mode on immobilized columns. Diehl et al. have presented applications using 10% chloroform and 35% ethyl acetate in CO_2 on the commercially available Chiralpak IA and Chiralpak IB columns [209]. In the same study, chloroform/ethanol and dichloromethane/methanol in various ratios were used as SFC modifiers on proprietary immobilized cellulose trifluoromethylchlorophenylcarbamate and cellulose 3,5-dichlorophenylcarbamate phases. Cox also studied the elution of 39 neutral compounds in SFC mode on Chiralpak IA, IB, and IC columns using methanol, isopropranol, acetonitrile, tetrahydrofuran, methyl *tert*-butyl ether, chloroform, methylene chloride, and methyl acetate [210]. In this study, 93% of the compounds were at least partially resolved. Based on this study and other work, the following solvents are recommended for primary SFC screening on immobilized polysaccharide-based CSPs to obtain the highest success rate: tetrahydrofuran (25% in CO_2) > methyl *tert*-butyl ether/5% methanol (25% in CO_2) > methanol (20% in CO_2) > 2-propanol (20% in CO_2). If these solvents do not yield satisfactory separations, the following solvents may be considered for secondary SFC screening: chloroform, dichloromethane, acetonitrile, and methyl or ethyl acetate. Figure 12 illustrates an example of extended-range solvents being applied to chiral SFC separations. Note that methyl *tert*-butyl ether, tetrahydrofuran, and dichloromethane have all been used successfully to separate enantiomers. All these solvents may offer improved solubility for sparingly soluble compounds, reducing the risk of sample precipitation during chromatography

7.5 Recent Developments and Trends in SFC with Polysaccharide-Based CSPs

Reviews [194–196] of the relatively recent SFC literature on chiral separation technology reveal several trends, such as efforts to establish innovative high-throughput screening strategies, coupling of complementary columns to resolve multi-isomer mixtures, the use of gradients in SFC screening, and hyphenation of SFC instruments with mass sensitive detection systems. SFC can readily be coupled to mass spectrometry detectors using either electrospray or atmospheric pressure chemical ionization (APCI) ionization. Recent applications along this line for method development [211] and drug monitoring in biological matrices have been reported (Figure 13) [212,213]. However, of particular interest are innovative rapid screening strategies in which mixed pools of samples are evaluated with standard columns and solvents, and then detected with mass spectrometry [207]. For example, up to six samples might be pooled and subjected

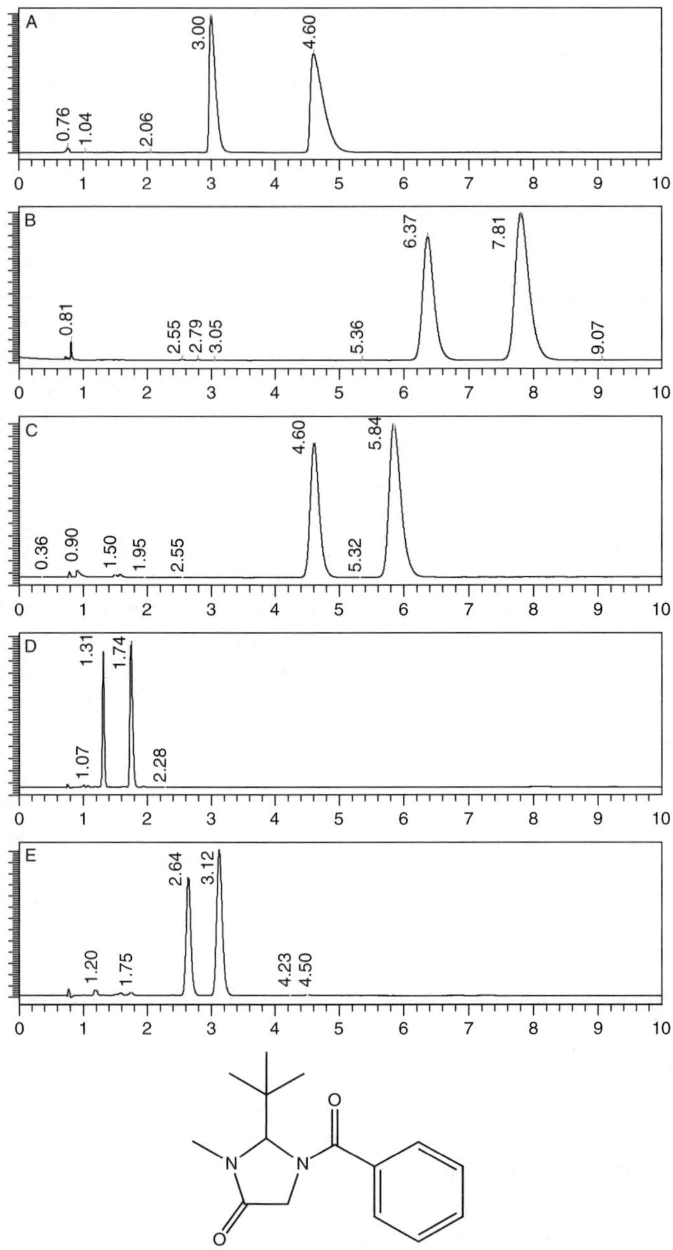

1-benzoyl-2-*t*-butyl-3-methyl-4-imidazolidinone

Figure 12 SFC chromatograms on immobilized columns with extended-range solvents. (A) Chiralpak IA 250 × 4.6 mm 25% (MTBE 20% MeOH) in CO_2; 4 mL/min, 150 bar, 35°C, 210 nm; (B) Chiralpak IC 250 × 4.6 mm 25% (MTBE + 20% MeOH) in CO_2; 4 mL/min, 150 bar, 35°C, 210 nm; (C) Chiralpak IC 250 × 4.6 mm 25% THF in CO_2; 4 mL/min, 150 bar, 35°C, 210 nm; (D) Chiralpak IA 250 × 4.6 mm 25% (CH_2Cl_2 + 20% MeOH) in CO_2; 4 mL/min, 150 bar, 35°C, 250 nm; (E) Chiralpak IC 250 × 4.6 mm 25% (CH_2Cl_2 + 20% MeOH) in CO_2; 4 mL/min, 150 bar, 35°C, 250 nm.

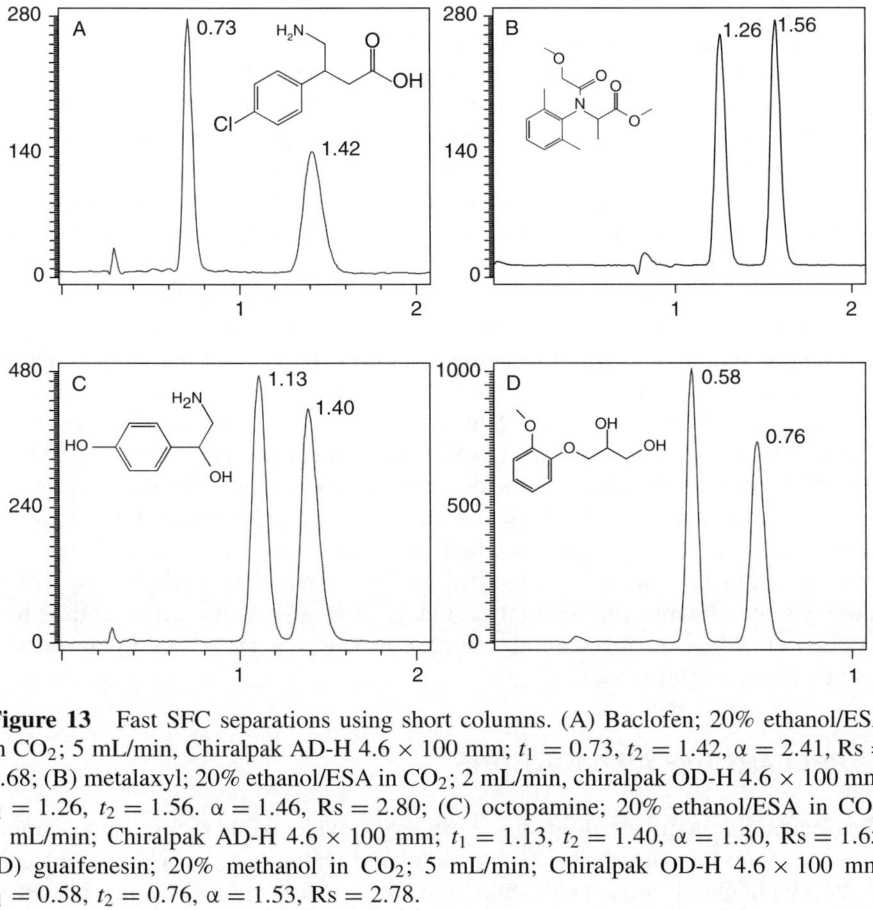

Figure 13 Fast SFC separations using short columns. (A) Baclofen; 20% ethanol/ESA in CO_2; 5 mL/min, Chiralpak AD-H 4.6 × 100 mm; $t_1 = 0.73$, $t_2 = 1.42$, $\alpha = 2.41$, Rs = 4.68; (B) metalaxyl; 20% ethanol/ESA in CO_2; 2 mL/min, chiralpak OD-H 4.6 × 100 mm; $t_1 = 1.26$, $t_2 = 1.56$. $\alpha = 1.46$, Rs = 2.80; (C) octopamine; 20% ethanol/ESA in CO_2; 5 mL/min; Chiralpak AD-H 4.6 × 100 mm; $t_1 = 1.13$, $t_2 = 1.40$, $\alpha = 1.30$, Rs = 1.65; (D) guaifenesin; 20% methanol in CO_2; 5 mL/min; Chiralpak OD-H 4.6 × 100 mm; $t_1 = 0.58$, $t_2 = 0.76$, $\alpha = 1.53$, Rs = 2.78.

to screening of a specified set of CSPs and mobile-phase conditions. For each of these runs, extracted ion chromatograms (EICs) are obtained at the expected *m/z* ratio of every enantiomeric pair, and then displayed as six EICs on the same page. This process sorts out all of the results such that each EIC displays the chromatogram for only one particular pair of enantiomers and essentially eliminates potential interferences from achiral impurities. The authors of this study claimed that this procedure would allow for at least a fourfold improvement in throughput.

With respect to chiral gradient SFC screening technology, White has implemented a highly efficient strategy in which short 10-cm SFC screening columns are combined with 5-mL/min flow rates and 2.5-min gradients [214]. Using this procedure, one sample can be screened on the four standard columns with methanol, ethanol, and isopropanol modified CO_2 within 80 minutes. As an alternative approach to fast SFC screening, Zeng et al. have established a parallel column strategy in which samples are injected onto the four standard polysaccharide-based CSPs operating simultaneously [215]. The four column effluents are combined at a MUX electrospray interface into the mass spectrometer. Using an automated program designated as intelligent parallel optimization for chiral SFC separation, the software interprets each ion screening chromatogram selected and selects the next experiments for optimization autonomously. As with the other techniques, a large increase in sample throughput is realized using this approach. The technique was also adapted for rapid determination of the enantiomeric excess of each enantiomeric pair in two large compound libraries [216].

The inherently low viscosity of SFC mobile phases readily allows for the serial coupling of two or more SFC columns to engineer separation systems with unique chiral and achiral selectivity profiles. Phinney et al. coupled achiral columns together with Chiralcel OD or Chiralpak AD columns in SFC mode to separate complex mixtures containing one or more pairs of enantiomers [217]. Barnhart et al. coupled Chiralpak AD-H and Chiralcel OD-H columns together in SFC mode to separate a mixture of four stereoisomers that could not be separated on either column alone [218]. Welch et al. designed a chiral method development strategy for complex mixtures in which columns from two sequential banks of columns could be selected for tandem column screening in SFC mode [219]. Although in principle one column could be achiral and the other chiral, best results were obtained for several complex mixtures when two polysaccharide-based CSPs were utilized.

8 PREPARATIVE APPLICATIONS

The sequential stages of drug discovery and development require increasingly larger amounts of pure materials for biological testing purposes. For drug candidates with chiral centers, all enantiomers and diastereomers must be tested individually for effectiveness, toxicity, side effects, and alternative applications. Nowadays, in each of these development stages, preparative chiral chromatography plays a vital role in isolating target enantiomers or key intermediates in the synthetic pathway to the end product. A survey of the literature on preparative chiral chromatography indicates that polysaccharide-based CSPs are the primary or sole phases employed for this purpose, and occasionally are used as a production tool at the process level [5,220–224]. Industrial-scale enantiomer separation becomes an especially attractive option if the undesired enantiomer can easily be racemized and recycled to greatly improve the yield of target enantiomer. Such an increase in yield can make a significant improvement in cost savings and may tip the balance in favor of preparative chromatography vs. other techniques of enantiomer resolution.

8.1 Preparative Method Development on Polysaccharide-Based CSPs

It is important to realize that compared to analytical method development, method development for preparative chiral chromatography involves a number of additional considerations. The ultimate goal of preparative chiral chromatography is not only to establish information on the enantiomeric or diastereomeric composition of given mixtures, but to isolate sizable amounts of stereoisomers of defined purity in the shortest possible time at the lowest cost. Essentially, this challenge means identifying the separation conditions providing the highest productivity for a given problem. Productivity is defined as the amount of enantiomer that can be processed per weight of stationary phase per unit time typically expressed in units of kilograms of enantiomer per kilogram of CSP per day. Polysaccharide-based CSPs are well established in the preparative chromatography community as the enantioselective adsorbents that provide the broadest range of applications in combination with the highest productivity levels. Next, we discuss in detail the multiple criteria and individual steps necessary for the identification of highly productive preparative enantiomer separation.

8.2 Preparative Method Development Screening

The initial step in method development for preparative enantiomer separation is generally a broad screening program to identify experimental conditions providing useful selectivity levels, typically with a selectivity factor of 1.5 or greater. All of the method development schemes described previously (Figures 5,6,7, and 11) can be employed for initial preparative screening. Chiral preparative screening philosophy is a matter of personal or group preference; some groups screen exclusively in SFC mode, some groups screen with immobilized CSPs first, others screen with four standard polysaccharide-based CSPs and a small number of nonpolar and polar organic mobile phases, and still others screen in HPLC and SFC modes with as many primary and secondary columns as possible to develop several optional separations for further consideration. As the amount of material to be processed (and therefore the cost) go up, which is typically the case in the later stages of drug development, more care is given to the selection of a final method, for which considerable resources will be applied.

8.3 General Solvent Considerations

Apart from selectivity, sample solubility in the mobile phase is the most important factor in preparative chiral chromatography. In general, the highest loading and best productivity will be achieved when injections containing the highest stable concentration of solute in mobile phase are made onto the column. Larger volume injections at a lower concentration will result in volume overload, compromising the achievable resolution by excessive band broadening. It is always tempting to resort to the rich "box of dirty tricks" to enhance sample solubility, such as dissolving a sample in a "good" solvent different than the mobile phase or using heating to enhance poor solubility at ambient temperature. However, any action that produces an unstable solution may cause the sample to "crash" out of solution and cause havoc with the preparative chromatography. A sample that is loaded in good solvents and then injected into a stream of a "less good" mobile phase will sooner or later precipitate, blocking partially, or even completely, the flow path of mobile phase through the column. This will affect the selectivity, peak shapes, and resolution negatively, and in extreme cases may even damage the hardware of the chromatographic preparative instruments. However, even in the cases of excellent sample solubility in the mobile phase, a compromise needs to be established between sample solubility and the viscosity of the injection solution. Highly concentrated sample solutions may cause malfunction of the injection devices, lead to pressure spikes upon injection, and may even give cause to severe band dispersion on the column inlet due to viscous fingering phenomena. The viscosity of the mobile phase is a further crucial factor to be considered because most preparative separations are run at the fastest possible flow rates within the allowable pressure limits for a given CSP. For example, in some cases, for a particular column and solute combination, ethanol may yield a higher selectivity separation than a competing separation utilizing methanol or acetonitrile. However, the latter solvents may still yield a higher productivity separation than ethanol, due to their lower viscosity, allowing operation at high flow rates with reduced peak broadening. Preparative solubility problems can be solved with common aprotic solvents such as methyl *tert*-butyl ether, chloroform, dichloromethane, acetone, terahydrofuran, and ethyl acetate and immobilized polysaccharide-based CSPs, and this combination provides an excellent solution for use in both HPLC and SFC applications.

8.4 General Considerations: Loading Capacity

One of the most attractive features of polysaccharide-based CSPs in context with preparative applications is their high loading capacity. The loading capacity of a given chromatographic material is generally expressed as saturation concentration, which describes the maximum amount of compound that can be adsorbed per weight of CSP. Polysaccharide-based CSPs derive their favorable loading capacities from a relatively thick polymer surface layer supported by macroporous silica gel, with the chiral polymer content accounting for as much as 20% of the weight of entire adsorbents. A comparison of the average saturation capacities of the most popular classes of CSPs are given in Figure 14 [122,225], indicating that polysaccharide-based CSPs are by far the materials with the most favorable saturation characteristics.

8.5 Loading Studies

It is important to acknowledge that there exists no general relationship between the enantioselectivity observed and the loading capacity of a given CSPs. To establish information on the inherent productivity of a given CSP/mobile phase combination for a specific chiral analyte, the dynamic loading capacity must be measured under strictly controlled experimental conditions [220]. For this purpose, a sample solution at the highest stable concentration is prepared in mobile phase, and in a series of injections, successively larger volumes of this solution are applied to the chosen CSPs until the two peaks to be separated

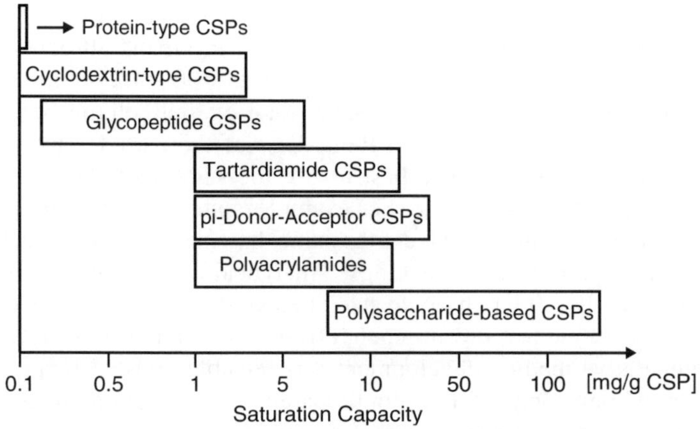

Figure 14 Typical saturation capacity of the most used commercially available CSPs. (Adapted from [225].)

start to merge at the baseline. Considering the total sample amount injected, the flow rate employed, and the time required to elute the peak pair under these "touching band conditions," experimental productivity figures can easily be calculated. These figures can then be transferred to preparative columns packed with identical CSP material by scaling the sample amount and flow rate to the cross-sectional area of the preparative column. Typically, the shapes for larger peaks will become distorted as the CSP becomes overloaded. This is the result of the compound-specific adsorption isotherm, which describes the uptake of sample as increasing amounts are applied to the stationary phase. A more detailed description of adsoprtion isotherms is beyond the scope of this work. Interested readers are referred to articles on preparative chromatography [222].

8.6 Particle Size for Preparative Chiral Separations and Column Formats

Particle size is a variable to consider to optimize productivity in preparative chiral separations [222]. For complex separations and relatively small sample amounts (milligrams to grams) it may be best to choose small-diameter particles in semipreparative (10 to 30-mm diameter) columns. Thus, 5-μm particles are frequently used in semipreparative columns of 25 cm or less column length. This is especially true of SFC applications, in which higher flow rates can be used, due to the lower viscosity of the mobile phase. For the preparative separation of substantial amounts of material (hundred grams to kilograms to tons), CSPs with larger particle sizes, frequently 20 μm, are a more favorable choice, as they allow for higher HPLC flow rates. Columns of 5 cm or larger diameter are typically used with longer column beds in prepacked columns (e.g., 50 × 500 mm or 100 × 500 mm). However, many users also employ dynamic axial compression columns prepared with bulk 20-μm packing. In the special case of SMB, 20-μm columns are also likely to be used; however, in almost all cases for SMB, the individual column length may be reduced to 100 mm. In this case each individual column is a segment of a longer total overall column bed composed of four or more individual columns. Regardless of the particle size used in the ultimate preparative process, it is best to use the same particle size in method development at the analytical scale. For example, some vendors offer preparative method development columns that are the same dimensions as analytical columns, but with 20-μm particle size packings. Analytical sized columns of 4.6 mm diameter packed with 20-μm particle size of each of these phases are available to support preparative method development efforts and loading studies. Conversely, for SFC preparative method development it is sensible to use 5-μm particle size for both method development and semipreparative (1 to 3-cm-diameter columns) or preparative (5-cm-diameter columns) separations.

Polysaccharide-based CSPs have been used prominently with several special chromatographic process modes to enhance productivity in large-scale enantiomeric separations [5,221,226–235]. These techniques include stacked injections on batch columns, recycling and peak shaving, steady-state recycling (SSR), and simulated moving-bed (SMB) chromatography. All of these techniques seek to overcome the inherent inefficiency in elution batch chromatography for which complete elution of a loaded sample is required before the next injection is made. However, under these conditions only a portion of the chromatographic bed is engaged in the separation process, often resulting in modest productivity. Productivity can be boosted by process design, taking advantage of continuous loading of a racemic feed solution and withdrawal of pure fractions of partially resolved chromatographic bands. The most productive implementation of this concept is the SMB process, which simulates a countercurrent movement of the stationary and mobile phases, producing two product streams: the raffinate (the first eluting component) and the extract (the second eluting component). Although a detailed treatment is outside the scope of this chapter, it should be emphasized that the SMB process allows for significant improvements in productivity over batch elution chromatography. Specifically, compared to single-column-batch chromatography, SMB-based enantiomer separation processes have considerably lower requirements concerning CSP and mobile phases and can be scaled easily to accommodate industrial-scale production volumes. On the other hand, SMB method development tends to be more involved to achieve robust and reproducible operation. In addition, the implementation of SMB processes requires substantial investment in dedicated equipment and operation infrastructure. Several important commercial pharmaceuticals are now manufactured with SMB as a major production step [5]. Novasep has introduced Varicol, an advanced multicolumn continuous process, to further improve productivity for the preparative separation of enantiomers [236]. Also, recent reports suggest that further improvement in productivity might be possible by the implementation of gradient features in SMB operations [5].

8.7 Preparative SFC for Chiral Separations

The inherent benefits of polysaccharide-based CSPs (high loading capacity, unparalleled broad enantioselectivity) and SFC [high efficiency at high flow rates, environmentally benign nature of super (or sub)-critical carbon dioxide, low requirements of organic solvents, and relative ease of product recovery] have established this winning combination as the preferred preparative tool for the isolation and purification of drug candidates and synthetic intermediates in early-stage drug development. Indeed, the recent literature [196,195,237,238] suggests that preparative enantiomer separation by SFC is performed almost exclusively using polysaccharide-based CSPs. Another specific advantage of SFC for preparative chiral separations of acids is that acidic additives are generally not required for separation. This can be a crucial advantage, as the presence of strongly acidic modifiers in an alcohol-containing mobile phase can potentially lead to esterification of the acid enantiomers. The high speed of SFC separations [194,239], coupled with the concept of stacked injections, can lead to rapid enantiomer production. Figure 15 illustrates a rapid semipreparative SFC production run of 10 g of a proprietary research compound. From the inset, it can be seen that the cycle time for the stacked injections is 1.7 min, and the entire run was completed in approximately 7 h.

However, solubility issues remain an often underestimated issue in SFC. As there are currently no practical procedures for determining the solubility of a compound in the supercritical mobile phase prior to processing, inexperienced users tend to overestimate the possible sample loads. Precipitation of sample during chromatography can block inlet frits and columns, thus causing a dramatic increase in pressure drop across the column. In extreme cases, high-pressure drop acting on an inlet can actually bend the inlet frits and thus destroy the entire column. As a minimal precaution, use of a precolumn filter is recommended in semipreparative or preparative SFC, and sample loading may need to be revised lower at the first sign of pressure increase in the column.

8.8 Preparative Applications with Immobilized Columns

The much broader solvent compatibility and extended scope of applications of immobilized polysaccharide-based CSPs can be exploited effectively for preparative applications in both HPLC and SFC modes. Diehl et al. have presented a series of example separations in SFC mode in which both proprietary and commercial immobilized columns were used with extended range solvents to obtain preparative chiral separations [209]. For example, an aromatase inhibitor had very poor solubility in nonpolar mobile phases (90 : 10 heptane/IPA; 3 mg/mL) but excellent solubility (>40 mg/mL) in terahydrofuran, dichloromethane, and 9 : 1 chloroform/methanol. The sample was well separated on an immobilized cellulose tris(3,5-dichlorophenylcarbamate) CSP using a mobile phase of 70% CO_2, 30% 9 : 1 dichloromethane/ethanol. Zhang et al. compared the SMB preparative

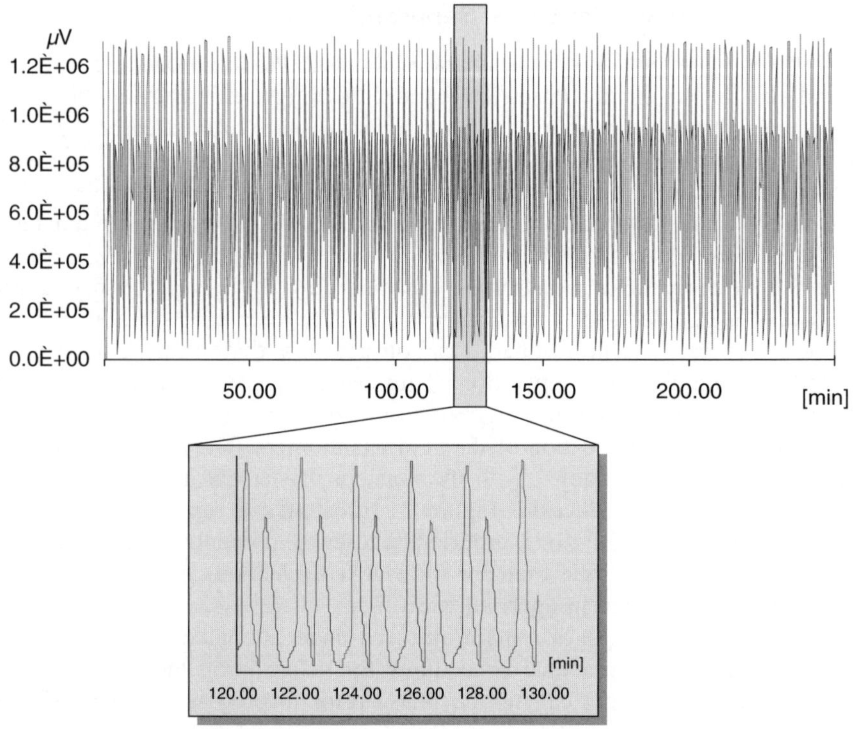

Figure 15 Stacked injection chromatograms using rapid semipreparative SFC. Ten-gram production run of proprietary research compound. Column: Chiralcel OJ-H 25 × 2.1 cm ID; sample: 40 mg/injection; cycle time: 1.7 min; mobile phase: 5% methanol; flow rate: 80 g/min; production: 1.4 g/h.

separation of a calcium-sensitizing drug on 20-μm Chiralpak AD CSP with ethanol as mobile phase with the separation achievable using 20-μm immobilized Chiralpak IA CSP eluted with 70 : 30 dichloromethane/tetrahydrofuran [176]. The solubility of the Ca-sensitizing drug was 45.5 g/L in the DCM/THF mobile phase, whereas it was 8 g/L in ethanol. The Chiralpak IA separation was also characterized by higher selectivity and lower retentivity, with the net result that productivity on the Chiralpak IA column was 2.8 g enantiomer/kg CSP per day vs. 0.4 for the Chiralpak AD-H separation, accounting for an improvement in productivity by a factor of 6.5. A similar study compared the preparative separation of glutethimide enantiomers on 20-μm Chiralpak AD and IA [99]. With all of the standard mobile phases (alkane/alcohols, methanol, ethanol, and acetonitrile), glutethimide was too strongly retained on both Chiralpak AD and IA. However, by using ethyl acetate with the IA column, dramatic improvements in productivity could be realized. The combination of IA and ethyl acetate afforded excellent enantioselectivity ($\alpha = 4.11$) and resolution ($Rs = 16.39$), with an elution time under 10 minutes. Further, the solubility of glutethimide in ethyl acetate

TABLE 3 Enantioseparations of Glutethimide[a]

Mobile Phase	t_1	t_2	k'_1	k'_2	α	Rs
Chiralpak AD						
Methanol	12.64	57.09	3.21	18.03	5.61	13.22
Acetonitrile	13.26	43.2	3.42	13.4	3.92	7.18
Chiralpak IA						
Methanol	10.12	37.82	2.37	11.61	4.9	9.14
Acetonitrile	8.18	25.33	1.73	7.44	4.3	7.91
Ethyl acetate	**4.32**	**8.43**	**0.44**	**1.81**	**4.11**	**16.39**

Source: Adapted from [99], with permission.
[a]Common conditions: flow rate, 1 mL/min; temperature, 25°C; particle size, 5 μm; column size: 250 × 4.6 mm. Using SMB conditions, it was possible to obtain a productivity of 7116 g enantiomer/kg CSP per day for glutethimide eluted from a 20-μm IA column with ethyl acetate.

is outstanding (solubility ~ 370 g/L), and ethyl acetate is a single solvent with an intermediate viscosity (ν = 0.45 cP at 20°C. SMB modeling predicts that this separation will have productivity greater than 7 kg enantiomer/kg CSP per day (Table 3). As the immobilized polysaccharide-based CSPs become more widely employed in preparative chiral separations, it is expected that many examples of high-productivity chiral separations will emerge employing these columns and extended-range solvents in both SFC and HPLC modes.

8.9 Additives in Preparative Mobile Phases

As for analytical applications, trifluoracetic acid (TFA) and diethylamine (DEA) are the most popular additives for most acids and bases, respectively, for preparative HPLC separation on polysaccharide-based CSPs [239]. Although TFA can convert acid salts in situ to free acids, the presence of the corresponding counterion tends to compromise the loading capacity. It is recommended that the preparative separation of acidic compounds always be processed as free acids. A similar conclusion can be reached with basic compounds, which should also be processed as free bases for preparative purposes. Additives with low boiling points, such as TFA, DEA, or butylamine, can readily be removed by evaporation; however, other effective additives with high boiling points, such as ethanolamine and ethanesulfonic acid, are difficult to remove by evaporation and may require additional manipulations for product isolation, such as extraction or crystallization steps. In contrast to HPLC, amine salt (e.g., propranolol-HCl, thioridazine-HCl, and tramadol-HCl on Chiralpak AD-H) enantiomers can be separated preparatively on polysaccharide-based CSPs under SFC conditions without prior conversion to the free base [204,205]. Such separations are accomplished using methanol in the absence of additives and may be considered as an interesting preparative option for compounds difficult to obtain or to handle as free bases.

8.10 Stability and Other Special Preparative Considerations

Having selected the lowest cost or best productivity separation available, a few final aspects must be demonstrated before the preparative operation proceeds. The sample should undergo a stability study in which it is stored for 12 to 24 hours in mobile phase (including additives) at a moderately elevated temperature resembling that likely to be encountered in the solvent removal step. Following this storage step, the sample should be reanalyzed to ensure that it has not racemized, decomposed, or otherwise reacted chemically. For large-scale separations it is also important to check the purity of the starting mobile-phase components to ensure that no contaminants will be concentrated into the final product upon solvent removal.

9 SPECIAL TOPICS

9.1 Reversal of Enantiomer Elution Order

On occasion, in both analytical and preparative enantiomer separation applications, the need may arise to reverse the elution order of a chiral separation developed on a polysaccharide-based CSP. The most common need to reverse the elution order in analytical applications arises when being challenged to measure a trace amount of impurity of a late-eluting enantiomer in the main early-eluting enantiomer. In this situation, the peak of the enantiomer of interest is often obscured by the tail of the major peak, making precise integration and quantification difficult or even impossible. The same is true for preparative separation with strongly tailing peak shapes; in this situation it will be possible to isolate (at least partially) pure fractions of the early-eluting peak, while the late-eluting peak may be contaminated throughout with various amounts of the opposite enantiomer. With the late-eluting enantiomer being the target, reversal of the elution order would be required to achieve access to the desired compound. Unfortunately, there is no general strategy when using polysaccharide-based CSPs to predictably obtain a reversal in elution order. However, based on a number of reports in the literature for which elution-order reversals have been observed, there are several method development strategies that can be pursued that should have a reasonable chance of success. In this brief discussion it is not possible to cite all of the reports for which reversal in elution has occurred on polysaccharide-based CSPs, some of which are mentioned only briefly in passing. However, a few examples may serve to illustrate strategies that have resulted in elution-order change.

Thus, frequent reversal of elution order has been observed upon switching from cellulose-based columns to amylose-based columns, such as between Chiralcel OD and Chiralpak AD columns [87,88]. Although far from being a general rule, there appears to be some type of chiral recognition complementarity between the cellulose and amylose backbones. For a certain number of compounds that resolve on both types of columns, there will be opposite preferences in chiral recognition. This phenomenon is also seen with immobilized polysaccharide-based CSPs,

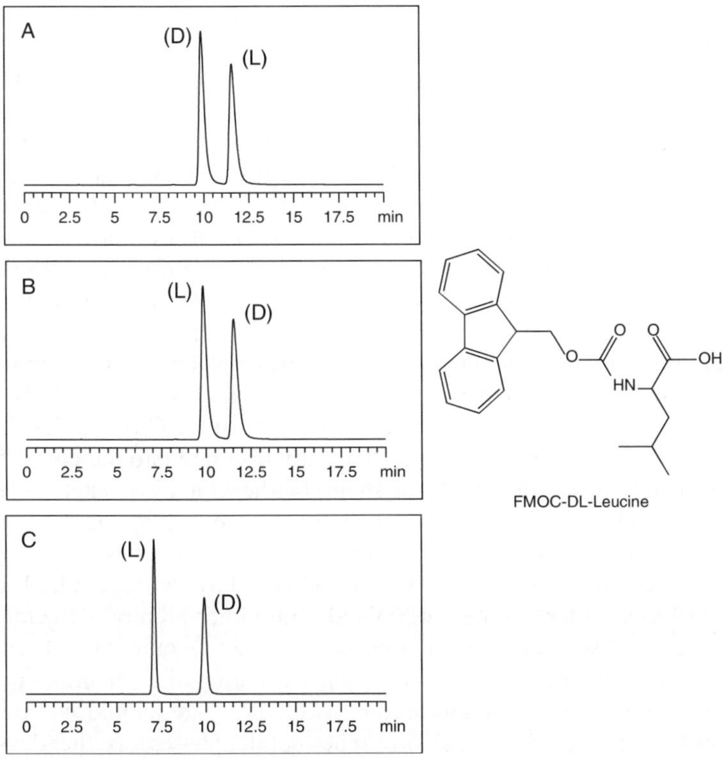

Figure 16 Reversal of elution order on immobilized columns. Common mobile phase: 90 : 10 : 0.1 hexane/isopropanol/TFA. (Adapted from Chiralpak IC launch presentation developed by Tong Zhang, Chiral Technologies, Europe, Illkirch, France.)

as illustrated in Figure 16. In this particular case, a reversal of elution order for FMOC-D,L-leucine was observed when switching from the amylose-based Chiralpak IA column to cellulose-based Chiralpak IB and IC columns under otherwise identical chromatographic conditions.

Occasionally, reversal of elution order with a given polysaccharide-based CSP can be enforced by appropriate changes in the mobile-phase conditions [79,87,88,240–243]. Thus, changes in elution order were seen in alkane/alkane mobile phases upon changing from a shorter or less bulky alcohol to a longer or more bulky alcohol. For example, an elution order change may occur for hexane/ethanol vs. hexane/isopropanol; other changes have occurred when using C_3 to C_8 alcohol modifiers in alkane. It was also noted that a reversal in elution order can be effected by changing the water level in an alkane/alcohol mobile phase between 0 and 1600 mg H_2O/L [242]. Similar observations were made for the elution order for three of four profen NSAIDs, for which reversals occurred upon switching SFC modifiers from methanol to 2-propanol, 2-butanol, and 2-pentanol [244]. A third strategy might involve changing the temperature of the separation [242,244]. Some changes in elution order in SFC experiments were noted on Chiralpak AD by varying the temperature between 15 and 45°C. Similar changes were able to effect a normal-phase reversal on a Chiralcel OD column by varying the temperature from 8 to 55°C.

It does not follow logically that switching from normal-phase to reversed-phase mobile phases on the same chiral selector will necessarily lead to a reversal in enantiomer elution order. However, in our laboratories we were recently able to reverse the elution order of a pair of enantiomers of a proprietary compound by switching from normal phase to reversed phase on different polysaccharide-based CSPs. The molecular mechanisms triggering reversal of elution orders on polysaccharide-based CSPs are largely unknown. A hypothesis often cited for solvent-driven elution-order changes is that they are caused by "conformation changes" of the chiral polymer [2,240]. In addition, Wang and others attribute this behavior to changes in the "steric environment" of the CSP [2,87,88]. Thus, if solute molecules of different size and shape are absorbed into the chiral polymer, they may change to some extent the shape of the binding pocket where chiral discrimination takes place and, consequently, the selectivity of the separation. Obviously, obtaining a reversal of enantiomer elution order is a matter of trial and error, so the chances of success will be enhanced by trying standard screening protocols. Elution order can be determined with either a chiral detector such as a polarimeter or by preparing an unequal (e.g., 2 : 1) ratio of two individual enantiomers if they are available, or even by noting the retention order if a racemic mix and a single enantiomer are available. Care should be taken when using a polarimeter detector and different mobile phases, as the direction of rotation may change in different solvents.

9.2 Fast Chiral Separations with 3-μm CSPs

Many analytical laboratories are faced with high volumes of enantiomeric samples to resolve, and as a consequence, mastering high-speed, high-throughput analyses is becoming a routine requirement in enantioselective chromatography. To a certain extent, these needs can be met by implementation of parallel chromatography and (gradient) SFC, eventually in combination with shorter column formats. However, outside the enantiomeric separation realm, much progress has been made recently with fast separations using sub-2-μm particles and UPLC/HPPLC techniques. Fast analyses generally require short, high-efficiency columns packed with small particles. Polysaccharide-based CSPs generally work best with wide-pore silica, and there may be pragmatic trade-offs among smaller particle size, higher pressure, phase effectiveness, and mechanical stability. To meet the needs for faster, higher-efficiency columns, two vendors have recently introduced polysaccharide-based CSPs in a 3-μm particle size. Van Deemter curves for 3- and 5-μm particle size CSPs show that 3-μm particles can be operated at significantly higher linear velocities and flow rates than 5-μm CSPs without significant loss in efficiency, as shown in Figure 17 [245].

Analytical columns of 4.6 mm diameter that are normally operated at 1 to 1.5 mL/min with 5-μm particles can be operated readily at 4 to 5 mL/min if the

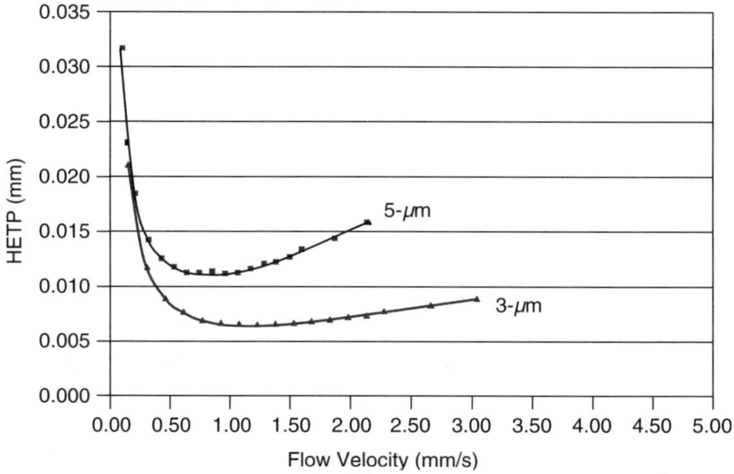

Figure 17 Van Deemter curves for 3- and 5-μm chiral stationary phases in HPLC mode. 5 μm: Chiralpak AD-H; 3 μm: Chiralpak AD-3; data for peak1: *trans*-stilbene oxide eluted with 9 : 1 hexane/isopropanol.

columns are packed with 3-μm particles having a narrow particle size distribution. Stability studies conducted on 3-μm Chiralpak AD-3 columns introduced recently by Daicel Chemical Industries show that these columns can readily tolerate system pressures up to the 400-bar limit that is generally associated with conventional HPLC equipment. High-efficiency columns also typically require low extra-column system volume to maintain high efficiency, and these criteria can usually be met without modification on conventional HPLC equipment. Apart from high-speed applications, the higher-efficiency aspects of these columns should not be overlooked. When a series of 13 chiral probes was injected onto standard 5-μm AD-H columns and the new AD-3(μm) columns, the latter showed 40 to 60% higher efficiency. Figure 18 illustrates several fast normal-phase separations on 4.6 × 50 mm Chiralpak AD-3 columns. On the other hand, the potential to master very difficult separations with these high-efficiency columns needs to be emphasized. Figure 19 illustrates the very difficult separation of all four metolachlor isomers on a series of 2 4.6 × 150 mm Chiralcel OD-3 columns. Very fast separations can also be achieved in SFC mode, using short (50 × 4.6 mm) 3-μm AD-3 columns, as illustrated in Figure 20. However, full exploitation of this potential will have to await the advent of optimized commercial SFC instrumentation with significantly reduced extra-column volume.

10 CONCLUSIONS, PERSPECTIVES, STATE OF THE ART, AND FUTURE DIRECTIONS

Over the last three decades, chiroscience, encompassing asymmetric synthesis technology, biotransformation, enantioselective catalyst methodology, directed

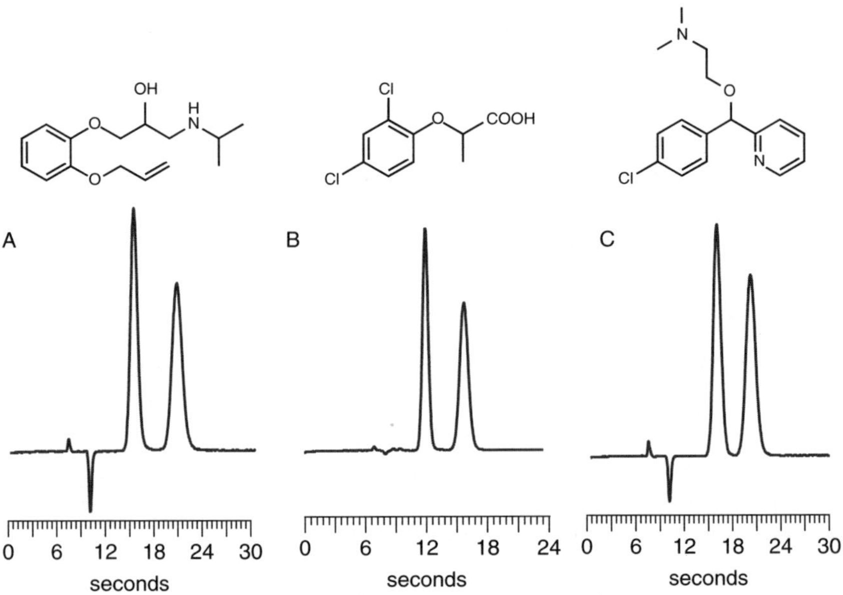

Figure 18 Fast HPLC separations using 3-μm chiral stationary phases. A = Oxprenolol; B = dichlorprop; C = carbinoxamine; Chiralpak AD-3 50 × 4.6 mm ID; for oxprenolol and carbinoxamine: n-hexane/2-propanol/diethylamine 90:10:0.1 (v/v/v); for dichlorprop: n-hexane/2-propanol/trifluoroacetic acid 90:10:0.1 (v/v/v); flow rate: 5.0 mL/min; UV detection: 235 nm; temperature: 25°C.

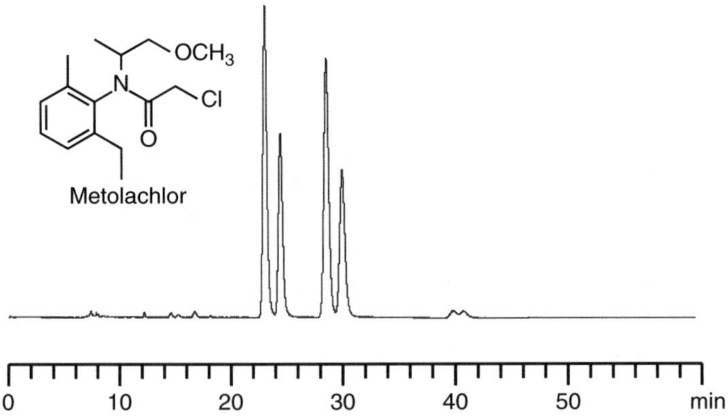

Figure 19 Mastering difficult separations with 3-μm chiral stationary phases. Serial coupled columns: 2× Chiralcel OD-3 150 × 4.6 mm ID; n-hexane/2-propanol 97.5 : 2.5 (v/v); 0.5 mL/min; 235 nm; 15°C; $\alpha_{1,2} = 1.06$; $\alpha_{3,4} = 1.05$; Rs $1, 2 = 2.09$; Rs $3, 4 = 1.73$; $N_1 = 19,969$; $N_2 = 20,304$; $N_3 = 20,433$; $N_4 = 19,657$.

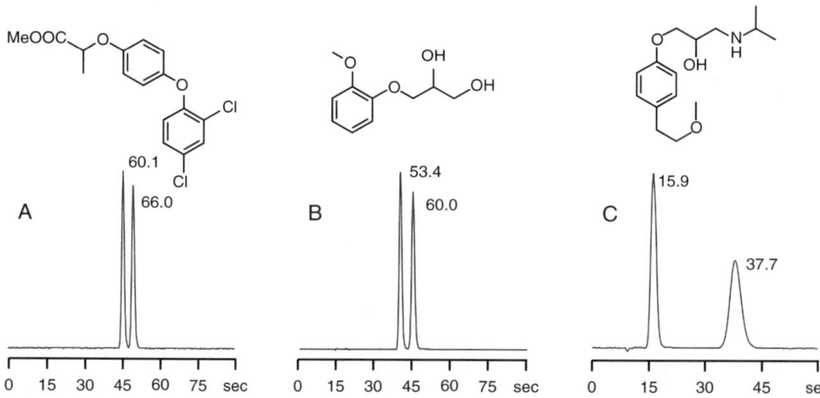

Figure 20 Very fast 3-μm SFC separations. (A) Diclofop methyl Chiralcel OD-3 150 × 4.6 mm ID; 20% MeOH in CO_2, 5 mL/min, 100 bar, 220 nm, 35°C; (B) guaifenesin Chiralpak AD-3 150 × 4.6 mm ID; 20% IPA in CO_2, 5 mL/min, 100 bar, 220 nm, 35°C; (C) metoprolol tartrate Chiralcel OD-3 50 × 4.6 mm ID; 20% MeOH + 1% DEA in CO_2, 5 mL/min, 100 bar, 220 nm, 35°C.

evolution of tailor-made enzymes, and all disciplines of enantiomer separation have made tremendous progress. It is fair to state that all these developments have profited greatly from the contributions of enantioselective chromatography, especially from the extremely versatile chiral recognition capacity of polysaccharide-based CSPs. With chirality aspects moving increasingly into the center of all fields of chemical and biological science, polysaccharide-based CSPs will continue to support efficient research and discovery toward new technologies and materials. From the viewpoint of enantioselective chromatography, challenges in the near future will include establishing robust and generally applicable screening platforms to support the high-throughput philosophy prevailing in drug development materials science circles. Also, there will be a need to address new chiral separation problems resulting from changing philosophies in the pharmaceutical industry. These challenges will favor fast separation technology, such as SFC, eventually in combination with columns packed with CSPs of small particle sizes.

In the future we can also expect to see higher-molecular-weight, more polar pharmaceutical candidates based on peptides, nucleosides, or designed drug delivery platforms. These emerging separation challenges will require enhancing the current repertoire of established chromatographic materials, by adding new CSPs with complementary chiral recognition profiles. These needs will be met efficiently by using the more robust immobilized CSPs in combination with extended-range solvents. Much progress has been made in understanding the mechanisms by which polysaccharide-based chiral stationary CSPs are able to discriminate between pairs of enantiomers. However, to date, the majority of this mechanistic work has focused on unrealistic operation scenarios, largely in the absence of mobile-phase environments. Further work addressing polysaccharide-based CSPs under realistic operation conditions will be needed to arrive at a refined understanding, ultimately allowing the establishment of predictive models truly supportive of method development efforts.

REFERENCES

1. E. Francotte. Presentation at ISCD-14, Hamburg, Germany, July 2002.
2. R. W. Stringham. In *Advances in Chromatography*, Vol. 44, E. Gruska, N. Grinberg, Eds. Taylor & Francis, Boca Raton, FL, 2006, pp. 257–290.
3. X. M. Chen, C. Yamamoto, Y. Okamoto. *Pure Appl. Chem*., 79:1561, 2007.
4. Y. Okamoto, T. Ikai. *Chem. Soc. Rev*., 37:2593, 2008.
5. S. Abel, M. Juza. In *Chiral Separation Techniques*, 3rd ed., G. Subramanian, Ed. Wiley-VCH, Weinheim, Germany, 2007, pp. 585–600.
6. M. Kotake, T. Sakan, N. Nakamura, S. Senoh. *J. Am. Chem. Soc*., 73:2973, 1951.
7. C. Dalgliesh. *J. Chem. Soc*., 3940, 1952.
8. G. Hesse, R. Hagel. *Chromatographia*, 6:277, 1973.
9. G. Blasche. *Angew. Chem. Int. Ed. Engl*., 19:13, 1980.
10. T. Shibata, I. Okamoto, K. Ishii. *J. Liq. Chromatogr*., 9:313, 1986.
11. Y. Okamoto, Y. Kaida. *J. Chromatogr. A*, 666:403, 1994.
12. E. Yashima, Y. Okamoto. *Bull. Chem. Soc. Jpn*., 68:3289, 1995.
13. K. Ogumi, H. Oda, A. Ichida. *J. Chromatogr. A*, 694:91, 1995.
14. Y. Okamoto, E. Yashima. *Angew. Chem. Int. Ed*., 37:1020, 1998.
15. G. Hesse, R. Hagel. *Chromatographia*, 9:62, 1976.
16. G. Hesse, R. Hagel. *Liebigs' Ann. Chem*., 996, 1976.
17. E. Francotte, R. M. Wolf. *Chirality*, 2:16, 1990.
18. Y. Okamoto, M. Kawashima, K. Yamamoto, K. Hatada. *Chem. Lett*., 739, 1984.
19. H. Yuki, Y. Okamoto, I. Okamoto. *J. Am. Chem. Soc*., 102:6356, 1980.
20. M. Nakazaki, K. Yamamoto, M. Maeda. *J. Org. Chem*., 46:1985, 1981.
21. Y. Okamoto, I. Okamoto, H. Yuki. *Chem. Lett*., 835, 1981.
22. Y. Okamoto, S. Honda, I. Okamoto, M. Yuki, S. Murata, R. Noyori, H. Takaya. *J. Am. Chem. Soc*., 103:6971, 1981.
23. A. Ichida, T. Shibata, I. Okamoto, Y. Yuki, H. Namikoshi, Y. Toga. *Chromatographia*, 19:280, 1984.
24. Y. Okamoto, M. Kawashima, K. Hatada. *J. Am. Chem. Soc*., 106:5357, 1984.
25. Y. Okamoto, R. Aburatani, K. Hatada. *J. Chromatogr*., 389:95, 1987.
26. E. Francotte. *J. Chromatogr. A*, 576:1, 1992.
27. K. H. Rimbock, F. Kastner, A. Mannschreck. *J. Chromatogr*., 351:346, 1986.
28. E. Francotte, R. M. Wolf. *Chirality*, 3:43, 1991.
29. Z. Juvancz, K. Grolimund, E. Francotte. *Chirality*, 4:459, 1992.
30. E. Francotte, R. M. Wolf. *J. Chromatogr*., 595:63, 1992.
31. Y. Okamoto, M. Kawashima, K. Hatada. *J. Chromatogr*., 363:173, 1986.
32. B. Chankvetadze, E. Yashima, Y. Okamoto. *Chem. Lett*., 617, 1992.

33. B. Chankvetadze, E. Yashima, Y. Okamoto. *J. Chromatogr. A*, 670:39, 1994.
34. E. Yashima, C. Yamamoto, Y. Okamoto. *Polym. J.*, 27:858, 1995.
35. (a) G. B. Cox, N. M. Maier. Presentation at the Pittsburgh Conference, Chicago, Mar. 2009. (b) H. Ates, D. Mangelings, Y. Vander Heyden. *J. Chromatogr. B*, 875:57, 2008.
36. Y. Okamoto, Y. Kaida. *J. Synth. Org. Chem. Jpn.*, 51:59, 1993.
37. Y. Okamoto, T. Ohashi, Y. Kaida, E. Yashima. *Chirality*, 5:616, 1993.
38. B. Chankvetadze, E. Yashima, Y. Okamoto. *J. Chromatogr. A*, 694:101, 1995.
39. Y. Kaida, Y. Okamoto. *J. Chromatogr.*, 641:267, 1993.
40. Y. Kaida, Y. Okamoto. *Chem. Lett.*, 85, 1992.
41. Y. Kaida, Y. Okamoto. *Chirality*, 4:122, 1992.
42. T. Shimizu, Y. Yamakazi, H. Taka, N. Namagita. *J. Am. Chem. Soc.*, 119:5966, 1997.
43. T. Kubota, C. Yamamoto, Y. Okamoto. *J. Am. Chem. Soc.*, 122:4056, 2000.
44. U. Vogt, P. Zugenmaier. Presentation at the European Science Workshop on Specific Interaction in Polysaccharide Systems, Uppsala, Sweden, 1983.
45. U. Vogt, P. Zugenmaier. *Ber. Bunsenges. Phys. Chem.*, 89:1217, 1985.
46. H. Steinmeier, P. Zugenmaier, *Carbohydr. Res.*, 164:97, 1987.
47. I. Ali, Y. Aboul-Enein. In *Chiral Separation Techniques*, 3rd ed., G. Subramanian, Ed. Wiley-VCH, Weinheim, Germany, 1997, pp. 29–97.
48. T. Shibata, K. Mori, I. Okamoto. In *Chiral Separations by HPLC*, A. M. Krstulovic, Ed. Ellis Horwood, Chichester, UK, 1989, pp. 336–349.
49. E. Yashima, M. Yamada, C. Yamamoto, M. Nakashima, Y. Okamoto. *Enantiomer*, 2:225, 1997.
50. E. Yashima, C. Yamamoto, Y. Okamoto. *Synlett*, 344, 1997.
51. C. Yamashima, E. Yashima, Y. Okamoto. *Bull. Chem. Soc. Jpn.*, 72:1815, 1999.
52. E. Yashima. *J. Chromatogr. A*, 906:105, 2001.
53. T. O'Brien, L. Crocker, R. Thompson, K. Thompson, P. H. Toma, D. A. Conlon, B. Feibush, C. Moeder, G. Bicker, N. Grinberg. *Anal. Chem.*, 69, 1997.
54. Y. Bereznitski, R. LoBrutto, N. Variankaval, R. Thompson, K. Thompson, P. Sajonz, L. S. Crocker, J. Kowal, D. Cai, M. Journet, T. Wang, J. Wyvratt, N. Grinberg. *Enantiomer*, 7:305, 2002.
55. H. Ding, N. Grinberg, R. Thompson, D. Ellison. *J. Liq. Chromatogr. Relat. Technol.*, 23:2641, 2000.
56. S. Ma, S. Shen, H. Lee, N. Yee, C. Senanayake, L. A. Nafie, N. Grinberg. *Tetrahedron: Asymmetry*, 19:2111, 2008.
57. C. Yamamoto, Y. Okamoto. *Bull. Chem. Soc. Jpn.*, 77:227, 2004.
58. E. Francote, R. W. Wolf, D. Lohmann, D. Mueller. *J. Chromatogr.*, 347:25, 1985.
59. H. Staudinger, W. Döhle. *J. Prakt. Chem.*, 161:219, 1942.
60. K. Ikeda, T. Hamasaki, H. Kohno, T. Ogawa, T. Matsumoto, J. Sakai. *Chem. Lett.*, 1089, 1989.
61. A. Ishikawa, T. Shibata. *J. Liq. Chromatogr.*, 16:859, 1993.
62. H. Hopf, W. Grahn, D. G. Barret, A. Gerdes, J. Hilmer, J. Huckler, Y. Okamoto, Y. Kaida. *Chem. Ber.*, 123:841, 1990.

63. T. Shibata, T. Sei, H. Nishimura, K. Deguchi. *Chromatographia*, 24:552, 1986.
64. I. W. Wainer, M. C. Alembic. *J. Chromatogr.*, 358:85, 1986.
65. I. W. Wainer, M. C. Alembic, E. Smith. *J. Chromatogr.*, 388:65, 1987.
66. I. W. Wainer, R. M. Stiffin, T. Shibata. *J. Chromatogr.*, 411:139, 1987.
67. E. Yashima, M. Yamada, Y. Okamoto. *Chem. Lett.*, 579, 1994.
68. E. Yashima, M. Yamada, Y. Kaida, Y. Okamoto. *J. Chromatogr. A*, 694:347, 1995.
69. E. Yahima, H. Fukaya, P. Sahavattanapong, Y. Okamoto. *Enantiomer*, 1:193, 1996.
70. E. Yashima, C. Yamamoto, Y. Okamoto. *J. Am. Chem. Soc.*, 118:4036, 1996.
71. T. Booth, I. W. Wainer. *J. Chromatogr. A*, 737:157, 1996.
72. T. D. Booth, W. J. Lough, M. Saeed, T. A. G. Noctor, I. W. Wainer. *Chirality*, 9:173, 1997.
73. T. D. Booth, K. Azzaoui, I. W. Wainer. *Anal. Chem.*, 69:3879, 1997.
74. D. T. Witte, F. J. Bruggeman, J. P. Franke, S. Copinga, J. M. Jansen, R. A. De Zeeuw. *Chirality*, 5:545, 1993.
75. S. A. Matlin, M. E. Tiritan, A. J. Crawford, Q. B. Cass, D. R. Boyd. *Chirality*, 6:135, 1994.
76. E. Francotte, T. Zhang. *J. Chromatogr. A*, 718:257, 1995.
77. H. Y. Aboul-Enein, I. Ali. *J. Biochem. Biophys. Methods*, 48:175, 2001.
78. E. P. Sousa, M. E. Tiritan, R. V. Oliveira, C. M. M. Afonso, Q. B. Cass, M. M. M. Pinto. *Chirality*, 16:279, 2004.
79. C. Yamamoto, E. Yashima, Y. Okamoto. *J. Am. Chem. Soc.*, 124:12583, 2002.
80. N. Enomoto, S. Furukawa, Y. Ogasawara, H. Akano, Y. Kawamura, E. Yashima, Y. Okamoto. *Anal. Chem.*, 68:2798, 1996.
81. R. M. Wenslow, Jr., T. Wang. *Anal. Chem.*, 73:4190, 2001.
82. K. Matsumoto, C. Yamamoto, E. Yashima, Y. Okamoto. *Anal. Commun.*, 35:63, 1998.
83. K. Matsumoto, C. Yamamoto, E. Yashima, Y. Okamoto. *Rapid Commun. Mass Spectrom.*, 13:2011, 1999.
84. C. Yamamoto, E. Yashima, Y. Okamoto. *Bull. Chem. Soc. Jpn.*, 72:1815, 1999.
85. T. Wang, Y. W. Chen. *J. Chromatogr. A*, 855:411, 1999.
86. T. Wang, Y. W. Chen, A. Vailaya. *J. Chromatogr. A*, 902:345, 2000.
87. M. J. Gidley, S. M. Bociek. *J. Am. Chem. Soc.*, 10:3820, 1988.
88. T. Wang, R. M. Wenslow, Jr. *J. Chromatogr. A*, 1015:99, 2003.
89. R. Wirz, T. Brgi, A. Baiker. *Langmuir*, 19:785, 2003.
90. R. B. Kasat, C. Y. Chin, K. T. Thompson, E. I. Franses, N.-H. L. Wang. *Adsorption*, 12:405, 2006.
91. R. B. Kasat, Y. Zvinevich, H. W. Hillhouse, K. T. Thompson, N.-H. L. Wang, E. I. Franses. *J. Phys. Chem. B*, 110:14114, 2006.
92. R. B. Kasat, N.-H. L. Wang, E. I. Franses. *Biomacromolecules*, 8:1676, 2007.
93. R. B. Kasat, N.-H. L. Wang, E. I. Franses. *J. Chromatogr. A*, 1190:110, 2008.
94. R. B. Kasat, S. Y. Wee, J. X. Lian, N.-H. L. Wang, E. I. Franses. *J. Chromatogr. B*, 875:81, 2008.
95. S. Sun, E. R. Bernstein. *J. Phys. Chem.*, 100:13348, 1996.
96. R. Wirz, D. Ferri, A. Baiker. *Anal. Chem.*, 80:3572, 2008.

97. P. Franco, A. Senso, L. Oliveros, C. Minguillón. *J. Chromatogr. A*, 906:155, 2001.
98. G. Felix. *J. Chromatogr. A*, 906:171, 2001.
99. T. Zhang, P. Franco. In *Chiral Separation Techniques*, 3rd ed., G. Subramanian, Ed. Wiley-VCH, Weinheim, Germany, 2007, pp. 99–134.
100. T. Ikai, C. Yamamoto, M. Kamigaito, Y. Okamoto. *J. Chromatogr. B*, 875:2, 2008.
101. Y. Okamoto, R. Aburatani, S. Miura, K. Hatada. *J. Liq. Chromatogr.*, 10:1613, 1987.
102. E. Yashima, H. Fukaya, Y. Okamoto. *J. Chromatogr. A*, 677:11, 1994.
103. X. Chen, H. Zou, Q. Zhang, J. Ni, Z. Zhang. *J. Chromatogr. Sci.*, 40:315, 2002.
104. X. Chen, Y. Liu, F. Qin, L. Kong, H. Zou. *J. Chromatogr. A*, 1010:185, 2003.
105. F. Ling, E. Brachmachary, M. Xu, F. Svec, J. M. J. Fréchet. *J. Sep. Sci.*, 26:1337, 2003.
106. K. Kimata, R. Tsuboi, K. Hosoya, N. Tanaka. *Anal. Methods Instrum.*, 1:23, 1993.
107. K. Kimata, R. Tsuboi. U.S. patent 5,302,633, 1994.
108. K. Kubota, T. Kusano, C. Yamamoto, E. Yashima, Y. Okamoto. *Chem. Lett.*, 724, 2001.
109. K. Kubota, Y. Yamamoto, Y. Okamoto. *Chirality*, 15:77, 2003.
110. K. Kubota, Y. Yamamoto, Y. Okamoto. *J. Polym. Sci. A*, 42:4704, 2004.
111. X. Chen, F. Qin, Y. Liu, X. D. Huang, H. Zou. *J. Chromatogr. A*, 1034:109, 2004.
112. F. Qin, X. Chen, Y. Liu, L. Kong, H. Zou. *J. Sep. Sci.*, 27:109, 2004.
113. X. Chen, F. Qin, Y. Liu, L. Kong, H. Zou. *Electrophoresis*, 25:2817, 2004.
114. L. Oliveros, C. Minguillón, P. López. French patent 2,714,671, 1994.
115. L. Oliveros, P. López, C. Minguillón, P. Franco. *J. Liq. Chromatogr.*, 18:1521, 1995.
116. C. Minguillón, P. Franco, L. Oliveros, P. López. *J. Chromatogr. A*, 728:407, 1996.
117. C. Minguillón, P. Franco, L. Oliveros. *J. Chromatogr. A*, 728:415, 1996.
118. L. Oliveros, C. Minguillón, B. Serkiz, F. Meunier, J. P. Volland, A. Cordi. *J. Chromatogr. A*, 729:29, 1996.
119. L. Oliveros, P. Franco, A. Senso, C. Minguillón. *Am. Biotechnol. Lab.*, 14:34, 1996.
120. P. Franco, C. Minguillón, L. Oliveros. *J. Chromatogr. A*, 791:37, 1997.
121. Y. Okamoto, N. Enomoto, S. Furukawa, Y. Oasawara, H. Akano, Y. Kuwamura. European patent 0 625 524 B1, 1994.
122. E. Francotte. *J. Chromatogr. A*, 906:379, 2001.
123. E. Francotte. PCT international patent application WO 96/27615.
124. E. Francotte, T. Zhang. PCT international patent application WO 97/04011.
125. S. Mayer, X. Briand, E. Francotte. *J. Chromatogr. A*, 876:331, 2000.
126. E. Francotte, D. Huyhn. *J. Pharm. Biomed. Anal.*, 27:421, 2002.
127. E. Francotte. PCT international patent application WO 97/49733.
128. R. Duval. French patent 2,784,108, 2000.
129. R. Duval. U.S. patent 6,346,616, 2002.
130. X. Chen, Y. Liu, F. Qin, L. Kong, H. Zou. *J. Chromatogr. A*, 1010:185, 2003.
131. T. Ikai, R. Muraki, C. Yamamoto, M. Kamigaito, Y. Okamoto. *Chem. Lett.* 33:188, 2004.

132. P. Franco. doctoral thesis, Barcelona, 1997.
133. T. Ikai, C. Yamamoto, M. Kamigaito, Y. Okamoto. *J. Sep. Sci.*, 30:971, 2007.
134. T. Ikai, C. Yamamoto, M. Kamigaito, Y. Okamoto. *J. Chromatogr. A*, 1157:151, 2007.
135. H. G. Breitinger. *Tetrahedron Lett.*, 43:6127, 2002.
136. B. Chankvetadze, T. Ikai, C. Yamamoto, Y. Okamoto. *J. Chromatogr. A*, 1042:55, 2004.
137. N. Kasuya, Y. Kusaka, N. Habu, A. Ohnishi. *Cellulose*, 9:263, 2002.
138. P. Piras, C. Roussel, J. Pierrot-Sanders. *J. Chromatogr. A*, 906:443, 2001.
139. P. Jandera. *J. Chromatogr. A*, 965:239, 2002.
140. M. L. de la Puente, C. T. White, A. Rivera-Sagredo, J. Reilly, K. Burton, G. Harvey. *J. Chromatogr. A*, 983:101, 2003.
141. M. L. de la Puente. *J. Chromatogr. A*, 1005:55, 2004.
142. L. Zhou, C. Welch, C. Lee, X. Gong, V. Antonucci, Z. Ge. *J. Pharm. Biomed. Anal.*, 49:964, 2009.
143. H. A. Wetli, E. Francotte. *J. Sep. Sci.*, 30:1255, 2007.
144. J. M. Stevens, M. Crawford, Z. Wang. Presentation at the Pittsburgh Conference, New Orleans, LA., Mar. 2008.
145. E. Francotte. *G.I.T. Lab. J.*, 46, Mar. 2006.
146. P. Sajonz, X. Gong, W. R. Leonard, Jr., M. Biba, C. J. Welch. *Chirality*, 18:803, 2006.
147. T. Huybrechts, G. Torok, T. Vennekens, R. Sneyers, I. Somer. *Pharm. Technol. Eur.*, Oct. 1, 2007.
148. K. M. Kirkland. *J. Chromatogr. A*, 718:9, 1995.
149. L. R. Snyder, J. J. Kirkland, J. L. Glajch. *Practical HPLC Method Development*, 2nd ed. Wiley, New York, 1997, pp. 537–615.
150. C. Perrin, V. A. Vu, N. Matthijs, M. Maftouh, D. L. Massart, Y. Vander Heyden. *J. Chromatogr. A*, 947:69, 2002.
151. M. E. Andersson, D. Aslan, A. Clarke, J. Roeraade, G. Hagman. *J. Chromatogr. A*, 1005:83, 2003.
152. Q. B. Cass, F. Batigalhia. *J. Chromatogr. A*, 987:445, 2003.
153. F. Geiser, J. Lee, T. Yun, W. L. Champion, Jr. Presentation at the 215th American Chemical Society National Meeting, Dallas, TX, Mar. 1998.
154. C. Suteu. Presentation at the 12th International Symposium on Chirality, Chamonix, France, 2000.
155. B. Chankvetadze, I. Kartozia, C. Yamamoto, Y. Okamoto. *J. Pharm. Biomed. Anal.*, 27:467, 2002.
156. N. Matthijs, M. Maftouh, Y. V. Heyden. *J. Chromatogr. A*, 1111:48, 2006.
157. N. Matthijs, M. Maftouh, Y. V. Heyden. *J. Sep. Sci.*, 29:1353, 2006.
158. K. G. Lynam, R. W. Stringham. *Chirality*, 18:1, 2006.
159. B. Chankvetadze, C. Yamamoto, Y. Okamoto. *J. Chromatogr. A*, 922:127, 2001.
160. A. M. Krstulovic, G. Rossey, J.-P. Porziemsky, D. Long, I. Chekroun. *J. Chromatogr.*, 411:461, 1987.
161. C. Weinz, G. Blaschke, H.-M. Schiebel. *J. Chromatogr. B*, 690:233, 1997.

162. M. Schulte, R. Ditz, R. M. Devant, J. N. Kinkel, F. Charton. *J. Chromatogr. A*, 769:93, 1997.
163. L. Miller, C. Orihuela, R. Fronek, J. Murphy. *J. Chromatogr. A*, 865:211, 1999.
164. K. Tachibana, A. Ohnishi. *J. Chromatogr. A*, 906:127, 2001.
165. Y. Okamoto, M. Kawashima, R. Aburatani, K. Hatada, T. Nishiyama, M. Masuda. *Chem. Lett.*, 1237, 1986.
166. Y. Okamoto, R. Aburatani, Y. Kaida, K. Hatada. *Chem. Lett.*, 1125, 1988.
167. R. J. Bopp F. Geiser. Presentation at the 9th International Symposium on Chiral Discrimination, Nagoya, Japan, 1997.
168. Y. K. Ye, B. Lord, R. W. Stringham. *J. Chromatogr. A*, 945:139, 2002.
169. R. W. Stringham, K. G. Lynam, B. S. Lord. *Chirality*, 16:493, 2004.
170. L. Tang, D. Xiang, J. A. Blackwell. *Enantiomer*, 5:345, 2000.
171. Y. Hatefi, W. G. Hanstein. *Biochemistry*, 8:1129, 1969.
172. C. Perrin, N. Matthijs, D. Mangelings, C. Granier-Loyaux, M. Maftouh, D. L. Massart, Y. Vander Heyden. *J. Chromatogr. A*, 966:119, 2002.
173. T. Zhang, D. Nguyen, P. Franco. Presentation at the 20th International Symposium on Chromatography, Geneva, 2008.
174. L. Thunberg, J. Hashemi, S. Andersson. *J. Chromatogr. B*, 875:72, 2007.
175. T. Zhang, C. Kientzy, P. Franco, A. Ohnishi, Y. Kagamihara, H. Kurosawa. *J. Chromatogr. A*, 1075:65, 2005.
176. T. Zhang, M. Schaeffer, P. Franco. *J. Chromatogr. A*, 1083:96, 2005.
177. T. Zhang, D. Nguyen, P. Franco, T. Murakami, A. Ohnishi, H. Kurosawa. *Anal. Chim. Acta*, 557:221, 2006.
178. A. Ghanem, H. Y. Aboul-Enein. *J. Liq. Chromatogr. Relat. Technol.*, 28:2669, 2005.
179. A. Ghanem, L. Naim. *J. Chromatogr. A*, 1101:171, 2006.
180. A. Ghanem, H. Hoenen, H. Y. Aboul-Enein. *Talanta*, 68:602, 2006.
181. A. Ghanem. *J. Sep. Sci.*, 30:1019, 2007.
182. J. Y. Jin, W. Lee, J. H. Park, J. J. Ryoo. *J. Liq. Chromatogr. Relat. Technol.*, 30:1, 2007.
183. R. Cirilli, V. Orlando, R. Ferretti, L. Torchetto, R. Silvestri, G. De Martino, F. La Torre. *Chirality*, 18:621, 2006.
184. A. Ghanem, H. Y. Aboul-Enein. *Anal. Chim. Acta*, 548:26, 2005.
185. C. W. Amoss, B. Coryell. *LCGC Appl. Notebook*, June 2004.
186. P. Franco, T. Zhang. *J. Chromatogr. B*, 875:48, 2008.
187. T. Zhang, D. Nguyen, P. Franco. *J. Chromatogr. A*, 1191:214, 2008.
188. T. Zhang, D. Nguyen, P. Franco. *J. Sep. Sci.*, 29:1517, 2006.
189. T. Zhang, D. Nguyen, P. Franco, Y. Isobe, T. Michishita, T. Murakami. *J. Pharm. Biomed. Anal.*, 46:882, 2008.
190. B. S. Lord, R. W. Stringham. *LCGC Appl. Notebook*, Feb. 2004.
191. P. Petersson, K. E. Markides. *J. Chromatogr. A*, 666:381, 1994.
192. K. L. Williams, L. C. Sander. *J. Chromatogr. A*, 785:149, 1997.
193. G. Terfloth. *J. Chromatogr. A*, 906:301, 2001.
194. K. W. Phinney, R. W. Stringham. In *Chiral Separation Techniques*, 3rd ed., G. Subramanian, Ed. Wiley-VCH, Weinheim, Germany, 2007, pp. 135–154.

195. L. T. Taylor. *J. Supercrit. Fluids*, 47:566, 2009.
196. D. Mangelings, Y. Vander Heyden. *J. Sep. Sci.*, 31:1252, 2008.
197. Y. K. Ye, K. G. Lynam, R. W. Stringham. *J. Chromatogr. A*, 1041:211, 2004.
198. K. Anton, J. Eppinger, L. Frederiksen, E. Francotte, T. A. Berger, W. H. Wilson. *J. Chromatogr. A*, 666:395, 1994.
199. L. Siret, N. Bargmann, A. Tambute, M. Caude. *Chirality*, 4:252, 1992.
200. G. Peytavin, F. Gimenez, B. Genissel, C. Gillotin, A. Baillet, I. W. Wainer, R. Farinotti. *Chirality*, 5:173, 1993.
201. N. Bargmann-Leyder, C. Sella, D. Bauer, A. Tambute, M. Caude. *Anal. Chem.*, 67:952, 1995.
202. N. Bargmann-Leyder, A. Tambute, M. Caude. *Chirality*, 7:311, 1995.
203. R. W. Stringham. *J. Chromatogr. A*, 1070:163, 2005.
204. F. Geiser, M. Schultz, L. Betz, M. Shaimi, J. Lee, W. Champion, Jr. *J. Chromatogr. A*, 865:227, 1999.
205. F. Geiser, R. Shah. *Chirality*, 16:263, 2004.
206. M. S. Villenueve, R. J. Anderegg. *J. Chromatogr. A*, 826:217, 1998.
207. Y. Zhao, G. Woo, S. Thomas, D. Semin, P. Sandra. *J. Chromatogr. A*, 1003:157, 2003.
208. M. Maftouh, C. Granier-Loyaux, E. Chavana, J. Marini, A. Pradines, Y. Vander Heyden, C. Picard. *J. Chromatogr. A*, 1088:67, 2005.
209. G. Diehl, A. Meishammer, D. Huynh, E. Francotte. Presentation at SPICA 2006, Innsbruck, Austria, Oct. 2006.
210. G. Cox. *LC/GC Appl. Notebook*, Feb. 2007.
211. M. Garzotti, M. Hamdam. *J. Chromatogr. B*, 770:53, 2002.
212. J. Chen, Y. Hsieh, J. Cook, R. Morrison, W. A. Korfmacher. *Anal. Chem.*, 78:1212, 2006.
213. R. A. Coe, J. O. Rathe, J. W. Lee. *J. Pharm. Biomed. Anal.*, 42:573, 2006.
214. C. J. White. *J. Chromatogr. A*, 1074:163, 2005.
215. L. Zeng, R. Xu, D. B. Laskar, D. B. Kassel. *J. Chromatogr. A*, 1169:193, 2007.
216. D. B. Laskar, L. Zeng, R. Xu, D. B. Kassel. *Chirality*, 20:885, 2008.
217. K. W. Phinney, L. C. Sander, S. A. Wise. *Anal. Chem.*, 70:2331, 1998.
218. W. W. Barnhart, K. H. Gahm, S. Thomas, S. Notari, D. Semin, J. Cheetham. *J. Sep. Sci.*, 28:619, 2005.
219. C. J. Welch, M. Biba, J. R. Gouker, G. Kath, P. Augustine, P. Hosek. *Chirality*, 19:184, 2007.
220. C. Suteu. In *Preparative Enantioselective Chromatography*, G. B. Cox, Ed. Blackwell, Oxford, UK, 2005, pp. 78–109.
221. S. Anderrson. In *Chiral Separation Techniques*, 3rd ed., G. Subramanian, Ed. Wiley-VCH, Weinheim, Germany, 2007, pp. 585–600.
222. G. B. Cox. In *Preparative Enantioselective Chromatography*, G. B. Cox, Ed. Blackwell, Oxford, UK, 2005, pp. 78–109.
223. E. Francotte. *Chirality*, 10:492, 1998.
224. C. M. Kraml, D. Zhou, N. Byrne, O. McConnell. *J. Chromatogr. A*, 1007:108, 2005.

225. E. Francotte. In *Preparative Enantioselective Chromatography*, G. B. Cox, Ed. Blackwell, Oxford, UK, 2005, pp. 48–77.
226. J. Dingenen, J. N. Kinkel. *J. Chromatogr. A*, 666:627, 1994.
227. G. Nadler, C. Dartois, D. S. Eggleston, R. C. Curtis, M. Martin. *Tetrahedron: Asymmetry*, 9:4267, 1998.
228. C. M. Grill, L. Miller, T. Q. Yan. *J. Chromatogr. A*, 1026:101, 2004.
229. L. Miller, C. Orihuela, R. Fronek, D. Honda, O. Dapremont. *J. Chromatogr A*, 849:309, 1999.
230. T. A. Berger, J. Smith, K. Fogelman, K. Kruluts. *Am. Lab.*, 34:14, 2004.
231. L. Miller, C. Grill, T. Yan, O. Dapremont, E. Huthmann, M. Juza. *J. Chromatogr. A*, 1006:267, 2003.
232. J. Dingenen. In *Preparative Enantioselective Chromatography*, G. B. Cox, Ed. Blackwell, Oxford, UK, 2005, pp. 78–109.
233. C. M. Grill, L. M. Miller. In *Preparative Enantioselective Chromatography*, G. B. Cox, Ed. Blackwell, Oxford, UK, 2005, pp. 78–109.
234. M. Schulte, J. Strube. *J. Chromatogr. A*, 906:399, 2001.
235. J. H. Kennedy, M. D. Belvo, V. S. Sharp, J. D. Williams. *J. Chromatogr. A*, 1046:55, 2004.
236. J. Blehaut, Presentation at Chirality 2001, Orlando, FL, July 2001.
237. C. Kraml. *Am. Pharm. Rev.*, July–Aug. 2007.
238. D.-R. Wu, L. Leith. *Am. Pharm. Rev.*, Sept.–Oct. 2007.
239. C. White. *J. Chromatogr. A*, 1074:163, 2005.
240. K. Balmér, B.-A. Person, P.-O. Lagerström. *J. Chromatogr. A*, 660:269, 1994.
241. M. H. Gaffney, R. M. Stiffin, I. W. Wainer. *Chromatographia*, 27:15, 1989.
242. K. Balmér, P.-O Lagerström, B.-A. Persson. *J. Chromatogr.*, 592:331, 1992.
243. M. Okamoto, H. Nakazawa. *J. Chromatogr.*, 588:177, 1991.
244. O. Gyllenhaal, M. Stefansson. *Chirality*, 17:257, 2005.
245. N. M. Maier, G. B. Cox, B. Krueger, T. Zhang, D. Nguyen, P. Franco, A. Ohnishi, T. Shibata, T. Murakami. Presentation at HPLC 2008, Baltimore, 2008.

CHAPTER 5

Chiral Separations by Various Techniques

JIAN G. NING

Schering Plough Research Institute, Union, New Jersey

Chiralpak AD and Chiracel OD columns from Chiral Technology, Inc. have recently gained recognition in the separation of enantiomers of pharmaceutically important compounds [1–9]. Both columns have the same dimethylphenyl carbamate functionality; however, AD and OD are an amylose and a cellulose derivative, respectively. The Chiralcel OD column is marketed in both normal-phase mode (OD) and reversed-phase mode (OD-R). However, since the Chiralpak AD column was originally designed by the manufacturer to be used only in the normal-phase mode, there was very little literature on reversed-phase applications of Chiralpak AD columns in the early 1990s. The work presented here demonstrates the superior separations in the reversed-phase mode as compared to the normal-phase mode for Chiralpak AD columns.

1 PHASE-CONVERSION CHIRAL SEPARATIONS

Three case studies of separations of pharmaceutical chiral intermediates in the reversed phase are illustrated below. A Chiralpak AD column was converted from the originally designed normal-phase to reversed-phase mode. To convert a chiral column in normal phase to the reversed phase, the column was flushed adequately with 200 proof ethanol at 0.2 mL/min (not exceeding the pressure limit of the column), then equilibrated in an organic and aqueous mixture of the reversed mobile phase. The characteristics of this new reversed-phase application are compared with those of the normal-phase mode.

Case Study 1 The molecular structures of the first reversed-phase mode separation of pharmaceutical intermediates are as shown in Scheme 1. Compound

Chiral Separation Methods for Pharmaceutical and Biotechnological Products, Edited by Satinder Ahuja

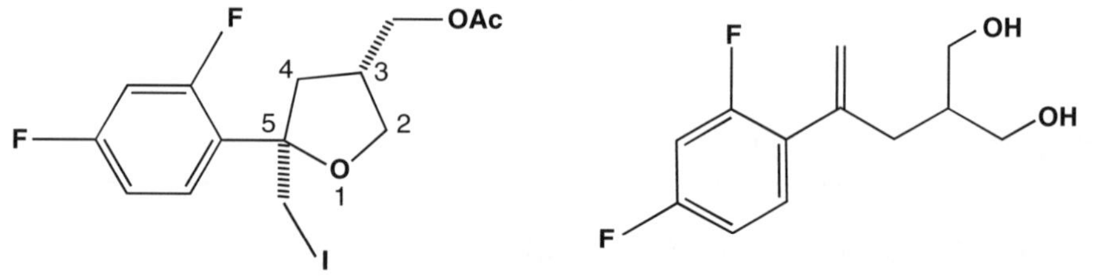

1 2

Scheme 1

1 has two chiral centers; the enantiomer desired has the *cis*-(3*S*,5*R*) configuration. Compound **2** is a potential achiral impurity. A solution of the mixture was prepared and chromatographed, first using the normal-phase mode, then the reversed-phase mode. Figure 1 shows that the elution order of five compounds is compound **1** *trans*-(*S,S*), compound **1** *trans*-(*R,R*) coeluting with *cis*-(*S,R*) and compound **2**, and compound **1** *cis*-(*R,S*).

The enantiomer compound desired, compound **1** *cis*-(*S,R*), was not separated from compound **2** and the compound **1** *trans*-(*R,R*) stereoisomer in the normal-phase mode. The mixture was then chromatographed using the reversed-phase mode after careful conversion from the normal phase. The chromatogram is shown in Figure 2. Although the conversion of the mobile-phase mode did not change the elution order of the **1** stereoisomers, **2** was completely separated from

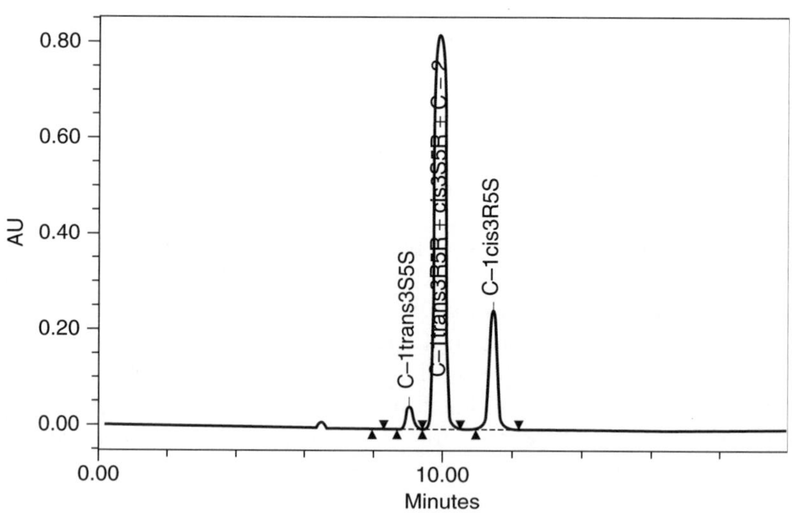

Figure 1 Separation of stereoisomers of compounds **1** and **2** by normal-phase chromatography. High-performance liquid chromatography (HPLC) parameters: Chiralpak AD 25 × 0.46 cm at 30°C. Mobile phase: 99% hexane/1% anhydrous ethanol (200 proof). Flow rate: 1 mL/min. Detector wavelength: 215 nm.

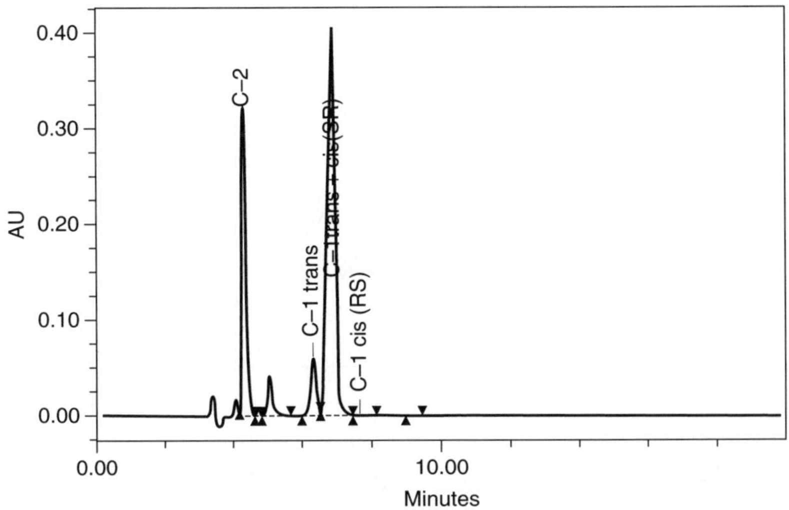

Figure 2 Compound **1** stereoisomers and **2** in the reversed phase. HPLC parameters: Chiralpak AD 25 × 0.46 cm at 40°C. Mobile phase: 80% methanol/20% H_2O. Flow rate: 1 mL/min. Detector wavelength: 215 nm.

the desired *cis*-(*S,R*) enantiomer. Since the coeluting *trans*-(*R,R*) is present in a negligible amount, the separation is adequate to calculate the enantiomeric purity of the compound 1 *cis*-(*S,R*) enantiomer.

As indicated in the example above, the conversion of the mobile-phase mode may not necessarily change the elution order of the stereoisomers, but it could have profound effects on achiral impurities that coelute with chiral isomers, because the polarity of the mobile phase is changed from one extreme to the other. The polar reversed mobile phase is a much stronger eluent than the normal mobile phase to polar compound **2**. This results in a much earlier elution of the achiral impurity and thus resolves completely from the compound **1** enantiomers.

Case Study 2 Conversion of the mobile-phase mode can sometimes even change the elution order of stereoisomers. The molecular structure of the compound is illustrated in Scheme 2. Compound **3** has two chiral centers; again the *cis*-(3*S*,5*R*) is the desired stereoisomer.

A chromatogram of the normal-phase mode is shown in Figure 3. The chromatogram on the top of the overlay is of the racemate, and the one at the bottom is of a typical sample. Notice that the minor impurity *cis*-(3*R*, 5*S*) rides on the tail of the major peak *cis*-(3*S*,5*R*). This peak is hardly resolved and would be very difficult to quantitate even though the resolution (1.66) between two *cis* enantiomers in the racemate standard is fairly good.

In the reversed mobile phase mode, not only the resolution but also the elution order of the stereoisomers has changed favorably, as shown in Figure 4. As indicated by the results above, the reversed-phase mode not only offers better

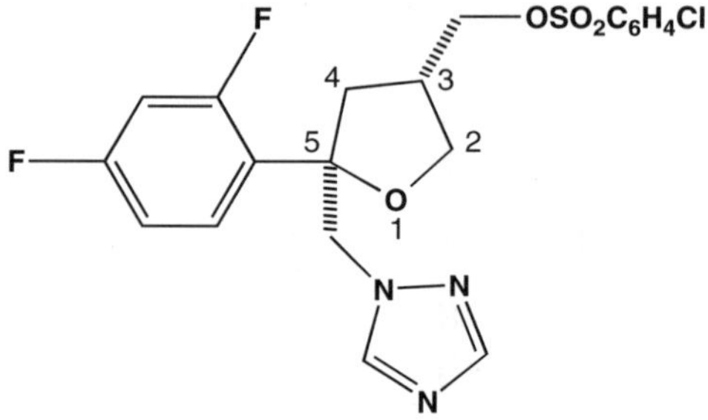

3

Scheme 2

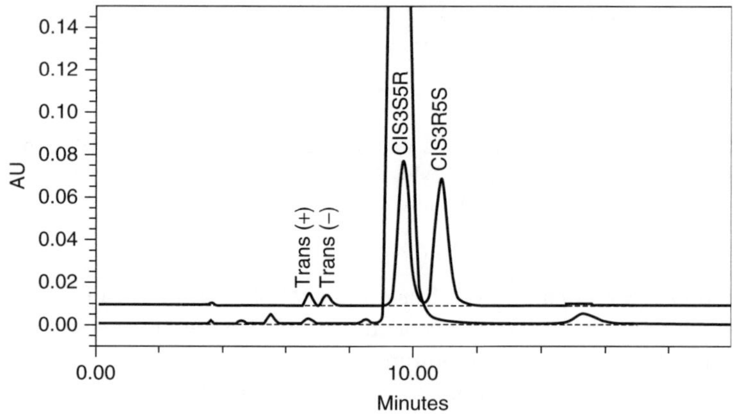

Figure 3 Overlay plot of compound **3** racemate standard and **3** *cis*-(*S,R*) sample in the normal-phase mode. HPLC parameters: Chiralpak AD 25 × 0.46 cm at 30°C. Mobile phase: (200 proof) anhydrous ethanol. Flow rate: 1 mL/min. Wavelength: 233 nm.

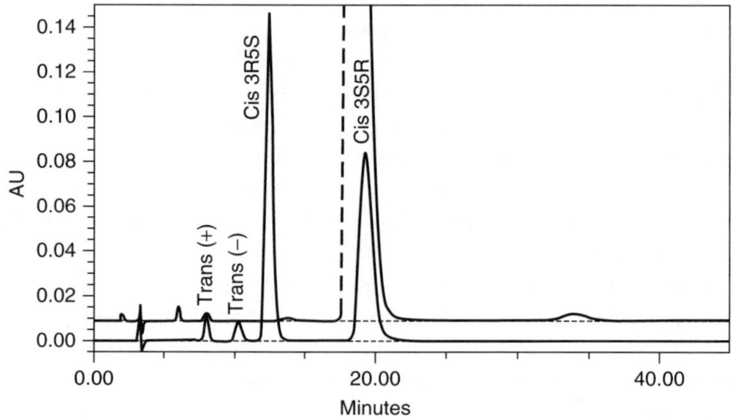

Figure 4 Overlay plot of compound **3** racemate standard and **3** *cis*-(*S,R*) sample in the reversed-phase mode. HPLC parameters: Chiralpak AD 25 × 0.46 cm at 40°C. Mobile phase: 90% methanol/10% H_2O. Flow rate: 1 mL/min. Wavelength: 233 nm.

resolution for each of four isomers, but also places the minor impurity *cis*-(3*R*, 5*S*) before the major peak *cis*-(3*S*,5*R*). However, the elution order for the *trans* enantiomers remained unchanged. The reversed-phase mode enables much more reliable quantitation of the minor chiral impurities. The sensitivity of the detection is also improved considerably. The limit of detection of *cis*-(3*R*, 5*S*) impurity in the system is obtainable to a level of 0.1% of the *cis*-(3*S*,5*R*).

Case Study 3 Converting to the reversed phase greatly expands the scope of chiral chromatography of the Chiralpak AD column. Some chiral compounds that could not be separated on the normal phase can now be adequately resolved in the reversed-phase mode. The D-L-Boc-*N*-Arg(NO_2)-lactams are the compounds that previously failed to resolve in the normal-phase mode but separate well in the reversed-phase mode. The molecular structure of the Boc-*N*-Arg(NO_2)-lactam is shown in Scheme 3.

After many failed attempts to resolve the enantiomers in the normal-phase mode, the reversed-phase mode was finally tried. With very little effort, the compounds were separated with ample resolution. This is shown in Figure 5. Note that the minor impurity D enantiomer elutes before the major peak L enantiomer. This elution order is desirable for quantitation, as there is no interference from the

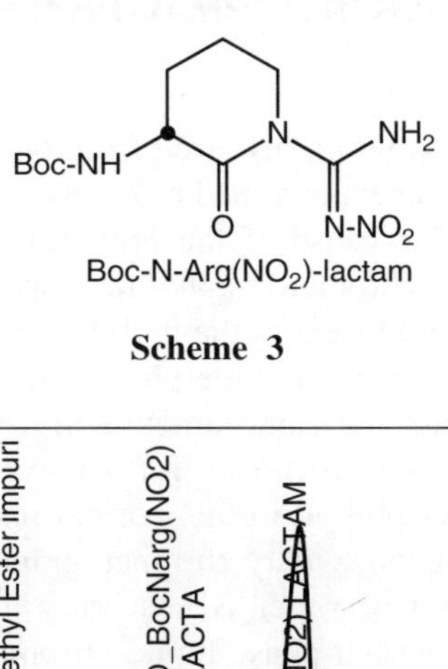

Boc-N-Arg(NO_2)-lactam

Scheme 3

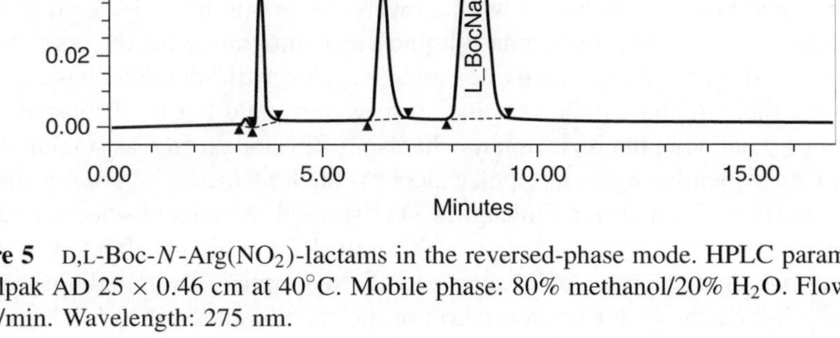

Figure 5 D,L-Boc-*N*-Arg(NO_2)-lactams in the reversed-phase mode. HPLC parameters: Chiralpak AD 25 × 0.46 cm at 40°C. Mobile phase: 80% methanol/20% H_2O. Flow rate: 1 mL/min. Wavelength: 275 nm.

tail of the major peak. After a year of operation in the reversed phase, there has been no evidence of deterioration of column efficiency. Although the peak shape in the reversed phase tends to be a little broader than that in the normal phase, this is characteristic from the beginning for many compounds. The conversion does not impair the integrity of the column.

Conversion from the normal phase to the reversed-phase mode of Chiralpak AD gives much more freedom to design new separation patterns for those chiral compounds that are difficult to resolve using the normal phase. As indicated above, the reversed-phase mobile phases employed here are very simple, efficient, and robust. The reversed-phase mode also surpasses the inherent limitations of conventional normal-phase chiral HPLC; thus buffers, and other additives can easily be employed, if necessary. There are many other chiral columns on the market that were originally intended to be used in the normal phase. This work offers chromatographers another dimension to tackle the onerous tasks of trial-and-error efforts. One word of caution: Before converting your expensive chiral column to the reversed phase, study the column chemistry and contact the manufacturer to make sure that the column will not collapse in the reversed-phase mode.

2 WATER AS A MODIFIER IN NORMAL-PHASE CHIRAL SEPARATIONS

When a chiral HPLC method was transferred from one lab to another, the resolution between the two enantiomers obtained previously in the method development lab was completely lost. The cause of the problem was discovered by analyzing mobile phases using gas chromatography/mass spectrometry (GC-MS). There was a small amount of water present in the mobile phase (A), which had resolved the enantiomers. The freshly made mobile phase (B), which failed to resolve the enantiomers, did not contain the same amount of water as was present in the mobile phase (A). It was then proved by further experiments, contrary to conventional wisdom that water present in this normal-phase liquid chromatography is not useful for obtaining good-quality chromatograms, which led us to the conclusion that a small amount of water is necessary and critical to separate the chiral intermediate in this normal-phase liquid chromatograph.

Macaudiere and colleagues published an interesting paper in 1987 [10] that demonstrated how he and his co-workers were the first to investigate polar modifiers (such as water) in normal-phase liquid chromatography for the resolution of racemic amides and phosphine oxides on a β-cyclodextrin-bonded phase column. However, their findings indicated that since water could not be displaced easily by solutes (analytes), the inclusion mechanism of cyclodextrin was somehow hindered. Consequently, water and polar alcohols such as methanol are not the best polar modifiers. Balmér and colleagues [11] studied the enantioselective resolution of a series of amino alcohols on a Chiralcel OD column affected by water in a normal mobile phase of hexane and diethylamine. Their findings suggest that with increasing water concentration in the mobile phase, the k' value of one enantiomer decreases whereas the k' of the other enantiomer is not influenced. Therefore, changing water content in the mobile phase affects the resolution and even the elution order of two enantiomers of amino alcohols.

This study describes the work of enantiomeric resolution of a 2-oxo-3-piperidine acetamide analog with a Chiralcel OD column and a normal-phase chiral separation using water as a modifier. The molecular structure of a 2-oxo-3-piperidineacetamide analog is shown in Scheme 4, where R_1 and R_2 are different alkyl functional groups. The mobile phase developed for chiral separation of the compound consists of 92.5% hexane, 7.5% ethanol, and 0.1% trifluoroacetic acid. The chromatogram of the original separation is shown in Figure 6. However, when the method was transferred to a different laboratory environment, the method did not work as expected, resulting in the chromatogram shown in Figure 7.

Initial investigations were tried on different lots of columns, using different HPLC apparatus and with freshly made mobile phase, without much success. Further investigation utilized GC-MS to compare the mobile phase of (A) with (B). The results were very informative. It was determined that the amount of water present in mobile phase A was more than that in mobile phase B.

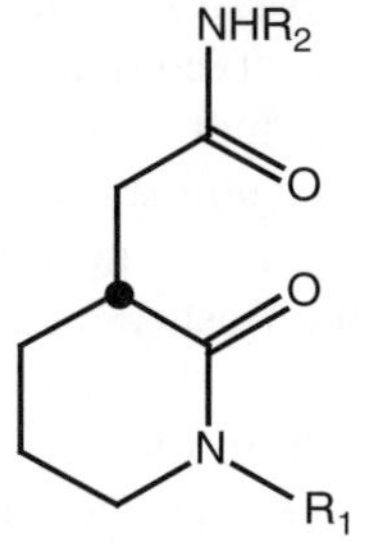

Scheme 4

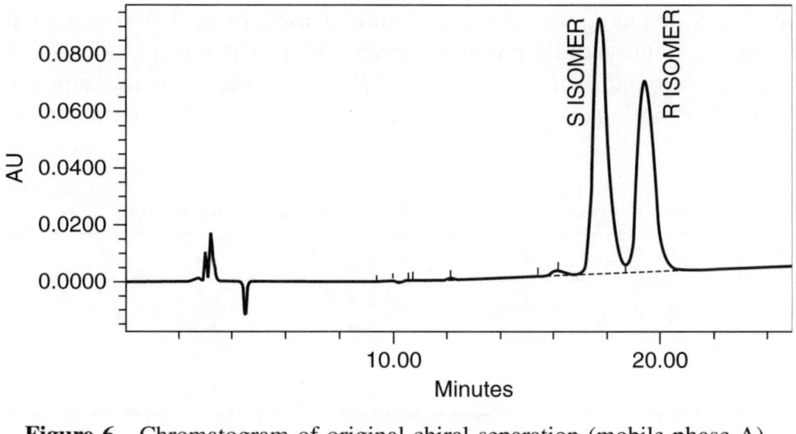

Figure 6 Chromatogram of original chiral separation (mobile phase A).

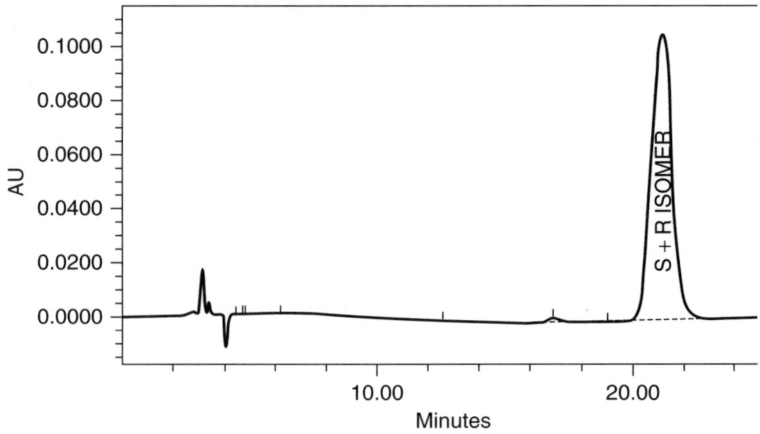

Figure 7 Chromatogram from another laboratory (mobile phase B).

The water content of each mobile phase was further quantified by Karl Fisher titration. There was 0.20% of water in mobile phase A, which has been recycling in HPLC for about a week, while the freshly made mobile phase B had less than 0.01% water content. It was also observed that moisture from the air can increase the water content of a recirculating mobile phase up to 0.1% overnight. At this level of water content, the HPLC system started to differentiate the enantiomers. All these findings indicate that water was necessary to separate the chiral intermediate in mobile phase A. But the freshly made mobile phase was "too dry" to separate the chiral intermediate.

Our next task was to find out how much water is needed to achieve chiral separation. Bearing in mind that we were dealing with a normal mobile phase, the maximum amount of water was limited by the solubility of water in the mobile phase. Water is the stronger eluent in the normal phase. The more water that is present in the mobile phase, the faster the compound elutes. Up to 0.4% of water was added to the normal mobile phase. Table 1 shows that as the water content of the mobile phase increases, the k' value of the (S)-enantiomer decreases more rapidly than that of the (R)-enantiomer, and the separation of (S)- and (R)-enantiomers thus improves. With 0.38% of water in the mobile

TABLE 1 Effect of Water Content in Mobile Phase[a] on Resolution

Water Content (%)	k'_S	k'_R	α	Resolution
0.01	3.4	3.4		
0.18	2.9	3.2	1.10	1.15
0.28	2.7	3.1	1.15	1.33
0.38	2.6	2.9	1.12	1.45

[a] 90% hexane, 10% ethanol, 0.1% TFA, H_2O % varies.

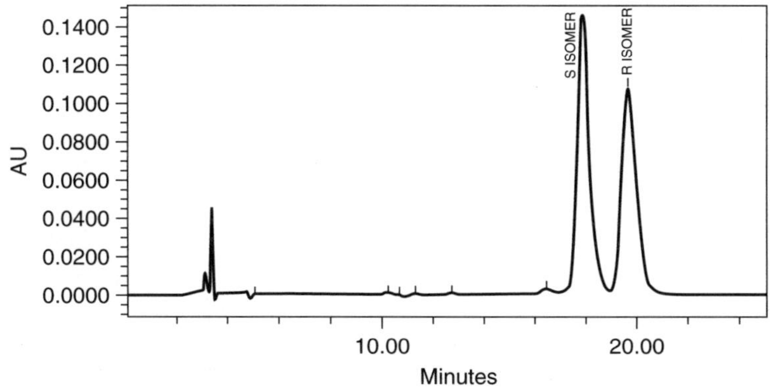

Figure 8 Optimized chromatogram with 0.25% water in the normal mobile phase. HPLC parameters: Chiralcel OD 25 × 0.46 cm at 30°C. Mobile phase: 0.25% H_2O in 93% hexane, 7% ethanol, 0.1% TFA. Flow rate: 1 mL/min. Wavelength: 210 nm.

phase, a resolution factor of 1.45 was attained between (*S*)- and (*R*)-isomers. However, as more water was added to the normal mobile phase, it became more difficult to mix water with the normal mobile phase. Also, the baseline noise of the chromatogram was getting worse. The mobile phase was further optimized; the chiral separation of the 2-oxo-3-piperidineacetamide analog was achieved with a resolution factor of 1.76, as demonstrated in Figure 8.

The critical role of water as a polar modifier in the normal phase was demonstrated in this work. Although in normal-phase liquid chromatography, the interaction between the chiral stationary phase and the mobile phase are hydrophobic in nature, the interactions of water with the analyte and the chiral stationary phase are inherently hydrophilic. Ethanol in this normal mobile phase acts as a polar component for increasing eluent strength, but ethanol alone does not work well with respect to the hydrophilic interactions necessary to achieve resolution. Water is a much more polar solvent than ethanol. A small amount of water present in this normal mobile phase greatly increases the hydrophilicity of the mobile phase and creates a more polar environment for both the analytes and the chiral stationary phase. Thus, two enantiomers can interact differentially with the chiral stationary phase, resulting in this successful enantiomeric resolution.

3 OTHER POPULAR CHIRAL SEPARATION TECHNIQUES

3.1 Chiral GC

Chiral GC offers several advantages (see Chapter 11) for the resolution of small molecules:

- It is valuable for the detection of compounds with poor ultraviolet absorbance.

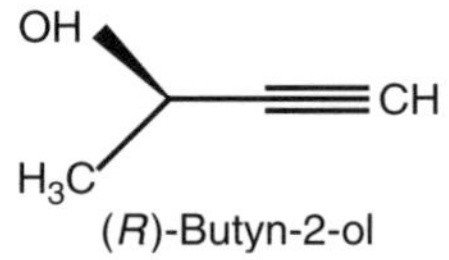

Scheme 5

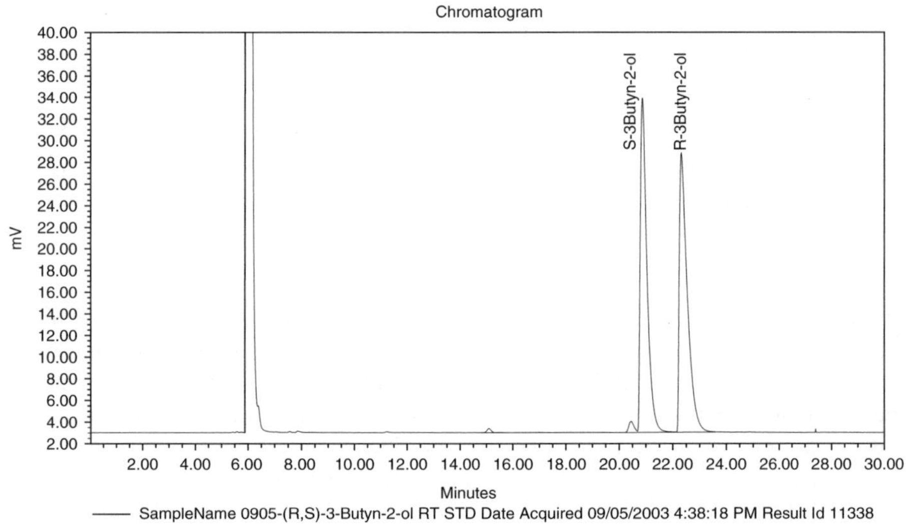

Figure 9 Chiral GC separation of butyn-2-ol. GC parameters: Agilent 6890-N GC. Column: Astec Chiraldex G-PN, 30m × 0.25 mm. Flow: 0.7 mL/min, split 100 mL/min. Oven temperature 40°C, FID 250°C, injection temperature 110°C; sample preparation: 20 mg/mL TBME.

- It is useful for small molecules, as they are generally volatile.
- Derivatization techniques to alter polarity and volatility and/or enantioselectivity, such as acylation with trifluoroacetic anhydride, are available for amine/alcohol compounds.

An example is shown in Scheme 5 and Figure 9.

3.2 Chiral Capillary Electrophoresis

Chiral capillary electrophoresis offers some major advantages (see Chapter 12):

- It is a speedy screening method.
- There are abundant chiral selectors and separation modes to select.
- It is a fully automated process.
- The cost of analysis is low.

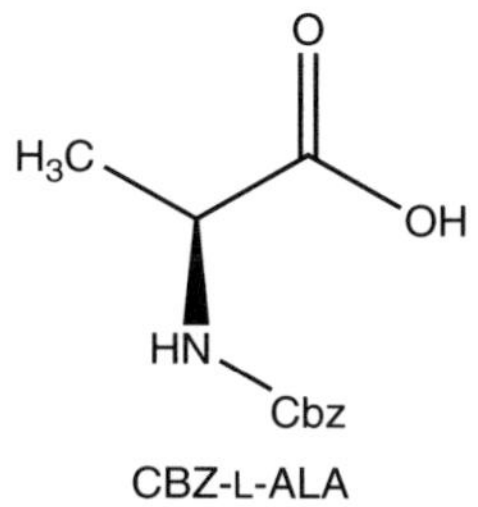

Scheme 6

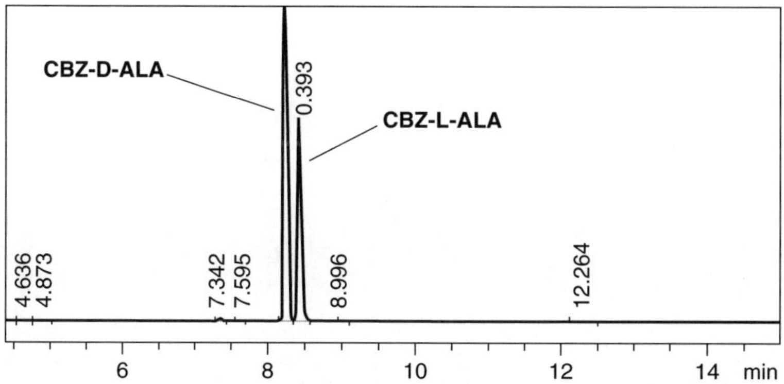

Figure 10 Chiral capillary electrophoresis separation of Cbz-ALA. CE parameters: HP 3D-CE. Column: Fused silica capillary 40 cm × 50 μm. Electrolytes: borate buffer 25 mM, pH 9.3; heptakis-6-sulfato-β-cyclodextrin 15 mM. Potential: +30 kV at inlet current limit 100 μA.

An example is shown in Scheme 6. This structure represents Cbz (carbobenzoxy)–L-ALA (alanine) (Figure 10).

3.3 Chiral Supercritical Fluid Chromatography

Some of the excellent features offered by chiral SFC (see Chapter 10) are as follows:

- SFC runs as normal phase; therefore, it is complementary to reversed-phase HPLC.
- It uses the same columns as in HPLC but offers faster analysis (three to five times higher flow rate).
- It frequently offers higher efficiency in less time (more separation in less time).
- Reequilibration is very rapid (speedy method development).

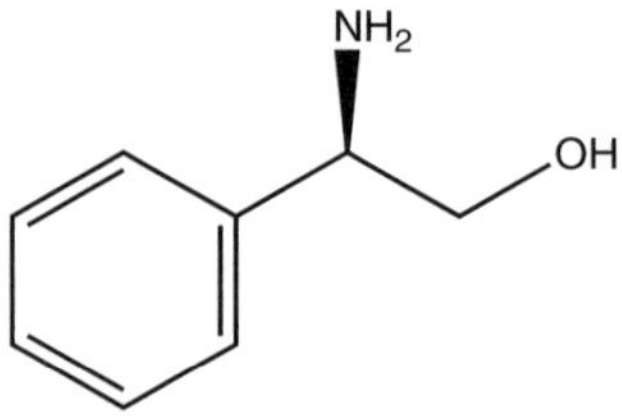

(R + S)-2-phenyglycinol

Scheme 7

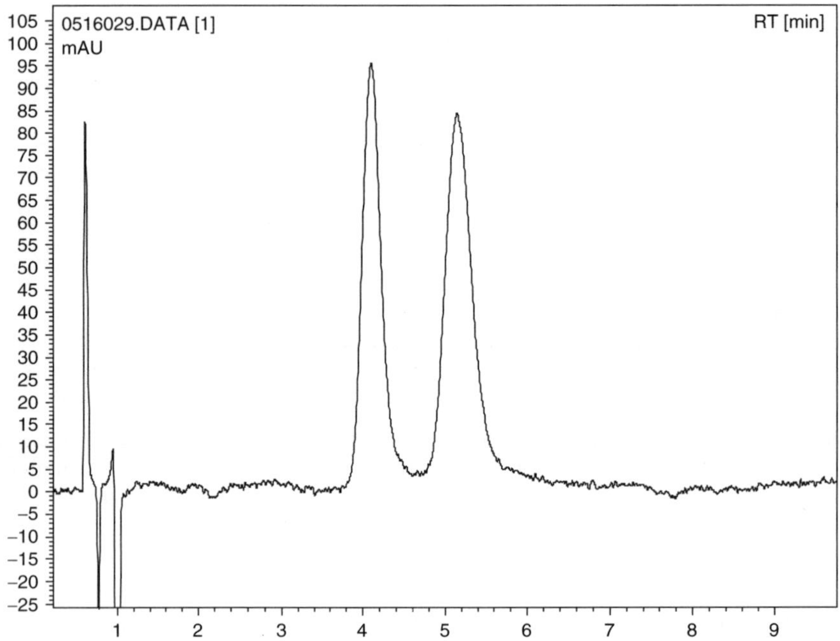

Figure 11 Chiral SFC separation of ($R + S$)-2-phenyglycinol. SFC parameters: Berger SFC. Chiralpak ADH 25 cm × 4.6 mm × 5 μm. 100 bar. 10% EtOH (0.2% DEA), 5 mL/min.

- It is easy to remove solvent, environmentally friendly, and easy to scale up to preparative scale.

An example is shown in Scheme 7. This structure represents ($R + S$)-2-phenyglycinol (Figure 11).

3.4 Separation of a Rotamer with an LC Sample Recycler

The LC sample recycler is designed to analyze the sample by eluting the analyte repeatedly between two exactly configured HPLC columns and a detector in an

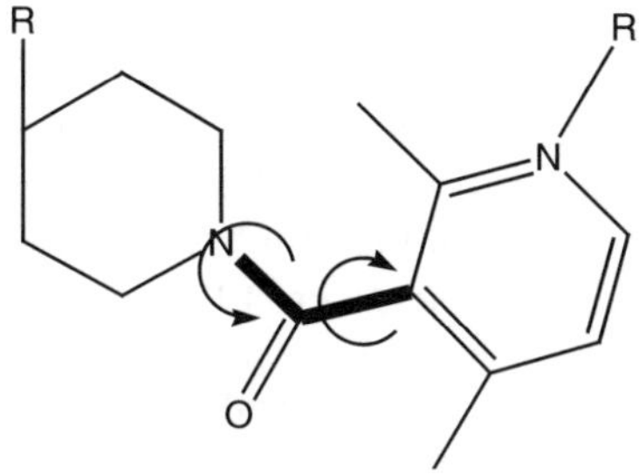

Scheme 8

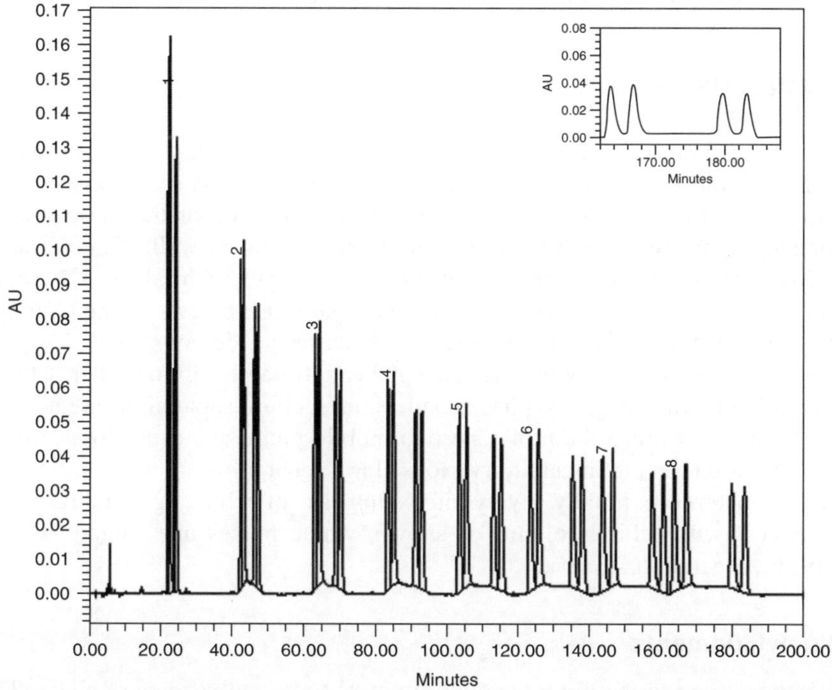

Figure 12 Separate rotamers with an LC sample recycler. HPLC parameters: LC sample recycler: Chiralizer Services CS 3080. HPLC system: Waters Alliance 486 variable-wavelength detector. HPLC columns: 2 × YMC Pro C18 25 cm × 4.6 mm × 5 μm. Mobile phase: 50% acetonitrile/50% 10 mM KH_2PO_4. pH 6.4. Flow rate: 0.7 mL/min. Column temperature: 30°C.

HPLC system via an automatic switching device. Analyte eluted from the first column and the detector is switched onto the second column, continuing the process of chromatographic separation in a recycling loop. The recycle can be repeated many times, until adequate separation is achieved or the first peak starts to overlay the last peak. An example of a rotamer is shown in Scheme 8.

Analyte was recycled or chromatographed in a system eight times, which is indicated by the numbers above all the first peaks in Figure 12. There are four well-defined peaks in the rotamer that are well separated at the eighth elution, with a minimum resolution factor of 1.8. The partial chromatogram insert at the upper right corner shows the resolution of the rotamer at the eighth cycle/elution.

It is noted that the last peak of rotamer at the seventh elution is getting really close to overlap the first peak of the rotamer at the eighth elution. At that point, the recycling of analytes reaches its limit and analytes are diverted to the waste. The area percentages of four peaks of the rotamer at the last elution are 26.6, 26.1, 24.1, and 23.2%, respectively. This separation of rotamers did not involve any chiral selector in the mobile or stationary phases, which is the prerequisite for all other chiral separation techniques. Is this separation chiral, or not?

4 CONCLUSIONS

Chiral separation remains one of the most expensive and challenging chromatographic separations today. It is hoped that this chapter will shed some light on the mysteries of chiral separations. Remember that some of the basic rules of chromatographic separation apply to chiral separations as well. Three cases of chiral separations based on phase conversion of a popular Chiralpak AD column are presented. Examples of successful chiral separations by converting Chiralpak AD column from normal phase to reversed phase are demonstrated. By phase conversion, some of the compounds changed enantiomeric elution order, whereas others did not. Advantages of phase conversion in chiral separations are also discussed. Improper preparation of a normal mobile phase can cause loss of chiral resolution observed previously for various chiral separations, thus leading to poor method transference. Finally, a very interesting case of achieving chiral resolution on rotamers with achiral columns is shown, which makes one wonder whether or not the separation is chiral.

Acknowledgments

The author is very grateful for the opportunity to work with a team of high-caliber scientists led by Robert Strack in Schering Plough, and is obliged to Alicia Duran for her great GC-MS work.

REFERENCES

1. R. Carr, G. Caille, A. Ngoc, R. Foster. *J. Chromatogr. B*, 668(1):175, 1995.
2. R. Ferretti, B. Gallinella, F. La Torre, C. Villani. *J. Chromatogr. A*, 704(1):217, 1995.
3. S. Caccamese, G. Principato, R. Ottana, T. Previtera, C. Zappala. *J. Chromatogr. A*, 694(2):355, 1995.
4. J. McCarthy. *J. Chromatogr. A*, 685(2):349, 1994.
5. A. Kot, P. Sandra, A. Venema. *J. Chromatogr. Sci*., 32(10):439, 1994.

6. T. Chen, R. Mills. *J. Chromatogr.*, 659(2):321, 1994.
7. S. Thomas, B. Surber, M. Fitzgerald. *J. Chromatogr.*, 632(2):390, 1992.
8. E. Mikami, Y. Fujii, S. Yamada, J. Hayakawa, H. Nakazawa, S. Suzuki. *Yakuzaigaku*, 51(4):241, 1991.
9. T. Hollenhorst, G. Blaschke, *J. Chromatogr.*, 585(2):329, 1991.
10. P. Macaudiere, M. Caude, R. Rosset. *J. Chromatogr.*, 405:135–143, 1987.
11. K. Balmér, P. Lagerström, B. Persson, G. Schill. *J. Chromatogr.*, 592:331–337, 1992.

CHAPTER 6

Chiral Discrimination Study for Polysaccharide-Based Chiral Stationary Phases

YUN K. YE

Analytical Research and Development, Bristol-Myers Squibb Company, New Brunswick, New Jersey

1 INTRODUCTION

Although chirality was discovered more than a century ago, chiral separation is still perceived as one of the most challenging tasks for scientists in the separation field. In fact, significant progress has only been made within the last two decades, due to the intense levels of interest in the development and application of chiral separation techniques in biological research fields, especially in the pharmaceutical industries [1–5]. Chiral separation within the pharmaceutical area is essential for investigation of the pharmacokinetical, physiological, toxicological, and metabolic activities of both enantiomers. There is a broad range of examples concerning the means by which the stereoisomers of drugs show differences in terms of their bioavailability, distribution, metabolism, and excretion behavior. Examples also exist in which stereochemical parameters have a fundamental significance in their action and disposition in biological systems. Currently, evaluation of chiral purity is used routinely to monitor the stereoselective synthesis, pharmaceutical research and development, and quality control for the pharmaceutical production of finished chiral drugs.

Chiral HPLC is undoubtedly the most important technique for the separation and quantitation of enantiomers with respect to its availability, versatility, simplicity, and ability to scale up easily from an analytical scale to semipreparative and preparative scales [6,7]. Direct separation of enantiomers by high-performance liquid chromatography (HPLC) has become one of the most useful procedures

Chiral Separation Methods for Pharmaceutical and Biotechnological Products, Edited by Satinder Ahuja

in fields that deal with drugs, natural products, and agrochemicals. This process not only determines an enantiomer's optical purity, but also obtains optical isomers on a larger scale [8,9]. The preparation of a chiral stationary phase (CSP) capable of effective chiral recognition is the key step in the chiral HPLC method. Therefore, many CSPs for HPLC have been prepared, with over 100 being commercialized [10].

Nowadays, the most widely used CSPs are the derivatives of polysaccharides developed by Okamoto and co-workers. They are one of the most useful CSPs in regard to their versatility, durability, and in particular, their loadability on preparative-scale chromatography [11]. More than 10 CSPs of this type are commercially available, with different derivative side groups using either amylose or cellulose. Figure 1 represents the structures of the CSPs. Most of the CSPs are physically coated onto silica for support, and, more recently, three bonded polysaccharide polymer CSPs have become available [12].

Although the polysaccharide-based CSPs have been commercialized for more than two decades, the chiral discrimination mechanisms are still unclear at the molecular level. The polysaccharide-based CSPs might show chiral recognition that depends on the higher-ordered structures of the polymers, which makes it even more difficult to understand the chiral recognition mechanism. Problems often arise with regard to the selection of appropriate systems from the polysaccharide-based CSPs available with fitting mobile phases. Unfortunately, no selector–select and combinations or reliable chiral recognition models have been developed to allow for predictions with respect to separability, magnitude of enantioselectivity, elution order, and suitable chromatographic conditions. Often, the identification of suitable CSPs for a specific pair of enantiomers requires a considerable amount of experimentation, time, material, and labor. One way to solve this dilemma is through the blind screening method. This is achieved by combining commercially available CSPs with different mobile phases [13,14]. However, due to the large selections of CSPs and different combinations of mobile-phase modifiers and additives, the screening approach also demands a great amount of time and resources. Only a limited number of combinations of CSPs and mobile-phase screening is feasible for the interested analytes, and satisfactory enantioseparation is not always guaranteed. In the case of partially separated enantiomers, further method development is required. Moreover, the screening approach does not provide any insight into how and where chiral recognition by a chiral selector molecule occurs. Evidently, there is a need for empirical and/or rational strategies to facilitate this tedious selection procedure in order to lessen the occurrence of the old-fashioned trial-and-error approach.

A lack of understanding of the chiral discrimination mechanism also prevents an explanation of the observed and unexpected effects of separation conditions on chiral resolution using polysaccharide-based CSPs. To date, among the CSPs commercially available, unusual effects reported most frequently have come from polysaccharide-based CSPs. These have stemmed mainly from studies in the dominant normal-phase mode. Commonly encountered and unexplained abnormal effects, such as different organic modifiers, temperatures, additives, and so on,

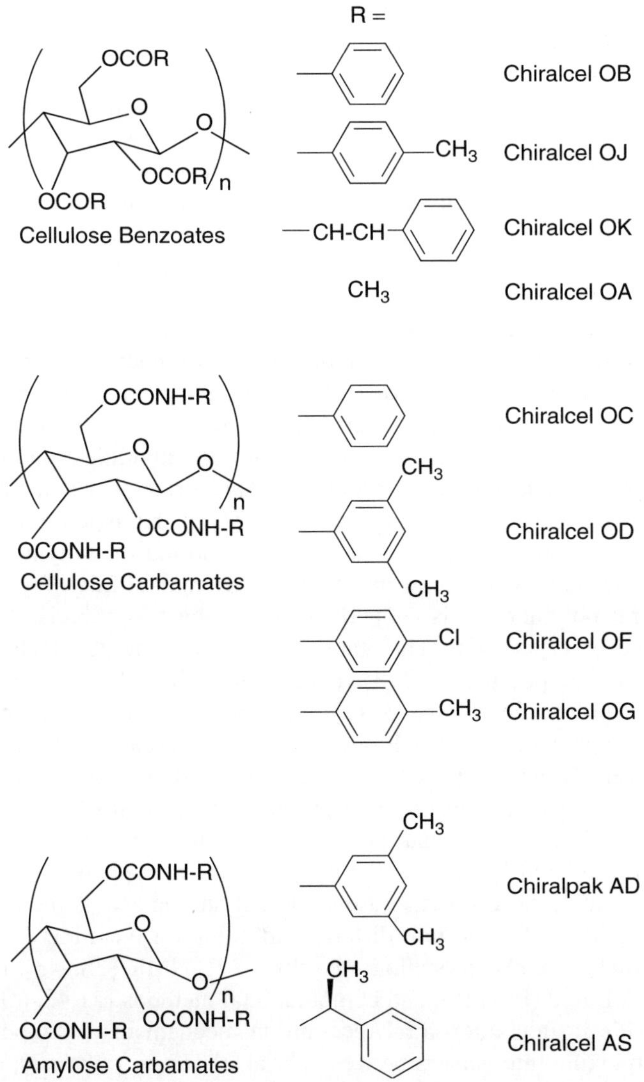

Figure 1 Commercially available polysaccharide-based CSP structures.

caused a great deal of bewilderment; this significantly hindered systematic chiral method development [15–25].

The lack of understanding of the chiral separation mechanism also hindered the development of newer and better polysaccharide-based CSPs. The current CSPs were usually developed by a variation of the side groups or by random substitutions on the phenyl ring without theoretical guidance, in hopes that one of the changes would produce better CSPs.

2 STRUCTURE CHARACTERIZATION FOR POLYSACCHARIDE-BASED CSPS

In the past decade, many attempts to clarify the mechanism of chiral recognition on CSPs for liquid chromatography have been made by means of chromatography, nuclear magnetic resonance (NMR) spectroscopy, x-ray analysis, and computational methods. Compared with the number of chiral selectors available, relatively few detailed studies on enantioseparation mechanisms are accessible. Among them, most investigations have been performed on low-molecular-mass selectors, including π-donors and acceptors, crown ethers, and cyclodextrins. The most popular strategy of establishing chiral recognition models for a given selector involves the collection of a representative body of chromatographic enantioseparation data with a series of analytes displaying incremental structural modifications. A systematic interpretation of these data may provide some mechanistic information on the contributions of the individual structure elements to enantioselective selector–analyte binding. This strategy was most successful with small-molecule CSPs, although only a few mechanistic studies on chiral discrimination at a molecular level have been reported on polymeric CSPs. A number of different interaction sites with a different affinity for enantiomers exist on chiral polymers; it is difficult to determine their precise structures in the solid state or in solution. This causes our understanding of chiral recognition mechanism of polymeric CSPs to be challenging. The chiral recognition mechanism of polysaccharide-based CSPs has been investigated based primarily on chromatographic methods; it has not yet been elucidated satisfactorily at a molecular level. Recently, spectroscopic [26–30] and computational approaches [31,32] have been carried out on polysaccharide-based CSPs.

Insight into chiral discrimination at the molecular level for polysaccharide-based CSPs is hindered by the complexity of the polymer, such as the exact stereochemical structure, the geometry of the interaction, the accessible binding sites, and the multiplicity of sites with different affinities for enantiomers. Numerous techniques, such as x-ray crystallography [33–37], NMR [38–43], calorimetric studies [44], infrared (IR) [45], and computational methods [41,42,46], have been used to provide insight into chiral recognition mechanisms for other CSPs. To grasp the perplexing interactions between chiral stationary polymers, understanding the chiral stationary-phase structures, chiral cavities, and surface properties are the basic foundations. Some progress was made recently for the structural characterizations of polysaccharide-based CSPs.

2.1 Structure Analysis for Polysaccharide-Based CSPs

Vogt and Zugenmaier reported on structural models for some liquid-crystalline cellulose derivatives [47]. Models for helicoidal cholesteric structures are proposed for cellulose derivatives. The pitch of these structures is determined by independent methods as a function of temperature, concentration, solvent, and degree of polymerization. The results were reported on the basis of available theoretical considerations. These structural results formed the foundation for further structural characterization of polysaccharide-based CSPs.

Yashima et al. reported the first computational structural model for cellulose tris(4-trimethylsilylphenylcarbamate) (CTSP) [48] based on Vogt and Zugenmaier's report on the structure [47]. The optimization of CTPC was performed by a full energy minimization of 1 unit of CTPC, containing CH_3O groups at the 1- and 4-positions of a glucose unit, and the optimized units were allowed to construct an octamer with a left-handed threefold (3/2) helix according to the structure of CTPC reported by Vogt and Zugenmaier [47] on the basis of x-ray analysis of a CTPC fiber. The dihedral angles defined by $H_1-C_1-O-C'_4$ (ϕ) and $H'_4-C'_4-O-C_1$ (δ) were fixed to be 60° and 0°, respectively. Then the 8 units of CTPC as a starting structure were optimized by the steepest-descent method. During the end of the minimization, a metastable structure was obtained. The 8-mer has hydrogen bonds between the NH protons of the carbamate moieties at the 6-positions and the carbonyl oxygens at the 2-positions. The distance between the NH proton of the 6-position and the carbonyl oxygen at the 2-position is 2.634 Å. Then the minimization was further performed by using the conjugate gradient followed by the adopted-basis Newton–Raphson methods. Molecular dynamic (MD) simulation was applied to the optimized 8-mer of CTPC. The structures with lower energies were extracted from the trajectory files obtained from MD simulation, and molecular mechanical calculations as described were performed again for these extracted structures. However, significant changes were not observed before and after MD simulation. Figure 2 represents the structure. It is the first time that a polysaccharide-based CSP structure has been proposed.

Okamato's group reported another computational structure of the CSP, cellulose tris(5-fluoro-2-methylphenylcarbamate) [29]. The polymer model was constructed using the crystalline structure of cellulose tris(phenylcarbamate) [47]. First, a full energy minimization of a repeating unit of cellulose tris(5-fluoro-2-methylphenylcarbamate) containing CH_3O groups at the 1- and 4-positions of a glucose unit was performed. Then the monomeric unit of cellulose tris(5-fluoro-2-methylphenylcarbamate) was allowed to construct a trimer with a left-handed threefold (3/2) helix according to the structure of CTPC. The trimer was placed into a simulation cell using three-dimensional periodic boundary conditions. The energy minimization of the periodic structure was then performed. The resulting

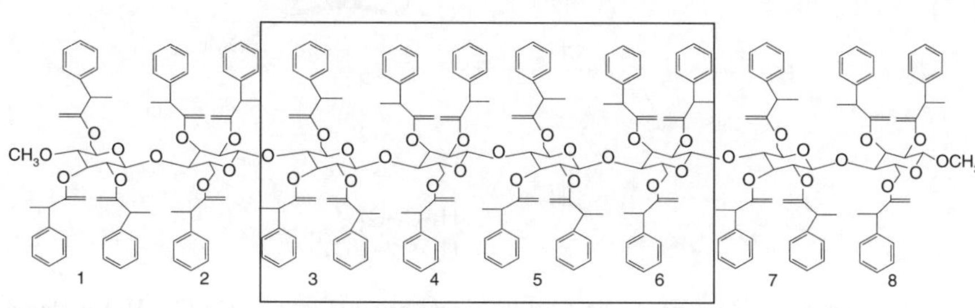

Figure 2 Octamer of CTPC reported by Yashima et al. [48].

optimized trimer in the unit cell was connected to give a nanomer (9-mer) as the model polymer. However, both proposed polysaccharide-based CSPs were based on the computational study without direct experimental data.

One of the most diversified and widely used polysaccharide-based chiral stationary phases is amylose tris(3,5-dimethylphenylcarbamate) (ADMPC), which is available commercially as the Chiralpak AD chiral column. Recent reports [30] demonstrate that low-molecular-weight (DP ~ 100) ADMPC, prepared by the enzymatic polymerization of α-D-glucose 1-phosphate dipotassium, is soluble in chloroform, which makes NMR investigation of the chiral discrimination possible for this polysaccharide CSP. The structural analysis of ADMPC was performed by NMR spectroscopy using a sample with a lower degree of polymerization in order to understand the structure of ADMPC CSP [30]. The two-dimensional NOESY NMR, combined with the computer modeling study, revealed the detailed three-dimensional structure of the ADMPC polymer.

The NOESY spectrum of ADMPC reveals that a number of NOE cross peaks were observed in the region of glucose–glucose, NH of carbamate residues–methyl on the phenyl group, NH–glucose, and NH–NH proton resonances. The chemical shifts of the glucose protons (H_1 to H_6) of ADMPC were assigned from the COSY experiment. The assignments of the methyl protons on the phenyl groups and NH proton resonances at the 2-, 3-, or 6-position of a glucose unit were attained by comparing the NMR data of the regioselectively carbamoylated model polymers, amylose 6-(3,5-dichlorophenylcarbamate)-2,3-bis(3,5-dimethylphenylcarbamate) and amylase 2,3-bis(3,5-dichlorophenylcarbamate)-6-(3,5-dimethylphenylcarbamate). Figure 3 shows the dimer structure of ADMPC with atom and angle assignments.

To propose a structure for ADMPC, the interproton distances of the glucose protons (e.g., H_1 to H_4') have to be determined by measuring the peak volumes

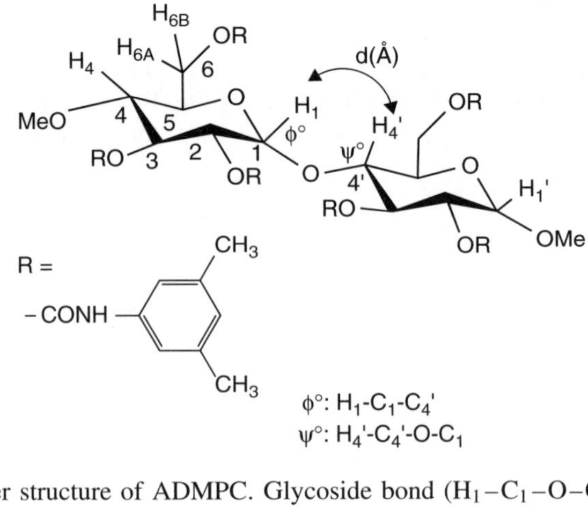

Figure 3 Dimer structure of ADMPC. Glycoside bond (H_1–C_1–O–C_4'–H_4') is defined by two dihedral angles, ϕ and ψ.

of the cross and diagonal peaks at different short mixing times. These distances can be related to the torsion angle about the glycoside bond defined by two dihedral angles, $H_1-C_1-O-C'_4$ (ϕ) and $H'_4-C'_4-O-C_1$ (ψ) (Figure 3). The peak volumes of the cross peaks in the glucose proton resonances of ADMPC were first estimated using the two-dimensional NOESY method acquired at different short mixing times (25 to 50 ms). The distance of the $H_1-H'_4$ protons was estimated to be 2.83 to 2.96 Å, using the proton distances H_2-H_4 and H_3-H_4 in the same glucose unit as the internal reference together with the peak volume.

To obtain information about the glycoside bond geometry between two adjacent glucose rings, the dimer model of ADMPC shown in Figure 2 was constructed, and the $H_1-H'_4$ distance and energy profiles depending on the two dihedral angles defined by ϕ and ψ were calculated; ϕ and ψ were rotated individually at 6° intervals, and the distance of the $H_1-H'_4$ protons and the potential energy were estimated. There are numerous possible dihedral angles that satisfy the estimated $H_1-H'_4$ distance (2.83 to 2.96 Å). However, the computational total energy contour map estimated from the dimer model of ADMPC indicates that the lowest total energy of the dimer is (−68.5°, −42.0). This dihedral angle leads to the left-handed 4/3 helical structure for ADMPC.

The polymer model of ADMPC was then constructed using the dihedral angles obtained and optimized under three-dimensional periodic boundary conditions [29,48]. The full structure of ADMPC is shown in Figure 4A and B; the main chain is shown in Figure 4C and D. The glucose residues are regularly arranged along the helical axis. A chiral helical groove with polar carbamate groups exists along the main chain. The polar carbamate groups are preferably located inside, and the hydrophobic aromatic groups are placed outside the polymer chain so that polar enantiomers may interact predominantly with the carbamate residues in the groove through hydrogen-bond formation. The ADMPC polymer proposed, one of the most widely used CSPs in the chiral separation field, is the most reliable three-dimensional structure so far. It is the first time that reliable NMR results were incorporated into the computation as the necessary constraints to obtain a meaningful computational optimization.

Recently, we carried out further computational optimization of ADMPC polymer based on the structure proposed by Okamoto's group [30], demonstrated in Figure 4. The approach to minimizing the ADMPC structure is built on the method developed by Okamoto et al. The coordinates for the ADMPC structure published [30] were provided by Okamoto's group. This is based on the $H_1-H'_4$ distance (2.83 to 2.96 Å) of the two neighboring sugar rings, as determined by two-dimensional NMR and molecular modeling using 12 sugar units. We used these results as the initial coordinates for further minimization and introduced three new conditions for the minimization. First, methoxy groups were substituted for the hydroxyl groups of the two end sugar units of the 12-mer to represent ADMPC as a polymer, preventing unusual conformational arrangements of the phenyl side chains on the two end units.

Additionally, a periodic box (40, 40, 50 Å) for the ADMPC 12-mer was built, and this was bonded to the neighbor ADMPC 12-mer in another periodic box

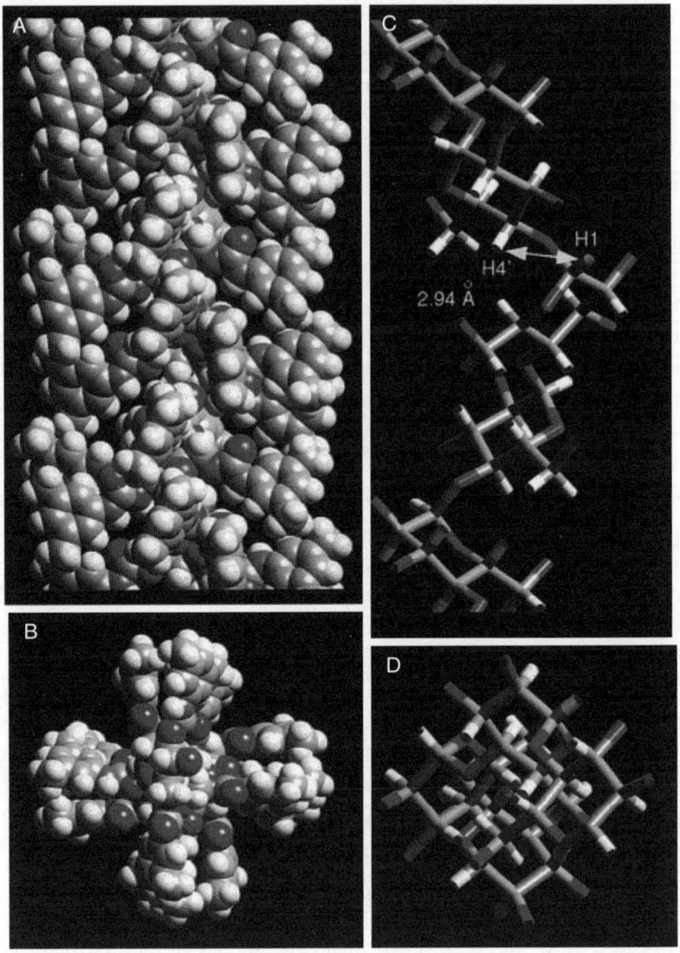

Figure 4 Optimized structure (A and B) and the main chain (C and D) of ADMPC. The viewpoints are along the chain axis (A and C) and perpendicular to the chain axis (B and D).

with a 1,4′ sugar linkage to form one long chain polymer. These steps virtually eliminate the end-unit effects in the minimization. Second, while the polysaccharide backbone atoms were fixed to the positions determined by Okamoto et al. [30] in accord with their NMR results, we allowed the phenyl side chains to move freely. Third, molecular dynamic simulation was also used to obtain a better structure for the ADMPC polymer, which was not employed in Okamato's study. We used quenched annealing to reduce the possibility of a local minimum. With quenched annealing, the initially minimized complexes were simulated with molecular dynamics at 500 K, 20 different trajectories were sampled, and each conformation was then minimized to find the lowest energy configuration.

TABLE 1 Energies of ADMPC Polymer Before and After Minimization (kcal/mol)

	After Minimization	After Quenched Anealing
Total potential energy	−1151.1	−1224.4
Internal	−593.9	−584.3
Bond	116.0	103.0
Angle	198.7	171.8
Torsion	−636.1	−602.5
Out-of-plane	2.6	1.9
Cross	−275.1	−258.4
Nonbond	−555.2	−638.2
Van der Waals	−10.1	−109.5
Repulsive	1925.7	1961.0
Dispersive	−1935.8	−2070.5
Electrostatic	−545.1	−528.7

The component energy terms for minimization of the AMPDC structure are listed in Table 1, including those for Okamato's initial structure, the minimized structure after reducing end effects (as described later), and the final minimized structure after quenched annealing. These results show that reducing end effects reduced the energy to −1151.1 kcal/mol, which is much lower than the 1116 kcal/mol of the initial ADMPC structure [30]. The quenched annealing, which allows the polymer to search for different minima by raising the temperature temporarily, further reduced the total energy to −1224.4 kcal/mol. Almost all of this reduction of energy as shown by Table 1 is due to internal energy; less than 15% of the energy reduction is from van der Waals or electrostatic interactions.

The original and final energy-minimized structures are pictured in Figure 5. This shows that the ADMPC polysaccharide is a left-handed helix with the sugar units forming the backbone for the polymer. Beyond that, this structure differs from that of Okamoto et al. [30] in the arrangement of the phenyl side groups, which now wind along the backbone helix to line the chiral groove along the main chain axis. By virtue of being significantly lower in energy, this structure for ADMPC is expected to be a more insightful starting point for understanding the enantioselectivity observed experimentally.

The effects of backbone and side chain on the molecular environments of chiral cavities in polysaccharide-based polymers were reported recently by Kasat et al. [49]. The backbone and side-chain chiral cavities were studied using three commercially available polysaccharide-based CSPs, cellulose tris(3,5-dimethylphenylcarbamate) (CDMPC), amylose tris(3,5-dimethylphenylcarbamate) (ADMPC), and amylose tris[(*S*)-*R*-methylbenzylcarbamate] (ASMBC). Attenuated total reflection infrared spectroscopy (ATR-IR), x-ray diffraction (XRD), ^{13}C cross-polarization/magic-angle spinning (CP/MAS) and MAS solid-state NMR, and density functional theory (DFT) modeling methods were used for the study.

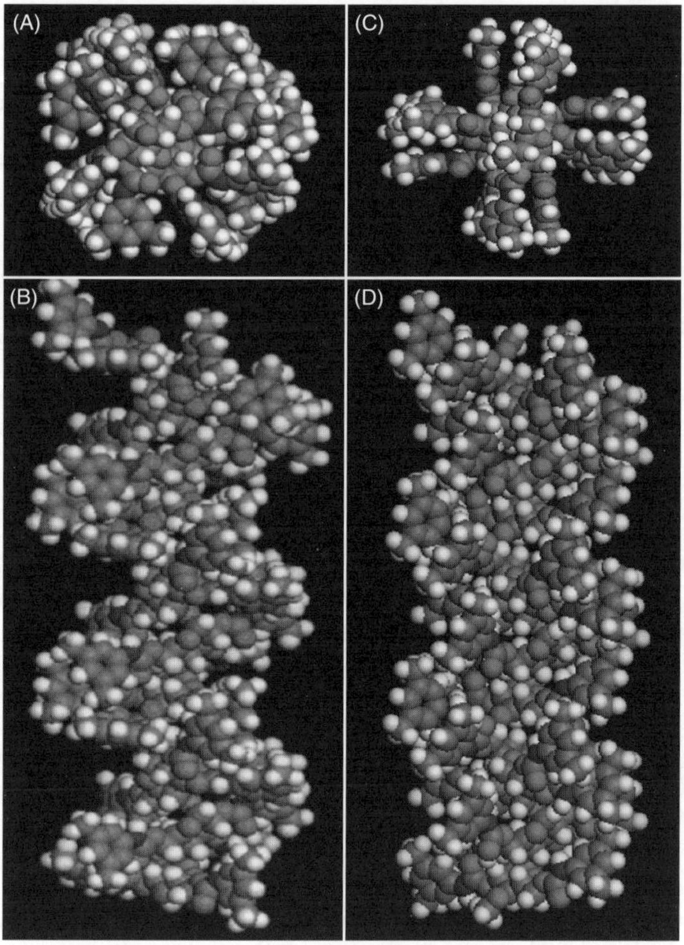

Figure 5 MM-minimized ADMPC structures. The viewpoints are (A) along the chain axis and (B) perpendicular to the chain axis. (C) and (D) are the original ADMPC structure.

It is reported that the strengths of H-bonds of the C=O and NH groups in the chiral cavities of these polymers are significantly different [49]. For CDMPC, the cavities are slightly bigger than for ADMPC, due primarily to the differences in the polymer backbones, resulting in weaker intrapolymer H-bonds in the CDMPC. For ASMBC, the H-bond strengths of the C=O groups are stronger than for ADMPC, even though they have the same backbones, because of the interactions with the neighboring CH groups in the same side chains. The strengths of the H-bonds of C=O groups for these polymers are inferred to be in the following order: ASMBC > ADMPC > CDMPC. The strengths of the H-bonds of the NH groups are inferred in the order ADMPC > CDMPC > ASMBC. The coupling of the NH and CH_3 groups with the phenyl ring results in much higher IR intensities in ADMPC and CDMPC than in ASMBC. XRD results show different d-spacings between the polymer chains in the three polymers, resulting in variations in the packing arrangements, and different intra- and interpolymer interactions between the C=O and NH groups. From the NMR chemical shifts of the C_1 carbon, it is inferred that the backbone glycosidic bond conformations and helicities are similar for ADMPC and ASMBC but different for ADMPC and CDMPC. DFT simulations predict that the ADMPC and CDMPC side chains have a planar conformation, while the ASMBC side chain has nonplanar, and possibly multiple, conformations. Hence, the C=O, NH, and phenyl groups in the chiral cavities of these polymers have different molecular environments, which may be a major factor affecting the selectivities of chiral solutes.

2.2 Solvent Effects of the Structure of Polysaccharide-Based CSPs

Since HPLC chiral separation involved different solvent systems and different selectivities were observed with different mobile-phase composition using the same CSP and chiral analytes, the solvents used in chiral separation must also have played an important role. Different solvents may have had a significant impact on the structures of chiral polymers. Structural differences of chiral polymer may exist in the presence of different solvents.

Recently, Wenslow and Wang examined the structure analysis of ADMPC by solid-state NMR ($^{1}H/^{13}C$ CPMAS) [50]. The CPMAS NMR spectrum of the CDMP polymer flushed with hexane was compared to the spectrum of dry ADMPC dry polymer (Figure 6). The hexane carbons are clearly present in the NMR spectra. These solvent peaks represent incorporation of solvent molecules into the structure of the ADMPC chiral polymer. Sharper peaks for hexane were observed due to the solvent polarity. Weak-polarity solvent incorporation produces relatively high solvent mobility. This high mobility is displayed in the CPMAS spectrum through the sharpness of the hexane peaks. The hexane mobility, however, is slow enough to produce ^{13}C signals mediated through the $^{1}H/^{13}C$ dipole coupling.

More interesting results were reported for the ethanol and iropropanol solvent incorporation comparing to hexane and dry ADMPC spectra (Figure 7). As ethanol and iropropanol content in the flush solvent, the resolution (line narrowing) increased for the ^{13}C peaks representing ADMPC polymer carbons on both the amylose backbone and the dimethylphenylcarbamate side chain. This increased resolution is accompanied by obvious peak multiplicities in the region of side-chain carbons. With hexane incorporation, ADMPC polymer displayed broadened ^{13}C CPMAS peaks due to either numerous conformation geometries or a more amorphous nature. When ethanol and iropropanol replaced hexane in the ADMPC polymer, these broadened ^{13}C peaks were replaced by relatively sharp, resolved ^{13}C peaks. The peak multiplicities may be a result of either multiple geometrical conformations of the side chain becoming “locked in” or crystal packing interactions. Either of these effects was induced by ethanol and iropropanol solvent incorporation, which created multiple crystallographically unique ^{13}C sites

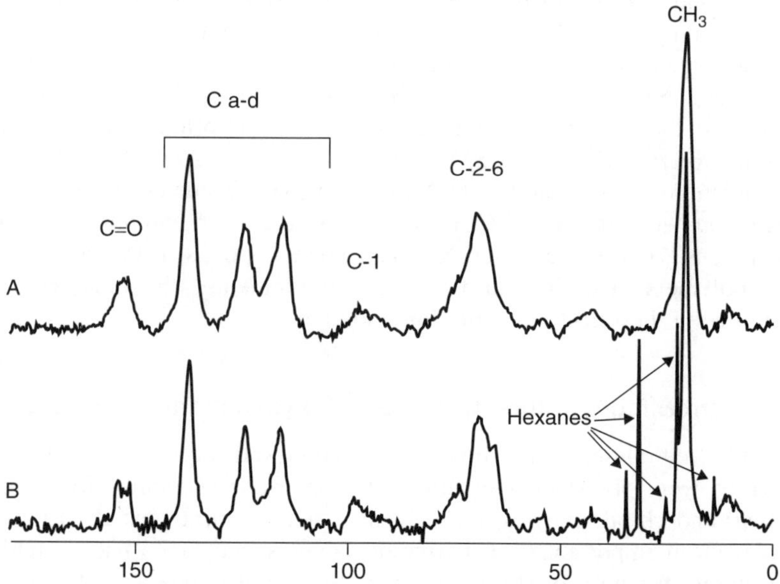

Figure 6 $^1H/^{13}C$ CPMAS spectra of (A) dry ADMPC polymer and (B) ADPMC polymer flushed with hexane.

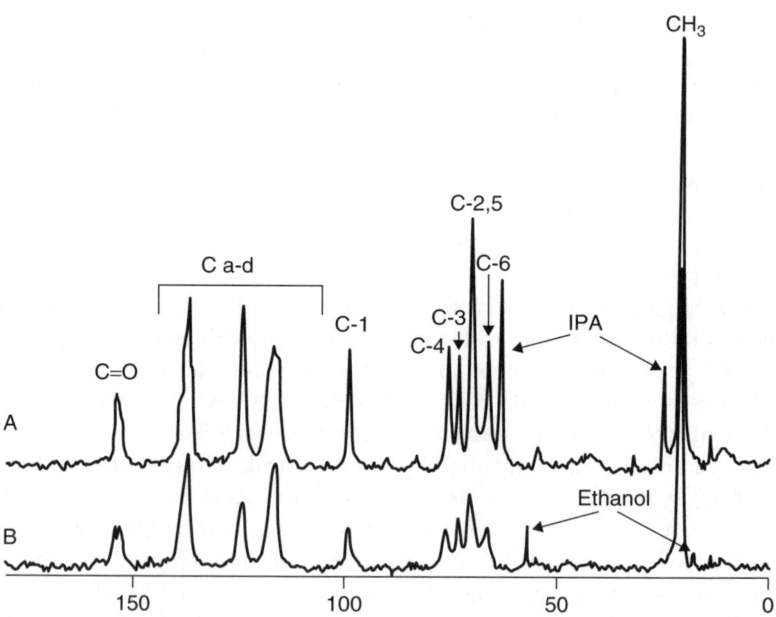

Figure 7 $^1H/^{13}C$ CPMAS spectra of ADMPC polymer flushed with (A) 10 : 90 IPA/hexane and (B) 20 : 80 ethanol/hexane. Both spectra are scaled to match the intensity of the peak at 138.0 ppm in part A.

in the ADMPC side chain. The line narrowing of solid-state NMR signals can generally be attributed to increased crystallinity, increased mobility, or both.

The ^{1}H T_1 experiments were performed on dry ADMPC as well as ADMPC flushed with 100% hexane, 90 : 10 hexane isopropylamine (IPA), and 60 : 40 hexane/ethanol to determine the mobility effects of the ADMPC backbone and side chain. The ^{1}H T_1 values for dry ADMPC polymer and for the other three solvent systems were all about the same (i.e., 3 s). These results demonstrated that increased mobility was not the responsible driving force for the peak sharpening observed, and the increased crystallinity may be the cause of differing solvent incorporation into ADMPC polymer.

As displayed by the shape and number of the ADMPC ^{13}C peaks (Figure 7), incorporated ethanol induced significantly less crystallographically distinct ^{13}C resonances than did incorporated IPA. It is clear from the presence of IPA and ethanol peaks in the CP/MAS spectra in Figure 6a and b, respectively, that incorporation of the polar alcohol modifier occurred in both solvent systems. In comparing ADMPC backbone carbons (66 to 100 ppm), IPA incorporation clearly yielded sharper, more intense ^{13}C resonances, indicating an increased crystalline character of ADMPC in the presence of IPA as compared to ethanol. Additionally, by contrasting the region of ADMPC side-chain carbon (21, 116 to 209 ppm) in Figure 6, it is clear that IPA incorporation produced more distinctly inequivalent ^{13}C resonances in the side-chain region, indicating an increased number of crystallographically unique ^{13}C sites in the side chain. Both observations demonstrate that significantly different conformational distributions can exist in ADMPC, depending on whether IPA or ethanol is in the mobile phase. This difference in conformational distribution, induced by different alcohol modifiers, is probably due to the difference between the bulkiness of the IPA and ethanol molecules.

The polysaccharide polymer–solvent interactions for ADMPC with five commonly used solvents—hexane, methanol, ethanol, IPA, and acetonitrile (ACN)—are studied using ATR-IR for thin sorbent films, XRD for thin films, CP/MAS and MAS solid-state NMR for ADMPC polymer-coated silica beads, and DFT modeling [51]. The polymer forms helical rods with intra- and interrod hydrogen bonds. The IR spectrum of the dry ADMPC polymer (prepared from DMF) changes to a varying extent upon sorption of different solvents (Figure 8). The IR spectra results indicate that all the NH in the polymer is H-bonded. Since most of the NH groups are found to be H-bonded by IR, and since approximately half are predicted to form intrarod H-bonds, it is inferred that the remaining H-bonds are formed from interrod H-bond interactions.

Upon absorption of excess hexane by the polymer, the polymer spectrum in the in situ experiments remains essentially unchanged. These results agree with the solid-state NMR results described above [50]. With the absorption of methanol by the polymer, the ATR-IR results indicate that the methanol evidently interacts with the polymer, mainly via strong H-bonding interactions: namely, C=O (polymer)$\cdots$ HO (methanol) and NH (polymer) $\cdots$ OH (methanol). In the presence of methanol, some portion of the C=O $\cdots$ HN H-bonds in and between the

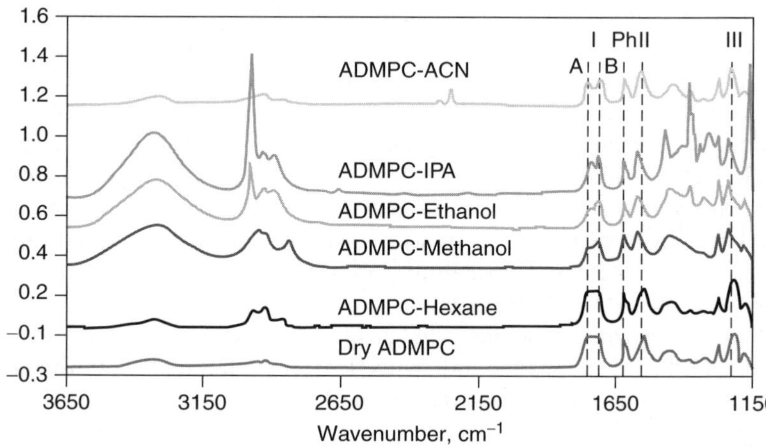

Figure 8 IR spectra of ADMPC polymer in the dry state and upon equilibration with various solvents.

rods of the dry polymer are broken, and new, stronger H-bonds are formed with the solvent molecules changing the bulk structure of the polymer. The results imply that the weakly H-bonded C=O $\cdots$ HN groups between the polymer rods are the ones mainly affected by the solvent molecules. The absorption of ethanol, with slight spectral changes compared to methanol, is observed, suggesting that ethanol–polymer interactions are qualitatively the same but have some quantitative differences. Similar changes are also observed upon the absorption of IPA. All the alcohols tested are incorporated into the structure of the polymer and interact with the polymer via H-bond interactions strong enough to change the polymer structure. The small quantitative differences observed in the amide I band of ADMPC polymer may be related to changes in the polymer structure upon absorption of different alcohols.

Results from the absorption of ACN and ATR-IR indicate that ACN also formed H-bonds with ADMPC polymer but to a lesser extent than for the alcohols. It is understandable that ACN is a weaker H-bond donor and acceptor than are the alcohols, resulting in smaller shifts in the IR wavenumbers due to weaker NH(polymer)$\cdots$ NC–(ACN) H-bond interactions. The IR spectra changes observed upon absorption of ACN may result from significant dipole–dipole interactions of the amide group with ACN.

Various nanometer-sized cavities are formed between the polymer side chains and rods. Changes in the H-bonding states of the C=O and NH groups of the polymer upon absorption of each of the solvents are clearly evident in ATR-IR spectra results. The IR wavenumbers, the H-bonding interaction energies, and the H-bonding distances of the polymer side chains with each of the solvent molecules are predicted using DFT theory. With hexane, the interaction energies are less than 1 kcal/mol, the energies of polymer–methanol range from 5 to 6 kcal/mol, and with ethanol and IPA, the energies are also about 5 to 6 kcal/mol.

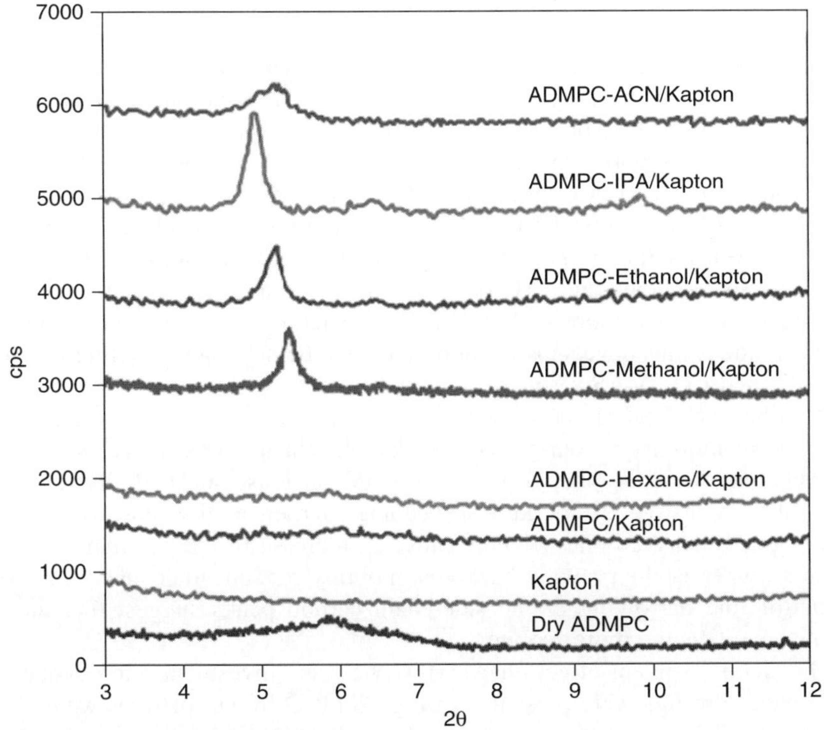

Figure 9 XRD patterns of ADMPC polymer in the dry state and upon equilibration with various solvents. The Kapton film, examined for reference, shows no peaks in this region.

With ACN, the interaction energies are lower than those with the alcohols, ranging from 3 to 4 kcal/mol, indicating weaker H-bond interactions between the polymer and ACN compared to the polymer with alcohols.

Changes in the polymer crystallinity upon absorption of each solvent are also evident in XRD (Figure 9). A very broad peak indicates that the polymers are substantially amorphous. Upon equilibration with hexane, the XRD pattern of the polymer remains almost unchanged. This shows that the polymer film retains the structure that it has in the dry state and remains partially ordered. With methanol absorption, one sharp peak appears, indicating that the long-range crystalline order, or the higher-order structure, of the polymer increases substantially. Evidently, methanol, ethanol, and IPA enter and change the polymer structure, consistent with the IR results. These results imply that alcohols form H-bonds with polymer C=O groups (as H-bond donors) and NH groups (as H-bond acceptors) and alter the interrod interactions, allowing for a more ordered arrangement of the polymer rods. Other interesting results are the d-spacing results. From methanol to ethanol to IPA, the d-spacing is increasing with enlarging solvent molecular size. The spacing observed could be the repeat distance between the stacked polymer rods. The polymer rods are reorganized upon solvent absorption, and the distance between the rods increases with the increase in solvent molecular size. These results have implications for understanding the role of the solvent in modifying the structure and behavior of the polymer CSPs.

The changes in the polymer crystallinity and the H-bonding states of C=O groups have also been probed with ^{13}C CP/MAS solid-state NMR, and agree with the results reported by Wenslow and Wang [50]. The interesting results were shown in ^{13}C MAS solid-state NMR. The ^{13}C MAS spectra (Figure 10) look quite different from the CP/MAS spectra (Figures 6 and 7). The polymer backbone peaks in the dry polymer are absent, indicating no substantial molecular mobility. Some phenyl and CH_3 peaks are observed in Figure 10, indicating that the side chains have more mobility than the backbone. It is concluded that the polymer side chains have substantial mobility in the dry state, resulting in intense aromatic and CH_3 MAS resonances.

With hexane, the CH_3 peak of the polymer becomes slightly sharper, indicating a slight mobility increase. With methanol, ethanol, IPA, or ACN, the peak intensities of the three phenyl peaks (II to IV) increase substantially relative to the peak I intensity. The peaks also become sharper, as does the CH_3 peak. It is concluded that the polar solvents cause an increase in the polymer side-chain mobility, whereas the polymer backbone mobility remains unchanged. The MAS results of line narrowing of the side-chain carbon peaks suggest that the side chains have become more mobile.

The alcohol solvent effect on ADMPC was also investigated in our lab. The XRD pattern results were presented in Figure 11. Different patterns were clearly observed as IPA and ethanol incorporated into the ADMPC polymer. Crystallinity increased as ethanol and IPA were added, which agrees with results reported previously [50]. However, comparing Figures 9 and 11, different XRD patterns were

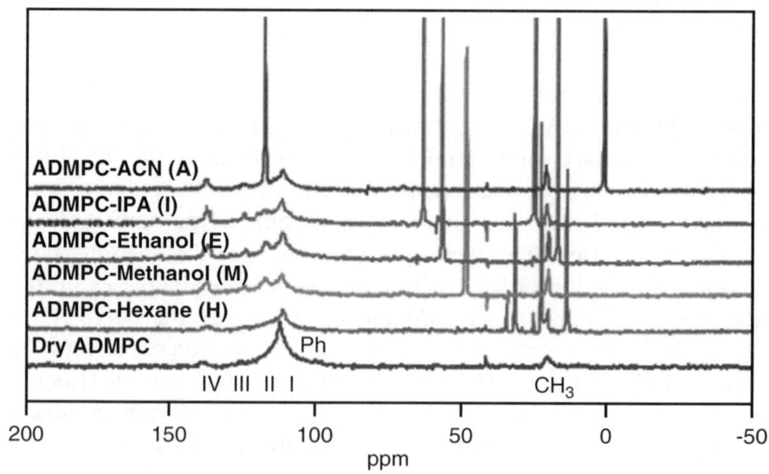

Figure 10 ^{13}C MAS solid-state NMR spectra of ADMPC polymer (inside silica beads, Chiralpak AD) in the dry state and upon equilibration with an excess of various solvents.

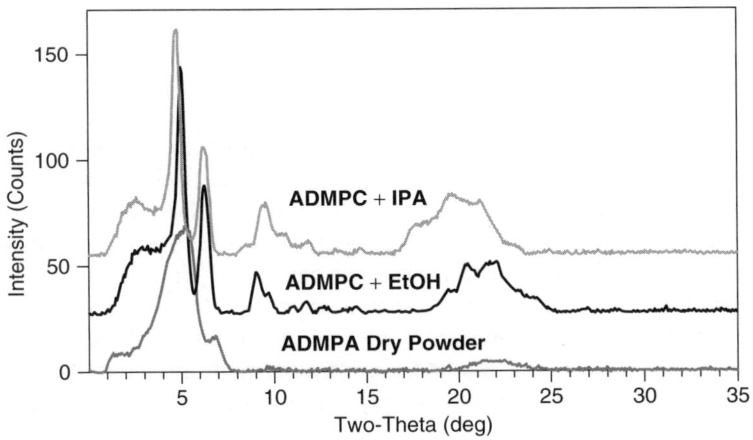

Figure 11 XRD patterns of ADMPC polymer in the dry state and upon equilibration with IPA and ethanol.

observed for both dry ADMPC powder and with ethanol and IPA incorporation. Multiple peaks were observed for all XRD patterns in our lab results, but only one peak was seen in Figure 9. Furthermore, a new peak at about 9θ was observed after alcohol incorporation. These differences may rise from a different source of ADMPC polymer, such as the history of the ADMPC polymer and what solvents were used to produce the polymer.

2.3 Additive Effects of the Structure of Polysaccharide-Based CSPs

The use of mobile-phase additives is quite routine in chiral HPLC employing polysaccharide-based CSPs under normal-phase conditions. In fact, up to a maximum of 1% (v/v), acidic and basic additives are frequently incorporated into the mobile phase for polar analytes as a tailing reducer. Severe tailing is often observed with compounds having amine groups, which require basic mobile-phase additives, such as triethylamine and diethylamine, to be incorporated into the typical hexane–alcohol mobile phase to obtain satisfactory chiral selectivity.

For acidic compounds, acidic additives such as trifluoroacetic acid and acetic acid are required in any mobile-phase combinations for the purpose of both efficiency and elution in normal-phase mode. The benefit of achiral acidic or basic additives is generally considered only to improve efficiency and peak shape for the polar analytes in chiral HPLC. Other important functions of the acidic and basic additives, such as the effect on retention and selectivities, are overlooked for polysaccharide-based CSPs. In fact, improved selectivities were found in many cases when achiral acidic or basic mobile-phase additives were used in mobile phase for chiral HPLC. Numerous studies have been reported in macrocyclic antibiotic-based, protein-based, and Pirkle-type CSPs using achiral acidic and/or basic additives to control the chiral selectivity. Our recent work [52–55] demonstrated that the acidic and basic additives could have a dramatic effect on both efficiency and chiral selectivity for amino acids using polysaccharide-based CSPs.

Recently, we reported a large chiral selectivity achieved for *O-tert*-butyltyrosine allyl esters using the ADMPC polymer as the chiral stationary phase [56]. In this study, acidic additives were used in the mobile phase to achieve the unusually huge selectivity. High enantioselectivity is observed generally for amino acid esters with this CSP, where the enantioselectivity is switched on when the mobile phase is acidic. The high enantioselectivity, combined with the large effect of acidity, make this system a particularly interesting one to study.

Chromatograms of the D- and L-enantiomers of *O-tert*-butyltyrosine allyl esters using the ADMPC polymer as the chiral stationary phase are shown in Figure 12, with and without the addition of ethanesulfonic acid. This indicates that ethanesulfonic acid activates a very large chiral selectivity in which the

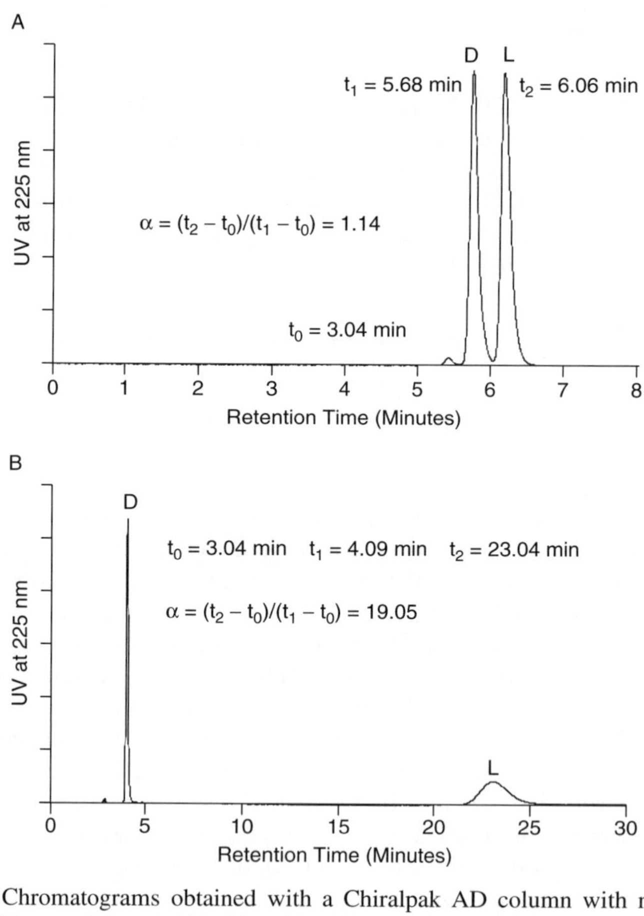

Figure 12 Chromatograms obtained with a Chiralpak AD column with mobile phase consisting of hexane/ethanol (85 : 15, v/v): (A) no acidic additive ESA; (B) 0.2 vol% ESA added to the mobile phase.

selectivity goes from 1.14 to 19.05 upon the addition of ethanesulfonic acid. The retention of the less retained D-enantiomer is decreased slightly upon addition of the acid, while retention of the L-enantiomer is increased markedly. The decreased retention of the D-enantiomer could be due either to a change in the hydrogen bonding of the amino group on the ester, since it has changed from a base to an acid, or to a structural change of the chiral polymer to reduce the number of binding sites. The large increase in retention of the L-enantiomer is probably due to new interactions that are enabled by the amino group becoming acidic. Such interactions cannot readily be visualized without the help of other methods; hence, NMR was used to investigate the nature of these highly selective interactions [57].

The proton NMR survey spectra of ADMPC with and without ethanesulfonic acid (ESA) in chloroform are shown in Figure 13. No dramatic changes were noted for glucose protons H_2 to H_6 when ESA was added. However, the chemical shifts in H_1 move downfield and merge with the signal from H_3 as ESA is added, as demonstrated in Figure 14. This observation indicates noticeable local chemical environmental changes in the H_1 proton. The fact that H_1 shifts downfield with chemical shift changes in the glucose unit in the presence of ESA implies that the complex formed causes slightly reduced electron densities around the H_1 protons. However, the H_1 chemical shift change was reversed with a small addition of D_2O (10 μL). It is noted that the merged H_1 and H_3 signals are well separated and the glucose proton shows its natural state in the absence of ESA with D_2O addition. This one-dimensional NMR result indicates normal adsorption and desorption of ESA onto ADMPC polymer.

Okamoto et al. [30] point out that the internuclear distances in H_1 and H'_4 play a critical role in the tertiary structure of ADMPC in solution because these distances can be related to the torsion angles about the glycoside bond defined by two dihedral angles, $H_1{-}C_1{-}O{-}C'_4$ and $H'_4{-}C'_4{-}O{-}C_1$. To answer the question of whether the tertiary structure of ADMPC is changed with the addition of ESA, it would be useful if the distance of H_1 and H_4 (H'_{14}) can be estimated before and after the introduction of ESA in the ADMPC solution. To measure such a distance, the two-dimensional NOESY technique of NMR spectroscopy is utilized. It is required to measure the NOE buildup as a function of mixing time in a set of the two-dimensional NOESY spectra as a cross-peak volume. The initial slope of such a build-up curve is the cross-relaxation rate between spins i and j. Under the initial rate approximation, the NOE enhancement between two spins i and j at a mixing time τ will be proportional to the cross-relaxation rate (σ_{ij}), which in turn depends on the sixth power of the internuclear distance r_{ij}^{-6}:

$$\eta_i(j) = k\sigma_{ij}\tau = k' r_{ij}^{-6}\tau \tag{1}$$

If a known internal internuclear distance r_k exists, one has

$$\frac{\sigma_{ij}}{\sigma_k} = \frac{r_k^6}{r_{ij}^6} \tag{2}$$

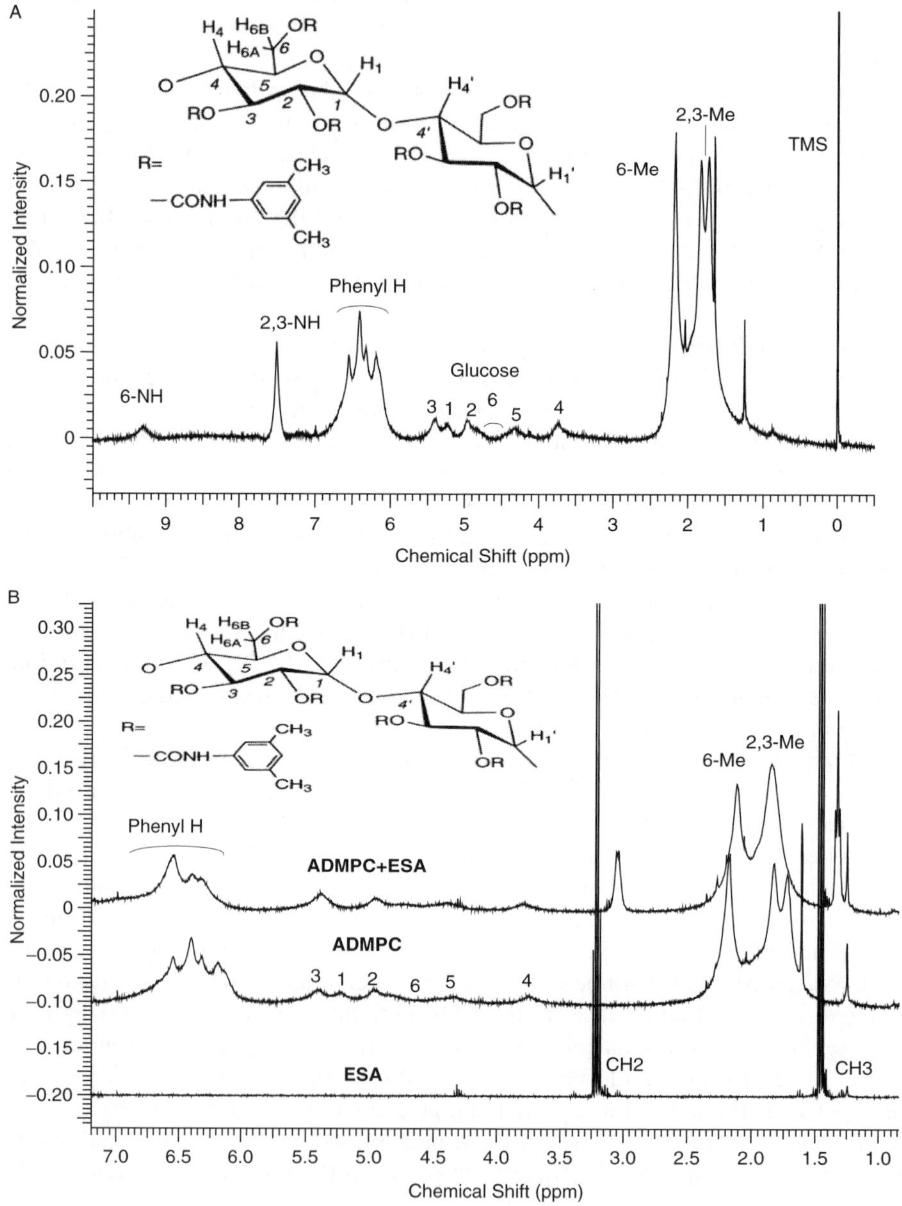

Figure 13 (A) NMR spectrum of ADMPC polymer. (B) NMR spectra of ESA, and ADMPC polymer in the presence of ESA. The concentration of the ADMPC polymer is 22 mg/mL in $CDCl_3$.

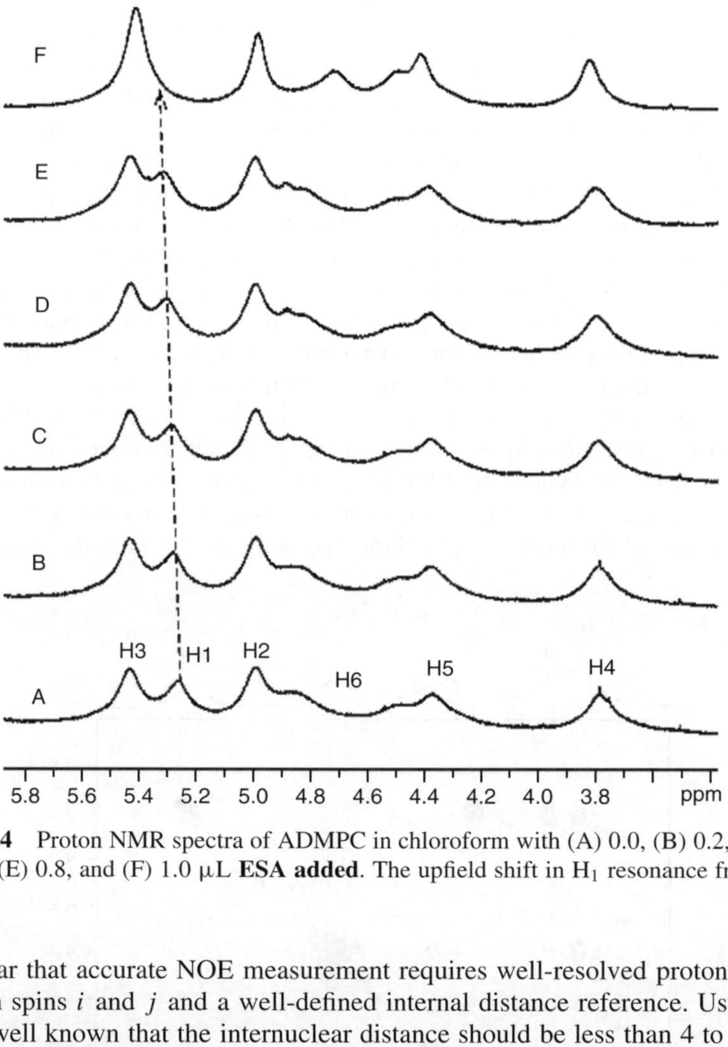

Figure 14 Proton NMR spectra of ADMPC in chloroform with (A) 0.0, (B) 0.2, (C) 0.4, (D) 0.6, (E) 0.8, and (F) 1.0 μL **ESA added**. The upfield shift in H_1 resonance frequency is noted.

It is clear that accurate NOE measurement requires well-resolved proton signals between spins *i* and *j* and a well-defined internal distance reference. Usually, it is also well known that the internuclear distance should be less than 4 to 5 Å for reasonable distance estimation.

In this case, the internuclear distance between protons H_1 and H_4 in the same glucose unit is larger then 4 Å; therefore, any cross-peak measured between protons H_1 and H_4 can be attributed to protons H_1 and H_4' from the consecutive units. Furthermore, the internuclear distances between H_2 and H_4 (2.63 Å) and H_3 and H_4 (2.98 Å) of amylose and amylose ester derivatives serve as a convenient internal distance reference. For a neat ADMPC solution, the proton H_1 and H_3 signals are well resolved, as demonstrated by Okamoto et al. [30]. With the addition of ESA, however, overlap between protons H_1 and H_3 (cf. Figure 14B) prevents the independent measurement of cross-peak volume that is related to proton H_1. To overcome this problem, an ESA titration experiment was conducted and the results are shown in Figure 14. As shown in Figure 14A to F, the chemical shift of H_1 shifts downfield as ESA is added, until overlap of signals H_1 and H_3 is observed. The ADMPC with the addition of 0.8 μL of neat ESA (Figure 14E) was chosen to conduct two-dimensional NOESY for the internuclear distance $H_1-H'_4$ estimations. At this concentration, a significant shift (0.1 ppm) of H_1 signal is observed, which indicates that sufficient ESA is in the solution to take effect in changes in the chemical shifts. In addition, the protons H_1 and H_3 are still resolvable for the NOE cross-peak measurements at the concentration.

For the purpose of calibration, a neat ADMPC solution is used for the NOESY measurement. Figure 15 demonstrates the results. A set of NOESY spectra with mixing time between 15 and 100 ms was recorded. Using internal distance references of H_{24} (2.63 Å) and H_{34} (2.98 Å), along with the initial NOE buildup slopes, the distance between H_1 and H'_4 was determined to be between 2.73 and 2.97 Å, which is comparable with those (2.83 to 2.96 Å) measured by Okamoto et al. [30]. With the same method and experimental conditions, the two-dimensional NOESY spectra of the ADMPC with ESA were recorded at a set of mixing times ranging from 15 to 100 ms. Typical NOESY spectra of a glucose section with a mixing time of 60 ms are shown in Figure 14. As anticipated, the cross-peaks H'_{14} and H_{34} are well resolved for independent volume measurements. Under these conditions

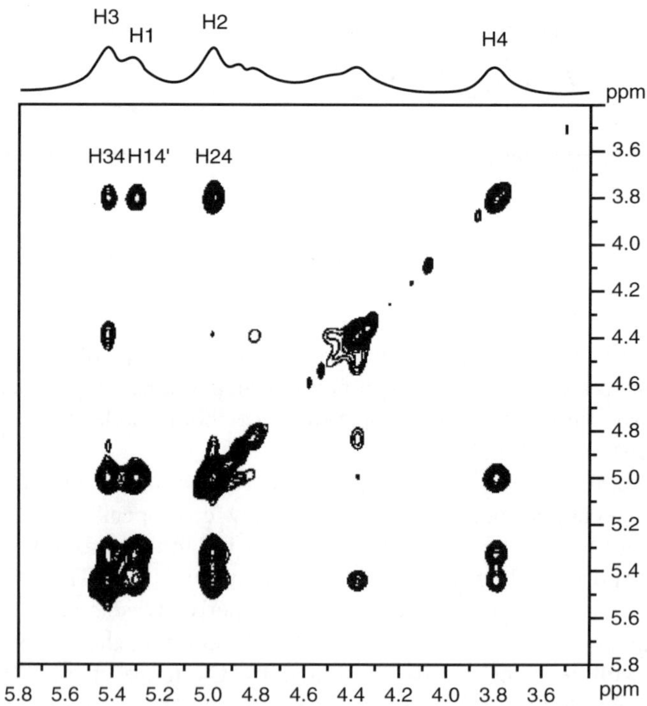

Figure 15 Glucose proton section of the NOESY spectrum of ADMPC in chloroform with the addition of ESA (0.8 μL). Resolvable H_{34} and H'_{14} cross-peaks are labeled in the spectrum.

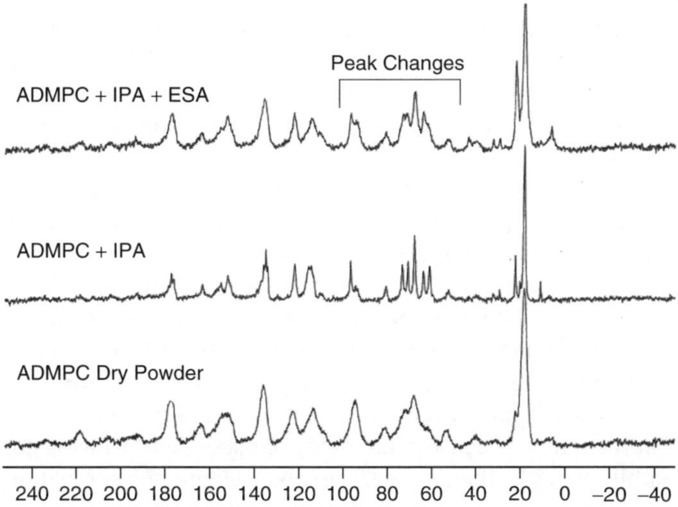

Figure 16 Solid-state NMR results with ESA addition.

and the same internal distance references, the internuclear distance between H_1 and H'_4 is estimated at between 2.72 and 2.93 Å. The distance data indicate that the internuclear distance H'_{14} of ADMPC does not alter in the presence of ESA. As mentioned before, this distance is the critical factor that determines the torsion of glycoside bond. Therefore, the tertiary structure of the ADMPC in solution did not change with the introduction of ESA. The NMR results reported seemed to be consistent with the LC results reported previously [56].

Solid-state NMR was also used to study the effect of ESA on ADMPC in our lab. Figure 16 shows the results. IPA is clearly incorporated into ADMPC polymer, and NMR peak narrowing and splitting were also observed, which agree with the results reported previously [50,51]. An interesting phenomenon was observed once ESA was added. The NMR peaks broadened with less splitting for ADMPC polymer in the ADMPC + IPA + ESA system compared to the ADMPC + IPA system. These results indicate that the crystallinity was decreased after ESA incorporation. The effect of the ESA is completely opposite to that of a polar solvent such as alcohol or ACN. Furthermore, the ESA is only 1% of the amount of IPA used in the experiment, but it almost completely reverses the IPA effect with regard to peak splitting and narrowing. Further study is needed to clarify these results.

3 CHIRAL DISCRIMINATION MECHANISM STUDY BY POLYSACCHARIDE-BASED CSPS

In the past decade, many attempts to clarify the mechanism of chiral recognition on CSPs for liquid chromatography have been made using computational methods. The most successful studies were performed with small-molecule CSPs; only a few mechanistic studies on chiral discrimination at the molecular level have been reported for polymeric CSPs. A number of different interaction sites with different affinities for enantiomers exist on chiral polymers. Analysis of their precise structures in both the solid state and in solution is not easy. This makes understanding the chiral recognition mechanism of polymeric CSPs difficult. The chiral recognition mechanism of the polysaccharide-based CSPs had been investigated based mostly on chromatographic methods, but it has not yet been elucidated satisfactorily at a molecular level.

As explained earlier, NMR is a powerful tool used to investigate the chiral recognition mechanism of CSPs, but most phenylcarbamates of the polysaccharides with high resolving abilities are soluble only in polar solvents such as pyridine, tetrahydrofuran (THF), and dimethylformamide (DMF). In such cases the interaction between the solvent and the phenylcarbamates is too strong to detect the interaction between the phenylcarbamates and a racemate. For these $CDCl_3$-insoluble phenylcarbamate derivatives, a computer simulation involving molecular mechanical and molecular dynamic calculations is a very useful and effective approach to elucidating the mechanism of chiral recognition, which might allow a prediction of the elution order of enantiomers.

Yashima et al. reported computational studies on the chiral discrimination mechanism of cellulose trisphenylcarbamate (CTPC) [48]. CTPC shows a high chiral resolving ability as a CSP for HPLC. The calculation of interaction energies between CTPC and enantiomers of (±)-*trans*-stilbene oxide and (±)-*trans*-1,2-diphenylcyclopropane was performed using Quanta/CHARMm and Molecular Interaction programs to gain an insight into the chiral recognition mechanism of the phenylcarbamate derivatives of cellulose. The structure of CTPC was optimized with the CHARMm force field based on the proposed structure of CTPC by x-ray analysis, as described in Section 2.1.

In chromatographic enantioseparation on CTPC, (±)-*trans*-stilbene oxide was resolved completely ($\alpha = 1.46$) and the (*R*,*R*)-(+)-isomer eluted first, followed by the (*S*,*S*)-(−)- isomer, but (±)-*trans*-1,2-diphenylcyclopropane was not resolved ($\alpha \sim 1$). The results of the calculation of interaction energies between CTPC and the enantiomers (±)-*trans*-stilbene oxide suggested that the most important adsorbing site of CTPC for (±)-*trans*-stilbene oxide may be the NH protons of the carbamate moieties at the 3-position of glucose units, and the (*S*,*S*)-(−)-isomer of (±)-*trans*-stilbene oxide may interact more closely than the (*R*,*R*)-(+)-isomer with CTPC. In contrast, there was little difference in the minimum interaction energies between the enantiomers (±)-*trans*-1,2-diphenylcyclopropane.

These calculations agreed with the results observed for chromatographic resolution on CTPC, although the difference in the interaction energy calculated between the two enantiomers of *trans*-stilbene oxide is large compared with that estimated by the chromatographic enantioseparation. This may be due to the uncertainty in the interaction energy values calculated, which may be dependent on the structure of CTPC.

The interaction energy calculations between CTPC/cellulose tris(3,5-dimethylphenylcarbamate) (CDMPC) and *trans*-stilbene oxide/benzoin were performed using various methods with different force fields [58]. In a chromatographic enantioseparation, *trans*-stilbene oxide is resolved completely on both CSPs ($\alpha = 1.46$ for CTPC CSP and $\alpha = 1.68$ for CDMPC CSP), with the reversed elution order of enantiomers; the (*R*,*R*)-isomer elutes first on CTPC CSP, and the (*S*,*S*)-isomer on CDMPC CSP. On the other hand, benzoin is separated completely on CDMPC CSP ($\alpha = 1.58$) but is not separated on CTPC CSP. The results of the calculations matched the chromatographic resolution data for *trans*-stilbene oxide and benzoin. The interaction energy between CTPC and (*S*,*S*)-*trans*-stilbene oxide obtained in the calculation was lower than that between CTPC and (*R*,*R*)-*trans*-stilbene oxide, whereas an opposite enantiomer preference was observed for CDMPC and *trans*-stilbene oxide. Although the same calculation was carried out for the benzoin and CTPC complex, almost no difference in the interaction energies was observed for the enantiomers. The clear interaction energy differences between the enantiomers was recognized only when the enantiomers were generated inside CTPC and CDMPC, indicating that the polar carbamate residues of the cellulose carbamates may be the most important adsorbing and chiral recognition site for polar racemates.

The (*S*,*S*)- *trans*-stilbene oxide is bound in the chiral groove to form a hydrogen bond between the ether oxygen of *trans*-stilbene oxide and the NH proton of the carbamate group of CTPC. Each phenyl group appears to interact with the phenyl groups of CTPC through $\pi-\pi$ interactions. In the calculations above, the interaction energies are calculated between a single polysaccharide chain and an enantiomer because polar racemates can preferentially come into the chiral groove of the polymer chain with the polar carbamate residues. Besides these polar interactions, hydrophobic interactions such as the $\pi-\pi$ interaction between the phenyl group of the phenylcarbamate derivatives and the aromatic groups of an enantiomer may contribute to the chiral recognition, especially under reversed-phase conditions. In this case, the hydrophobic chiral space created by the phenyl moieties of the phenylcarbamates may also play an important role in chiral recognition.

NMR is well established as the most powerful tool for the elucidation of chiral recognition mechanisms. The changes in chemical shifts and line widths as well as signal splitting of the enantiomers in the proton NMR spectrum provide information about which protons are affected by the interaction between the chiral selector and each enantiomer. In addition, two-dimensional NOESY measurements reveal proximities of protons in the CSP and the analyte to provide structural information for the CSP–analyte complex. These two tools have been used to study chiral recognition for polysaccharide-based CSPs with different analytes.

One example is reported by Okamato's group [9]. Chromatographic separation of 1,1′-bi-2-naphthol has been performed on cellulose tris(5-fluoro-2-methylphenylcarbamate) as a chiral stationary phase for HPLC. The complete

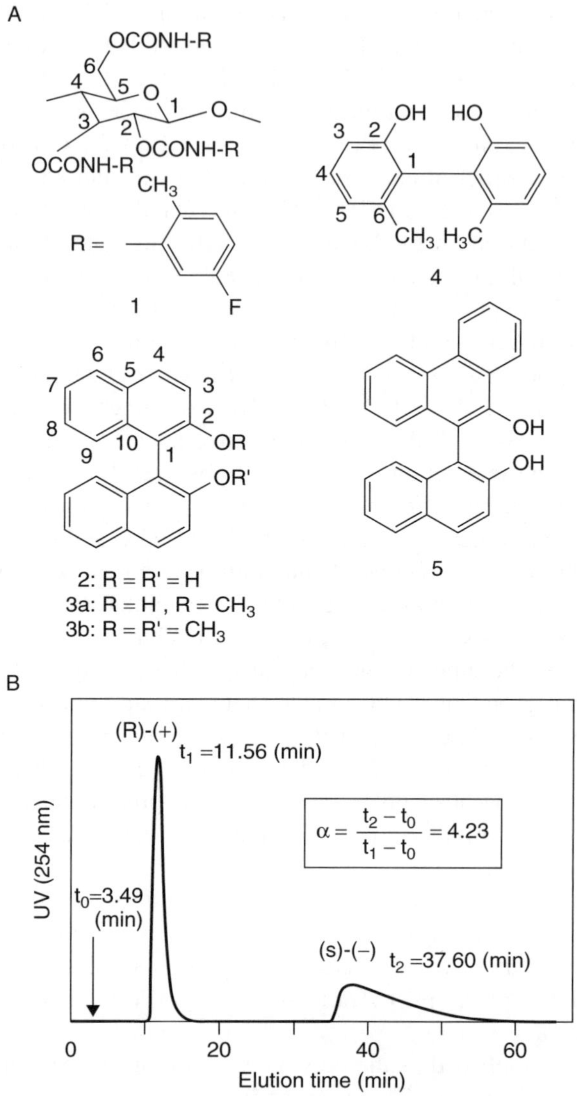

Figure 17 (A) Structures used in the study. (B) Chromatogram of the enantioseparation of (R, S)-1,1′-bi-2-naphthol with hexane-2-propanol (90 : 10) as the eluent at 20°C. Column, 25 × 0.46 cm ID; flow rate, 1.0 mL/min.

baseline separation of 1,1′-bi-2-naphthol was achieved with the elution order with (R)-enantiomer eluted first, followed by the (S)-enantiomer (Figure 17).

The cellulose derivative CSP dissolved in chloroform also exhibited a chiral discrimination for 1,1′-bi-2-naphthol in ^{1}H and ^{13}C NMR spectroscopies as well as in HPLC. The hydroxy and some aromatic protons and carbon resonances of 1,1′-bi-2-naphthol were clearly separated into a pair of peaks, due to enantiomers

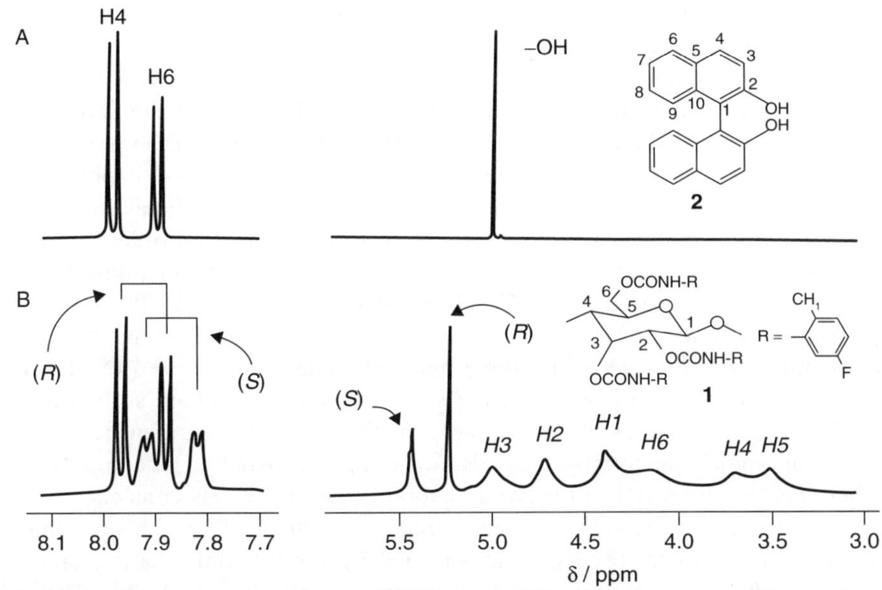

Figure 18 ^{1}H-NMR spectra of a selected region of (*R*, *S*)-2 (19.4 mM) in the absence (A) and presence (B) of 1 (36.1 mM) glucose unit) in $CDCl_3$ at 23°C. The assignments were performed with two-dimensional COSY and NOESY experiments.

in the presence of CSP polymer, as shown in Figure 18. Significantly, each of the hydroxy and naphthyl protons (H_4 and H_6) of the enantiomers of 1,1′-bi-2-naphthol was separated into two peaks in the presence of CSP polymer. It was found that the hydroxy protons [(*S*)-2-OH] were shifted more largely downfield, accompanied by line broadening, than was the corresponding (*R*)-2-OH, whereas the (*S*)-2-H_4 and (*S*)-2-H_6 resonances exhibited upfield shift and broadening, indicating that (*S*)-1,1′-bi-2-naphthol interacts more strongly with CSP polymer. The downfield shift of the OH resonances is ascribed to hydrogen-bond effects, and the upfield shifts for the aromatic protons of 1,1′-bi-2-naphthol are probably due to a π-stacked or shielding effect by a neighboring aromatic ring of CSP polymer. The significant broadening of these proton resonances of (*S*)-1,1′-bi-2-naphthol indicates that exchange rates between the free and bound forms of 1,1′-bi-2-naphthol to CSP polymer is slow compared with that on the NMR time scale. The larger chemical shift movement and broadening of the (*S*)-1,1′-bi-2-naphthol resonances than (*R*)-1,1′-bi-2-naphthol observed with ^{1}H-NMR is associated with the chromatographic elution order of the enantiomer. Similar splittings into enantiomers in the presence of CSP polymer were also observed in ^{13}C NMR spectroscopy.

For both enantiomers, the association constants with the CSP polymer were determined based on complexation-induced shifts; the corresponding $\Delta\Delta G_0$ value was also calculated. A comparison of the $\Delta\Delta G_0$ values derived from

HPLC measurements, however, revealed that the $\Delta\Delta G_0$ value obtained by NMR was substantially larger.

The binding geometry and dynamics between CSP polymer and the enantiomers of 1,1′-bi-2-naphthol were investigated on the basis of the spin-lattice relaxation time, ^{1}H-NMR titrations, and intermolecular NOEs in the presence of CSP polymer. Clear NOE signals were observed between 1,1′-bi-2-naphthol and CSP polymer, which provided the foundation for the computational study. The NMR data are fully consistent with the chromatographic elution order.

The initial model of CSP polymer was constructed using three-dimensional periodic boundary conditions in CERIUS2 starting from the CTPC, which is reported to have a left-handed threefold (3/2) helical structure, and then was energy-minimized by a Dreiding force field. The energy-minimized (*S*)-1,1′-bi-2-naphthol was placed manually in the groove of the main chain so that all of the NMR data, including the intermolecular NOEs and the titration results, as well as intermolecular hydrogen bonds, are visually satisfied. The complex was further energy minimized to relieve unfavorable van der Waals contacts.

Figure 19 shows the lowest-energy structure of the 1-(*S*)-1,1′-bi-2-naphthol complex. The polymer possesses a left-handed 3/2 helical conformation, and the glucose residues are arranged regularly along the helical axis. A chiral helical groove, or ditch, with polar carbamate residues exists along the main chain. The polar carbamate groups are located inside, and hydrophobic aromatic groups are placed outside the polymer chain so that polar enantiomers can get in the groove to interact with the carbamate residues via hydrogen-bonding formation.

Although chromatographic retention behavior for (*R*)-1,1′-bi-2-naphthol in HPLC and NMR data is available, it is difficult to propose an analogous model for the ADMPC polymer–(*R*)-1,1′-bi-2-naphthol complex because of lack of NOE data. No computational study was performed for the (*R*)-enantiomer. The HPLC, NMR, and computational results illustrated that (*S*)-1,1′-bi-2-naphthol binds site-selectively to the phenylcarbamoylated cellulose derivative CSP polymer through multiple interactions, including intermolecular hydrogen bonding and $\pi-\pi$ and/or CH$-\pi$ interactions to afford a 1 : 1 complex. Molecular modeling on the basis of chromatography and NMR data reveals the chiral discrimination rationale for a cellulose phenylcarbamate derivative.

Chiral interactions between ADMPC and a series of analytes were also investigated by Okamato's group [30]. The $CHCl_3$ was used for both HPLC and NMR studies, which facilitated direct comparison of the results in HPLC with those in NMR. HPLC results using chloroform as the eluent shows that (*S*)-1-(9-anthryl)-2,2,2-trifluoroethanol interacts more strongly with ADMPC than does (*R*)-1-(9-anthryl)-2,2,2-trifluoroethanol. The ^{1}H-NMR titrations of ADMPC with (*S*)- and (*R*)-1-(9-anthryl)-2,2,2-trifluoroethanol were performed to obtain information with respect to the binding sites of ADMPC in the complexation. The H_1, H_2, and H_4 proton resonances of a glucose unit of ADMPC (Figure 3 shows the H assignment) were affected significantly by the addition of (*S*)-1-(9-anthryl)-2,2,2-trifluoroethanol and shifted upfield, whereas the other glucose proton resonances moved only slightly. The significant upfield shifts of the three proton resonances

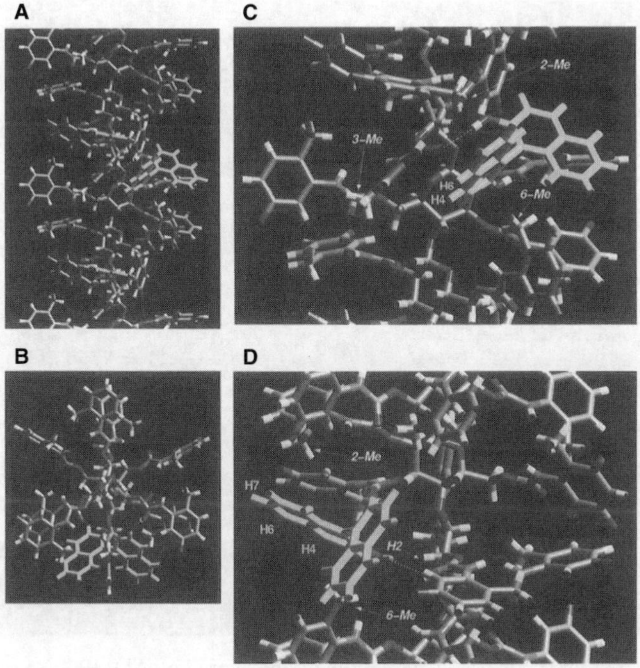

Figure 19 Computer-generated depiction of the complex of 1-(*S*)-1,1′-bi-2-naphthol. The (*S*)-1,1′-bi-2-naphthol is shown in yellow. Dashed lines correspond to hydrogen bonds. (A) Configuration: along the helix axis and (B) perpendicular to the chain axis. (C and D) Expanded region of the same structural model showing the interactions of (*S*)-1,1′-bi-2-naphthol with CSP polymer.

indicate that an anthryl ring of (*S*)-1-(9-anthryl)-2,2,2-trifluoroethanol may be located closely above the H_1, H_2, and H_4 protons on the same side of the glucose ring.

The two-dimensional NOESY spectra of the ADMPC–(*S*)-1-(9-anthryl)-2,2,2-trifluoroethanol complex were measured under various conditions, and clear intermolecular NOEs for the complex have not been observed. A precise model for the complex was not proposed. Only a simple computational minimization was performed by placing (*S*)-1-(9-anthryl)-2,2,2-trifluoroethanol manually in the groove of the main chain of ADMPC using the ^{1}H-NMR titration results, and the intermolecular hydrogen bonds were visually satisfied. The complex was energy-minimized further to relieve the unfavorable van der Waals contacts. Figure 20 shows the lowest-energy structure of the ADMPC–(*S*)-1-(9-anthryl)-2,2,2-trifluoroethanol complex. The OH proton of (*S*)-1-(9-anthryl)-2,2,2-trifluoroethanol forms hydrogen bonding with the carbonyl oxygen of the carbamate group at the 2-position. The distance between the hydrogen and oxygen is 1.968 Å, which is short enough for hydrogen bonding. The anthryl ring is positioned favorably above the H_1, H_2, and H_4 protons, and

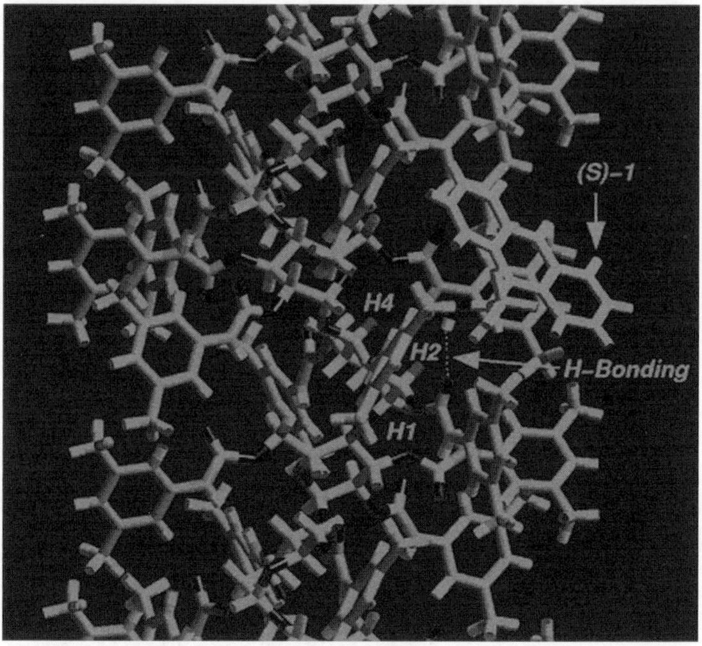

Figure 20 Computer-generated depiction of the complex ADMPC–(*S*)-1-(9-anthryl)-2,2,2-trifluoroethanol [(*S*)-1 in the figure]. The glucose carbon atoms and the H_1, H_2, and H_4 protons of the glucose residue of ADMPC are shown in green and orange, respectively. The dashed line corresponds to a hydrogen bond.

the interaction model explains the upfield shift of the H_1, H_2, and H_4 protons of the glucose residue in the presence of (*S*)-1-(9-anthryl)-2,2,2-trifluoroethanol.

The interaction between ADMPC and *O*-*tert*-butyl tyrosine allyl ester by NMR and computational study was reported recently [57]. An acidic additive (ESA) played a key role in the separation. A significantly improved selectivity was observed with ESA addition in the HPLC mobile phase [56]. Therefore, an NMR study both with and without ESA was performed. The interactions of each ester enantiomer with ADMPC polymer were studied by obtaining ^{1}H-NMR spectra for each ester in the presence of ADMPC polymer in $CDCl_3$ solution. Figure 20 shows the results of ^{1}H-NMR spectra of ADMPC polymer and D and L enantiomers without ESA added. Figure 21A is an ^{1}H-NMR spectrum of *O*-*tert*-butyl tyrosine allyl ester without ADMPC polymer; Figure 21B is a ^{1}H-NMR spectrum of the D enantiomer of *O*-*tert*-butyl tyrosine allyl ester with ADMPC polymer; Figure 21C is a ^{1}H-NMR spectrum of the L enantiomer of *O*-*tert*-butyl tyrosine allyl ester with ADMPC polymer; and Figure 21D is ADMPC polymer without *O*-*tert*-butyl tyrosine allyl ester added. The peaks are much broader for the polymer, as expected, due to its slower rate of tumbling. The results indicate no discernible interaction between the enantiomers or the chiral polymer in the absence of ESA. These results agree with the HPLC results (Figure 12).

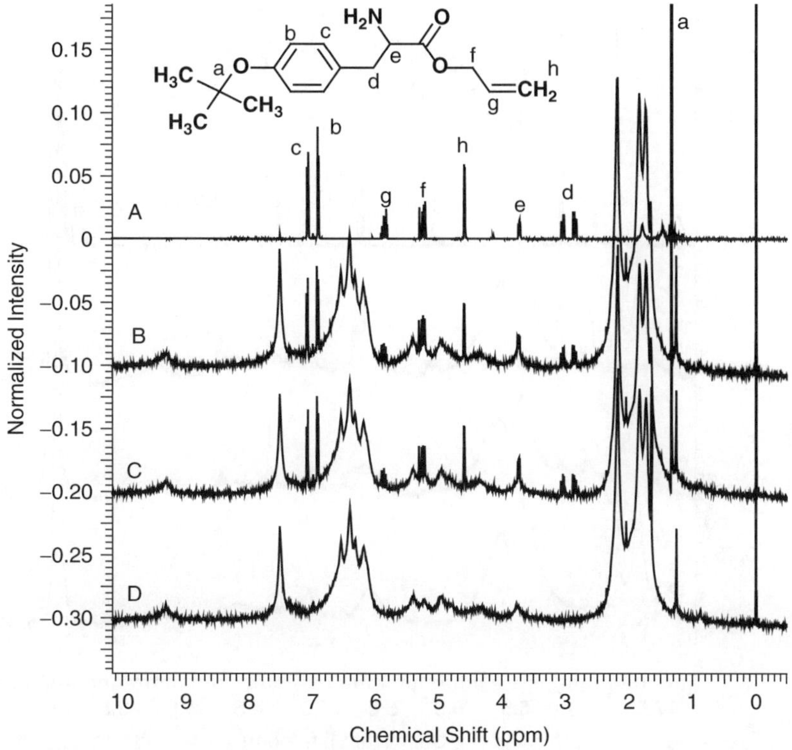

Figure 21 ^{1}H-NMR spectrum of ADMPC polymer with D- and L-enantiomers without ESA. The concentration of the ADMPC polymer is 22 mg/mL in $CDCl_3$. The concentration for both D and L enantiomers is 1 mg/mL in $CDCl_3$. (A) ^{1}H-NMR spectrum of *O-tert*-butyl tyrosine allyl ester without ADMPC polymer; (B) ^{1}H-NMR spectrum of the D-enantiomer of *O-tert*-butyl tyrosine allyl ester with ADMPC polymer; (C) ^{1}H-NMR spectrum of the L-enantiomer of *O-tert*-butyl tyrosine allyl ester with ADMPC polymer; and (D) ADMPC polymer without *O-tert*-butyl tyrosine allyl ester added.

Figure 22A shows the ^{1}H-NMR spectra of the analyte, *O-tert*-butyl tyrosine allyl ester in $CDCl_3$ in the presence of ethanesulfonic acid. The structure of the analyte is shown at the top, and the spectral assignments are shown with reference to the labels on the structure. The peaks are sharp, as one would expect for a small molecule in liquid. Figure 22D shows the spectrum for ADMPC in $CDCl_3$ solution and in the presence of ESA. The peaks are much broader for the polymer, as expected, due to its slower rate of tumbling. The assignments, provided with reference to the structure of the monomer unit of ADMPC, are the same as those in Figures 3 and 14. Figure 22B and C show the NMR spectra for the D- and L-enantiomers, respectively, in the presence of the ADMPC and ESA. These show that the peaks for the D-enantiomer broadened slightly in the presence of polymer, while those for the L-enantiomer broadened to a greater degree. This broadening of the peaks of the enantiomers in the presence of ADMPC is

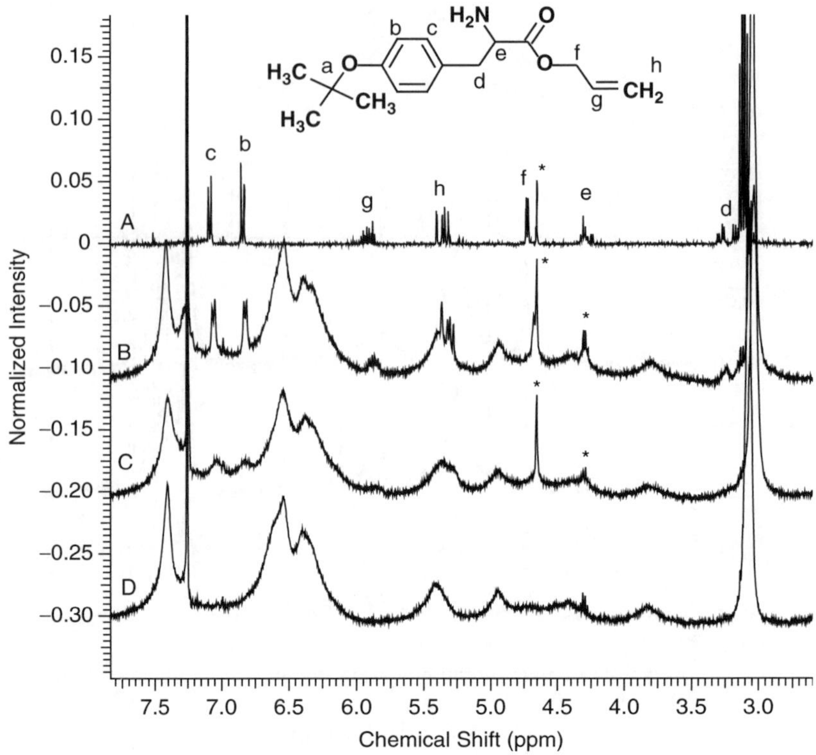

Figure 22 ^{1}H NMR spectra for samples indicated, each with ESA added. (A) *O-tert*-butyl tyrosine allyl ester alone; (B) D enantiomer of *O-tert*-butyl tyrosine allyl ester plus ADMPC polymer; (C) L-enantiomer of *O-tert*-butyl tyrosine allyl ester plus ADMPC polymer; and (D) ADMPC polymer alone. Asterisks represent impurity peaks.

attributed to interactions with the polymer, which slow the rate of tumbling of the small esters. The much greater broadening for the L enantiomer indicates much greater interaction with the polymer. In the absence of ethanesulfonic acid, there was no discernible broadening in the NMR spectrum of either enantiomer with ADMPC. Both of these results are consistent with the chromatography: much greater retention for the L enantiomer, and the necessity of acid to achieve high enantioselectivity (Figure 12).

Figure 22 reveals an upfield shift for the H_3 and H_5 protons of ADMPC glucose in the presence of each enantiomer, and the shift is larger for the L-enantiomer. The shifts for H_2 and H_4 of the ADMPC glucose are negligible. This indicates that both enantiomers have shielding effects for the glucose H_3, demonstrating that both the D- and L-enantiomers interact with ADMPC primarily on the H_3 and H_5 side of glucose, since H_2 and H_4 are on the opposite side of the sugar ring. The larger shift for the L-enantiomer indicates that, on average, it is closer to H_3 and H_5 than is the D-enantiomer. Interestingly, these results are contrary to the results reported by Okomato's group described above for 1-(9-anthryl)-2,2,2-trifluoroethanol with ADMPC polymer [30]. In their study, the 1-(9-anthryl)-2,2,2-trifluoroethanol was on the H_2 and H_4 side of the ADMPC glucose chiral pocket.

The two-dimensional NOESY spectra collected and presented at exactly the same conditions for D- and L-enantiomers in the presence of ADMPC with ESA are shown in Figure 23 in an overlaid mode. As expected, negative NOEs were observed for ADMPC as well as those between each of the enantiomers and ADMPC, and these are attributable to slowed molecular tumbling. The major difference between the D- and L-enantiomers involves the phenyl protons, cH and bH, of the tyrosine moiety. For the D-enantiomer, intermolecular NOEs

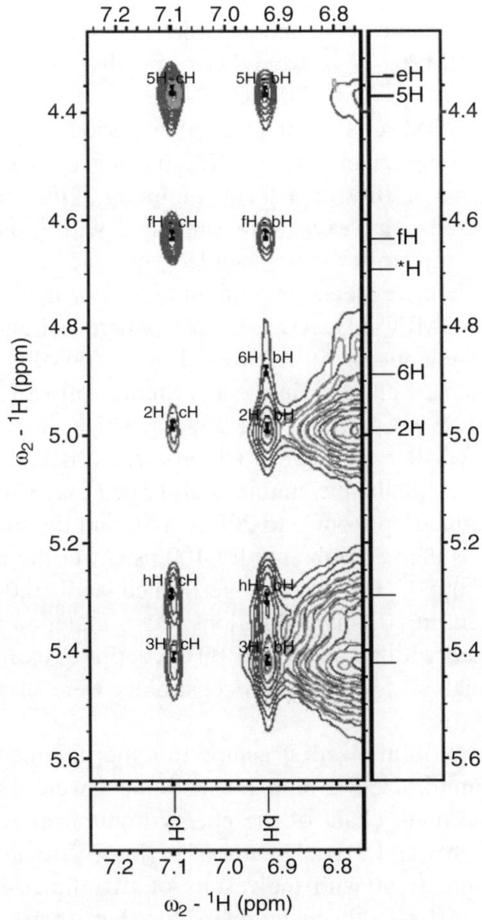

Figure 23 Two-dimensional NOESY NMR spectra of the ADMPC and ESA mixed with the L-enantiomer (red) and D-enantiomer (green). (*See insert for color representation of the figure*.)

were observed only between the cH of the tyrosine phenyl ring and the 5H of the AMDPC sugar ring, as well as the bH and 3H of these same moieties, which indicates that the phenyl and glucose rings are in close proximity. For the L-enantiomer, more intermolecular NOEs were observed for even more internuclear pairs between its tyrosine phenyl protons and the glucose protons of ADMPC, bH–2H, bH–3H, bH–5H, cH–2H, cH–3H, and cH–5H. It is apparent that phenyl and glucose rings are closer, on average, for the L-enantiomer than for the D-enantiomer. In addition to the ^{1}D proton NMR results discussed previously, the NOE data presented here provide direct experimental evidence that shows preferences of L- and D-enantiomers toward ADMPC, specifically the aromatic protons of the *O-tert*-butyl tyrosine ester with the glucose protons of ADMPC. These distance data provide a valuable starting point for molecular modeling studies.

The molecular mechanics calculations were performed using the COMPASS force field for the ADMPC–*O-tert*-butyl tyrosine allyl ester complex. A periodic box (40 Å, 40 Å, 50 Å) for the ADMPC 12-mer was built, and this was bonded to the neighboring ADMPC 12-mer in another periodic box with a 1,4′ sugar linkage to form one long-chain polymer. The quenched annealing approach was used to reduce the possibility of a local minimum. With quenched annealing, the initially minimized complexes were simulated with molecular dynamics at 500 K, 20 different trajectories were sampled, and each conformation was then minimized to find the lowest energy configuration. For the MD simulations, the energy-minimized ADMPC-LH, ADMPC-DH (where LH and DH represent the protonated L- and D-enantiomers), ADMPC-L, and ADMPC-D complexes were simulated using Discover building in the MS Studio software of Accelrys. These dynamic simulations were performed at 298 K, which is the same temperature as that used in the NMR experiments. Chloroform was included by setting the dielectric constant to 4.8. In the initial equilibrated stage, the minimized four complexes were warmed up from 0 to 298 K first, and the first MD was done in 1 ps. The system was then equilibrated for 100 ps. After the initial equilibration, a production MD run of 100 ps was performed and 10,000 structures were sampled uniformly during 0.1-ns simulations. Postsimulation analysis of the MD results was performed with an additional BTCL script to extract different energy terms for further analysis. No distance constraints were used in the molecular dynamics simulations.

To compute energy-minimized structures for the polymer–enantiomer complexes, the energy-minimized D- and L-enantiomers were each placed initially in the groove of the main chain of the energy-minimized ADMPC so that the positions of the 3H–bH and 5H–cH pairs of protons were about 5 Å apart. This initial position is consistent with the results of two-dimensional NMR, which showed NOE signals for both enantiomers for these pairs of protons, and it treated each enantiomer similarly. Initially, the amino group for each enantiomer was directed away from the ADMPC. A distance constraint subset was created for the molecular mechanics minimization to constrain the 5H–cH and

3H–bH distances to be within 5 Å for both enantiomers, in accord with the two-dimensional NOESY spectra. No distances other than the 5H–cH and 3H–bH protons were constrained. The structures of the complexes of ADMPC with the L- or D-enantiomer were then minimized with respect to energy using the quenched annealing method.

Figure 24 shows the energy-minimized structures of the ADMPC-LH and ADMPC-DH complexes. The ADMPC structure changed slightly upon binding for each enantiomer to achieve minimum energy, which was expected. The ADMPC structure upon binding each enantiomer is indicated in Figure 24 with a

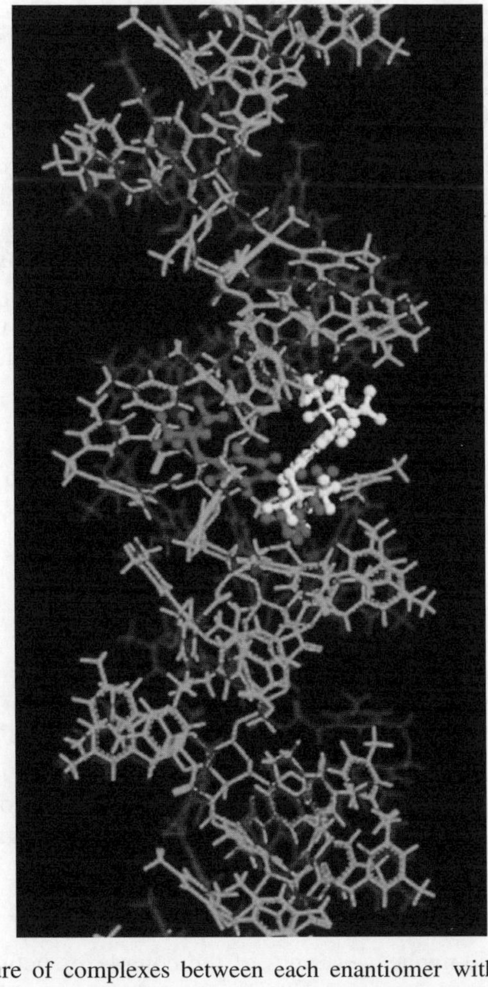

Figure 24 Structure of complexes between each enantiomer with the chiral ADMPC polymer. The enantiomers perturb the structure of the polymer differently. The images shows the L-enantiomer (red) interacting with ADMPC polymer (blue), and the D-enantiomer (yellow) interacting with the ADMPC polymer (green). (*See insert for color representation of the figure*.)

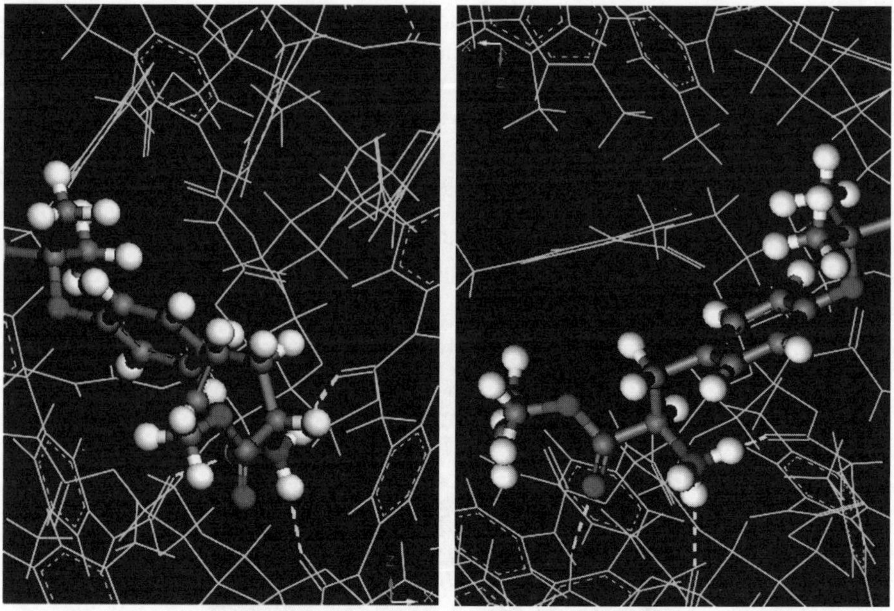

Figure 25 Close-up views of each enantiomer interacting with the chiral ADMPC polymer: (A) L-enantiomer and (B) D-enantiomer. The yellow dashed lines show the three hydrogen-bonding interactions between the enantiomer and the polymer with the polymer. (*See insert for color representation of the figure*.)

different color to distinguish them. The structures of the complexes show that the two enantiomers bind quite differently to ADMPC. The L-enantiomer is aligned lengthwise within the chiral groove of the polymer, whereas the D-enantiomer is aligned across the chiral groove. Since protonation has such a large effect on chiral discrimination, the hydrogen bonding is of particular interest. Figure 25 shows the structures of the two complexes in close-up detail to indicate hydrogen bonding. For the L-enantiomer, three hydrogen bonds are formed between the protonated amino group and three carbamate oxygens in this minimum-energy structure to provide strongly favorable interaction energy. Two hydrogen bonds are formed between the protonated amino group of the D-enantiomer and two carbamate oxygen groups, and one more hydrogen bond is formed between the ester oxygen of the D-enantiomer and the carbamate –NH group of the ADMPC. Both enantiomers thus have considerable hydrogen-bonding interaction with ADMPC. How much these interactions impart enantioselectivity can be judged from the energies computed.

The minimized component energies of the complexes between ADMPC and each protonated enantiomer are summarized in Table 2, in which the differences in minimized energies between ADMPC-DH and ADMPC-LH complexes are given by $\Delta E_{\text{L-D}}$. These data show that binding to either enantiomer increases the internal energy of the system, which is largely a consequence of the structural

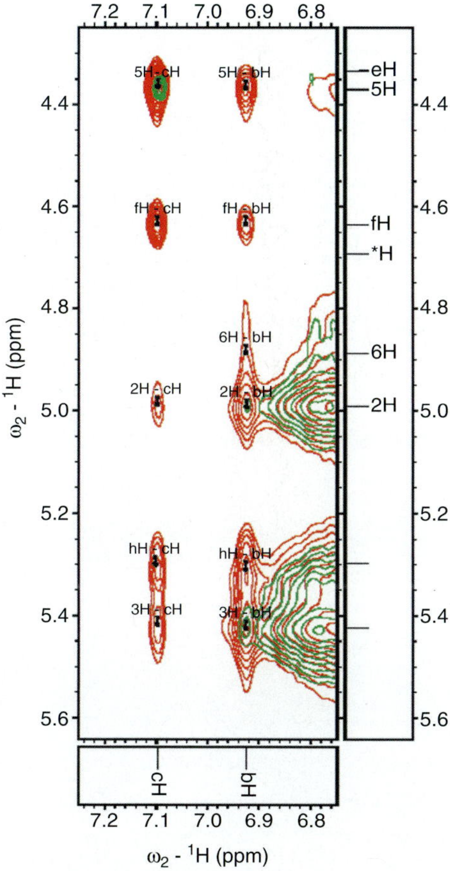

Chapter 6, Figure 23 See page 179.

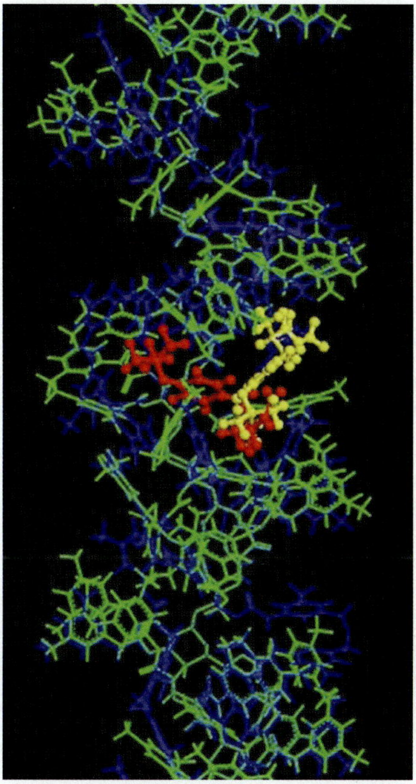

Chapter 6, Figure 24 See page 181.

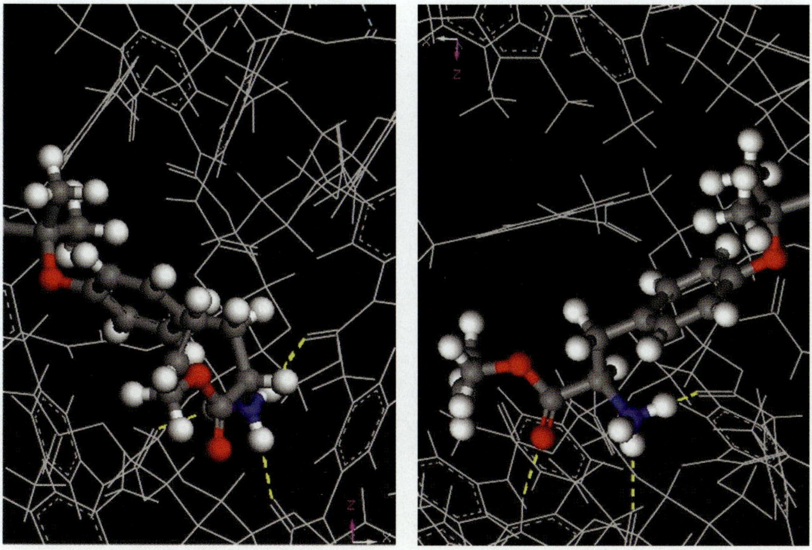

Chapter 6, Figure 25 See page 182 for full caption.

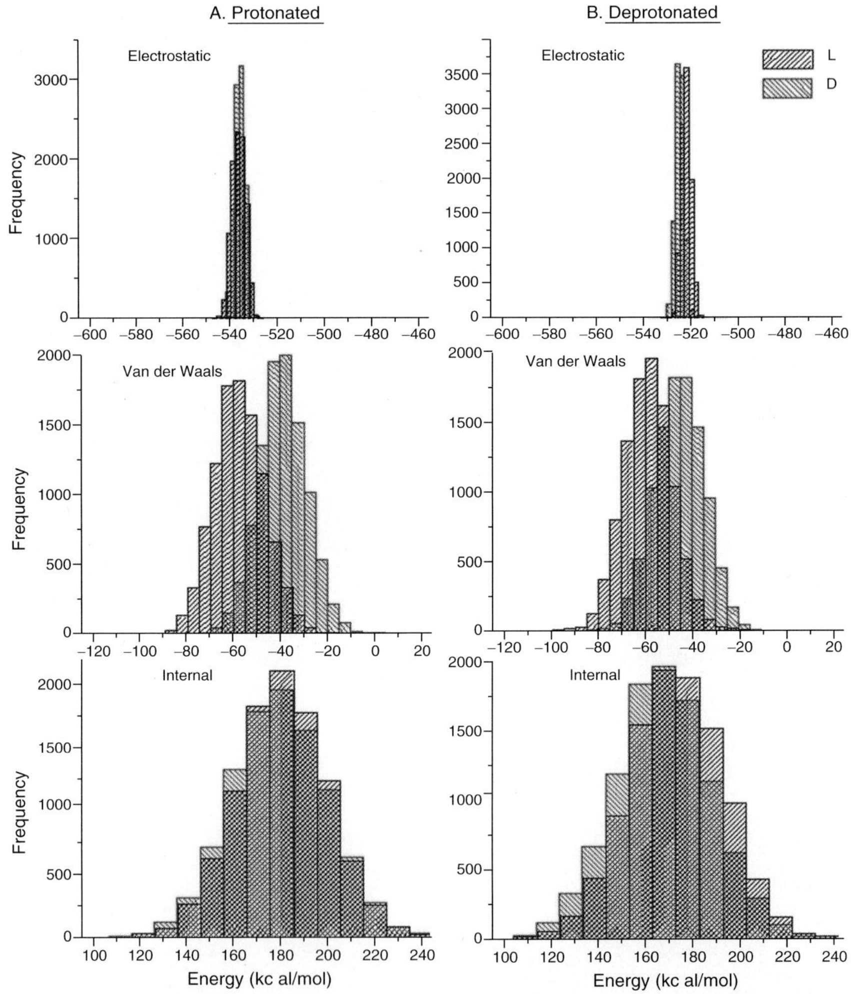

Chapter 6, Figure 26 Component energy distributions from MD simulations of the ADMPC complex with each enantiomer at 298 K in chloroform, after 100 ps of equilibration time. (A) Distributions for the case of the protonated enantiomer; (B) distributions upon deprotonation, starting with the energy-minimized position for the protonated enantiomer. Each plot ranges over 150 kcal/mol.

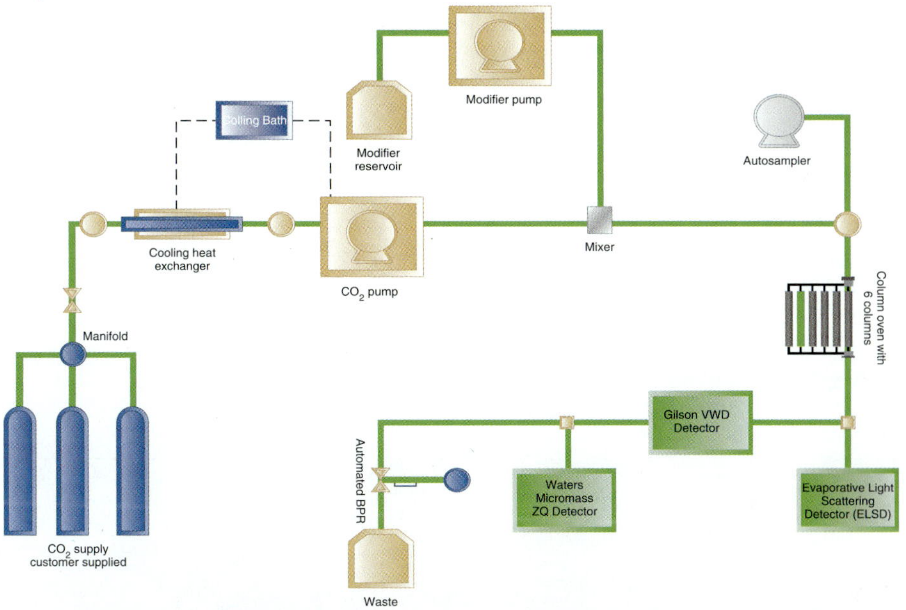

Chapter 10, Figure 3 See page 303 for full caption.

Chapter 15, Figure 1 See page 444 for full caption.

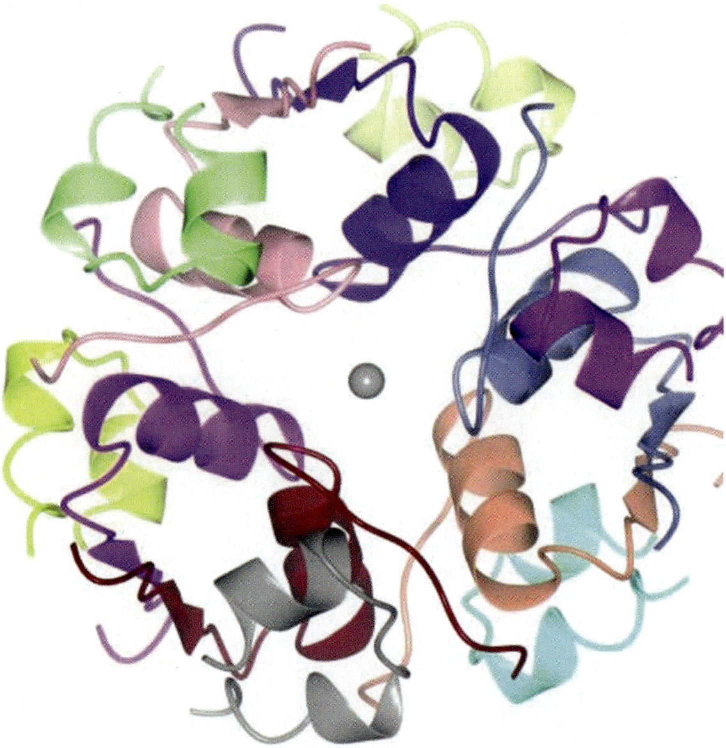

Chapter 15, Figure 5 Insulin hexamer. Each of the 12 chains is shown in a different color.

TABLE 2 Potential and Component Energies Obtained from Simulations for ADMPC with D- and L-Enantiomer Interactions (kcal/mol)

Energy Components	ADMPC-D	ADMPC-L	ΔE
Total energy	−1207.50	−1212.90	−5.30
Internal	−563.70	−553.30	10.40
Bond	121.70	121.50	−0.20
Angle	200.10	192.60	−7.50
Torsion	−604.50	−589.90	14.60
Out-of-plane	2.80	3.00	0.20
Cross	−283.60	−280.50	3.10
Nonbond	−641.70	−657.40	−15.70
Vdw	−81.90	−96.30	−14.40
Repulsive	2114.60	2156.00	41.40
Dispersive	−2196.40	−2252.30	−55.90
Electrostatic	−559.80	−561.10	−1.30

perturbation of the polymer by binding to an enantiomer. The effect is greater for the LH-enantiomer, which is consistent with the expectation that it binds more tightly. The van der Waals interactions are more favorable upon binding to either enantiomer, and the effect is greater for the LH enantiomer. This is again consistent with the expectation that LH binds more tightly. The electrostatic interaction is favorable upon binding either enantiomer, and the effect is nearly equal. The energy-minimized structures thus point to van der Waals interactions as being the primary enantioselective interactions.

One caveat in studying the energy-minimized structures is that these could be entropically very unfavorable (i.e., they could be improbable compared to structures of only slightly higher energies that are well within kT). It is also possible that the lack of solvent could skew the results. MD simulations using minimized structures as a starting point are useful for introducing solvents and exploring the entropy through thermal distributions of positions. MD simulations were run for 100 ps, and after preequilibration over 100 ps, beginning at the energy minimum for each protonated enantiomer. The average component energies for each reactant (the polysaccharide and analyte) are summarized in Table 3A, and the component energies for the complexes of each enantiomer are summarized in Table 3B. The effect of solvent and room temperature on the polysaccharide itself is to greatly increase the internal energy, as shown by comparing Table 3 (quenched annealing) with Table 3A (ADMPC). The van der Waals interactions are reduced significantly with solvent and temperature, while the electrostatic interactions are reduced only moderately. For the complexes, comparing Table 1 with Table 3B shows that the general trends remain, in which the van der Waals interactions still contribute the main difference in energy between the two enantiomers. This conclusion holds no matter how the data are viewed, including the overall reaction energies, listed in Table 3C, which again show that the differences between D and L are in the van der Waals interactions.

TABLE 3 Potential and Component Energies Obtained from MD Simulations for ADMPC with D- and L-Enantiomer Interactions (kcal/mol)

	A. Reactants		B. Complexes			C. Energies of Reaction		
	ADMPC	LH	ADMPC-DH	ADMPC-LH	$\Delta E_{D\text{-}L}$	ADMPC-DH Formation	ADMPC-LH Formation	$\Delta E_{L\text{-}D}$
Total energy	−416.7	44.1	−390.9	−409.4	−18.5	−18.3	−36.8	−18.5
Internal	148	34.9	183.9	185	1.1	1	2.1	1.1
Bond	382.3	17.2	400.6	397.4	−3.2	1.1	−2.1	−3.2
Angle	471	29	508	497.1	−10.9	8	−2.9	−10.9
Torsion	−485.2	−6	−495.5	−485.3	10.2	−4.3	5.9	10.2
Out-of-plane	33.2	1	34.3	34.1	−0.2	0.1	−0.1	−0.2
Cross	−253.4	−6.2	−263.5	−258.3	5.2	−3.8	1.4	5.2
Van der Waals	−37.8	9.1	−39	−57.7	−18.7	−10.3	−29	−18.7
Repulsive	1887.1	62	1990.1	2001.3	11.2	41	52.2	11.2
Dispersive	−1924.9	−53	−2029.1	−2059	−29.9	−51.3	−81.2	−29.9
Electrostatic	−526.8	0.1	−535.8	−536.7	−0.9	−9	−9.9	−0.9

Histograms of the distributions of component energies provide an easier visual assessment than does a comparison of averages in Table 2. To obtain the thermal distributions of interaction energies for the two complexes, the energies of the configurations were sampled in the course of the molecular dynamics simulations to create histograms of the energy distributions. These histograms are shown in Figure 26A for the component energies of each ADMPC–enantiomer complex

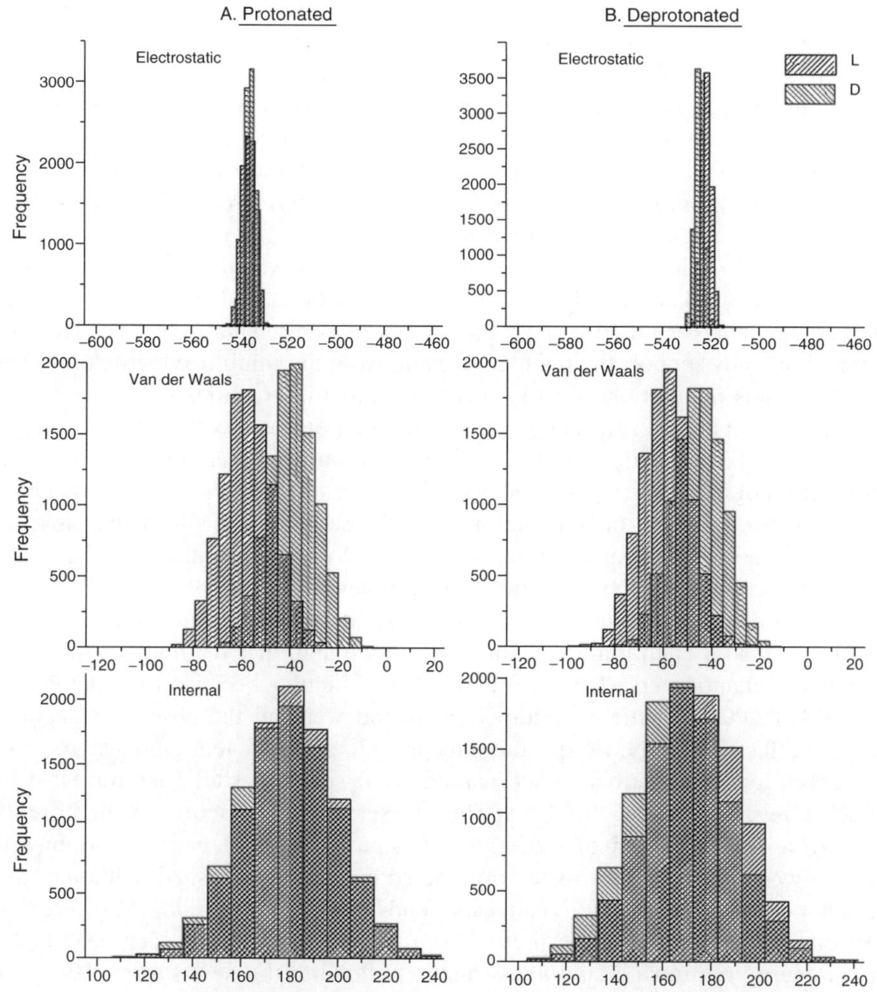

Figure 26 Component energy distributions from MD simulations of the ADMPC complex with each enantiomer at 298 K in chloroform, after 100 ps of equilibration time. (A) Distributions for the case of the protonated enantiomer; (B) distributions upon deprotonation, starting with the energy-minimized position for the protonated enantiomer. Each plot ranges over 150 kcal/mol. (*See insert for color representation of the figure.*)

for the case of protonated enantiomers, where the averages for these correspond to the first column of Table 3B. The histograms are plotted over a constant range of energies, 150 kcal/mol, to allow for an easy discernment of the relative enantioselectivities for each type of interaction. The histograms illustrate that there is a wide range of configurations greatly exceeding kT for a given component energy (i.e., internal, van der Waals, or electrostatic interaction energy). This indicates that the energy-minimized structures point to entropically reasonable candidates for the enantioselective sites. The distributions of internal energy for the D- and L-complexes are wide but virtually identical. The broad widths are attributed to the low energy of the individual internal modes. The electrostatic interactions remain strongly attractive, but the averages for the two enantiomers are almost identical in both magnitude and width. The narrow distribution is probably a consequence of the much higher energy of individual electrostatic interactions relative to kT. The distribution of energies corresponds to a distribution of structures, and in these, both enantiomers sometimes had their carbonyl groups interacting with NH_2 groups of the carbamate rather than the D-enantiomer only, which had been indicated by the minimized structure of Figure 24. The van der Waals interactions remain much weaker than the electrostatic interactions, but the large difference in average value for L and D show that these are very enantioselective. The MD simulations thus support the main conclusion from the minimized structures: The van der Waals interactions are the enantioselective interactions.

It may seem at first surprising that enatioselectivity occurs through differential van der Waals interactions rather than electrostatic interactions given that the protonation of the chiral compound acts virtually as a switch for turning on chiral selection. Specifically, no interactions were observed in NMR in the absence of ESA. Also, in the earlier chromatographic work, the addition of ESA had caused the chiral selectivity for the L-enantiomer to increase 12-fold, from 1.36 to 16 [56]. To explain this apparent paradox, it would be interesting to study the deprotonated enantiomers by the same computational methods as those for protonated enantiomers. This is fraught with difficulty because the interactions of the ADMPC with the enantiomers were too weak in the absence of acid to observe effects in the NMR spectra, thus providing insufficient guidance to serve as a check for computational identification of the selective site. Therefore, instead of claiming to represent the actual chirally selective interaction in the absence of acid, an investigation of the effect of deprotonation of each enantiomer in the energy-minimized sites of the protonated enantiomers ensued. Although this does not uncover the thermodynamics, it lends insight. MD simulations were thus performed after removing the proton from the amino group of each enantiomer in the energy-minimized complex structure, allowing the new system 100 ps to reequilibrate, and then monitoring energies for 100 ps. The resulting distributions are shown in Figure 26B. The energy distributions reveal that the primary effect of the deprotonation is to shift the electrostatic distribution to less favorable energies, but the effect is equal for the two enantiomers. The shift owes to the decreasing electrostatic interaction when the amino group is uncharged, and this affects D and L equally. The more insightful change upon deprotonation is that van der

Waals interactions become less selective. The van der Waals interactions of the L-enantiomer are minimally affected, but the D-enantiomer now begins to have more favorable van der Waals interactions upon deprotonation. A possible reason might be that the reduced influence of the electrostatic interaction, which constrained the position of the amino group, now allows the D-enantiomer to position itself for more favorable van der Waals interactions. Overall, for enantioselectivity in the presence of ESA, the results show that whereas the electrostatic interaction drives complexation, the van der Waals interactions direct the enantioselectivity.

The MD simulations also report on the positions of the protons in the thermal distribution, which allows a check for consistency with the two-dimensional NMR results. Figure 27 shows how the pair distribution functions for the same proton pairs were observed to couple in the two-dimensional NOESY measurements.

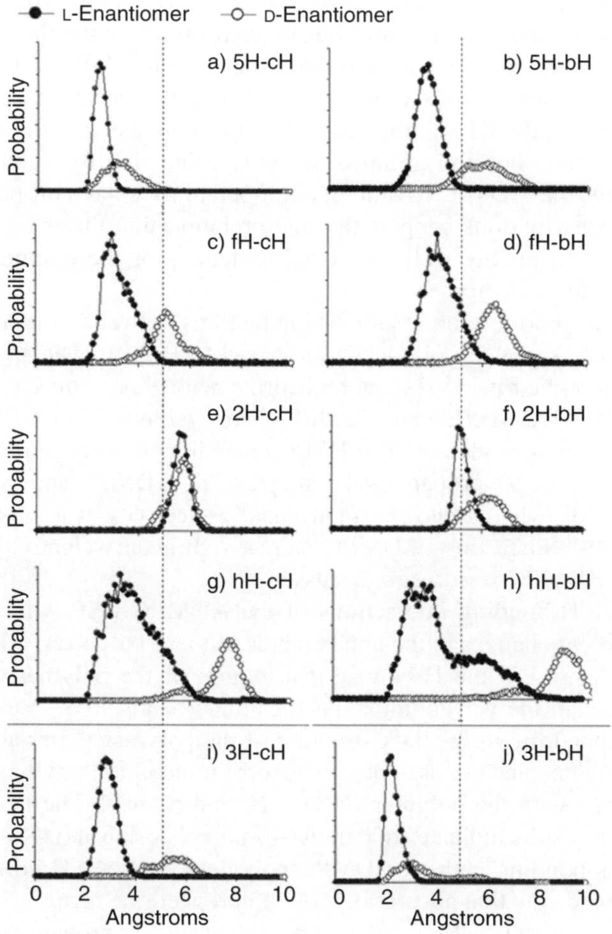

Figure 27 Pair distribution profiles over the 100 ps of molecular dynamics simulation for the protonated L- and D-enantiomers at 298 K with chloroform.

The graphs are arranged in the same order as in the two-dimensional NMR spectrum of Figure 23, to allow for convenient comparison. In Figure 27A and J, which show the distributions of distances of 5H–cH and 3H–bH, respectively, these are within 5 Å for both the L- and D-enantiomers. These are the only pairs that were experimentally within the NOE distance for the D-enantiomer as well as for the L-enantiomer; therefore, these agree with the NMR results. For the other intermolecular pair distances, the 5H–bH, 2H–bH, and 3H–cH of Figure 27B, F, and I, respectively, are within 5 Å for the L-enantiomer, in agreement with the two-dimensional NMR spectrum described above (Figure 23). The one disagreement is the 2H–cH distance in Figure 27E, which has the average distances for both enantiomers at 5.7 Å; this is somewhat outside the NOE distance. This coupling in the smallest one in the two-dimensional NMR spectrum, and 5 Å is an estimate rather than a fixed rule, so the disagreement is weak. For the intramolecular distances, the fH–cH, fH–bH, hH–cH, and hH–bH distances are all within 5 Å for the L-enantiomer but none within 5 Å for the D-enantiomer, which is also in agreement with the two-dimensional NMR data. This last set of intramolecular correlations is interesting because it means that the molecular minimization and MD simulations correctly predicted that the L enantiomer would fold to have the allyl group near the tyrosine ring for L but not D upon interaction with the ADMPC. Although computations today can be estimates at best, the pair distributions support the interpretation that these computed structures represent reasonably well the actual structures of the complexes between ADMPC and each enantiomer.

Franses et al. reported the results of interactions between norephedrine enantiomers and three types of polysaccharide-based CSPs [59]. The specific binding sites of (−)-norephedrine or (+)-norephedrine with three polysaccharide-based CSPs—ADMPC, ASMBC, and CDMPC—are studied using HPLC, ATR-IR, XRD, and molecular modeling. The HPLC data show that for norephedrine with 90 : 10 (v/v) *n*-hexane/2-propanol solvent at 298 K, ADMPC shows a significant enantioselectivity, whereas no or small enantioselectivity was observed on the other two CSPs. Since only ADMPC shows significant selectivity toward the analyte, only ADMPC results are discussed.

The specific H-bonding interactions of the ADMPC CSP with norephedrine cause the different changes in the amide bands of these polymers. ATR-IR spectra indicate the changes in the H-bonded distribution of the polymer C=O groups, and the changes in the wavenumbers of the amide bands are due to the incorporation of norephedrine in the bulk structure of the polymer. After absorption, the norephedrine molecules are assumed to adsorb in the chiral cavities of the polymer and interact with the polymer C=O, NH, and probably the phenyl groups.

The ATR-IR results indicates that both (+)-norephedrine and (−)-norephedrine form hydrogen bonding with an ADMPC polymer C=O or NH group. However, the IR spectra indicate that absorption of (−)-norephedrine forms a stronger C=O (of ADMPC polymer)· · · H_2N [of (−)-norephedrine] hydrogen-boding interaction, which was also demonstrated by molecular modeling results. The extra hydrogen bonding is thought to be the key. The XRD results show the absorption of norephedrine on ADMPC polymers, splitting the broad ADMPC polymer peak into two narrower peaks, with d-spacings of 14.6 and 18.9 Å, indicating an apparent increase in the crystallinity. Evidently, norephedrine absorption changes the polymer structure. It was proposed that PPA alters the interrod interactions and allows for a more ordered arrangement of the polymer rods.

MD simulation of an ADMPC polymer–norephedrine complex was also studied. The MD results suggest that the (−)-PPA enantiomer interacts with the polymer functional groups via three simultaneous interactions: two hydrogen-bonding interactioning NH (polymer) $\cdots$ OH [(−)-PPA], C=O (polymer)$\cdots H_2N$ [(−)-PPA], and one $\pi-\pi$ interaction. The MD-predicted elution orders for PPA enantiomers in ADMPC polymer agree with the HPLC results. The suggested mechanism of the significant enantioselectivity observed for (−)-norephedrine in ADMPC involves three simultaneous interactions: (−)-norephedrine forms two simultaneous hydrogen-bonding interactions, NH (polymer) $\cdots$ OH [(−)-norephedrine] and C=O (polymer) $\cdots$ H_2N [(−)-norephedrine], and one $\pi-\pi$ interaction, whereas (+)-norephedrine has only two interactions: one hydrogen-bonding interaction, OH of (+)-norephedrine with either the C=O or the NH group of the polymer, and one $\pi-\pi$ interaction.

An alternative strategy for obtaining a more detailed insight into chiral discrimination is the development of chemometrically driven predictions of retention and enantioselectivity: quantitative structure–enantioselective retention relationships (QSERRs). QSERRs combine quantitatively comparable retention data for a set of solutes with molecular descriptors reflecting the structural features of these solutes. The focus of the study is to develop statistically significant equations, which might prove to be meaningful in the understanding of solute and CSP interactions. Cellulose triacetate [60] and amylose tris(3,5-dimethylphenylcarbamate) [61,62] were studied using this method. More sophisticated statistical algorithms, such as multivariate regression analysis and neural networks, have been used successfully to model enantioselective chromatographic behavior as a function of nonempirical descriptors [62].

4 CONCLUSIONS

Chiral recognition exhibited by polysaccharide-based CSPs depends on the higher-ordered structures of the polymers, which makes it difficult to understand the chiral recognition mechanism. Insight into chiral discrimination at the molecular level for polysaccharide-based CSPs is hindered by the complexities of the polymer, such as the exact stereochemical structure, the geometry of the interaction, the accessible binding sites, and the multiplicity of sites with different affinities for enantiomers. Numerous techniques, such as x-ray crystallography, NMR, calorimetric studies, IR, and computational methods, have been used to provide insight into chiral recognition mechanisms for other CSPs. To better grasp the perplexing interactions between chiral stationary polymers, it is necessary to improve our understanding of the chiral stationary phase structures, chiral cavities, and surface properties.

REFERENCES

1. M. Simonyi. *Med. Res. Rev.*, 4:359, 1984.
2. E. J. Ariens. *Med. Res. Rev.*, 6:451, 1986.
3. E. J. Ariens, J. J. van Rensen, W. Welling. *Stereoselectivity of Pesticides*. Elsevier, Amsterdam, The Netherlands, 1988.
4. J. R. Powell, J. T. Ambre, T. I. Ruo. In *Drug Stereochemistry: Analytical Methods and Pharmacology*, I. W. Wainer, D. Drayer, Eds. Marcel Dekker, New York, 1988, Chap. 10.
5. I. W. Wainer. In *Drug Stereochemistry: Analytical Methods and Pharmacology*, I. W. Wainer, Ed. Marcel Dekker, New York, 1993, Chap. 6.
6. Special volume for chiral separations. *J. Chromatogr. A*, 666, 1994.
7. Special volume for chiral separations. *J. Chromatogr. A*, 906, 2001.
8. Y. Okamoto. *Chemtech*, 176, 1987.
9. Y. Okamoto, Y. Kaida. *J. Chromatogr. A*, 666:403, 1994.
10. C. Roussel, P. Piras, I. Heitmann. *Biomed. Chromatogr.*, 11:311, 1997.
11. Y. Okamoto, E. Yashima. *Angew. Chem. Int. Ed.*, 37:1020, 1998.
12. Brochure. Chiral Technologies, Inc., Exton, PA.
13. C. Perrin, V. A. Vu, N. Matthijs, M. Maftouh, D. L. Massart, Y. Vander Heyden. *J. Chromatogr. A*, 947:69, 2002.
14. C. Perrin, N. Matthijs, D. Mangelings, C. Granier-Loyaux, M. Maftouh, D. L. Massart, Y. Vander Heyden. *J. Chromatogr. A*, 966:119, 2002.
15. C. Rousell, J. L. Stein, F. Beauvais, A. Chemlal. *J. Chromatogr.*, 462:95, 1989.
16. E. Küsters, V. Loux, E. Schmid, P. H. Floersheim. *J. Chromatogr. A*, 666:421, 1994.
17. E. Küsters, C. Spöndlin. *J. Chromatogr. A*, 737:333, 1996.
18. R. J. Smith, D. R. Taylor, S. M. Wilkins. *J. Chromatogr. A*, 697:591, 1995.
19. K. Yaku, K. Aoe, N. Nishimura, F. Morishita. *J. Chromatogr. A*, 848:337, 1999.
20. M. H. Gaffney, R. M. Stiffin, I. W. Wainer. *Chromatographia*, 27:15, 1989.
21. M. Okamoto, H. Nakazawa. *J. Chromatogr.*, 588:177, 1991.
22. K. Balmér, B. A. Persson, P.-O. Lagerström. *J. Chromatogr. A*, 660:269, 1994.
23. T. Wang, Y. W. Chen. *J. Chromatogr. A*, 855:411, 1999.
24. K. Balmér, P.-O. Lagerström, S. Larsson, B.-A. Persson. *J. Chromatogr.*, 631:191, 1993.
25. K. Balmér, A. Persson, P.-O. Lagerström, B.-A. Persson, G. Schill. *J. Chromatogr.*, 553:391, 1991.
26. K. Balmér, P.-O. Lagerström, B.-A. Persson, G. Schill. *J. Chromatogr.*, 592:331, 1992.
27. T. Kubota, C. Yamamoto, Y. Okamoto. *Chirality*, 14:372, 2002.
28. E. Yashima, M. Yamada, Y. Okamoto. *Chem. Lett.*, 579, 1994.
29. E. Yashima, C. Yamamoto, Y. Okamoto. *J. Am. Chem. Soc.*, 118:4036, 1996.
30. C. Yamamoto, E. Yashima, Y. Okamoto. *J. Am. Chem. Soc.*, 124:12583, 2002.
31. E. Yashima, M. Yamada, Y. Kaida, Y. Okamoto. *J. Chromatogr. A*, 694:347, 1995.
32. C. Yamamoto, E. Yashima, Y. Okamoto. *Bull. Chem. Soc. Jpn.*, 72:1815, 1999.
33. T. Shibata, I. Okamoto, K. Ishii. *J. Liq. Chromatogr.*, 9:313–340, 1986.

34. H. Steinmeier, P. Zugenmaier. *Carbohydr. Res*., 164:97–105, 1987.
35. U. Vogt, P. Zugenmaier. *Ber. Bunsenges. Phys. Chem*., 89:1217–1224, 1985.
36. E. Francotte, R. M. Wolf, D. J. Lohmann. *J. Chromatogr*., 347:25–37, 1985.
37. T. Shibata, T. Sei, H. Nishimura, K. Deguchi. *Chromatographia*, 24:552–554, 1987.
38. J. Danhelka, M. Netopilik, M. Bohdanecky. *J. Polym. Sci. B*, 25:1801–1815, 1987.
39. E. Yashima, M. Yamada, Y. Okamoto. *Chem. Lett*., 3:579–582, 1994.
40. E. Yashima, M. Yamada, C. Yamamoto, M. Nakashima, Y. Okamoto. *Enantiomer*, 2:225–240, 1997.
41. E. Yashima, C. Yamamoto, Y. Okamoto. *J. Am. Chem. Soc*., 118:4036–4048, 1996.
42. C. Yamamoto, E. Yashima, Y. Okamoto. *J. Am. Chem. Soc*., 124(42):12583–12589, 2002.
43. Y. Okamoto, T. Ohashi, Y. Kaida, E. Yashima. *Chirality*, 5:616, 1993.
44. E. Francotte, R. Wolf, D. Lohmann, R. Mueller. *J. Chromatogr. A*, 347:25, 1985.
45. T. O'Brien, L. Crocker, R. Thompson, K. Thompson, P. H. Toma, D. A. Conlon, B. Feibush, C. Moeder, G. Bicker, N. Grinberg. *Anal. Chem*., 69:1999, 1997.
46. E. Yashima, M. Yamada, Y. Kaida, Y. Okamoto. *J. Chromatogr. A*, 694:347, 1995.
47. U. Vogt, P. Zugenmaier. *Ber. Bunsenges./Phys. Chem. Chem. Phys*., 89:1217, 1985.
48. E. Yashima, M. Yamada, Y. Okamoto. *Chem. Lett*., 579, 1994.
49. R. Kasat, N. L. Wang, E. Franses. *Biomacromolecules*, 8:1676, 2007.
50. R. M. Wenslow, Jr., T. Wang. *Anal. Chem*., 73:4190, 2001.
51. R. B. Kasat, Y. Zvinevich, H. W. Hillhouse, K. T. Thomson, N. H. L. Wang, E. I. Franses. *J. Phys. Chem. B*, 110:14114, 2006.
52. 52, Y. Ye, R. Stringham. *J. Chromatogr. A*, 927:47, 2001.
53. Y. Ye, R. Stringham. *J. Chromatogr. A*, 927:53, 2001.
54. Y. Ye, B. Lord, L. Yin, R. Stringham. *J. Chromatogr. A*, 945:147, 2001.
55. Y. Ye, B. Lord, R. Stringham. *J. Chromatogr. A*, 945:139, 2001.
56. Y. Ye, R. Stringham, M. Wirth. *J. Chromatogr. A*, 1057:75, 2004.
57. Y. Ye, S. Bai, S. Vyas, M. Wirth. *J. Phys. Chem. B*, 111:1189, 2007.
58. C. Yamamoto, E. Yashima, Y. Okamoto. *Bull. Chem. Soc. Jpn*., 72:1815, 1999.
59. R. Kasat, N. L. Wang, E. Franses. *J. Chromatogr. A*, 1190:110, 2008.
60. T. D. Booth, I. W. Wainer. *J. Chromatogr. A*, 737:157, 1996.
61. T. D. Booth, I. W. Wainer. *J. Chromatogr. A*, 741:205, 1996.
62. T. D. Booth, K. Azzaoui, I. W. Wainer. *Anal. Chem*., 69:3879, 1997.

CHAPTER 7

Comparison of Chiral Chromatography Columns for Pharmaceutical Method Development

GREGORY K. WEBSTER

Global Analytical Research and Development, Abbott Laboratories, Abbott Park, Illinois

1 INTRODUCTION

In today's pharmaceutical industry, companies are developing new chemical entities (NCEs) at a rapid pace, in search for the latest medical breakthrough and potential blockbuster drug. The relatively simple small molecules of past decades are being replaced by more complex molecules, often with a specific stereochemistry targeted. The specific stereochemistry requirement of these substances affects their biological activity as well as the commercial viability of the potential new drug. Thus, the separation sciences are being driven to developing new and more efficient separation strategies for isolating and testing these analytes. These strategies often play a vital role in the development of these entities into commercial products, especially in the early drug development stages, where many companies delay developing stereospecific synthesis routes until the NCE has proven itself to be a potential drug in the clinic. In this strategy, preparative chiral chromatography is critical in isolating the active pharmaceutical ingredient (API). Today, liquid chromatography (LC) remains the stalwart technique for chiral analytical testing and quality control, while the cost and efficiency advantages of supercritical fluid chromatography (SFC) is advancing it with preparative-scale chiral isolations. Analytical SFC is becoming more and more commonplace for chiral applications in the pharmaceutical laboratory, but it does not have the significant cost and green chemistry advantages found with the preparative-scale platform. Apart from the separation technique employed,

Chiral Separation Methods for Pharmaceutical and Biotechnological Products, Edited by Satinder Ahuja

developing optimum method conditions for new NCEs can be a complicated and prolonged process.

Unlike achiral column chromatography, chiral chromatography does not tend to keep to traditional method development strategy [1]. Chiral method development begins with finding a stationary phase that exhibits enantiomeric recognition for the molecule of interest. With this stationary phase, the mobile phase is further optimized to enhance this recognition and, ultimately, chromatographic resolution. Commercial systems are available to evaluate various chiral LC columns and mobile phases efficiently in an unattended mode. Employing such a strategy has enabled improved sample analysis time through rapid method development. Many chiral screening approaches have been applied to the major separation techniques available to the laboratory chemist. Elaborate screening procedures for assisted method development for liquid [2–7] and supercritical fluid [8–14] chromatography can be found in the literature. Capillary electrophoresis has been studied for chiral applications but is used to a lesser extent in the pharmaceutical industry, due primarily to cost and sensitivity concerns [15–32]. Chiral gas chromatography is used principally for raw materials and early intermediates, as most drug entities are not volatile [33,34]. The approaches for all these methods have been reviewed and optimized into a harmonized orthogonal approach to chiral screening [35,36].

2 CHROMATOGRAPHIC COLUMN CHIRAL SCREENING IN THE PHARMACEUTICAL INDUSTRY

Column chiral screening in the pharmaceutical industry centers primarily on normal-phase separations for liquid and supercritical fluid chromatography method development to enable preparative chromatographic collection of the enantiomers. These screening systems have similar chromatographic conditions. The use of polysaccharide-based stationary phases (Diacel Chiralpak AD and AS and Chiralcel OD and OJ) have been very effective in developing suitable chiral methods for pharmaceuticals. Although these stationary phases have been shown to be useful, it is important to evaluate these phases continually as molecular entities continue to evolve. Thus, pharmaceutical laboratories often supplement their column selections not only with polysaccharide stationary phases but with those using macrolide antibiotic and Pirkle stationary phases [37–40] platforms as well.

With reversed-phase achiral applications dominating pharmaceutical chromatographic analysis, it is surprising at first to note that few publications dedicate chiral method development screens to reversed-phase chiral chromatography [3,35,41]. Traditionally, better chromatographic performance is found with chiral separations in the normal phase for most column manufacturers. Normal-phase separations are typically used with isomer separations. It is the easier solvent removal for isolating the collected enantiomer that probably drove the industry to normal-phase chromatography for chiral applications.

Reversed-phase chiral separations are desired simply for efficiency in generating results from laboratories whose instrumentation is routinely configured to run in reversed- and not normal-phase modes. Normal-phase conditions are less attractive to the analytical chemist for this reason and deter laboratory efficiency. Reversed-phase chiral chromatography is described widely in the literature, and commercial columns for this application are readily available [42–51].

3 COMPARISON OF STATIONARY-PHASE EFFECTIVENESS FOR PHARMACEUTICAL APPLICATIONS

The pharmaceutical industry addresses its need for enantiomeric selectivity through stereospecific synthesis, preparative chiral isolations, or a combination of both. The need for stereopecific chiral separations propelled the chromatography industry and academia to deliver chiral stationary phases to meet these requirements. This ultimately yielded the cyclodextrin, macrocyclic glycopeptide, polysaccharide, and Pirkle-type stationary phases available today. Chiral chromatography differs from achiral chromatography in that the preparative scale drives the technique instead of analytical applications.

Chiral method development screening strategies are determined largely on the basis of whether the method is to be applied to analytical or preparative applications. Preparative chromatography tends to use volatile modifiers and mobile phases for easy removal after sample collection. Sample isolation capabilities have led to the advancement of chiral supercritical fluid chromatography since CO_2 is easily driven off. Chiral separations in preparative chromatography need to be optimized only to a point where pure samples can be collected. Conversely, while analytical chromatography can use a greater variety of modifiers and mobile phases, the separation needs to be optimized to a point where suitable resolution ($Rs \geq 1.5$) has been achieved between the enantiomers of interest.

It is advantageous to the chiral chromatographer that the majority of commercially available normal-phase LC chiral stationary phases and modifiers can be used on both LC and SFC instrumentation. This flexibility allows methods developed using one mode to be transferred to the other with minor modifications. LC and SFC are both operating in a normal-phase mode and with the same columns. A significant difference in chromatography to a point where only one mode shows method stereoselectivity using the same column is not generally expected.

4 DIRECT COLUMN COMPARISON

To compare chiral chromatography columns effectively, the results of several studies using similar chromatographic conditions and study samples are used for direct comparison of the stationary phases employed [35,36,40,41]. The criteria set for the samples used were representative of the current molecules of interest to the pharmaceutical industry. This was accomplished using a mix of commercially available and proprietary compounds. Specific compounds used to make up the study library are listed in Table 1. Literature screens cannot address all the molecules in development. To increase the complexity of the library, samples currently in development were used. However, to protect the intellectual property of the business concern, the true nature of these molecules cannot be disclosed.

In the following sections, the data tables containing a "Y" result indicate that at least one chromatographic condition in the specified platform posted a chromatographic separation with a resolution of $Rs \geq 1.5$. An "MD" result means that at least one chromatographic condition in this platform demonstrated chiral recognition but with a resolution of $Rs < 1.5$. Finally, an "N" result means that no resolution was achieved in the screen for this compound. An "MD" result enables the chromatographer to note that the stationary phase exhibits limited chiral recognition of the enantiomers and that the chromatographer can employ further optimization techniques to improve resolution. In a chiral screening application, the actual resolution result of the chromatographic profile is of less importance than an overall determination as to whether or not the stationary phase exhibits stereoselectivity.

5 INITIAL NORMAL-PHASE LC COLUMN SCREENS

Several of the major chiral stationary phases used by the pharmaceutical industry today are used in the first initial column screen [36]. The columns used are listed in Table 2. Noticeably absent from this study are macrolide antibiotic columns. This is because earlier normal-phase testing showed that these columns were largely ineffective for this particular sample set [36]. It is important to note again that columns chosen for screens need to be optimized for compounds that each business is developing. As shown in Table 3, 95% of the compounds produced stereospecific recognition to the study library, with 19 of the 22 compounds tested fully resolved.

The polysaccharide stationary phases (AD, OD, AS, and OJ) alone enabled suitable stereospecificity for nearly all chiral compounds typically analyzed in our laboratories. This result is consistent with other investigations [2,4,5]. In this study the Whelk-O Pirkle phases resolved only 25% of the compounds, and 50% of the study library should reach achievable resolution with further method development.

The chiral phases studied from Kromasil resolved the study library 20% of the time, and 40% should reach achievable resolution with further method development. The Whelk-O and Kromasil phases did not resolve any compound in this library that the polysaccharide stationary phases did not also resolve. Thus, for the compounds in this study library, running these additional column phases did not produce significant improvement in the chiral screen and would probably be removed to shorten run times.

TABLE 1 Study Library[a]

Label	Compound Identification	Compound Structure
M1	Econazole nitrate (racemate)	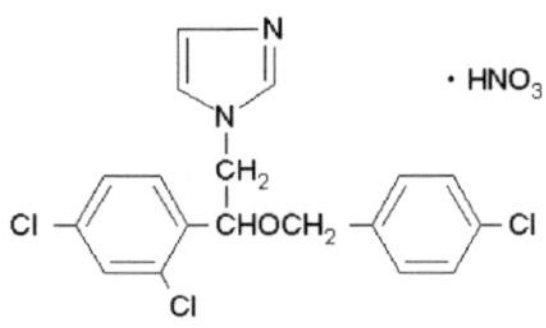
M2	Promethazine-HCl (racemate)	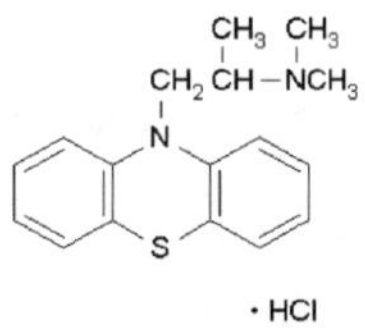
M3	Bupivacaine-HCl (racemate)	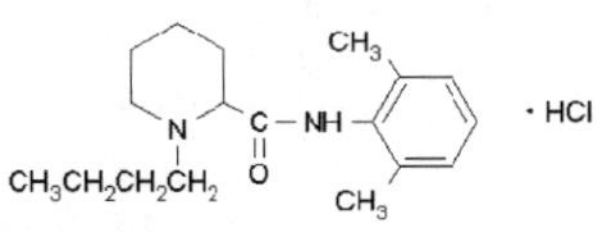
M4	Bendroflumethiazide (racemate)	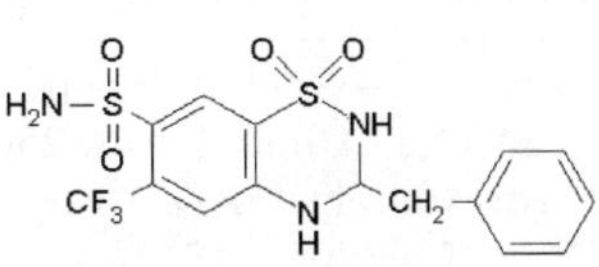
M5	Propanolol-HCl (racemate)	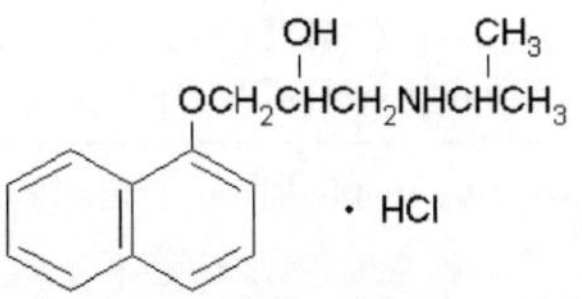
M6	Guaiacol glyceryl ether (racemate)	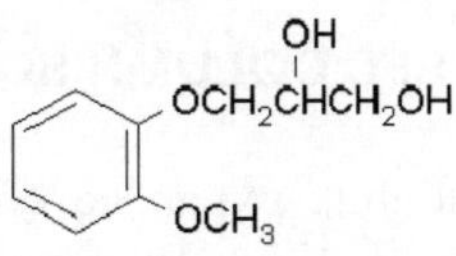
M7	Ketoprofen (racemate)	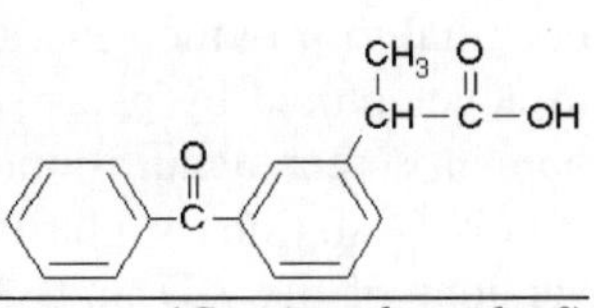

(*Continued overleaf*)

TABLE 1 ***(Continued)***

M8	Ibuprofen (racemate)	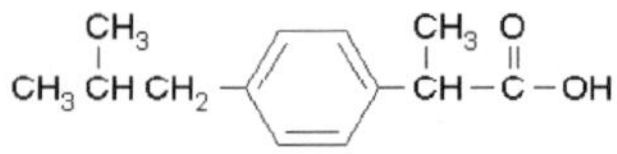
M9	Prilocaine-HCl (racemate)	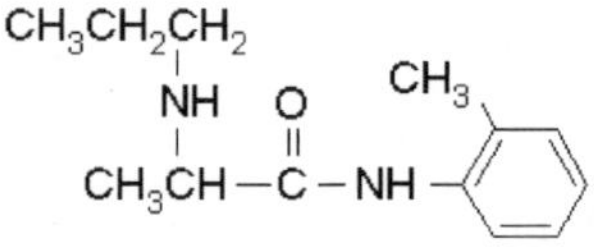
M10	Warfarin (racemate)	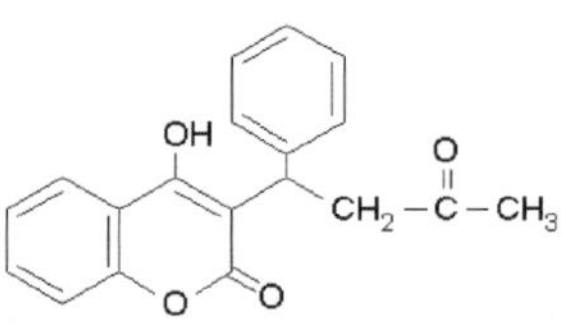

[a]M11 through M20 are proprietary structures and not listed.

TABLE 2 LC Chiral Columns

Regis (*S*,*S*)-Whelk-O, 10 μm, 250 × 4.6 mm, Regis Technologies (Morton Grove, IL)
Regis (*S*,*S*)-Ulmo, 5 μm, 250 × 4.6 mm, Regis Technologies
Diacel Chiralpak AD, 10 μm, 250 × 4.6 mm, Chiral Technologies (West Chester, PA)
Diacel Chiralpak AS, 10 μm, 250 × 4.6 mm, Chiral Technologies
Diacel Chiralcel OD, 10 μm, 250 × 4.6 mm, Chiral Technologies
Diacel Chiralcel OJ, 10 μm, 250 × 4.6 mm, Chiral Technologies
Kromasil CHI-TBB, 5 μm, 150 × 4.6 mm, Phenomenex (Torrance, CA)
Kromasil CHI-DMB, 5 μm, 150 × 4.6 mm, Phenomenex

Source: Adapted from [36].

6 SFC COLUMN SCREEN

A slightly expanded column set was used to obtain SFC results for the study samples [35]. The columns are listed in Table 4. Again in Table 5, the data illustrate that under SFC conditions, polysaccharide stationary phases have the potential to provide separations for most compounds representative of molecules in development by pharmaceutical laboratories. In this study, the method development screen again found no significant advantage in adding the Chirobiotic or Whelk-O stationary phases as a "first-tier" setup. The IA and IB immobilized versions of the AD and OD polysaccharide stationary phases did not present a resolution advantage as first-tier columns as well.

TABLE 3 Normal-Phase LC Screening Results

Mobile Phase	trans-Stilbene Oxide	Econazole nitrate (M1)	Promethazine HCl (M2)	Bupivacaince HCl (M3)	Bendroflumethiazide (M4)	Propanolol (M5)	Guaiacol Glyceryl Ether (M6)	Ketoprofen (M7)	Ibuprofen (M8)	Prilocaine HCl (M9)	Warfarin (M10)	Proprietary (M11)	Proprietary (M12)	Proprietary (M14)	Proprietary (M15)	Proprietary (M16)	Proprietary (M17)	Proprietary (M20)	Atenolol	Mianserin	Verapamil	Proprietary (M21)
Hept/EtOH w/ 0.1%ESA																						
AD	R	SP	SP	NR	NR	SP	SP	SP	NR	SP	NR	NE	SP	NR	NE	SP	NR	NR	NR	SP	NR	SP
Whelko	SP	NR	NR	NR	R	NE	NE	SP	NE	NE	NE	NR	NR	NR	NE	NR	NR	NR	NR	SP	NR	SP
OD	R	NR	NR	R	SP	R	R	NR	NR	SP	NR	NR	NR	R	NR	NR	NR	NR	NR	SP	SP	R
OJ	R	SP	NR	R	R	SP	NR	R	NR	NR	NR	SP	SP	SP	NR	NR	NR	NR	SP	R	NR	SP
AS	NR	SP	NR	NR	NR	SP	NR	NR	NR	NR	R	NR	SP	SP	NR	R	NR	NR	NR	SP	NR	SP
Ulmo	NR	NR	NR	NE	SP	NR	NR	NR	NR	SP	R	NR	NR	NR	NR	SP	R	NR	NR	NR	NR	SP
TBB	NR	NR	NR	NR	NR	R	NR	NR	NR	NR	NR	NR	NR	NR	NR	NR	R	NR	SP	NR	NR	NR
DMB	NR	NR	NR	NR	NR	R	NR	NR	NR	SP	R	NR	NR	NR	NR	NR	R	NR	SP	NR	NR	SP
Hept/EtOH																						
AD	R	SP	SP	NR	NR	SP	SP	SP	NR	SP	NR	R	SP	NR	NR	NR	NR	NR	NR	SP	NR	SP
Whelko	R	NR	NR	NR	R	NR	NR	SP	SP	R	R	NR	NR	SP	NR	NR	SP	NR	NR	SP	NR	SP
OD	R	NR	NR	R	SP	R	R	NR	NR	SP	NR	NR	NR	R	NR	NR	NR	NR	R	SP	SP	R
OJ	R	SP	SP	R	SP	SP	NR	R	R	NR	NR	SP	R	SP	NR	NR	NR	NR	SP	R	SP	SP
AS	NR	SP	NR	SP	NR	SP	NR	NR	NR	SP	R	NR	SP	SP	NR	NR	NR	NR	NR	SP	NR	SP
Ulmo	NR	NR	NR	NR	SP	NR	NR	NR	NR	NR	R	NR	NR	NR	NR	NR	SP	NR	NR	NR	NR	SP
TBB	NR	NR	NR	NR	NR	NR	NR	NR	NR	NR	R	NR	NR	NR	NR	NR	NR	NR	NR	NR	NR	NR
DMB	NR	NR	NR	NR	NR	NR	NR	NR	NR	SP	R	NR	NR	NR	NR	NR	NR	NR	NR	NR	NR	NR
Hept/EtOH w/ 0.1% TFA																						
AD	R	SP	NR	NR	SP	SP	SP	SP	NR	SP	NR	NR	SP	NR	NR	NR	NR	NR	NR	SP	NR	NR
Whelko	R	NR	NR	NR	R	NR	NR	SP	SP	NR	R	NR	NR	SP	NR	NR	NR	NR	NR	SP	NR	SP
OD	R	NR	NR	R	R	R	R	NR	NR	SP	R	NR	NR	R	NR	NR	NR	NR	R	R	SP	R
OJ	R	NR	NR	R	SP	SP	NR	R	R	NR	NR	SP	R	SP	NR	NR	NR	NR	R	R	SP	SP
AS	NR	SP	NR	SP	NR	SP	NR	NR	NR	SP	R	NR	SP	SR	NR	NR	NR	NR	SP	SP	NR	SP
Ulmo	NR	NR	NR	NR	SP	NR	NR	NR	NR	NR	R	NR	NR	NR	NR	SP	NR	NR	NR	NR	NR	SP
TBB	NR	NR	NR	NR	NR	NR	NR	NR	NR	NR	R	NR	NR	NR	NR	SP	NR	SP	NR	NR	NR	NR
DMB	NR	NR	NR	NR	NR	NR	NR	NR	NR	SP	R	NR	NR	NR	NR	NR	NR	NR	NR	NR	NR	NR
Hept/EtOH w/ 0.1% IPAm																						
AD	R	SP	SP	NR	NR	NR	SP	NR	NE	NR	SP	NR	NE	NE	NR	NR	NR	NR	NR	SP	SP	NR
Whelko	R	NR	NR	NR	R	NR	NR	SP	R	NR	SP	NR	NR	SP	NR	NR	NR	NR	NR	SP	NR	NR
OD	R	NR	NR	R	R	R	R	NR	NR	SP	NE	NR	NR	R	NR	NR	NR	NR	R	SP	SP	R
OJ	R	NR	SP	R	SP	SP	NR	NR	NR	NR	R	NR	R	SP	NR	NR	NR	NR	R	R	R	SP
AS	NR	SP	NR	NR	NR	SP	NR	NR	NR	NR	NR	NR	NR	SP	NR	NR	NR	NR	NR	NR	SP	SP
Ulmo	NR	NR	NR	NR	SP	NR	R	NR	NR	NR	SP	NR	NR	NR	NR	NR	NR	SP	NR	NR	NR	SP
TBB	NR	NR	NR	NR	NR	NR	NR	NR	NR	NR	NR	NR	NR	NR	NR	NR	NR	SP	NR	NR	NR	NR
DMB	NR	NR	NR	NR	SP	NR	NR	NR	SP	NR	NR	NR	NR	NR	NR	NR	NR	NR	NR	NR	NR	NR
Hept/EtOH w/ 0.1% IPAm; 0.1% TFA																						
AD	R	SP		SP	SP	NR	NR	NR	NR	NR	NR	NR	NR	NE	NR	NR	NR	NR	SP	SP	NR	NR
Whelko	R	NR		R	R	SP	NE	SP	R	NR	R	NR	NR	SP	NR	NR	NR	NR	R	NR	NR	SP
OD	R	NR		R	R	R	NE	NR	SP	NR	SP	NR	SP	NR	NR	NR	NR	NR	R	SP	SP	R
OJ	R	R		SP	SP	R	R	SP	SP	NR	NR	NR	R	SP	NR	NR	NR	NR	SP	R	SP	R
AS	NR	NR		NR	NR	NR	NR	NR	R	NR	R	NR	SP	SP	NR	NR	NR	NR	NR	NR	NR	R
Ulmo	NR	NR		NR	NR	NR	NE	NR	NR	SP	SP	NR	NR	NR	NR	NR	NR	NR	NR	NR	NR	SP
TBB	NR	NR		NR	NR	NR	NE	NR	NR	NR	NR	NR	NR	NR	NR	SP	NR	SP	NR	NR	NR	NR
DMB	NR	NR		NR	NR	NR	NE	NR	NR	R	NR	NR	NR	NR	NR	NR	NR	NR	NR	NR	NR	NR
Method? (Y/N/MD)	Y	Y	MD	Y	Y	Y	Y	Y	Y	Y	Y	Y	Y	Y	N	Y	Y	MD	Y	Y	Y	Y

NE - No Elution	Method:
NR - Not Resolved	Y- screen method posted suitable resolution
SP - Split Peak (R<1.5)	N-screen method did not indicate potential for resolution
R -Resolved (R> 1.5)	MD -screen method indicated potential for resolution, but R < 1.5

Source: [36].

7 PIRKLE STATIONARY PHASES

Pirkle chiral stationary phases use the same mobile-phase approaches as do polysaccharide stationary phases. However, since Pirkle phases tend to be bonded phases as opposed to coated polysaccharide stationary phases, these columns can be used with 1,2-dicholoroethane and tetrahydrofuran (THF) mobile phases. Many Pirkle stationary phases are available in multiple enantiomeric forms that

TABLE 4 SFC Chiral Columns

Regis (*S*,*S*)-Whelk-O, 10 μm, 250 × 4.6 mm, Regis Technologies (Morton Grove, IL)
Regis (*S*,*S*)-Ulmo, 5 μm, 250 × 4.6 mm, Regis Technologies
Diacel Chiralpak AD, 10 μm, 250 × 4.6 mm, Chiral Technologies (West Chester, PA)
Diacel Chiralpak AS, 10 μm, 250 × 4.6 mm, Chiral Technologies
Diacel Chiralcel OD, 10 μm, 250 × 4.6 mm, Chiral Technologies
Diacel Chiralcel OJ, 10 μm, 250 × 4.6 mm, Chiral Technologies
Kromasil CHI-TBB, 5 μm, 150 × 4.6 mm, Phenomenex (Torrance, CA)
Kromasil CHI-DMB 5 μm, 150 × 4.6 mm, Phenomenex
Diacel Chiralpak IA, 10 μm, 250 × 4.6 mm, Chiral Technologies
Diacel Chiralpak IB, 10 μm, 250 × 4.6 mm, Chiral Technologies
Chirobiotic V, 250 × 4.6 mm, Bodman Industries (Aston, PA)

Source: Adapted from [35].

provide the chromatographer with additional capabilities to optimize the separation. For example, with an (*S*,*S*)-stationary phase, the elution order can be reversed using the (*R*,*R*)-phase. This can be a significant advantage in peak collection with preparative separations.

Two Pirkle chiral stationary platforms were compared using the sample set library: (1) columns from Chirex and (2) columns from Regis [40]. The Pirkle stationary phases used are listed in Table 6.

7.1 Chirex Phases

Testing the study library series of compounds, the data presented in Table 7 illustrate that Chirex Pirkle stationary phases yielded eight separations with a resolution of Rs $\geq$ 1.5 for the enantiomeric pairs. An additional seven chromatographic profiles illustrate chiral recognition but need further optimization (Rs $<$ 1.5). The Chirex Pirkle column screen did not resolve compound M15. However, the polysaccharide stationary phases in a liquid chromatographic format did not resolve this compound either [36]. The results demonstrate that the Chirex Pirkle column platform can be an effective chiral screening tool to complement the polysaccharide column screen. Pirkle stationary phases can be used with SFC, as detailed by Holzheuer et al. [40].

7.2 Regis Phases

Unlike the Chirex platform of Pirkle stationary phases, the Regis Pirkle series can use amine modifiers, so this option was added to the screening sequence. As shown in Table 8, the Regis platform yielded four separations with a resolution of Rs $\geq$ 1.5 for the enantiomeric pairs [40]. Ten chromatographic profiles exhibited chiral recognition but needed further optimization (Rs $<$ 1.5). The Regis Pirkle

TABLE 5 Chiral SFC Screening Results

Mobile Phase	Econazole nitrate (M1)	Promethazine HCl (M2)	Bupivacaince HCl (M3)	Bendroflumethiazide (M4)	Propanolol (M5)	Gualacol Glyceryl Ether (M6)	Ketoprofen (M7)	Ibuprofen (M8)	Prilocaine HCl (M9)	Warfarin (M10)	M11	M12	M13	M14	M15	M16	M17	M18	M19	M20
0.1% IPAm in MeOH																				
AD15	R	SP	NR	R	R	SP	SP	SP	NR	R	SP	R	SP	NR	NE	SP	NR	R	NR	R
OD25	SP	NR	SP	SP	R	R	NR	NR	NR	R	NR	NR	R	R	R	NR	NR	R	NE	R
AS25	R	NR	NR	R	SP	NR	NR	NR	R	R	SP	SP	NR	SP	NE	NR	NR	R	R	R
OJ25	SP	SP	NR	R	SP	NR	SP	SP	NR	R	NR	NR	NR	NR	R	NR	NR	NE	NE	R
IA25	R	NR	R	NR	SP	NR	NR	NR	NR	R	NR	SP	NR	R	NE	NR	SP	NR	NE	R
OD25	SP	SP	SP	SP	R	R	NR	NR	SP	R	NR	SP	R	R	R	NR	NR	R	NE	R
IB25	NR	NR	SP	SP	R	R	NR	NR	NR	R	NR	NR	R	SP	R	NR	NR	SP	NE	R
Whelk-O	NR	NR	SP	R	NE	NR	R	R	NR	R	NE	NR	NR	SP	NE	R	NR	NR	NE	R
ChiroV	NR	NR	NR	NE	NE	NR	SP	NR	NR	SP	NE	NR	NR	NR	SP	NR	NR	NE	NE	R
0.1% TFA and 0.1% IPAm in Methanol																				
AD15	R	SP	NR	SP	SP	SP	SP	SP	NR	R	NR	R	SP	NR	NE	SP	R	NA	NA	R
OD25	NR	NR	R	SP	R	R	NR	NR	R	R	NR	NR	R	SP	R	SP	NR	NA	NA	R
AS25	R	NR	R	R	NR	NR	NR	SP	SP	R	SP	SP	NR	SP	NE	NR	NR	NA	NA	R
OJ25	R	NR	R	R	NR	NR	SP	NR	NR	R	NR	NR	NR	SP	R	SP	NR	NA	NA	R
IA25	NR	SP	NR	NR	NR	NR	NR	NR	SP	R	NR	NR	NR	R	NE	NR	SP	NA	NA	R
OD25	SP	SP	R	SP	R	R	NR	NR	R	R	NR	NR	R	R	NR	SP	NR	R	NR	R
IB25	NR	NR	R	SP	R	R	NR	NR	R	R	NR	NR	R	SP	R	NR	NR	NE	NE	R
Whelk-O	NA	NA	NA	NA	NA	NA	NA	NA	NA	NA	NA	NA	NA	NA	NA	NA	NA	NA	NA	NA
ChiroV	NR	NR	NR	NR	SP	NR	NR	NR	R	R	NR	SP	NR	NR	NR	NR	NR	NE	NE	SP
0,1% TFA in Methanol																				
AD15	NE	NE	NE	R	NE	SP	NR	SP	NE	R	NE	NE	R	NR	NE	SP	R	NA	NA	R
OD25	NR	NR	R	SP	R	R	NR	NR	SP	NR	SP	SP	R	R	NE	SP	NR	NA	NA	R
AS25	NE	NE	NE	R	NE	NR	NR	NR	NR	R	NR	NR	NR	SP	NE	R	NR	NA	NA	R
OJ25	R	NR	SP	R	NR	NR	SP	SP	NR	R	NR	NR	NR	NR	NE	SP	NR	NA	NA	NR
IA25	NR	SP	SP	NR	NR	NR	NR	SP	NR	R	NR	NR	NR	NR	NE	SP	SP	NA	NA	R
OD25	NR	SP	R	SP	R	R	NR	NR	R	NR	NR	NR	R	R	R	SP	NR	R	NR	R
IB25	NR	NR	R	SP	R	R	NR	NR	SP	R	NR	NR	R	SP	R	R	NR	NR	NR	R
Whelk-O	NA	NA	NA	NA	NA	NA	NA	NA	NA	NA	NA	NA	NA	NA	NA	NA	NA	NA	NA	NA
ChiroV	NR	NR	NE	NE	NR	NR	NR	NR	R	R	NR	NR	NR	NR	NR	R	NR	NR	NE	SP
0,1% EAS in Ethanol																				
AD15	R	R	R	SP	NR	R	SP	NR	NR	R	R	R	NR	NR	NE	SP	R	R	R	NR
OD25	NR	SP	R	SP	R	R	NR	NR	R	R	NR	NR	NR	NR	NE	R	R	NR	NR	NE
AS25	NR	SP	R	NE	SP	NR	R	NR	NR	R	NR	SP	SP	NR	NE	NR	NR	NR	NR	R
OJ25	NR	R	R	R	SP	NR	NR	SP	NR	R	SP	R	SP	NR	NE	SP	NR	NR	NR	NE
IA25	NR	NR	NR	NR	SP	SP	NR	NR	R	R	NR	R	SP	NE	NA	SP	SP	NR	NA	NE
OD25	NR	R	NE	SP	R	R	NR	NR	R	NR	NR	NR	R	SP	NR	SP	NR	SP	NR	R
IB25	NR	NE	NE	SP	R	R	NR	NR	SP	R	NR	NR	R	NR	NR	SP	NR	NR	NR	R
Whelk-O	NA	NA	NA	NA	NA	NA	NA	NA	NA	NA	NA	NA	NA	NA	NA	NA	NA	NA	NA	NA
ChiroV	NE	NR	NE	NE	NE	NR	NR	NE	NR	R	NE	NR	NR	NE	NE	NR	NR	NE	NE	NE
METHOD (Y/N//MD)	Y	Y	Y	Y	Y	Y	Y	Y	Y	Y	Y	Y	Y	Y	Y	Y	Y	Y	Y	Y

Method Status	
NE - No Elution	Method:
NR - Not Resolved	Y- screen method posted suitable resolution
SP - Split Peak (R<1.5)	N-screen method did not indicate potential for resolution
R -Resolved (R> 1.5)	MD -screen method indicated potential for resolution, but R < 1.5
NA - Not Analyzed	

Source: [35].

column screen did not resolve compounds M1, M12, and M12. For the LC conditions tested, the Chirex Pirkle screen seemed to give a slightly better response than the Pirkle columns from Regis.

Interestingly, the Regis columns ran much more effectively using SFC [40]. Samples M2, M5, M6, M11, M12, and M14 all improved in resolution. For the SFC conditions tested, the Regis Pirkle screen seemed to give a slightly better response than the Pirkle columns from Chirex.

TABLE 6 Pirkle Chiral Columns

Chirex Pirkle columns	(*R*)-NGLY/DNB, 10 μm, 250 × 4.6 mm, Phenomenex (Torrance, CA)
	(*S*)-ICA/(*R*)-NEA, 10 μm, 250 × 4.6 mm, Phenomenex
	(*S*)-Leu/(*R*)-NEA, 10 μm, 250 × 4.6 mm, Phenomenex
	(*S*)-PRO/(*R*)-NEA, 10 μm, 250 × 4.6 mm, Phenomenex
	(*R*)-PGLY/DNB, 10 μm, 250 × 4.6 mm, Phenomenex
	(*R*)-PGLY/DNAn, 0 μm, 250 × 4.6 mm, Phenomenex
Regis Pirkle columns	(*S*,*S*)-Whelk-O, 10 μm, 250 × 4.6 mm, Regis Technologies (Morton Grove, IL)
	(*S*,*S*)-Ulmo, 10 μm, 250 × 4.6 mm, Regis Technologies
	(*S*,*S*)-β-Gem, 10 μm, 250 × 4.6 mm, Regis Technologies
	(*S*,*S*)-α-Burke, 10 μm, 250 × 4.6 mm, Regis Technologies
	L-Leucine, 10 μm, 250 × 4.6 mm, Regis Technologies
	(*S*,*S*)-DACH, 10 μm, 250 × 4.6 mm, Regis Technologies

Source: Adapted from [40].

Using a dedicated platform for the Pirkle column formats significantly improved the impression of these columns over the results found with earlier screening investigations [35,36]. Instead of picking one or two Pirkle columns along with polysaccharide and antibiotic stationary phases in the earlier study, dedicating the screening to a series of Pirkle columns broadened the chemistry of the stationary phases to provide a better chance of "hits" in a variable column library. While polysaccharide column screens still seem to provide the best first-tier approach to finding chiral recognition conditions for stationary phases, the use of Pirkle columns surely warrants continued consideration as second-tier phases for pharmaceutical screening applications that require the Pirkle selectivity mechanism.

8 REVERSED-PHASE CHIRAL COLUMNS

Using the corresponding reversed-phase stationary phases as the polysaccharide phases used in normal-phase investigations as well as the Whelk-O column (as being representative of Pirkle stationary phases), chiral analysis of the study library using reversed-phase conditions was studied [41]. It is important to note that the mobile-phase modifiers shown to be effective in the normal-phase LC and SFC investigations were also most effective in the reversed-phase investigations [41]. The columns used are listed in Table 9. The data for reversed-phase screening is presented in Table 10. This screen yielded a suitable analytical method for 14 of 19 drug substances (Rs $\geq$ 1.5) and a potential method to be developed (Rs $\leq$ 1.5) for an additional five compounds. The data illustrate

TABLE 7 Chirex Pirkle Chiral Column Screening Results

Mobile Phase	trans-Stilbene Oxide	Econazole nitrate (M1)	Promethazine HCl (M2)	Bupivacaince HCl (M3)	Bendroflumethiazide (M4)	Propanolol (M5)	Guaiacol Glyceryl Ether (M6)	Ketoprofen (M7)	Ibuprofen (M8)	Prilocaine HCl (M(9)	Warfarin (M10)	Proprietary (M11)	Proprietary (M12)	Proprietary (M14)	Proprietary (M15)	Proprietary (M16)	Proprietary (M17)
Hept/EtOH w/ 0.1%ESA																	
(R)-NGLY / DNB	NR	NE	NR	NE	NE	NR	NE	NR	NE	SP	NE	NR	NR	NE	NE	NR	NR
(S)-ICA / (R)-NEA	NR	SP	R	SP	NE	SP	NR	R	NR	SP	NE	SP	NR	NE	NE	NR	SP
(S)-Leu / (R)-NEA	NR	NR	SP	R	NR	SP	NR	NR	NR	SP	NE	NR	SP	NR	NR	NR	SP
(S)-PRO / (R)-NEA	NR	NR	R	SP	NR	NR	SP	NR	NR	NR	NE	NR	NR	NE	NE	NR	R
(R)-PGLY / DNB	NR	NE	NR	SP	NE	NR	NE	NR	SP	SP	NE	NE	NR	NE	NE	NR	NR
(R)-PGLY / DNAn	NR	NE	NR	SP	SP	NR	NR	NR	NE	NR	NE	NE	NR	NE	NR	NR	NR
Hept/EtOH																	
(R)-NGLY / DNB	NR	NR	NR	NR	NE	NR	NE	NR	NR	SP	NE	NR	NR	NR	NR	NR	NR
(S)-ICA / (R)-NEA	NR	NE	SP	SP	NE	SP	NR	NR	NR	SP	NE	NR	NR	NR	NR	SP	NR
(S)-Leu / (R)-NEA	NR	NR	SP	R	NR	SP	NR	NR	NR	SP	NE	NR	SP	NR	NR	R	NR
(S)-PRO / (R)-NEA	NR	NR	R	SP	NE	NR	SP	NR	NR	NR	NE	NR	NR	NE	NE	NR	NR
(R)-PGLY / DNB	NR	NE	NR	SP	NE	NE	NR	NR	NR	SP	NR	NR	NR	NR	NR	NR	NR
(R)-PGLY / DNAn	NE	NE	NR	NE	SP	NE	NR	NR	NR	NR	SP	NE	NE	NR	NR	NR	NR
Hept/EtOH w/ 0.1% TFA																	
(R)-NGLY / DNB	NR	NR	NR	NR	NE	NR	SP	NR	NR	SP	NE	NR	NR	NR	NR	NR	NR
(S)-ICA / (R)-NEA	NR	SP	R	SP	NE	SP	NR	NR	NR	SP	R	NR	NR	NR	NE	NR	NR
(S)-Leu / (R)-NEA	NR	NR	SP	R	NR	SP	NR	NR	NR	SP	NR	NR	SP	NR	NR	NR	NR
(S)-PRO / (R)-NEA	NR	NR	R	SP	NE	NR	SP	NR	NR	NR	NE	NR	NR	NE	NE	NR	NR
(R)-PGLY / DNB	NR	NE	NR	SP	NE	NR	NR	NE	NR	SP	R	NR	NR	NR	NR	NR	NR
(R)-PGLY / DNAn	NR	NE	NR	NE	SP	NR	NR	NE	NR	NR	SP	NE	NE	NR	NR	NR	NR
Hept/EtOH w/ 0.1% NH4OAc																	
(R)-NGLY / DNB	NR	NR	NR	SP	NE	NR	SP	NE	NR	SP	NE	NR	NR	SP	NE	NE	NE
(S)-ICA / (R)-NEA	NR	R	R	SP	NE	SP	NR	NE	NR	SP	NE	SP	NR	SP	NR	NE	NE
(S)-Leu / (R)-NEA	NR	NR	SP	R	NR	SP	NR	NE	NR	SP	NE	NR	SP	NR	NR	NR	NE
(S)-PRO / (R)-NEA	NR	NR	R	SP	NE	NR	SP	NR	NR	NR	NE	NR	NR	NR	NR	NR	NR
(R)-PGLY / DNB	NR	NE	NR	SP	NE	NE	NR	NE	NE	NR	NR	NR	NR	NE	NE	NR	NR
(R)-PGLY / DNAn	NR	NR	NR	NE	SP	NE	NR	NE	NR	NR	NE	NR	NE	R	NR	NE	NR
Hept/EtOH w/ 0.1% TFA + 0.1% IPAm																	
(R)-NGLY / DNB	NR	NE	NR	SP	NE	NR	NE	NE	NE	NR	NE	NE	NE	SP	NE	NE	NR
(S)-ICA / (R)-NEA	NR	SP	SP	SP	NE	SP	NR	NE	NE	SP	NE	SP	NE	SP	NR	NE	NR
(S)-Leu / (R)-NEA	NR	NR	SP	R	NR	SP	NR	NE	NE	NR	NE	NE	NE	NR	NR	NE	NR
(S)-PRO / (R)-NEA	NR	NR	R	SP	NE	NR	SP	NE	NR	NR	NR	NR	NE	NR	NR	NE	NR
(R)-PGLY / DNB	NR	NR	NR	SP	NE	NE	NR	NE	NR	SP	NE	NE	NE	NR	NE	NE	NR
(R)-PGLY / DNAn	NR	NR	NR	SP	SP	NE	NE	NR	NE	NR	NE	NE	NE	NR	NE	NE	NR
Method? (Y/N/MD)	**N**	Y	Y	Y	**MD**	**MD**	**MD**	Y	**MD**	**MD**	Y	**MD**	**MD**	Y	**N**	Y	Y

NE - No Elution
NR - Not Resolved
SP - Split Peak (R<1.5)
R -Resolved (R> 1.5)

Method:
Y- screen method posted suitable resolution
N-screen method did not indicate potential for resolution
MD -screen method indicated potential for resolution, but R < 1.5

Source: [40].

that the same columns have stereoselectivity whether the chromatography was in either the normal- or reversed-phase mode. The reversed scheme was only slightly less effective than the equivalent column screen in the normal phase [36], where 19 of the 20 drug substances were fully resolved (Rs $\geq$ 1.5). The reversed-phase screen was as effective as Pirkle columns in the normal phase for the same library set [40]. Again, the reversed-phase conditions offer efficiency advantages for the analyst in not having to perform several system rinses in solvents in order to switch from their routine reversed-phase mode to the normal phase (or dedicating additional LC systems to normal-phase chiral applications).

TABLE 8 Regis Pirkle Chiral Column Screening Results

Mobile Phase	trans-Stilbene Oxide	Econazole nitrate (M1)	Promethazine HCl (M2)	Bupivacaince HCl (M3)	Bendroflumethiazide (M4)	Propanolol (M5)	Guaiacol Glyceryl Ether (M6)	Ketoprofen (M7)	Ibuprofen (M8)	Prilocaine HCl (M(9)	Warfarin (M10)	PHA-148575A (M11)	PHA-200578A (M12)	PF-557685 (M14)	PF-00190434/PF-00190431 (M15)	PD-0134298/PD-0134299 (M16)	SC-75416/SC-75417 (M17)	PHA-797804/PHA-797805 (M20)
Hept/EtOH w/ 0.1%ESA																		
Whelk-O	SP	NR	NR	NR	R	NR	NR	SP	NR	NR	NR	NR	NR	NR	NR	R	NR	NR
Ulmo	NR	NR	NR	NR	SP	NR	NR	NR	NR	SP	NR	NR	NR	NR	NR	NR	NR	NR
β-Gem	NR	NR	NR	NR	NR	NR	NR	NR	NR	NE	NR	NR	NR	NR	NR	NR	NR	NR
α-Burke	NR	NR	NR	R	NR	SP	SP	NR	NR	SP	NR	NR	NR	NR	NR	NR	SP	NR
Leucine	NR	NR	NR	R	NR	SP	SP	NR	NR	SP	NR	NR	NR	NR	NR	NR	NR	NR
DACH	NR	NR	NR	NR	NR	NE	NE	NR	NR	NR	NR	NR	NR	NR	NE	NR	NR	NR
Hept/EtOH w/ 0.1% TFA																		
Whelk-O	R	NR	NR	NR	SP	NR	NR	SP	NR	NR	R	NR	NR	SP	NR	NR	NR	NR
Ulmo	NR	NE	NR	NE	SP	NR	NR	NR	NR	NR	SP	NE	NR	NR	NR	NR	NR	NR
β-Gem	NR	NR	NR	NE	NR	NE	NE	NR	NE	NE	NR	NR	NR	NR	NR	SP	NR	NR
α-Burke	NR	NR	NE	NE	NR	NR	NE	NR	NR	NR	NR	NR	NR	NR	NR	NR	NE	NR
Leucine	NR	NE	NE	NE	NR	NE	NE	NR	NR	NE	NR	NE	NE	NR	NR	NR	NR	NR
DACH	NR	NE	NE	NR	SP	NR	NR	NR	NE	NR	NR	NE	NR	NR	NE	NR	NR	NR
Hept/EtOH w/ 0.1% NH4TFA																		
Whelk-O	R	NR	NR	NR	SP	SP	SP	SP	NR	NR	R	NR	NR	SP	NR	NR	NR	NR
Ulmo	NR	NR	NR	NR	SP	NR	SP	SP	NR	SP	SP	NR	NR	NR	NR	NR	NR	NR
β-Gem	NR	NR	NE	NR	NR	NR	NR	NR	NR	NR	NR	NR	NR	NR	NR	NR	NR	NR
α-Burke	NR	NR	NR	NR	NR	NR	NE	NR	NR	SP	NR	NR	NR	NR	NR	NR	NR	NR
Leucine	NR	NR	NE	R	NR	NR	SP	NR	NR	SP	NR	NR	NR	NR	NR	NR	NR	NR
DACH	NR	NR	NR	NR	SP	NR	NR	NR	NR	NR	SP	NE	NR	NR	NE	NR	NR	NR
Hept/EtOH w/ 0.1% NH4OAc																		
Whelk-O	R	NR	NR	R	SP	NR	NR	R	SP	NR	SP	NR	NR	SP	NR	NR	NR	NR
Ulmo	NR	NR	NR	NR	SP	NR	NR	NR	NR	NR	SP	NR	NR	NR	NR	NR	NR	NR
β-Gem	NR	NR	NR	NR	NR	NR	NR	NR	NR	NR	NR	NR	NR	NR	NR	NR	NR	NR
α-Burke	NR	NR	SP	SP	NR	SP	SP	NR	NR	NR	NR	NR	NR	NR	NR	NR	NR	NR
Leucine	NR	NR	NR	NR	NR	SP	SP	NR	NR	NR	SP	SP	NR	NR	NR	NR	NR	NR
DACH	NR	NR	NR	NR	SP	NR	NR	NR	NR	NR	NR	NE	NR	NR	NE	NR	NR	NR
Hept/EtOH w/ 0.1% IPAm																		
Whelk-O	R	NR	NR	R	SP	SP	SP	SP	NR	NR	SP	NR	NR	SP	NR	NR	NR	NR
Ulmo	NR	NR	NR	NR	SP	NR	NR	NR	NE	NR	SP	NR	NR	NR	NR	NR	NR	NR
β-Gem	NR	NR	NR	NR	NR	NR	NR	NR	NR	NR	NR	NR	NR	NR	NR	NR	NR	NR
α-Burke	NR	NR	NR	NR	NR	SP	SP	NR	NR	NR	NR	NR	NE	NR	NR	NR	NR	NR
Leucine	NR	NR	NR	NR	NR	NR	NR	NR	NR	NR	NR	NR	NE	NR	NR	NR	NR	NR
DACH	NR	NR	NR	NR	SP	NR	NR	NR	NE	NR	NR	NE	NR	NR	NE	NR	NR	NR
Hept/EtOH w/ 0.1% IPAm; 0.1% TFA																		
Whelko	R	NR		R	R	SP	NE	SP	R	NR	R	NR	NR	SP	NR	NR	NR	R
Ulmo	NR	NR		NR	NR	NR	NE	NR	NR	SP	SP	NR	NR	NR	NR	NR	NR	NR
Hept/EtOH																		
Whelk-O	R	NR	NR	R	SP	NR	NR	R	NE	NR		NR	NE	SP	NR	NR	SP	NR
Ulmo	NR	NE	NE	NR	SP	NE	NE	NR	NR	NR		NE	NE	NR	NR	NR	NR	NR
β-Gem	NR	NE	NR	NR	NR	NE	NE	NE	NR	NR		NE	NR	NR	NR	NR	NR	NR
α-Burke	NR	NE	NR	NR	NR	NE	NE	NR	NR	NR		NE	NR	NR	NR	NR	NR	NR
Leucine	NR	NE	NR	NE	NR	NE	NE	NR				NE	NR	NR	NR	NR	NR	NR
DACH	NR	NE	NR	NR	SP	NE	NE	NE				NE	NR	NR	NR	NE	NE	NR
Method? (Y/N/MD)	Y	N	MD	Y	Y	MD	MD	MD	MD	MD	Y	MD	N	MD	N	MD	MD	Y

NE - No Elution
NR - Not Resolved
SP - Split Peak (R<1.5)
R -Resolved (R> 1.5)

Source: [40].

9 CONCLUSIONS

Acceptable chiral resolution was obtained for all 20+ compounds in the screening library. Whether traditional normal phase or SFC is applied, and surprisingly, whether using normal phase or reversed phase, routine analytical chiral screening

TABLE 9 Reversed-Phase Chiral Columns[a]

Chiralcel OD-RH, 150 × 4.6 mm, Chiral Technologies (West Chester, PA)
Chiralpak AD-RH, 150 × 4.6 mm, Chiral Technologies
Chiralcel OJ-R, 150 × 4.6 mm, Chiral Technologies
Chiralpak AS-RH, 150 × 4.6 mm, Chiral Technologies
(*S*,*S*)-Whelk-O, 10 μm, 250 × 4.6 mm, Regis Technologies (Morton Grove, IL)

Source: Adapted from [41].
[a]Detection: UV at 254 nm.

TABLE 10 Reversed-Phase Chiral Column Screening Results

Mobile Phase	Econazole nitrate (M1)	Promethazine HCl (M2)	Bupivacaince HCl (M3)	Bendroflumethiazide (M4)	Propanolol (M5)	Guaiacol Glyceryl Ether (M6)	Ketoprofen (M7)	Ibuprofen (M8)	Prilocaine HCl (M9)	Warfarin (M10)	Proprietary (M11)	Proprietary (M12)	Proprietary (M14)	Proprietary (M15)	Proprietary (M16)	Proprietary (M17)	Proprietary (M20)	Naproxen	Lansoprazole
H_2O/MeOH																			
OD-RH	SP	SP	SP	SP		R	SP	NR	NE		NR	NR	R	R	SP	NR	NR	SP	SP
AD-RH	SP	SP	SP	SP		NR	NR	NR	NE		NR	NR	R	NR	R	SP	SP	SP	NR
OJ-R	R	NR	SP	SP		NR	NR	NE	NE		NR	NR	NR	NR	R	SP	SP	NR	NR
AS-RH	NR	NR	NR	SP		NR	SP	NR	NE		NR	NR	NR	NR	NR	SP	SP	SP	NR
Chiral Coat OD	NR	R	NR	NR		NR	NR	NE	NR		R	R	SP	SP	NR	NR	NR	NR	NR
Whelko	NR	NE	SP	SP		NR	SP	NR	SP		NR	NE	NE	NR	NR	SP	SP	R	NE
H_2O/MeOH w/ 0.1%ESA																			
OD-RH	NR	NR	NR	SP		R	NR	NR	NR		NR	NR	R	R	R	SP	NR	SP	SP
AD-RH	NR	NR	NR	NR		NR	NR	NR	NR		NR	NR	R	NR	R	SP	NR	SP	NR
OJ-R	NR	NR	NR	NR		NR	NR	SP	NR		NR	NR	R	NR	R	SP	NR	NR	NR
AS-RH	NR	NR	NR	NR		NR	NR	SP	NR		NR	NR	R	NR	NR	NR	NR	SP	SP
Chiral Coat OD	SP	R	NE	NR		NE	NR	NR	R		R	R	SP	SP	NR	NR	NE	NR	NR
Whelko	NE	NE	NR	NR		NR	NR	NR	NR		NR	NR	NR	NR	R	NR	NR	R	NE
H_2O/MeOH w/ 0.1%TFA																			
OD-RH	NR	NR	NE	NE	R	R	NR	NR	NE	R	NR	NR	R	R	SP	SP	NR	SP	NR
AD-RH	NR	NR	NR	NR	NR	NR	NR	NR	NR	SP	NR	NR	R	NR	R	SP	NR	SP	NR
OJ-R	NR	NR	NR	NR	SP	NR	NR	SP	NR	SP	NR	NR	R	NR	R	SP	NR	NR	NR
AS-RH	NR	NR	NR	NR	NR	NR	NR	NR	NE	SP	NR	NR	R	NR	NR	NR	NR	SP	NR
Chiral Coat OD	SP	R	NE	NE	NR	NE	NE	NR	NE	SP	R	R	SP	SP	NR	NE	NR	NR	NE
Whelko	NE	NE	NR	NR	R	NR	NR	NR	NR	R	NR	NR	NR	NR	R	NR	NR	R	NE
H_2O/MeOH w/ 0.1%TFA + 0.1% IPAm																			
OD-RH	NR	NR	SP	SP		SP	NR	NR	NE		NR	NR	R	R	SP	SP	SP	R	SP
AD-RH	NR	NR	NR	NR		SP	NR	SP	NR		NR	NR	R	NR	NR	NR	NR	R	SP
OJ-R	NR	NR	SP	NE		NR	NR	SP	NE		NR	NR	R	NR	SP	SP	SP	R	NR
AS-RH	NR	NR	NE	R		NR	NR	NR	NE		NR	NR	R	NR	NR	NR	NR	R	R
Chiral Coat OD	NR	R	NE	NE		NE	NE	NR	NE		R	R	SP	SP	NR	R	NR	NR	R
Whelko	NE	NE	SP	SP		NR	NR	NR	NR		NR	NR	NR	NR	SP	SP	NR	SP	NE
H_2O/MeOH w/ 0.1%DEA																			
OD-RH	NR	NR	NR	SP	NR	NR	NR	NR	NR	NR	NR	R	R	R	NR	NR	NR	SP	R
AD-RH	NR	NR	NR	SP	SP	NR	NR	NR	NR	SP	NR	NR	R	NR	NR	NR	SP	R	NR
OJ-R	NR	NR	NE	SP	NR	NR	NR	NR	NR	NR	NR	NR	R	NR	NR	NR	R	SP	NR
AS-RH	NR	NR	NR	SP	NR	NR	NR	NR	NR	NR	NR	NR	R	NR	NR	NR	R	NR	NR
Chiral Coat OD	NR	R	SP	NR	NR	NR	NR	NR	NR	NR	NR	SP		SP	NR	NR	SP	NR	NR
Whelko	NE	NR	SP	NR	R	SP	NR	SP	NR	NR	SP	SP	NR		SP	NR	SP	NR	NR
Method? (Y/N/MD)	Y	Y	MD	Y	Y	Y	MD	MD	MD	Y	Y	Y	Y	Y	Y	MD	Y	Y	Y

NE - No Elution	Method:
NR - Not Resolved	Y- screen method posted suitable resolution
SP - Split Peak (R<1.5)	N-screen method did not indicate potential for resolution
R -Resolved (R> 1.5)	MD -screen method indicated potential for resolution, but R < 1.5

Source: [41].

is very effective for representative chiral compounds using polysaccharide (AD, OD, AS, and OJ) chiral stationary-phase platforms.

Corresponding chiral method development screens using Pirkle stationary phases can provide resolution opportunities similar to those of the polysaccharide stationary phases prevalent in the pharmaceutical industry. While the literature suggests an 80% success rate with polysaccharide stationary phases, Pirkle stationary phases under similar conditions can provide positive results approximately 70% of the time. Using a series of screens based on Pirkle stationary-phase platforms warrants further investigation as complementary to the common polysaccharide-based screening system. Opportunities exist to take further advantage of the capabilities and solvent choices available to Pirkle stationary phases.

Reversed-phase chiral stationary phases can provide resolution opportunities similar to the polysaccharide stationary phases prevalent in the pharmaceutical industry running in the normal phase. Reversed-phase chiral systems have the advantage of not requiring the analyst to perform several system rinses when changing from their routine achiral reversed-phase mode of operation to a typical normal-phase mode for achiral investigations.

Of course, for preparative chromatography, normal and supercritical fluid chromatographic systems pose significant advantages in recovering enantiomers from the mobile phase. However, for analytical operations, analysts prefer overwhelmingly not to have to switch to normal-phase modes for chiral analysis. The primary focus, however, remains for chiral method development screening protocols to be optimized to predict the best opportunity for success in yielding a robust analytical method for the NCEs under investigation.

REFERENCES

1. L. R. Snyder, J. J. Kirkland. In *Introduction to Modern Liquid Chromatography*, 2nd ed., R. H. Thompson, A. S. Pereira, N. D. Meyer, Eds. Wiley, New York, 1980.
2. C. Perrin, V. A. Vu, N. Matthijs, M. Maftouh, D. L. Massart, Y. Vander Heyden. *J. Chromatogr. A*, 947:69, 2002.
3. C. Perrin, V. A. Vu, N. Matthijs, M. Maftouh, D. L. Massart, Y. Vander Heyden. *J. Chromatogr. A*, 966:119, 2002.
4. M. L. de la Puente, C. T. White, A. Rivera-Sagredt, J. Reilly, K. Burton, G. Harvey. *J. Chromatogr. A*, 983:101, 2003.
5. M. E. Andersson, D. Asclan, A. Clark, J. Roeraade, G. Hagman. *J. Chromatogr. A*, 1005:83, 2003.
6. E. Lipka, V. Glacon, G. Mackenzie, D. Ewing, C. Len, D. Postel, M. P Vaccher, J. P Bonte, C. Vaccher. *Anal. Lett*., 37:385, 2004.
7. C. K. Esser, R. M. Black, D. Von Langen. *Pharm. Discov*., 26, 2004.
8. Y. Zhang, W. Watts, L. Nogle, O. McConnell. *J. Chromatogr. A*, 1099:75, 2004.
9. T. A. Berger. *Packed Column SFC*. Royal Society of Chemistry, Cambridge, UK, 1995.
10. G. Terfloth. *J. Chromatogr. A*, 906:301, 2001.

11. K. K. Anton, C. Berger, Eds., *Supercritical Fluid Chromatography with Packed Columns*. Marcel Dekker, New York, 1998.
12. Y. Zhao, G. Woo, S. Thomas, D. Semin, P. Sandra. *J. Chromatogr. A*, 1003:157, 2003.
13. P. Borman, B. Boughtflower, K. Cattanach, K. Crane, K. Freebairn, G. Jonas, I. Mutton, A. Patel, M. Sanders, D. Thompson. *Chirality*, 15:S1–S12, 2003.
14. L. Di, O. J. McConnell, E. H. Kerns, A. G. Sutherland. *J. Chromatogr. B*, 809:235, 2004.
15. M. L. Riekkola, S. K. Wiedmer, I. E. Valko, H. Siren. *J. Chromatogr. A*, 735:57, 1996.
16. A. Guttman, S. Brunet, N. Cooke. *LCGC*, 14:36, 1996.
17. A. Guttman. *Electrophoresis*, 16:1900, 1995.
18. A. Guttman, S. Brunet, C. Jurado, N. Cooke. *Chirality*, 7:409, 1995.
19. A. Guttman, N. Cooke. *J. Chromatogr. A*, 680:157, 1994.
20. J. B. Vincent, A. D. Sokolowski, T. V. Nguyen, G. Vigh. *Anal. Chem.*, 69:4226–4233, 1997.
21. J. B. Vincent, D. Kirby, T. V. Nguyen, G. Vigh. *Anal. Chem.*, 69:4419, 1997.
22. H. Cai, T. V. Nguyen, G. Vigh. *Anal. Chem.*, 70:580, 1998.
23. J. Chapman, F.-T. A. Chen. *LCGC*, 19:427, 2001.
24. G. Terfloth. *J. Chromatogr. A*, 906:301, 2001.
25. S. Li, G. Vigh. *Electrophoresis*, 25:1201, 2004.
26. W. Zhu, G. Vigh. *Electrophoresis*, 24:130, 2003.
27. C. White. *J. Chromatogr. A*, 1074:163, 2005.
28. M. Maftouh, C. Granier-Loyaux, E. Chavanna, J. Marini, A. Pradines, Y. Vander Heyden, C. Picard. *J. Chromatogr. A*, 1080:67, 2005.
29. R. Szücs, I. Caron, K. A. Taylor, S. P. Gee, P. D. Ferguson, M. A. Kelly, J. V. Beaman, A. M. Lipczynski, P. A. Hailey. *J. Microcolumn Sep.*, 12:568, 2000.
30. D. K. Maynard, G. Vigh. *Electrophoresis*, 22:3152, 2001.
31. L. Liu, M. A. Nussbaum. *J. Pharm. Biomed. Anal.*, 19:679, 1999.
32. G. K. E. Scriba. *J. Pharm. Biomed. Anal.*, 27:373, 2002.
33. Chiraldex GC columns. In *A Guide to Using Cyclodextrin Bonded Phases for Chiral Separations by Capillary Gas Chromatography*. Advanced Separation Technologies, Whippany, NJ, (2002).
34. K. Ramig, A. Krishnaswami, L. A. Rozov. *Tetrahedron*, 52:319, 1996.
35. F. J. Akin, J. L. Antosz, K. F. Ausec, R. L. Greve, L.-E. Johnson, Magnusson, T. Ramstad, S. L. Secreast, D. S. Seibert, G. K. Webster. *Curr. Pharm. Anal.*, 3:53, 2007.
36. M. M. Wong, W. B. Holzheuer, G. K. Webster. *Curr. Pharm. Anal.*, 4:101–105, 2008.
37. J. H. Kennedy. *J. Chromatogr.*, 725:219–224, 1996.
38. B.-H. Shao, X.-Z. Xu, Q.-Z. Wu, J.-D. Lu, X.-Y. Fu. 28:63–80, 2005.
39. B.-H. Shao, X.-Z. Xu, Q.-Z. Wu, J.-D. Lu, X.-Y. Fu. *Huaxue Xuebao*, 61:1635–1640, 2003.
40. W. B. Holzheuer, M. M. Wong, G. K. Webster. *Curr. Pharm. Anal.*, 5:10–20, 2009.

41. W. B. Holzheuer, M. M. Wong, G. K. Webster. *Curr. Pharm. Anal.*, in press.
42. A. Ishikawa, T. Shibata. *J. Liq. Chromatogr.*, 16:859–878, 1993.
43. H. Y. Aboul-Enein, V. Serignese. *Biomed. Chromatogr.*, 8:22–25, 1994.
44. J. Mookdasanit, H. Tamura. *Food Sci. Technol. Res.*, 8:367–372, 2002.
45. A. L. Zhou, X. Y. Lu, Y. X. Xie, G. Q. Liu, H. X. Wang, X. S. Wang, R. Y. Gao. *Chromatographia*, 60:281–285, 2004.
46. L.-M. Yuan, Z.-G. Xu, P. Ai, Y.-X. Chang, A. K. M. F. Azam. *Anal. Chim. Acta*, 554:152–155, 2005.
47. F.-R. Ning, K.-K. Huang, S.-Q. Liu, F.-P. Jiao. *Fenxi Shiyanshi*, 25:1–5, 2006.
48. Y. Fu, X. Bai, C. Guo, Z. Wang. *Yaowu Fenxi Zazhi*, 26:830–832, 2006.
49. F. Liu, X. Cao, Q. Wang, D. Jin. *Ziran Kexueban*, 33:33–36, 2007.
50. K.-A. Lee, S. Yeo, K. H. Kim, W. Lee, J. S. Kang. *J. Pharm. Biomed. Anal.*, 46:914–919, 2008.
51. J.-H. Liu, B.-Y. Yu, Y.-J. Chen. *Chirality*, 20:51–53, 2008.

CHAPTER 8

Chiral Screening Methods for Pharmaceutical Analysis and Purification in an Industrial Laboratory

ROBERT DEPIANTA

Analytical Chemistry and Sample Logistics, Pfizer Global Research and Development, Groton, Connecticut

KENNETH DOUVILLE

Analytical Development, Vertex Pharmaceuticals, Inc., Cambridge, Massachusetts

BEVERLY NICKERSON

Analytical Development, Pfizer Global Research and Development, Groton, Connecticut

RICARDO E. BORJAS

Analytical Development, Vertex Pharmaceuticals, Inc., Cambridge, Massachusetts

1 INTRODUCTION

A large number of prescribed medicines are used as racemic mixtures of chiral drugs. It is known that the enantiomers that make up the racemate may have different pharmacological and toxicological properties [1–4]. These differences in enantiomer activities led to the thalidomide tragedy, in which its enantiomers had distinct drug activity: the (*S*)-enantiomer was a powerful teratogen, whereas the (*R*)-enantiomer provided the desirable effects of inducing sleepiness (sedative). Interesting and important new evidence has emerged claiming that the (*R*)-enantiomer may undergo chiral inversion (racemization) in vivo, thus casting doubt on the safety of the (*R*)-enantiomer [5–8]. It is

Chiral Separation Methods for Pharmaceutical and Biotechnological Products, Edited by Satinder Ahuja

also interesting to note that there is a resurgence of interest in thalidomide, due to its immunomodulatory and antiangiogenic properties [4]. In August 1998, thalidomide was approved for sale in the United States for the chronic treatment of erythema nodosum leprosum, a painful inflammatory dermatological reaction of lepromatus leprosy [9]. The incredible story of thalidomide is a product of advances in the chemical and biological sciences, in which analytical chemistry has a very important role. Analytical separation of the thalidomide enantiomers, and their characterization, led to deeper understanding of the pharmacological and toxicological properties of this once infamous drug [5–9].

The continuous search for increased efficiency and productivity in industrial laboratories has led to the use of technologies that allow the consolidation of tasks that are time consuming for scientific staff to perform. Method development and routine testing are two areas in which analytical scientists spend most of their time. Chiral chromatography method development remains a trial-and-error approach despite the advances made in an understanding of the mechanisms of enantioseparation. In our organizations, we have tackled chiral method development by using screens that automatically run a series of chromatographic conditions. These screens have evolved over time based on experience and scientific knowledge. They require minimal supervision, which allows multitasking of activities. The screens do generate large systematic files of data in which the selection of the "best method" remains a critical unautomated part of the process and requires considerable expertise.

Figure 1 shows typical stages in the development of a medicinal drug. The very early stage, from the later stage of lead seeking to candidate development in discovery, is characterized by analysis of a large number of samples; therefore, the productivity–efficiency–quality triangle becomes distorted, with an emphasis on productivity and efficiency. The resulting methods at this stage are *fit for purpose*; that is, the method used is sufficient to answer the question *at the time of need*, but it will probably change as the development progresses. As the candidate selected advances to the preclinical stage and eventually to clinical trials, the triangle distorts, emphasizing quality. At late stages of development, the concepts of robustness and ruggedness become important; consequently, more time is spent in the selection and optimization of methods, to ensure proper controls for quality and safety.

Our aim in this chapter is to review the fundamentals of chiral separation and to discuss chiral chromatographic screening models. These models are introduced generically; that is, the screening models presented are a subset of many possible variations; these variations come from the needs of the pharmaceutical development stage and the projects themselves. Few papers in the field of chiral chromatographic screens have been published [10–14]. We discuss multiplex techniques based on column and solvent switching valves and the parallel technology developed by Sepiatec and Combisep. We also review different separation modes that can be used in setting up the screens: supercritical fluid chiral chromatography; normal-, polar-, and reversed-phase chromatography; and chiral electrokinetic separation.

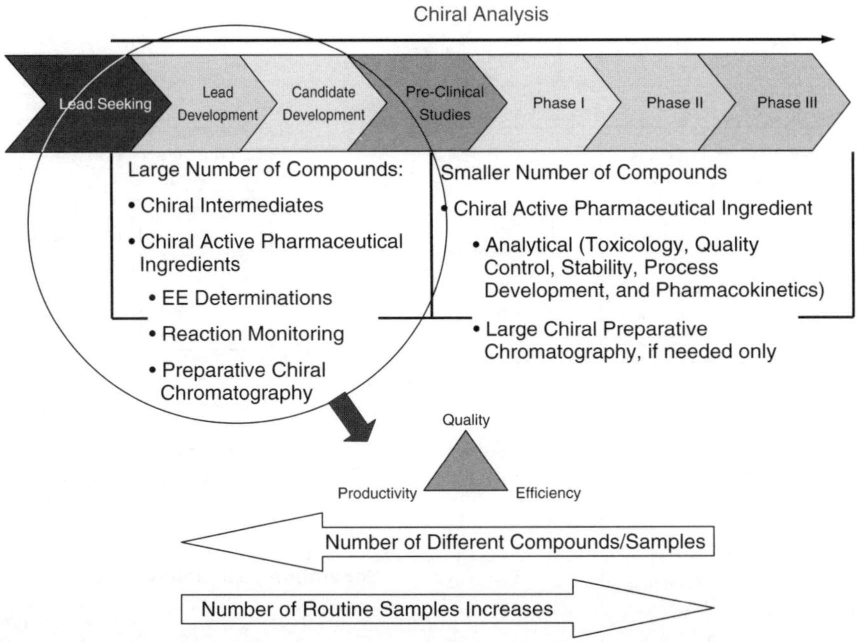

Figure 1 Typical pharmaceutical development stages. The work detailed in this chapter corresponds to the stages described within the circle. These stages are characterized by the need to analyze a large number of compounds and/or samples and the need for fast turnaround times, 2 to 24 hours for analytical samples and 1 day to 1 week for preparative-scale separations, depending on the scale. The quality–efficiency–productivity triangle is distorted at these stages, emphasing efficiency and productivity, and resulting in a method that is "fit for purpose." As the project progresses, the number of different compounds decreases but the number of routine samples per compound increases (e.g., samples for stability, releases, formulation support, and drug substance and formulation process development). Analytical methods at later stages of development are characterized by more rigorous development and validation procedures, and the quality–efficiency–productivity triangle emphasizes quality.

2 FUNDAMENTALS OF CHIRAL SEPARATION

Before introducing the fundamentals of chiral separation, it is important to point out that one critical goal in chiral drug development is to obtain enantiomerically pure compounds to perform all appropriate pharmacological and toxicological studies. There are different ways to address enantioselective preparations: *chiral pool synthesis, asymmetric induction, asymmetric catalysis, chiral resolution by crystallization*, and *chiral resolution by preparative chromatography*. Once the desired enantiomer is prepared, the next step is to determine its enantiomeric purity. To accomplish such a task, one should select the most appropriate technique for the compound in question. There are a considerable number of

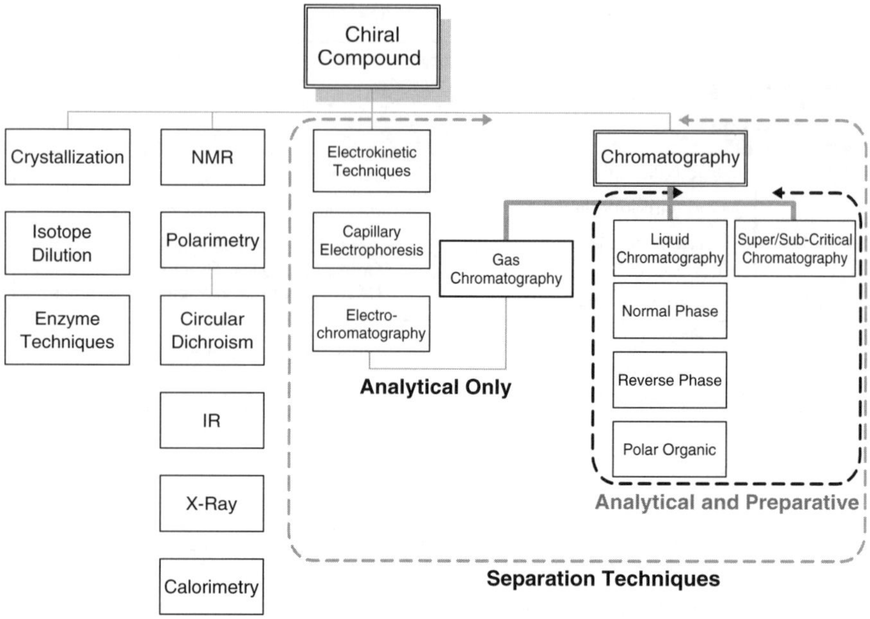

Figure 2 Techniques for the characterization and analysis of chiral compounds. The list of techniques shown is not comprehensive, but is intended to demonstrate the broad range of techniques available to the chemist, from chemical procedures to physical ones. This chapter deals exclusively with separation techniques.

techniques available, and some of them are shown in Figure 2. The techniques circled and labeled as separation techniques are discussed in this chapter.

2.1 Chiral Recognition

A model for the mechanism of chiral recognition was developed by Pirkle and Pochapsky [15]. In order for enantiomers to be chromatographically separated on a *chiral stationary phase* (CSP), each enantiomer must form transient diastereomeric complexes with the CSP, and the stabilities of these complexes must differ from each other to allow chromatographic separation to occur. The enantiomer that forms the less stable complex will be less retained and hence will elute earlier than will the enantiomer that forms the more stable complex. The concept of forming transient complexes implies the establishment of interactions between the chiral selector and the enantiomer; therefore, it is important to establish the minimum set of criteria required to form the diastereomeric complexes. Pirkle's model [15], also known as the *three-point rule*, states *that there must be three simultaneous interactions between the chiral selector and at least one of the enantiomers, and that one of these interactions must be stereochemically dependent*. It should be noted that the term *interaction*, as explained clearly by Maier and Lidner [1], *refers to intermolecular physical forces and their steric implications rather than to specific spatial relationships between the chiral selector and the enantiomers*.

2.2 Chiral Interactions Based on Intermolecular Physical Forces

The interactions that give rise to the formation of the diastereomeric complexes, in decreasing order of strength [16–17], are *hydrogen bonding, ion–dipole, dipole–dipole* (*Keesom*), *dipole-induced dipole* (*Debye*), *London dispersion forces, electron pair donor–acceptor* (π–π *interactions*), and *inclusion*. It should be noted that inclusion is not a type of intermolecular physical force, but such interactions help to establish stable diastereomeric complexes.

Hydrogen Bonding Hydrogen bonding is perhaps the most important interaction for many of the CSPs available. Hydrogen bonding takes place when an acidic proton comes within close proximity to an electron pair donor or a hydrogen-bond acceptor, such as oxygen (carbonyl, carboxyl, alcohols, and ethers), nitrogen (primary and secondary amines), halogens, and sulfur (thiols). Hydrogen-bonding energies can range from 20 to 150 kJ/mol, depending on the distance and the types of donor and acceptor groups. Hydrogen bonds are nearly linear, asymmetrical, and equal to the distance between the donor and the acceptor.

Ion–Dipole Ion–dipole is a coulombic attraction between an ionic and a polar compound. The potential energy of the ion–dipole interaction is defined mathematically as

$$E_{\text{i-d}} = \frac{-W^2 z\mu(\cos\alpha')}{\varepsilon r^2} \tag{1}$$

where z is the ion's charge, μ is the dipole moment of the polar molecule (which is defined as $\mu = qR$, where q is the charge magnitude and R is the separation between the charges). The angle between the ion and the dipole's center defines α'; ε is the relative electric permittivity and characterizes the ability of a solvent to pass a charge ($\varepsilon_{\text{water}}$ is 81 F/m, $\varepsilon_{\text{acetonitrile}}$ is 36.6 F/m, $\varepsilon_{\text{methanol}}$ is 33 F/m, $\varepsilon_{\text{ethanol}}$ is 24.3 F/m, $\varepsilon_{\text{1-propanol}}$ is 20.1 F/m, $\varepsilon_{\text{hexane}}$ is 2.3 F/m, and $\varepsilon_{\text{heptane}}$ is 1.92 F/m). W is a system constant and depends on the unit system applied, and r is the distance from the ion to the center of the dipole. Note that if the diastereomeric complex is based strongly on this interaction, the use of polar solvents (high ε) will decrease the E_{i-d} and, consequently, decrease retention.

Dipole–Dipole Dipole–dipole interaction occurs between two polar molecules possessing permanent dipole moments. The degree of attraction is dependent on the orientation of the two molecules, and the population of the multiple orientations that it can take follows the Boltzmann distribution, which is temperature dependent. The energy of the dipole–dipole interaction between molecules 1 and 2, E_{d-d}, is given by

$$E_{\text{d-d}} = -W^2\left(\frac{2}{3}KT\right)\frac{\mu_1^2\mu_2^2}{\varepsilon r^6} \tag{2}$$

All symbols are as in Eq. (1), except that K and T are the Boltzmann and the absolute temperature, respectively; therefore, low temperatures favor high dipole–dipole interaction.

Dipole–Induced Dipole Dipole–induced dipole interaction happens between a polar and a nonpolar molecule, in which the nonpolar molecule becomes polarized in the presence of a polar molecule that has a permanent dipole. As the nonpolar molecule is polarized, it generates a dipole (induced dipole). The interaction energy is given by

$$E_{\text{d-id}} = -W^2\left(\tfrac{1}{2}\right)(\alpha_1 E_2^2 + \alpha_2 E_1^2) \tag{3}$$

where α is the polarizability of molecules 1 and 2, and E_2 and E_1 are the electric fields generated by the polar and nonpolar molecules, respectively. E_{d-id} does not depend on temperature.

London Dispersion Forces London dispersion forces exist between all types of polar and nonpolar molecules. These forces arise from the concentration imbalance of the electron cloud around a molecule that induces a temporary dipole moment on a nearby molecule. The energy, E_{LD}, is given by

$$E_{\text{LD}} = -W^2\left(\frac{3}{2}\right)\left(\frac{I_1 I_2}{I_1 + I_2}\right)\frac{\alpha_1\alpha_2}{\varepsilon r^6} \tag{4}$$

where I_1 and I_2 are the ionization energies of the interacting molecules, and the other symbols are as defined earlier.

Electron Pair Donor–Acceptor ($\pi - \pi$ Interactions) Electron pair donor–acceptor interactions, also known as $\pi - \pi$ interactions, involve an electron pair from the donor higher energy level and a vacant acceptor lower energy level. Generally, π electrons from phenyl moieties are involved in $\pi - \pi$ interactions in chiral chromatography. This type of interaction is more pronounced in charged species than in neutral ones.

Inclusion In inclusion interaction an enantiomer diffuses into a chiral cavity to form diastereomeric inclusion complexes; this is the "host–guest" mechanism. There are two types of hosts, based on their cavity interior: hydrophobic and hydrophilic. The hydrophobic host favors inclusion of the hydrocarbon-rich part of molecules (e.g., cyclodextrin as a host), whereas the hydrophilic ones contains heteroatoms that allow bonding with an electron acceptor such as organic cations (e.g., crown ethers as a host). This enantiomeric separation mechanism is commonly used in gas chromatography and in electrokinetic-based separations.

Table 1 summarizes the classification of chiral phases, along with their description, chemistry, chiral separation mechanism based on the intermolecular interactions just described, and examples of these phases for each classification.

TABLE 1 Classification of Chiral Stationary Phases

Type	Description	Stationary-Phase Chemistry	Chiral Interaction Mechanism	Example
I	Brush (Pirkle), π-acidic or π-basic type	Various chiral selectors, ionic or covalent bonding	Attractive interactions H-bonding Charge transfer Dipole stacking	Whelk-O 1
II	Helical polymers	Cellulose/amylose derivatives Poly(methacrylates) Poly(acrylamides)	H-bonding Dipole interactions Inclusion	Chiralpak AD:amylose tris(3,5-dimethylphenylcarbamate) Chiralcel OJ: cellulose tris(4-methylbenzoate)
III	Cavity	Cyclodextrin Chiral crown ethers Macrocyclic glycopeptides	Inclusion $\pi-\pi$ Complexation H-bonding Anionic or cationic binding	Cyclobond columns Crownpak Chirobiotic columns
IV	Ligand exchange	Coordination complexes with metals	Solute is part of a diasteromeric metal complex	Davankov ligand-exchange chiral stationary phase
V	Protein	α_1-Acid glycoprotein Human serum albumin Cellobiohydrolase	Transient-diastereomeric complex based on combinations of hydrophobic interactions and polar interactions	Chiral: AGP HAS CBH

2.3 Thermodynamics of Enantiomer Separation

The chromatographic separation of enantiomers results from the reversible formation of a pair of diastereomer complexes. The formation can be written as

$$\mathrm{A}_R + \mathrm{CSP} \rightleftharpoons (\mathrm{A}_R \cdot \mathrm{CSP}) \tag{5}$$

$$\mathrm{A}_S + \mathrm{CSP} \rightleftharpoons (\mathrm{A}_S \cdot \mathrm{CSP}) \tag{6}$$

Here A_R and A_S represent enantiomers R and S in the mobile phase, $\mathrm{A} \cdot \mathrm{CSP}$ is the diastereomeric complex, and CSP is the chiral stationary phase. The difference in stability of the complexes controls the difference in retention times, wherein the enantiomer forming the less stable complex elutes first. The retention of the solute is given by the retention factor, k, and it represents the mass distribution of any given enantiomer between the mobile and stationary phases:

$$k = \frac{[(\mathrm{A} \cdot \mathrm{CSP})]_{\mathrm{SP}}}{[\mathrm{A}]_{\mathrm{MP}}} \tag{7}$$

$[(\mathrm{A} \cdot \mathrm{CSP})]_{\mathrm{SP}}$ and $[\mathrm{A}]_{MP}$ are the molar concentrations of the complex and the enantiomer, respectively. Expressing Eq. (7) in terms of moles, we obtain

$$k = \frac{[(\mathrm{A} \cdot \mathrm{CSP})]_{\mathrm{SP}}}{[\mathrm{A}]_{\mathrm{MP}}} \frac{V_{\mathrm{SP}}}{V_{\mathrm{MP}}} \tag{8}$$

V_{SP}/V_{MP} is the phase ratio, which will be denoted as ϕ. Therefore, Eq. (8) can be rewritten as

$$k = \frac{[(\mathrm{A} \cdot \mathrm{CSP})]_{\mathrm{SP}}}{[\mathrm{A}]_{\mathrm{MP}}} \phi \tag{9}$$

The ratio of the molar concentrations is the equilibrium distribution constant, K, and Eq. (9) simplifies to the form

$$k = K\phi \tag{10}$$

Since there is a k for each enantiomer, the enantioselectivity, α_{RS}, can be defined as

$$\alpha_{RS} = \frac{k_R}{k_S} \tag{11}$$

We can now express the enantioselectivity in terms of Gibbs free energy through its relation with the equilibrium constant, $\Delta G = -RT \ln K = -RT \ln(k/\phi)$. The enantioselectivity, α_{RS}, can then be written in terms of the Gibbs free energy,

$$\alpha_{RS} = \frac{k_R}{k_S} = \exp \sum \left(\frac{-\Delta G_R - \Delta G_S}{RT} \right) = \exp \Sigma \left(\frac{-\Delta\Delta G}{RT} \right) \tag{12}$$

The summation symbol, Σ, indicates the total molecular free energies of the (R)- and (S)-enantiomer interactions with the CSP. One must keep in mind that the enantioselectivity is given solely by $\Delta\Delta G$; hence, the column designer seeks ways to maximize $\Delta\Delta G$ but minimize the individual enantiomer's interaction Gibbs free energy, ΔG_R and ΔG_S. High individual Gibbs free energy translates to long retention times. Application of the Gibbs–Helmholtz equation to Eq. (12) gives

$$\ln \alpha_{R,S} = \frac{-\Delta\Delta H_{R,S}}{RT} + \frac{\Delta\Delta S_{R,S}}{R} \tag{13}$$

$\Delta\Delta H_{R,S}$ and $\Delta\Delta S_{R,S}$ correspond to the enthalpic and entropic components of the enantioselectivity, $\alpha_{R,S}$. The enthalpic contribution to the enantioselectivity tends to be negative, thus favoring chiral recognition, whereas the entropic contribution tends to be positive and counteracts the chiral recognition process. *Therefore, chiral recognition between the analyte and the CSP will be controlled enthalpically, and increasing temperature will decrease the enantioselectivity*.

CSPs based on macromolecules may show entropically controlled enantioselectivity because increasing temperature can lead to conformational changes in the backbone of the macromolecular CSPs. An important observation is that Eq. (12) assumes that all interactions are enantioselective in nature; however, the surfaces of the CSPs are heterogeneous in nature (supporting matrix, spacers, and linkers), which can lead to nonspecific interactions that in general are weak but higher in density than the enantioselective interactions. Consequently, the retention times observed reflect the sum of nonspecific and specific (enantioselective) interactions. Hence, the use of Eq. (13) in the form of van't Hoff's plots may produce apparent rather than true thermodynamic quantities, and the mechanistic interpretation of the data could be misleading.

3 HPLC CHIRAL SCREENS

A chiral screen consists of setting up a series of chromatographic methods that will be executed either sequentially or in parallel. The chiral screen can be used for method development or for scouting studies of samples. The goal of our work was to establish chiral screens that facilitate method development (e.g., methods to determine enantiomeric excess (ee) or to perform preparative-scale chiral isolations). Depending on the purpose of the method, the screen was parallel or sequential. Parallel screening supports moderate to high sample throughput, whereas sequential screening was used for low to moderate-sample-throughput applications.

3.1 Sequential Screen

The sequential screen consists of using solvent and column switching valves that allows automatic testing of columns with different media. The sequential screen is by nature slow, and the way to expedite the screen is through the use of fast HPLC methods. Keep in mind that the required speed of elution depends on the end goal of the screen. For moderate to high sample throughput, the HPLC methods should be short, with the goal of obtaining some level of separation on one or more columns. The best method can then be optimized off-line for its end use. The solvent switching valve is critical if the screen uses a binary pump to allow implementation of different desired mobile phases. The number of mobile phases commonly used in the screening method is greater than two. Switching valves and their operation are discussed below.

Switching Valve A switching valve usually consists of the following five parts:

1. A control unit that allows the valve selector to be rotated between the load and sample positions.
2. A connecting device that communicates the rotary movement to the rotor.
3. A valve body that houses the different ports necessary to provide connections to the mobile-phase supply, the column, the sample loop, the sample injection port, and a port to waste.
4. A rotor that actually selects the mode of operation of the valve and contains slots that can connect the alternative ports in the valve body to provide loading and sampling functions.
5. A preload assembly that furnishes adequate pressure between the faces of the rotor and the valve body to ensure a leaktight seal.

Column Switching Column switching is basically a method of column selection. The technique can be used pre- or postcolumn. The precolumn procedure, used in our work, is a valve-switching procedure that can be used to select a particular column or several columns on which to separate a particular sample. Postcolumn switching valve scan be used to allow the sample to flow to alternative detectors.

The technique of column switching increases the versatility of a liquid chromatograph significantly. A detailed example of a six-port valve arranged for six-column switching is shown in Figure 3, which is used in chiral reversed-phase HPLC screening system. In general, most switching valves operate similarly. The switching valve consists of 14 ports, including an "in" port, which allows the mobile phase to flow in the valve, and an "out" port, which allows the mobile phase to flow to the detector after it goes through the appropriate column. The remaining 12 ports are used to attach the six columns to the valve. The top six ports have a corresponding bottom port that the mobile phase flows into after going through the column. The valve assembly works through the six different columns by receiving an electronic signal from the controlling software (Empower by Waters or ChemStation by Agilent) at a scheduled time in the sample set. This signal rotates the rotor faceplate inside the valve counterclockwise, aligning the two holes on the rotor face with the appropriate column to proceed with the chromatographic run.

Ways to Speed Up Sequential Chromatographic Screens When working in an environment of moderate to high sample throughput, automation helps in

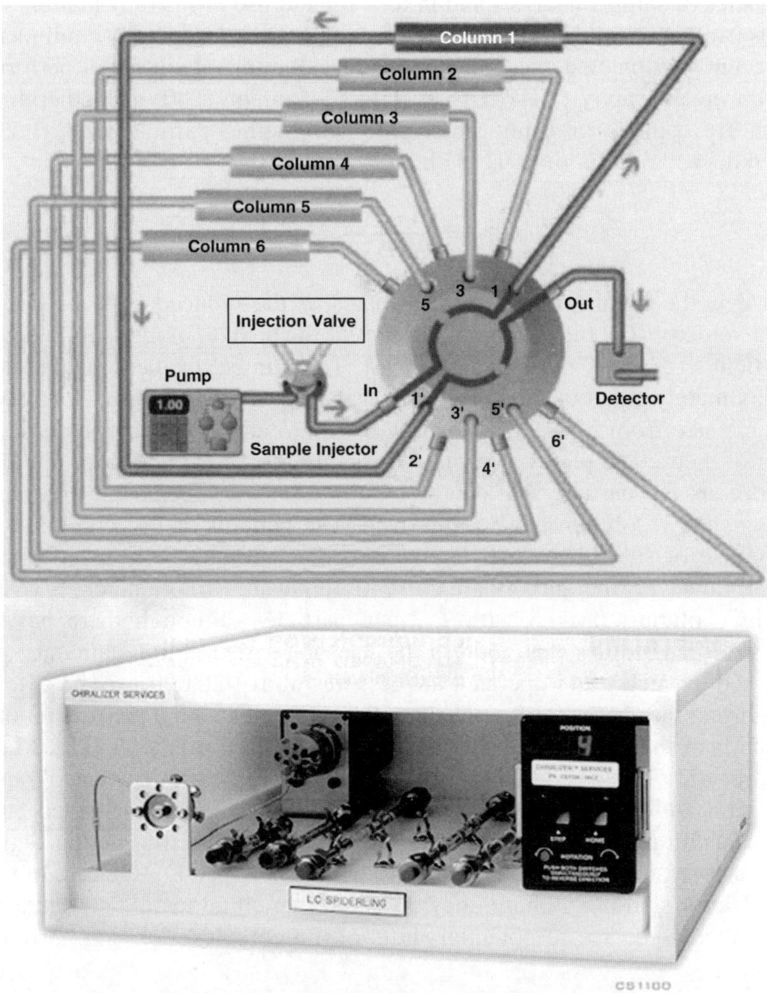

Figure 3 Use of six-column chiral screening value and the chiralizer. The top figure shows the six-column value used in the chiral RPLC screen. The front end of the value has the odd-numbered ports, and the back end has the even-numbered ports. The bottom of the valve is arranged similarly to the top but the bottom ports are the exit ports and are labeled with a number–prime. As an example, for the first two columns, column 1 is connected between ports 1 and 1′ and column 2 is connected between ports 2 and 2′. The mobile phase comes from a sample valve that enters through the "in" port that is adjacent to port 1′, and the detector is connected to the "out" *port* located between ports 1 and 6′. The sample route for column 1 follows the darker color; as the run completes the rotor changes to positions that allow column 2 to operate. (From Waters.) The bottom picture corresponds to the Chiralizer, which was also in our screens. (From ChiralizerServices.)

the logistics of sample analysis and in the efficient use of instrumentation, but the analysis itself is dependent on the method and the chromatography equipment. We have recently witnessed the commercial introduction of ultrahigh-performance liquid chromatography (UHPLC; i.e., UPLC, Agilent 1200, and others). These systems are capable of using small chromatographic particle sizes (<2 μm), which reduces analysis time, t_a [18]:

$$t_a = \frac{(1+k)Nh}{D_m v} d_p^2 \tag{14}$$

where N is the number of theoretical plates, h the reduced plate height, v the reduced velocity, D_m the diffusion constant of the analyte in the mobile phase, d_p the particle size, and k the retention factor. The number of theoretical plates, N, is approximately equal to $L/2d_p$. It is clear that for a given length, L, a reduction of particle size from 5 μm to 1.7 μm will lead to an appreciable increase in the number of theoretical plates. In fact, if L is constant, $N(1.7\ \mu\text{m})$ is $3 \times N(5\ \mu\text{m})$. Therefore, as the particle size decreases, the efficiency increases, time analysis decreases, and resolution is also improved. The benefits of fast chromatography with the use of sub-2-μm particle size have been discussed in recent literature [19–21]; however, the authors are currently unaware of commercially available chiral LC columns packed with < 2-μm particles, although there have been some published studies that address the use of bonded chiral stationary phases on < 2-μm particles and operated at pressures above 20,000 psi [22,23]. These ultrahigh-pressure procedures utilize capillary columns which provide relief from frictional heat that evolves when working at such high pressures [18]. Ma et al. [24,25] recently published the application of UPLC in chiral chromatography by using chiral additives of highly sulfated cyclodextrins. They achieved baseline enantioseparation in less than 3 min. Ringo and Evans [26] caution about the effects that pressure induces on CSPs even below 5100 psi. For the case of bonded β-*cyclodextrin* stationary phase, they observed that chiral resolution decreased as pressure increased, due to potential shifts of the complexation equilibrium with pressure.

Reduction in analysis time by reducing particle size allows fast sequential screens; however, the conversion of a UPLC method to conventional or to preparative LC systems may not be straightforward [27], whereas the reverse transfer is well documented [19–21]. Another approach to deliver fast LC methods is by using capillary columns, but at low inlet pressure (≪ 1000 psi). The abundant availability of commercial capillary LC systems allows greater access to the benefits of this technology, which provides greater mass sensitivity, reduction in analysis time, and cost-effectiveness.

The gain in mass sensitivity can be deduced easily from the chromatographic dilution factor, DF [28]:

$$\text{DF} = \frac{c_o}{c_{\max}} = \frac{\varepsilon_p \pi r^2 (1+k)\sqrt{2\pi L H}}{V_{\text{injection}}} \tag{15}$$

where c_o is the original compound concentration in a sample, $c_{\max}$ the final compound concentration at the peak maximum, r the column radius, L the column length, H the column plate height, ε_p the column porosity, $V_{\text{injection}}$ the injected sample volume, and π is the mathematical constant 3.1416. DF will increase or decrease proportionally to the square of the column radius and to the square root of the length and plate height of the column. Under further identical chromatographic and injection conditions, and assuming Gaussian peak shapes, $C_{\max}$ is inversely proportional to r^2. The decrease in analysis time is independent of the column diameter [29]. Several factors contribute to the decrease in analysis time upon reduction of column diameter, the critical one being the reduction in column volume, which translates into a reduction in the void volume, V_0, where $V_0 = 0.65V_c$, where V_c is the geometrical column volume of the LC column. Therefore,

$$V_R = V_0(1 + k) \tag{16}$$

The other factors that contribute to the decrease in analysis time are a faster equilibration time, which allows effective use of fast gradients, and reduced backpressure. The cost-effectiveness comes from the dramatic reduction in solvent consumption, and nowadays, environmental protection awareness is part of our job as chromatographers.

The use of sequential screens with conventional columns has the disadvantage of the equilibration time for gradient runs. The gradient retention factor, k^*, and resolution of the gradient elution are directly proportional to the gradient time, t_g, and the flow rate, F, but inversely proportional to the void volume, V_0, the change in organic fraction, $\Delta\phi$, and S, which is a constant for a given solute and experimental conditions and defined as $d(\log k)/d\phi$, where k and ϕ are the isocratic retention factor and the organic fraction, respectively. Mathematically [30],

$$k^* = \frac{t_g F}{1.15 V_0 S(\Delta\phi)} \tag{17}$$

Therefore, for a given or desired level of resolution, the gradient time can be shortened by increasing the flow rate and decreasing the void volume, both of which will lead to an increase in backpressure, especially if operating in the reversed-phase mode. One potential way to prevent the described dilemma is to operate the screen in a parallel fashion.

3.2 Parallel Chiral HPLC Screens

The concept of parallel chromatography arose from the need to perform high-throughput analysis with LC–mass spectrometry (MS) [31,32]. The instrumentation typically consisted of a gradient binary pump, a multiprobe autosampler (parallel autosampling device), and a parallel ultraviolet (UV) detector or mass spectrometer with a MUX interface (Figure 4). Using splitters makes the use of parallelization difficult. The approaches for parallelization vary from the use of a single pump to the use of multiple pumps, but regardless

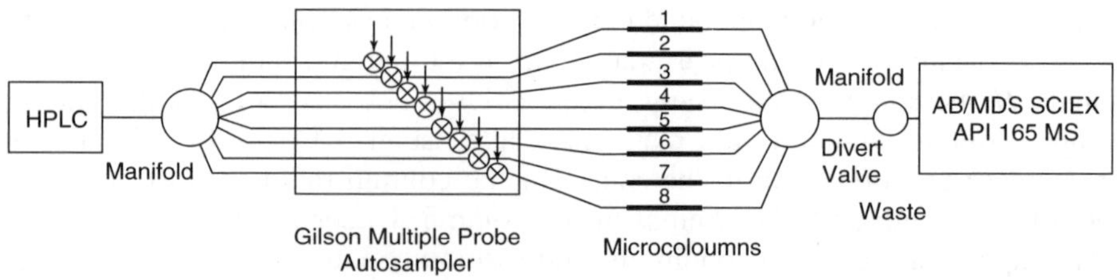

Figure 4 High-throughput eight-channel parallel LC/MS system. This system consists of a gradient HPLC pump, a Gilson 215 multiple-probe autosampler, a mass spectrometer equipped with an ion source, eight microbore columns, and a Valco switching valve. The volume of the sample loops is 20μL and full loop injections are used for all experiments. The total mobile-phase flow rate is 2.0 mL/min (0.25 mL/min for each column). During the early part of the gradient the flow is diverted to waste by a Valco switching valve. (From [31].)

of the approach, it is essential to have a flow controller that allows precise simultaneous flow through the parallel channels. The system described herein is manufactured by SEPIAtec GmbH and is based on a single gradient pump.

SEPIAtec Parallel HPLC System **[32–34]** The SEPIAtec parallel HPLC system does not control flow to each column of the array by individual pumps; rather, SEPIAtec's technique involves flow precision control from a single HPLC binary pump. This single pump delivers mobile phase to an array of columns by using a series of pressure feedback sensors. The flow controller follows Hagen–Poiseuille's law [35]; that is, the laminar flow (F) through a capillary tube (of length L) is directly proportional to the pressure drop (or pressure gradient, ΔP) between its ends, and η is the dynamic fluid viscosity. This is expressed mathematically as

$$F = \frac{\pi r^4 \, \Delta P}{8\eta L} \tag{18}$$

Operation of the flow controller consists of metering the pressure drop across a restrictor capillary, which is a fluid resistance, R_F. Fluid resistance is defined as the ratio of pressure difference to volume flow rate, $R_F = \Delta P/F$, and $\Delta P = P_{\text{total}} - P_{\text{restrictor}}$. It is clear from Eq. (18) that R_F is related directly to the viscosity as long as the geometry of the restrictor is constant; hence, R_F changes as the viscosity of the mobile changes during a gradient elution method. To circumvent this issue, the fluid controller measures a dimensionless variable, $Q_{\text{restrictor}}$, which is defined as $P_{\text{total}}/(P_{\text{total}} - P_{\text{restrictor}})$. $Q_{\text{restrictor}}$ is an instantaneous value that is compared to a certain set value, and any difference generates a signal that triggers a series of valves located after the restrictor to adjust the flow to keep them uniform throughout the columns. See Figures 5 and 6 for schematics of the SEPIAtech system (Sepmatix). The flow controller works with both analytical and preparative columns. It distributes the flow evenly to the columns, independent of the nature of each individual column.

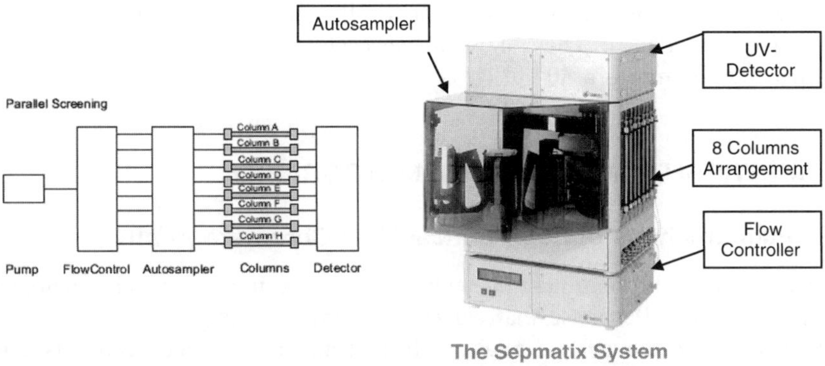

Figure 5 Sepmatix system. A single pump delivers the mobile phase to the parallel array of columns.

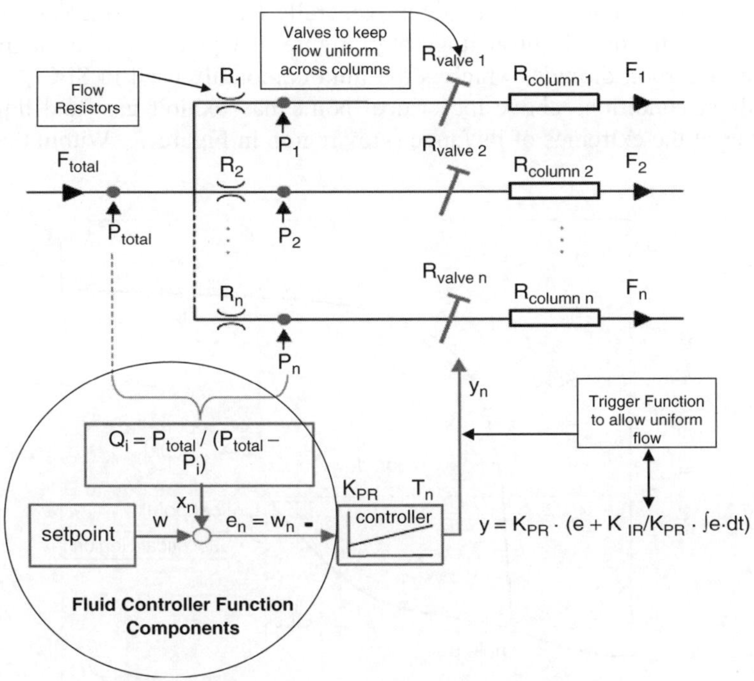

Figure 6 Sepmatix flow controller system. (From [32].)

It should be noted that there is commercially available an HPLC system with eight parallel capillary columns, in which each column is operated by an independent system manufactured by Eksigent. This system has the advantages of capillary columns cited in Section 3.1. Welch et al. [36] recently published a paper describing the use of microscale HPLC for piloting the large-scale preparative chromatographic resolution of the enantiomers of chiral pharmaceutical intermediates. They show a successful million-fold scale-up, from a 300-μm ID column to a 30-cm ID column [36].

4 SUPERCRITICAL FLUID CHROMATOGRAPHY

4.1 Fundamentals of Supercritical Fluid Chromatography

Like any other chromatographic technique, supercritical fluid chromatography (SFC) uses a mobile phase that carries the analyte through the column, except that in this case the mobile phase is heated and pressurized above its critical point to form a supercritical fluid. In practice, the temperature and pressure settings are generally just below the critical point. The differences among the various chromatographic techniques are determined by the properties of the fluids (i.e., density, polarity, diffusivity, and viscosity). Based on these fluid properties, supercritical fluid chromatography generally takes an intermediate position between gas and liquid chromatography. Figure 7 depicts a phase diagram for the case of carbon dioxide, which is the fluid commonly used in SFC.

Fluids at conditions above the critical point may exhibit gas- and liquidlike properties at the extremes of its range (gray region in Figure 7). Within the range

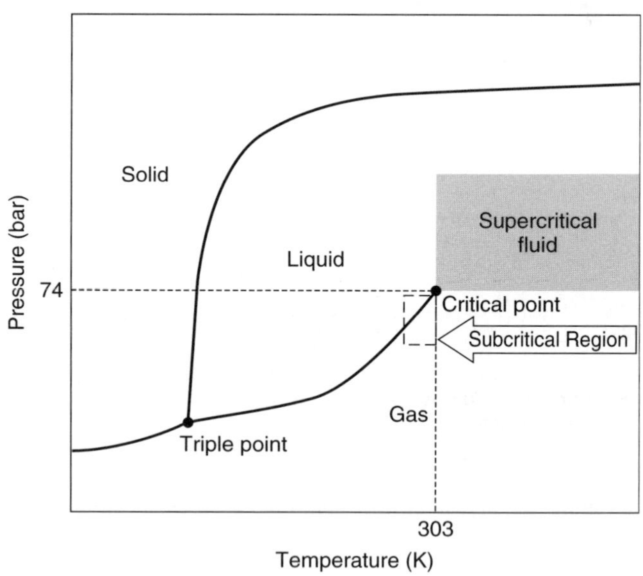

Figure 7 CO_2 phase diagram. The gray region corresponds to the supercritical fluid domain, whereas the dashed box is the subcritical domain where most of the SFC conditions reside. The ability to move along the lines of subcritical to supercritical allows SFC to have the advantageous attribute of solubility tuning. (From http://www.chem.leeds.ac.uk/People/CMR/whatarescf.html.)

described, the fluid can take a combination of those properties. An important point is that the viscosity, diffusivity, and solubilizing power of the supercritical fluid depend on its density, which itself is a function of the applied temperature and pressure. Essentially, this means that solubilizing power is tunable, and this gives enormous versatility to SFC. The solubility of analytes generally increases with density and becomes more pronounced near the critical pressure, where the rate of increase of the density, as a function of pressure, is greatest. The effect of pressure and temperature on solubility can be summarized as follows [37]:

- Polar solutes are most highly soluble in polar solvents.
- The maximum solubility increase vs. pressure usually occurs near the critical density; however, data are available that points out the need of greater pressures to increase the solubility of highly polar solutes in relatively nonpolar solvents.
- Within normal operating pressures and under isothermal conditions, solubility increases with pressure.
- If density is kept constant, solubility increases with temperature.
- Temperature increase under isobaric conditions results in a decrease in solubility when the pressure is set near or below the critical pressure, and solubility increases at pressures substantially above the critical pressure.

The benefits of SFC include:

1. *The analyte diffusion constant*, D_m, is three to five times larger than in HPLC; therefore, the time of analysis is decreased, as deduced clearly from Eq. (14).
2. *The viscosity of supercritical fluids* is approximately 1/20 of the mobile phases normally used in HPLC. Therefore, one can apply high flow rates without increasing considerably the pressure within the column. This also translates into a reduction of the time of analysis. Lower viscosity also translates into faster mass transfer; therefore, the C constant from the van Deemter equation remains fairly constant at a high flow rate; therefore, the efficiency remains constant.
3. *The cost is lower* since most of the solvent bulk is CO_2.
4. *For preparative SFC*, the cost of dry-down is lower since most of the solvent is CO_2. Also of great importance is the fact that since there is little solvent to remove, the recovery yield is greater.

The disadvantages and limitations of SFC include:

1. SFC has been considered to be applicable for nonpolar compounds; however, for all practical purposes, if the compound is soluble in methanol or in less polar organic solvent, it is a potential candidate for SFC. If the compound requires aqueous or buffers to be chromatograph, it may not be appropriate for SFC. This practical approach should not be interpreted

as meaning that the number of compounds amenable for SFC is narrow but as indicating a limitation that has reduced its impact as more experience and knowledge is gained. Berger [38] postulates that more than 95% of the small druglike molecules will be candidates for SFC because the drugs are less polar than water. Methanol, which can be considered a polar organic solvent, has an excellent probability of solubilizing the majority of the druglike compounds. Zheng et al. [39,40] have recently shown the ability to elute very polar/ionic compounds such as sulfonate salts when volatile ammonium salts were used as additives, hence expanding the range of compounds that can be chromatographed by SFC.

2. SFC systems are more complex than HPLC, mainly because of the need for two high-pressure pumps, one that allows liquefaction of CO_2 and the other to generate a CO_2 gradient with the polar additives. In addition, a backpressure regulator is needed to prevent expansion of the fluid into a gas, and detectors are designed to withstand higher pressures than normally experienced in HPLC. In the case of preparative SFC, the fraction collector is complex since it needs to prevent the loss of material due to aerosol generation at the outlet of the chromatograph.
3. There is a lack of familiarity of the principles and operation on the part of analysts.

4.2 Screening Methodology with SFC

The screening methodology with SFC is similar to the sequential HPLC approach: namely, use of switching valves for column scouting and mobile-phase selection.

5 ELECTROKINETIC SEPARATION: CAPILLARY ELECTROPHORESIS

5.1 Fundamentals of Capillary Electrophoresis [39–41]

Capillary electrophoresis (CE) is a separation technique that employs capillary columns with internal diameters on the order of 20 to 200 μm. CE can perform highly efficient separations of both large and small molecules. The major difference from a typical HPLC is that the separations are driven by an electrical field; hence, analytes are separated based on differences in their electrophoretic mobilities. The fundamental term is migration time, t_m, and it is the time that it takes an analyte to travel from the beginning of the capillary to the detector window. The migration time is related to other fundamental terms: electrophoretic mobility, μ_{ep}, electrophoretic velocity, v_{ep}, and electric field, E. The mathematical relationship is stated

$$\mu_{\text{ep}} = \frac{v_{\text{ep}}}{E} = \left(\frac{L_{\text{detector}}}{t_m}\right)\frac{L_{\text{total}}}{V} \tag{19}$$

Electrophoretic velocity is a measured term and corresponds to the term in parentheses in Eq. (19). L_{detector} is the length of the capillary to the detector. The last term on the right corresponds to the inverse of electric field strength, E^{-1}. L_{total} is the length of the entire capillary.

Equation (19) does not account for the presence of the electroosmotic flow (EOF), so it provides only the apparent mobility of the analyte. The EOF is the motion of liquid induced by a potential applied across a capillary column. The cause of electroosmotic flow is rooted in the formation of an electrical double layer at the capillary wall–solution interface. In CE, the capillary wall is generally made up of silica, and as in the case of HPLC, the walls are covered with silanol (Si–OH) groups. These silanol groups are ionized ($Si{-}O^-$) above pH 2; hence, the inner surface is negatively charged. In solutions containing ions, the cations will migrate to the negatively charged wall and partially neutralize them, forming a tightly adsorbed immobile layer of cations. The rest of the negative charge is neutralized by excess mobile phase, solvated cations located near the capillary wall. This arrangement of tightly and loosely adsorbed ions forms the electrical double layer. When an electrical potential is applied to the column, with an anode at one end of the column and a cathode at another, the cations will migrate toward the cathode. Since these cations are solvated and clustered at the walls of the column, they drag the rest of the solution with them, including neutral molecules and anions. This results in electroosmotic flow (see Figure 8).

The fundamental term for EOF is also the migration time, but corresponding to a neutral compound, and it is represented as t_0. The neutral migration time is related to the electroosmotic mobility, μ_{EOF}, electroosmotic flow, ν_{EOF}, and the electric field, E. The mathematic relationship is

$$\mu_{\text{EOF}} = \frac{\nu_{\text{EOF}}}{E} = \frac{L_{\text{detector}}}{t_0}\frac{L_{\text{total}}}{V} \tag{20}$$

The net electrophoretic mobility, $\mu_{\text{ep,net}}$, of an analyte will be $\mu_{\text{ep}} - \mu_{\text{EOF}}$. The electroosmotic mobility is defined by

$$\mu_{\text{EOF}} = \frac{\varepsilon\zeta}{4\pi\eta} \tag{21}$$

where ε is the dielectric constant, ζ the zeta potential (change of potential across the double layer), η the viscosity of the buffer, and e the total excess charge in solution per unit area. The zeta potential is defined as

$$\zeta = \frac{4\pi\delta e}{\varepsilon} \tag{22}$$

where δ is the double-layer thickness. Substitution of Eq. (22) into (21) results in

$$\mu_{\text{EOF}} = \frac{\delta e}{\eta} = \left(\frac{1}{3 \times 10^7 |Z| \sqrt{C}}\right)\frac{e}{\eta} \tag{23}$$

The expression in parentheses defines the zeta potential, where $|Z|$ is the number of valence electrons of the buffer and C is the buffer (electrolyte) concentration

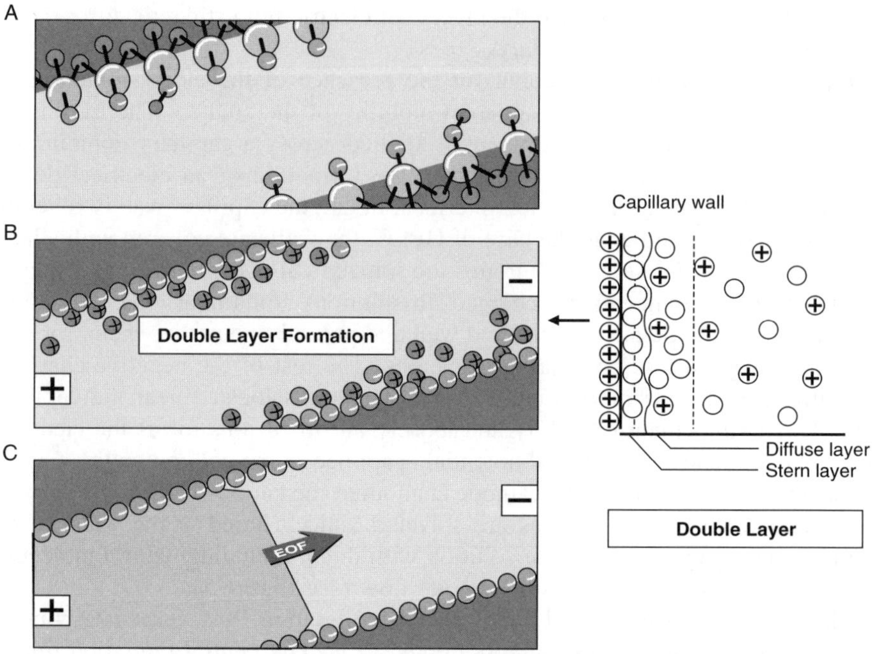

Figure 8 Formation of electroosmotic flow. (A) The fused-silica capillary surface is negatively charged, due to the SiO^- groups. (B) Solvated (hydrated) cations accumulate near the surface to develop the double layer (see the drawing on the right). (C) Upon an applied field, the flow moves toward the cathode carrying the cations and the solvent shell associated with it, thus generating electroosmotic flow (EOF). The EOF accelerates the travel of cations to the cathode and, depending on the magnitude of the EOF, it can also carry neutral molecules and anions to the cathode as well. + corresponds to the anode and − is the cathode. (From [53].)

in the bulk solution. From Eq. (23) it is clear that there are several practical ways to reduce the EOF contribution: Increase the concentration of the buffer (or electrolyte), increase the buffer viscosity, or reduce the total excess charge in solution. A decrease in the double-layer thickness is easily obtained by increasing the buffer concentration, thus making the double layer more compact (tighter cation adsorption) and making the cations more difficult to travel when voltage is applied. Changing the pH can decrease or increase the net charge at the capillary wall. For example, at $pH < 2$ the silanol groups are in their protonated form; therefore, there is no double layer (reduction of EOF), whereas at $pH > 2$ the double layer forms, due to deprotonation of the silanols. The use of neutral polymers in the buffer, such as polysaccharides, also reduces the EOF by increasing the viscosity in the double layer upon adsorption within the Stern layer (the compact region of the double layer). A similar result is observed when organic solvents such primary straight-chain alcohols are mixed into the electrolyte solution, as it is well known that alcohols increase viscosity when mixed with aqueous solutions [42]. However, even at concentrations larger than 50%, EOF continues to decrease even though at such concentrations η decreases. Closer inspection of Eq. (21) denotes that the addition of organic solvent affects the ratio ε/η, and it can increase upon addition of organic solvent [43,44]. Reduction of EOF due to the organic solvent is a complex matter which may include increase in the pK_a of the silanol groups, the buffer, and the analyte [41]. The control of EOF can lead to a situation in which it can change direction depending on the use of EOF modifiers (i.e., cationic or anionic surfactants) [45].

In capillary electrophoresis, as in the case of chromatographic separations, resolution (Rs) is a critical parameter that is closely related to the quality of a separation. Mathematically, Rs is defined as

$$\text{Rs} = \frac{\Delta\mu\sqrt{N}}{4(\mu_{\text{average}} + \mu_{\text{EOF}})} \tag{24}$$

where $\Delta\mu$ and μ_{average} are the difference and the average values of the electrophoretic mobilities of two adjacent peaks, respectively, and N is the number of theoretical plates, which are given by

$$N = \frac{\mu_{\text{ep}} V}{2D_{\text{analyte}}} \tag{25}$$

where μ_{ep} is the apparent mobility, as described in Eq. (19), D_{analyte} is the diffusion coefficient of an individual analyte, and V is the applied voltage. Large theoretical plate values are favored whenever the diffusion coefficient is low and the mobility and applied voltage are large. Substitution of Eq. (25) into (24) results in

$$\text{Rs} = 0.177\Delta\mu\sqrt{\frac{V}{(\mu_{\text{average}} + \mu_{\text{EOF}})D_{\text{analyte}}}} \tag{26}$$

Increasing voltage is not a very practical approach to increasing resolution. The best way to increase resolution is to maximize $\Delta\mu$. This is possible by reducing EOF or by changing its direction so that the “effective” length of the capillary is increased, but this occurs at the expense of increased runtime.

5.2 Parallel Capillary Electrophoresis

The parallel CE system used in our work houses a modular array of 96 capillaries (see Figure 9). Standard capillary dimensions are 50 or 75 μm in inner diameter (ID) and 200 μm in outer diameter (OD). Various capillary lengths can be used, depending on the analysis requirements (i.e., resolution and speed of analysis). The inlet side of the capillary array is arranged in an 8 × 12 format with electrodes placed side by side with each capillary, allowing for direct injection via vacuum or voltage from standard 96-well microtiter plates. The capillaries are aligned

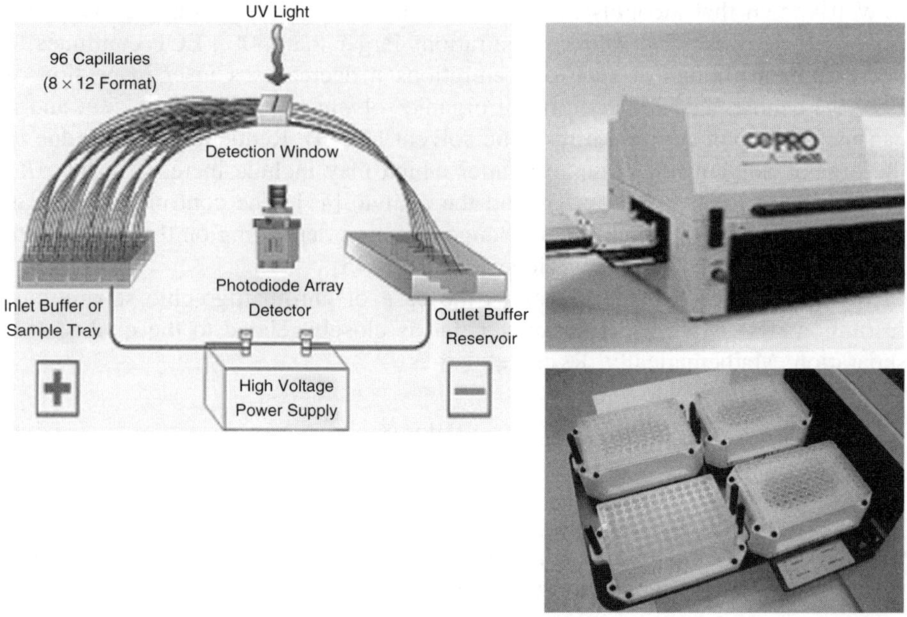

Figure 9 Parallel capillary electrophoresis system. Left: diagram of the CE Pro 9600. Right: CE Pro 9600 instrument and sample trays.

in parallel in the detection window, which is positioned between the UV light source and the detector assembly. The outlets of the capillaries are bundled to a common reservoir, where they are connected to ground. Room-temperature air is circulated around the capillaries to modulate the temperature during CE separation.

Detection consists of a UV light source, collimating lens, UV camera lens and a 1024-element linear photo diode array (PDA) detector. The UV light is provided by a line source consisting of a zinc (214 nm), cadmium (229 nm), or mercury (254 nm) lamp. Detection is performed at a fixed, single wavelength determined by the UV lamp and an interference filter. The capillary array is imaged continuously, providing real time, online UV detection for the 96 different CE separations.

A 20-kV 6-mA total current high-voltage power supply provides the appropriate voltage for electrokinetic sample injection and electrophoresis. The recommended operating limits are ±16 kV and 3 mA total current (30 μA per capillary). The system is also capable of polarity reversal. A syringe pump provides controlled pressure or vacuum to the outlet reservoir of the capillary array. Vacuum can be used for vacuum injection of sample, for introducing buffer solutions into the capillary array from the inlet tray, or for performing vacuum-assisted electrophoresis. A distribution valve allows up to five different buffer bottles and a waste bottle to be connected to the system simultaneously.

6 EXPERIMENT

Equipment Sequential HPLC screens consist of Agilent 1100 furnished with Chiralizer column-switching technology, except the reversed-mode screen setup, which consists of a Waters Alliance 2695 equipped with Waters column and solvent switchers. The Agilent systems are controlled with ChemStation software, whereas the Waters system is controlled with Empower software. Both the parallel HPLC system (the Sepmatix system) and the software are from Sepiatec. The chiral SFC system is a Berger analytical system from Mettler-Toledo, equipped with column and solvent switching valves and controlled with ChemStation software (note that Mettler-Toledo's SFC Division is now part of Thar, Inc.). The 96-channel parallel capillary electrophoresis instrument is from Advanced Technologies and is furnished with a 50-μm ID capillary array cartridge 55 cm in length. All HPLC screens are equipped with diode array detectors. The CE system is monitored at 214 nm. Once a method was selected, it was optimized in a separate HPLC, SFC, or CE system, utilizing the conditions obtained from the screens. The preparative modes were performed in an Agilent preparative HPLC and the Berger SFC II and SFC minigram systems.

Samples, Chemicals, and Solvents All solvents, buffers, and acids used for the HPLC and SFC screens are from Aldrich; water is from a Milli-Q system (Millipore). The samples from examples shown in this chapter are proprietary unless stated otherwise. Further details on the capillary electrophoresis screen can be found in a paper by Saavedra et al. [46].

Columns All Chiracel and Chiralpack columns are from Chiral Technologies, the Astec-Chirobiotic phases from Sigma-Aldrich, the Whelk columns from Regis, and the TBB and DBB phases from Kromasil. Sulfated cyclodextrins are from Beckman.

Screen Methods All methods are described in Tables 2 through 6. The injection volume for the reversed-phase screen is 10 μL, whereas it is variable (ranging form 5 to 20 μL) for the other screens, depending on the sample availability. Note also that all screens except the reversed-phase screen were kept at room temperature.

7 RESULTS

7.1 Chiral Normal and Polar Phase Screens (Conditions in Tables 2 and 3)

Prior to the implementation of the screens described in Table 2 (normal-phase screening methods), 1000 different chiral methods were developed using 14 different columns and three different mobile phases in support of our discovery portfolio. Of the 1000 methods developed, Figure 10 shows the percentage of these methods by column type. Based on these historical data, eight columns were

TABLE 2 Parallel Normal Phase Chiral LC Screen: Columns and Conditions[a]

Columns: ID = 046 cm, L = 5 cm, 5 μm	Gradient		
Chiralcel OD	Time (min)	% A	% B
Chiralpak AD	0	98	2
Chiralpak AS	2	98	2
Chiralcel OJ	20	60	40
Kromasil TBB	21	60	40
Kromasil DMB	21.5	98	2
Chirobiotic V			
S,S Whelk-O 1		Flow rate: 1 mL/min	
Mobile phase:	A (acid):	0.1% Trifluoroacetic acid in heptane	
	A (base)	0.2% Diethylamine in heptane	
	A (neutral):	Heptane	
	B (1):	Ethanol	
	B (2):	Propanol	

[a]Each A mobile phase is combined with B(1) and B(2) to total six different gradient compositions: A(acid):B(1), A(acid):B(2), A(base):B(1), A(base):B(2), A(neutral):B(1), A(neutral):B(2).

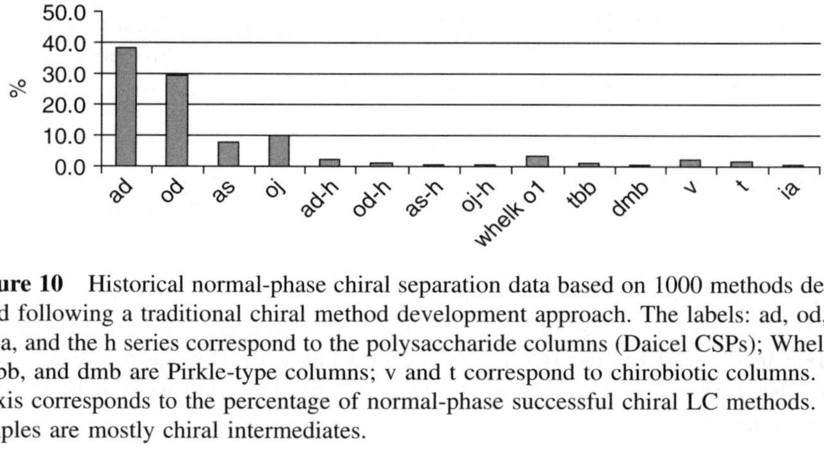

Figure 10 Historical normal-phase chiral separation data based on 1000 methods developed following a traditional chiral method development approach. The labels: ad, od, as, oj, ia, and the h series correspond to the polysaccharide columns (Daicel CSPs); Whelk-O 1, tbb, and dmb are Pirkle-type columns; v and t correspond to chirobiotic columns. The *y*-axis corresponds to the percentage of normal-phase successful chiral LC methods. The samples are mostly chiral intermediates.

selected to be used in our normal- and polar-phase chiral screens. The columns consist of Chiralcel OJ and OD, Chiralpak AD and AS, Chirobiotic V, Kromasil TBB, Kromasil DMB, and (*S*,*S*)-Whelk-O 1. Three mobile general phases are used for the normal-phase screen: neutral, acidic, and basic. The column selections have evolved somewhat through the years, but the polysaccharide-based CSPs remain the workhorse of chiral separations in our laboratories. Table 3 shows the polar-phase screening conditions, which were also selected based on experience; however, the historical data were fewer than for the normal-phase screens.

TABLE 3 Polar Organic Chiral LC Screen: Columns and Conditions[a]

Columns: ID = 0.46 cm, L = 5cm, 5 μm		Gradient 1	
Chiralpak AD	Time (min)	% A	% B
Chiralpak AS	0	95	5
Chirobiotic RN	30	5	95
Chirobiotic SN			
		Gradient 2	
	Time (min)	% A	% B
Chirobiotic T (L = 100 cm)	0	5	95
Chirobiotic V (L = 100 cm)	30	95	5
		Flow rate: 1 mL/min	
Mobile phase:	A: Acetonitrile B: Methanol		
		Chirobiotic T & V 100% Acetonitrile for 30 minutes	

[a]Chirobiotic T and V columns are screened isocratically with 100% acetonitrile. The other four columns are exposed to two different gradients, gradients 1 and 2.

As indicated in Section 1, there is a high-sample-throughput workload in the early stages of drug development—discovery to preclinical—where most of the samples are synthetic chiral intermediates that required enantiomeric excess (ee) evaluations and isolation by preparative chromatography. To improve the efficiency and productivity of chiral analysis further, we evaluated and implemented the Sepmatix (parallel) HPLC system from Sepiatec. The impact was immediate and significant:

- Multiple columns and mobile phases are evaluated in a fraction of the time required to screen them in series (via column switching). A sequential six-column screen with six different mobile phases takes approximately 900 min to run, whereas the parallel system takes 150 min for the same six different mobile phases. This is an 87% gain in efficiency! In addition, eight columns are screened in the parallel system instead of the six in the sequential screen.
- An optimum isolation method can be implemented sooner and the isolation/purification execution takes place sooner.
- Deadlines can be met when studies get moved up and time lines are compressed.

The CSP Chiralpak AD leads the frequency of success (ca. 35%) per column in the chiral NPLC screens, as shown in Figure 11. Note that regardless of the therapeutic area from which the samples originate, Chiralpak AD remains the leading CSP. The frequency of success for mobile phase changes between

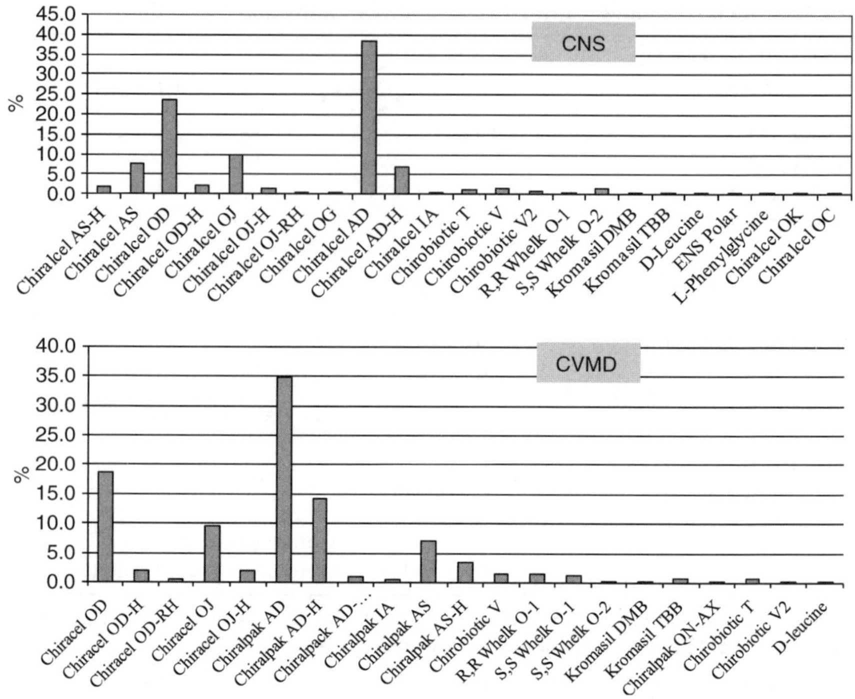

Figure 11 Historical outcome from the parallel (Sepmatix) normal-phase screen. The CSP Chiralcel AD gave the largest number of successful hits, regardless of therapeutic area.

acidic and neutral, depending on the nature of the compounds submitted to our laboratory. Chiralpak AS leads in frequency of success from the polar organic screen but is followed closely by the AD CSP (55% : 45%).

Figure 12 shows a representative output from the eight parallel columns. It is important to note that the quality of the chromatograms is not critical at the screen stage; the important point is to get a potential method for further optimization to accommodate high-quality chromatography. The rate of success is $\geq 95\%$, and in general about 95% of all our compounds are separated in either a normal-phase LC mode or SFC. The polar organic is a distant $< 5\%$. Figure 13 (p. 236) shows the optimized chromatogram for the sample shown in Figure 12.

7.2 Chiral Supercritical Fluid Chromatography Screen (Conditions in Table 4)

The selection of the columns for the chiral SFC screen followed the same course as that for our LC screens, namely, it was based on our early experience with SFC. The screen is a sequential one, as described in Table 4 (p. 237). The major benefit of SFC vs. normal-phase LC is the ability to use high flow rates without backpressure issues and, consequently, obtain faster analysis times. In addition,

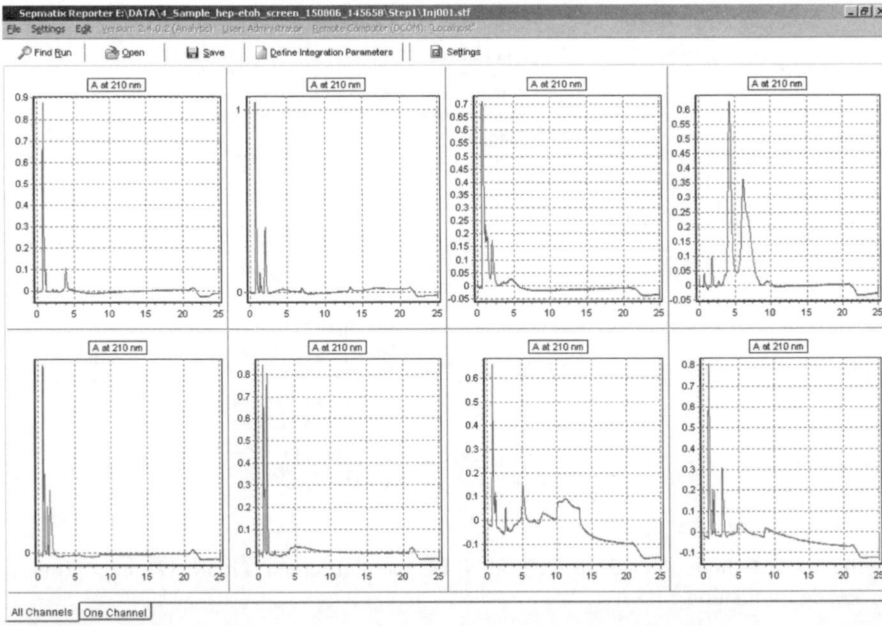

Figure 12 Parallel chiral HPLC screening output. Top channels from left to right correspond to Chiralcel OD, Chiralpak AD, Chiralpak AS, and Chiralcel OJ. Bottom channels from left to right correspond to Kromasil TBB, Kromasil DMB, Chirobiotic V, and S, S Whelk OI.

time is saved in mobile-phase preparations since they are based on methanol and additives, and mobile-phase consumption is minimal compared to liquid chromatography. Due to the benefits in cost, efficiency, and productivity, chiral SFC has now become our main chromatographic technique for chiral analysis, and the other liquid chromatography methods have become complementary to SFC.

The frequency of success for different SFC CSPs is shown in Figure 14 (p. 238). In this case, Chiralpak AD and Chiralcel OJ were the most successful columns based on more than 1200 methods developed. Figures 15 and 16 (pp. 238 and 239) show examples from the chiral SFC screen used to identify an analytical method for ee determination, and then the method is optimized for preparative work to isolate both enantiomers. Notice that the method selected is the AS column with the methanol/0.2% isopropylamine additives, and that it shows excellent resolution between the two enantiomers (Figure 16). The two small peaks correspond to a chiral impurity that differs from the main enantiomers by a methyl group around the chiral center. The selection of isopropyl amine vs. trifluoroacetic acid was the resolution of the latest eluting peak (retention time at 9.8 min) vs. the preceding one, since we needed to isolate the first two enantiomers and the referred impurity. The second best separation was from the OJ chiral phase. However, the peaks did not have enough resolution, even after optimization, to be transferred to preparative SFC.

Sample Information

117569-163-1

Chemists Amanda Nelson

Zone CVMD

Project Code:

Action EE Method Dev.

Scale-Up Amt

Date Rec. 8/14/2006

Date Comp. 8/17/2006

DAG Contact E. Pollard x14373

Method Information

Instrument HPLC 1051

Reference No. 081706-5

Column Chiralcel OJ-H

Dimensions 4.6mm x 15cm

Mobile Phase 70/30 Heptane/EtOH

Modifier None

Flow 1.5 ml/min

Notes

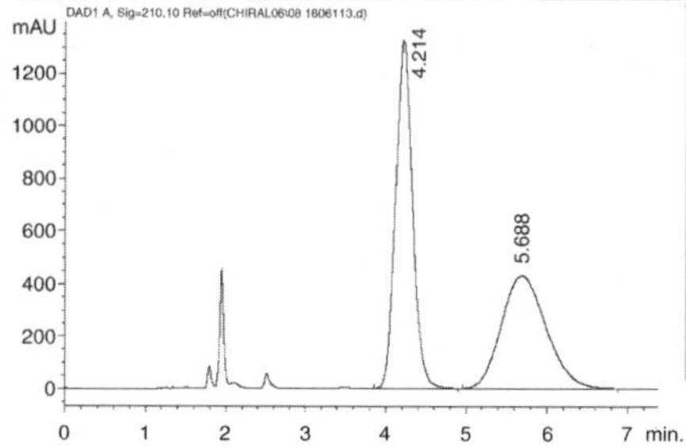

Results

Alpha 1.35

Ret. Time 1 4.214

Ret. Time 2 5.688

Enant. Excess 8.4%

Area 1 1951.64

Area 2 1649.89

UV 210

UV Max. 205

Figure 13 Method selected from chiral NPLC parallel screen (Figure 12) and optimized for EE determination.

7.3 Chiral NPLC vs. Chiral SFC

We performed several studies in which we compared the results from the chiral normal-phase screen with the results from the SFC screen. The result in most cases shows the following trend: (a) if the enantiomers are separated in the same CSP column for both modes, we observed peak reversal (change of selectivity) about 65% of the time; (b) if the enantiomers are separated in NPLC with Chiralpak AD, about 70% of the time we observed the same quality of separation in the SFC but the CSP was Chiralcel OJ and often with observed peak reversal; (c) we normally observed a 30 to 50% decrease in retention times when comparing SFC with NPLC, and the SFC peaks were sharper, due to the lower k'. Figure 17 (p. 239) shows an example of the differences typically observed between NPLC and SFC.

7.4 Chiral Reversed-Phase Liquid Chromatography Screen (Conditions in Table 5, pp. 240–241)

The chiral reversed-phase LC screen is dedicated to developing methods that will be used for quality control purposes for late-stage projects (phase 1 development and later), as these methods are very rugged, robust, and user friendly. An added practical advantage of using reversed phase lies in the fact that in many cases the mobile phases used in the achiral HPLC method can also be used in the chiral HPLC method, thus minimizing the need to prepare different mobile phases.

Use of the chiral reversed-phase LC screen to develop methods for early-stage projects (discovery stage) is minimal since when transferred to a preparative

TABLE 4 Chiral Supercritical Fluid Chromatography Screen: Columns and Conditions[a]

Columns: ID = 0.46 cm, L = 25 cm, 5μm	Modifier Program		
Chiralcel OD	Rate	%	Hold
Chiralpak AD		5	1
Chiralpak AS	7.0	40	1
Chiralcel OJ	35	5	2
Kromasil TBB			
Kromasil DMB	Supercritical fluid: CO_2		
Conditions:	Flow rate:	4 mL/min	
	Pressure:	100 bar	
	Temperature:	35°C	
Modifiers:	Methanol		
	Methanol (0.1% TFA or acetic acid)		
	Methanol (0.2% IPA)		
	Ethanol		
	Propanol		
	90 : 10 Methanol/propanol		
	Acetonitrile		

[a]The supercritical fluid is CO_2. There are six columns that are exposed to the following gradients: CO_2/methanol; CO_2/0.1% trifluoroacetic acid (TFA) or acetic acid in methanol; CO_2/0.2% isopropylamine (IPA) in methanol; CO_2/ethanol; CO_2/propanol, CO_2/methanol/propanol, and CO_2/acetonitrile. The rate is the slope of the gradient [Δ%/time (min)]. Runtime per method is 20 min.

HPLC, the resulting yields are generally low and there is a high risk of decomposition, due to the higher temperatures needed to remove the solvents, additives, and water; therefore, it is used as the last resort when nothing else gives positive results. Generally, the CSPs with higher success rate are the Chiralpak AD and AS. Figure 18 (p. 242) shows the outcome from the screen wherein a proprietary racemate is separated very well even when overloading the column.

7.5 Parallel Capillary Electrophoresis Screen (Conditions in Table 6, p. 243)

The parallel CE screen is still under development, and the evaluation study was published in 2008 [46]. In general, it has the same advantage as the Sepmatix system. That is, several runs happen simultaneously. For example, the separation of eight compounds with three HS-CDs (24 analyses) are completed in 100 minutes compared to a normal single-channel CE analysis, the same analysis takes about 1000 minutes. The current multiplex configuration does not allow for 24-h operation since every run needs to be set up manually before analyzing the samples. Joule heating plays a role as well: inducing nonlinear behavior that affects the elution time [46]. Figure 19 (p. 244) shows a typical output of the chiral CE.

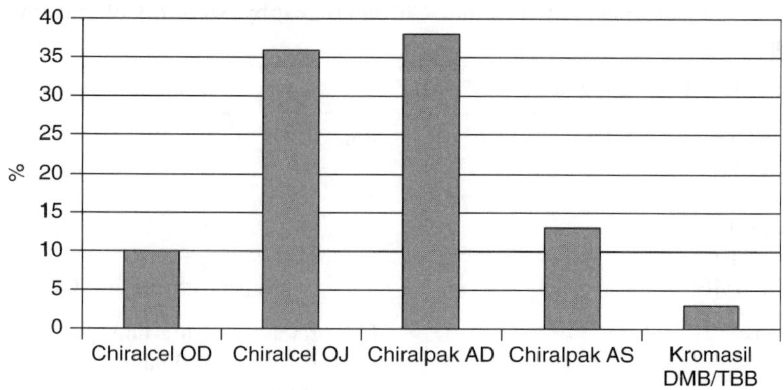

Figure 14 Historical outcome from the chiral SFC screen. The results are based on more than 1200 methods developed.

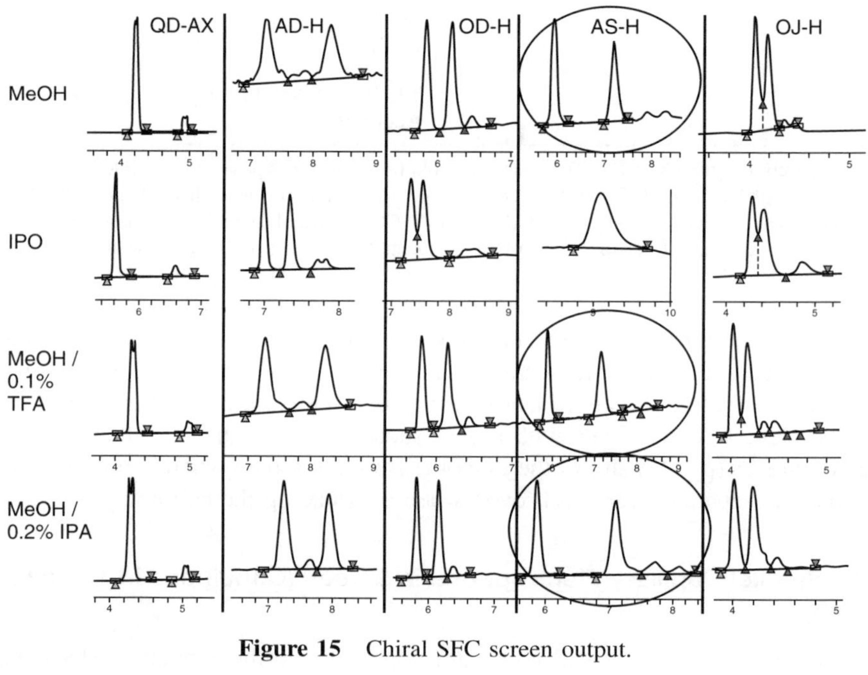

Figure 15 Chiral SFC screen output.

7.6 Chiral Preparative HPLC and SFC

In the early development (discovery) stage, the transfer from analytical- to preparative-scale chromatography is one of the most demanding aspects of chiral work because of the low availability of compounds at this stage. Therefore, it is critical that the analytical methodology be transferred seamlessly to the preparative system, and the yield is greater than 70%, with a minimum

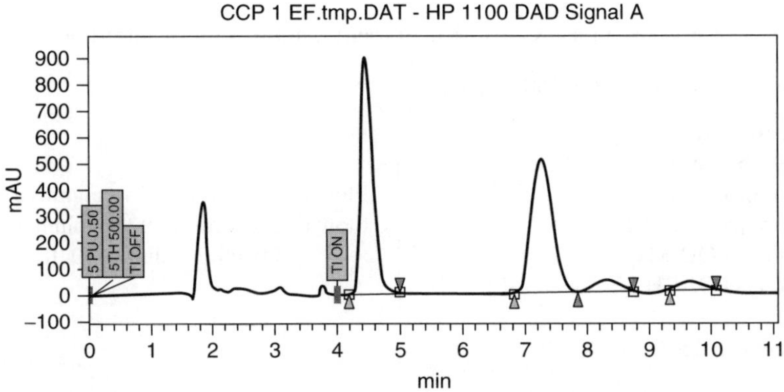

Figure 16 Optimized chiral SFC method from the method selected from Figure 15.

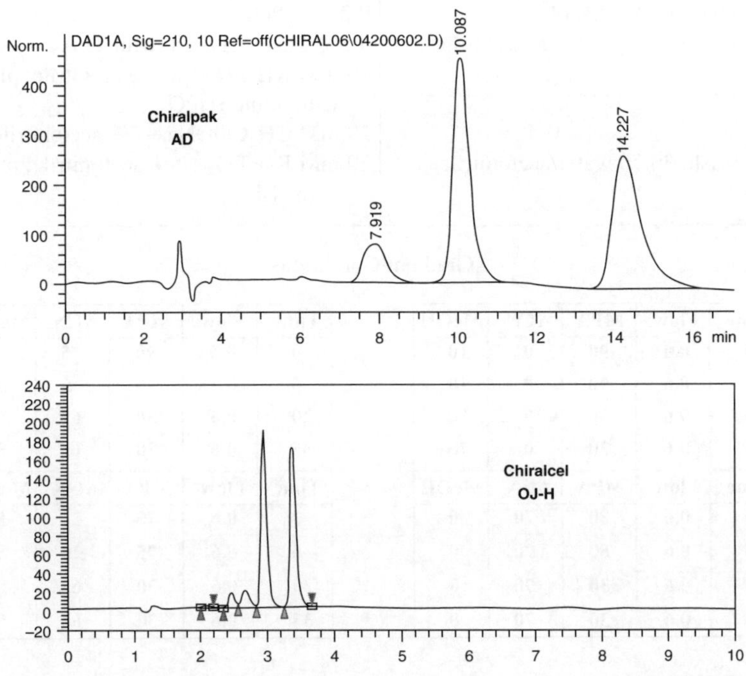

Figure 17 Comparison of chiral normal-phase LC (top) vs. chiral SFC (bottom). Both methods have been optimized for EE determinations. No peak reversal between NPLC and SFC was observed in this case.

chromatographic purity of about 75%. For samples that will undergo exploratory toxicology studies, a chromatographic purity above 85% may be required. So it is essential to have good resolution between the enantiomers in question in order to have good yields and acceptable purity. An important criterion is to develop the analytical method with the same stationary-phase particle size as the one

TABLE 5 Reversed-Phase Chiral LC Screen: Columns and Conditions

Columns: ID = 0:4 cm, L = 150 cm, 5 μm	
(1) Chirobiotic T (2) Chirobiotic V (3) Chirapack AD-RH (4) Chiralpack AS-RH (5) Chiracel OD-RH (6) Chiracel OJ-RH Temperature: 25°C and 40°C	Flow rate: 1 mL/min for columns 1 and 2 Flow rate: 0.5 mL/min for columns 3 to 6 Injection volume: 10 μL
Mobile Phases B, C, and D:	Mobile Phases A:
Mobile phase B: Acetonitrile (ACN) Mobile phase C: Methanol Mobile phase D: 95:5Water:Acetonitrile Needle wash: 30:70 water/acetonitrile	0.2% H_3PO_4/100 mM $NaOCl_4$ 0.2% H_3PO_4 0.1% TEA; pH ~ 3:0 with TFA 20 mM KH_2PO_4 + 5% acetonitrile; pH ~ 4:5 with dilute H_3PO_4 20 mM CH_3COONa + 5% acetonitrile 20 mM K_2HPO_4 + 5% acetonitrile; pH ~ 7:0 with H_3PO_4

Gradient Conditions

Time	Flow	MPA	ACN	MeOH
0	0.6	90	0	10
5	0.6	90	0	10
20	0.6	30	0	70
35	0.6	30	0	70
Time	**Flow**	**MPA**	**ACN**	**MeOH**
0	0.6	80	20	0
5	0.6	80	20	0
20	0.6	30	70	0
35	0.6	30	70	0

Top gradient: columns 1, 2, 3, 5, and 6
Bottom gradient: column 4

Time	Flow	MPA	ACN	MeOH
0	0.6	90	5	5
5	0.6	90	5	5
20	0.6	30	65	5
35	0.6	30	65	5
Time	**Flow**	**MPA**	**ACN**	**MeOH**
0	0.6	75	20	5
5	0.6	75	20	5
20	0.6	30	65	5
35	0.6	30	65	5

Top gradient: columns 1, 2, 3, 5, and 6
Bottom gradient: column 4

(*Continued overleaf*)

TABLE 5 (*Continued*)

Time	Flow	MPA	ACN	MeOH
0	0.6	90	0	10
5	0.6	90	0	10
20	0.6	30	0	70
35	0.6	30	0	70
Time	**Flow**	**MPA**	**ACN**	**MeOH**
0	0.6	80	0	20
5	0.6	80	0	20
20	0.6	30	0	70
35	0.6	30	0	70

Top gradient: columns 1, 2, 3, 5, and 6
Bottom gradient: column 4

Time	Flow	MPA	ACN	MeOH
0	0.6	95	5	0
5	0.6	95	5	0
20	0.6	35	65	0
35	0.6	35	65	0
Time	**Flow**	**MPA**	**ACN**	**MeOH**
0	0.6	85	15	0
5	0.6	85	15	0
20	0.6	35	65	0
35	0.6	35	65	0

Top gradient: columns 4 and 5
Bottom gradient: column 4

Time	Flow	MPA	ACN	MeOH
0	0.6	90	5	5
5	0.6	90	5	5
20	0.6	30	65	5
35	0.6	30	65	5
Time	**Flow**	**MPA**	**ACN**	**MeOH**
0	0.6	80	15	5
5	0.6	80	15	5
20	0.6	30	65	5
35	0.6	30	65	5

Gradient for mobile phases A: 4 and 5 with mobile phase D for all columns except for AS, which follows the bottom gradient

Time	Flow	MPA	ACN	MeOH
0	0.6	95	0	5
5	0.6	95	0	5
20	0.6	35	0	65
35	0.6	35	0	65
Time	**Flow**	**MPA**	**ACN**	**MeOH**
0	0.6	85	0	15
5	0.6	85	0	15
20	0.6	35	0	65
35	0.6	35	0	65

Gradient for mobile phases A: 4 and 5 with mobile phase C for all comumns except for AS, which follows the bottom gradient

for the preparative scale. A mismatch will cause difficulty in the transfer, and it can cause delay in the turnaround time. The transfer of the analytical method to the preparative method is done using scaling factors, since the transfer involves changes in column dimensions. The general scaling factor, Y, is defined as

$$Y = \frac{L_p}{L_a}\frac{d_p^2}{d_a^2} \tag{27}$$

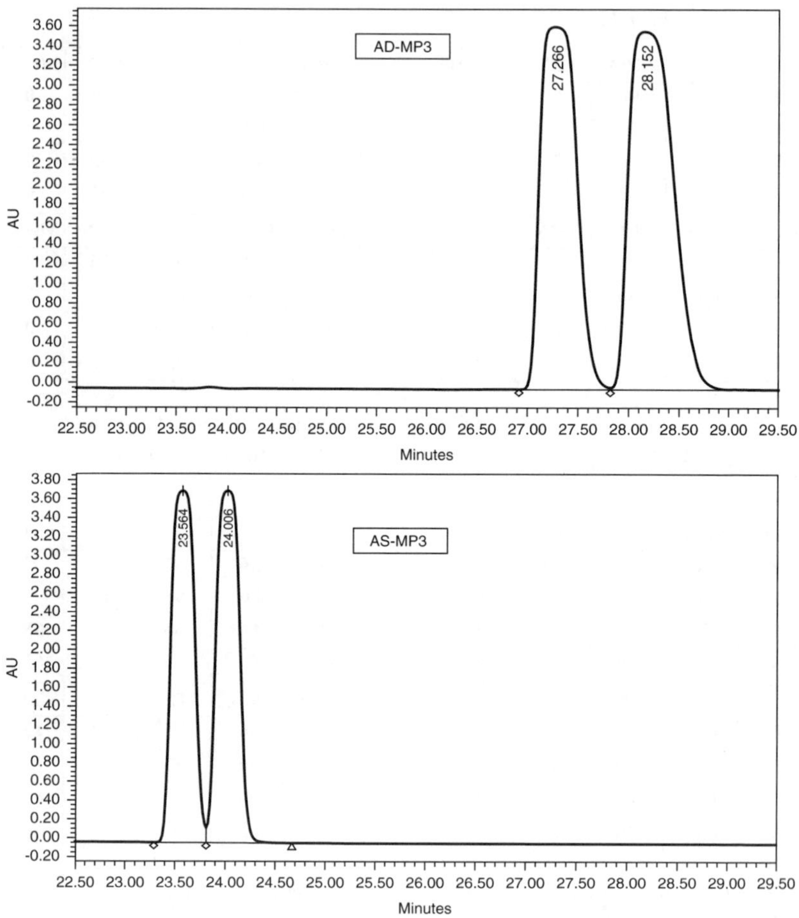

Figure 18 Best two methods from the chiral reversed-phase screen for a given sample.

where L and d are column length and diameter, respectively, and the subscripts a and p correspond to analytical and preparative columns, respectively. If one scales the mass load, M, from analytical to preparative, one employs $M_p = M_a Y$. The flow rate is scaled similarly, $F_p = F_a Y$. A major concern in preparative chiral chromatography is selectivity because optimization of preparative methods maximizes the sample (mass) loading on the preparative column. Maximizing the sample loading allows greater production rates [47]. For the case of touching-peak chromatographic separations, as seen in Figure 20 bottom picture, the complex α_{RS} vs M_x relationship can be simplified to $M_x = \psi[(\alpha_{RS} - 1)/\alpha_{RS}] \cdot M_s$ as long as α_{RS} is > 1.03, and ψ is a constant for the system [47]. M_x is the mass of the injected sample, and M_s is the column saturation capacity, which is defined as the mass of solute that will fill completely an adsorbed monolayer. M_s is specific to a particular column and depends on the nature of the solute and the chromatographic conditions [47]. The ratio, M_x/M_s, is the loading factor,

TABLE 6 Chiral Parallel CE Screen

Capillary Dimensions for a 96-Channel Cartridge	
Capillary length	55 cm from which 33 cm are effective
Capillary internal diameter	50 μm
Separation conditions	
Electric field	−100 V/cm
Applied voltage	−5.5 kV
Hydrodynamic injection	0.5 psi for 10 s
Runtime	100 min
Data collection	CEPro Data Manager
Buffer	50 mM Phosphate and adjust pH to desired value with triethylamine
Cyclodextrins	HS-α-CD, HS-β-CD, and HS-γ-CD, commercially available at 20% (w/v)
Background electrolyte	5% (w/v) of S-β-CD in 25 mM phosphate at pH 2.5 (adjusted with triethylamine)
System performance marker	1,3,6,8-Pyrentetrasulfonate at 250 μg/mL

Source: Adapted from [46].

$L_f (= \psi[(\alpha_{RS} - 1)/\alpha_{RS}])$; this indicates that as the demand for larger production rate increases, so does the selectivity and the loading factor. It simply indicates that as the amount of mass loading increases, so does the selectivity. However, the loadability of a column is not easy to predict since it depends on the competitive isotherms of the solutes on the CSP. The situation described gives the screen approach greater relevance because it normally provides more than one "best hit," which allows rapid adjustment to loss of selectivity due to loadability problems. Similarly, we have found that as the chiral column ages, the loadability decreases and the selectivity also decreases.

It should be noted that all the screens are performed under gradient elution, but the "best hit" is optimized isocratically whenever the method is to be transferred to a semipreparative scale. We have found that in the chiral preparative work the use of SFC maximizes the degree of success of the transfer. The examples shown in Figures 20 (p. 245) and 21 (p. 246) clearly demonstrate the point. In Figure 20 the tailing in the chiral NPLC method is pronounced, thus reducing the practical resolution between the two enantiomers. The retention times of the peaks of interest are past 20 min. If one now compares this with the profile obtained with chiral SFC in Figure 21, it is clear that the efficiency is greater with chiral SFC. Better resolution and much shorter retention times directly affect the turnaround time and the quality and yield of the final product. In addition, stacking is highly favored in SFC, and this allows repetitive injections of the sample. In the example shown in Figure 22 (p. 247), we stack 33 injections in about 55 min. For this we developed an isocratic method wherein the k' values are close to 1, and α_{RS} to 1.25. The injection volume was 3 mL of a sample solution of about 5 mg/mL, and the injections were 1.7 min apart. The result of the stacking was 92% yield of the desired enantiomer with 85% purity in less than 8 h (including drydown).

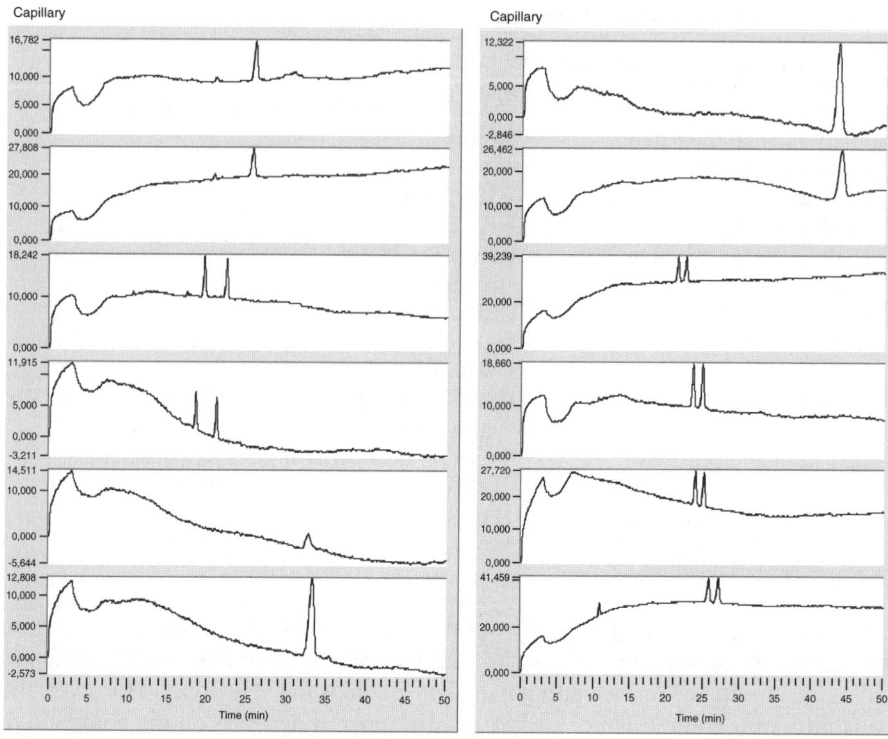

Figure 19 Chiral (highly sulfated cyclodextrins) capillary electrophoresis. The left side corresponds to the screen of atropine racemate. The right side corresponds to the screen of phenylalaninamide racemate. From top to bottom: HS-α-CD, HS-β-CD, and HS-γ-CD. Each sample is injected in duplicate in HS-CD.

8 DISCUSSION

The various screens presented in this chapter have a single purpose: to increase efficiency and productivity in identifying and then optimizing methods for analytical- and preparative-scale chiral separations. A critical point is that the screens must evolve continuously by analyzing historical trends, testing new technologies, and evaluating their pros and cons.

Chiral normal-phase chromatography remains very popular since most laboratories have the capability to perform them; hence, transfer of the method to other laboratories (internal or external) is straightforward. However, as shown in the results section, chiral SFC can be very successful and is becoming more accessible, and scientists are also becoming more familiar with the technique. In terms of performance, we have observed that the polysaccharide CSPs frequently generate the method for the separation of the enantiomers, and this is in accord with the literature. What makes theses phases so efficient? Many studies have been performed to try to determine the reasons for the success of these CSPs; these

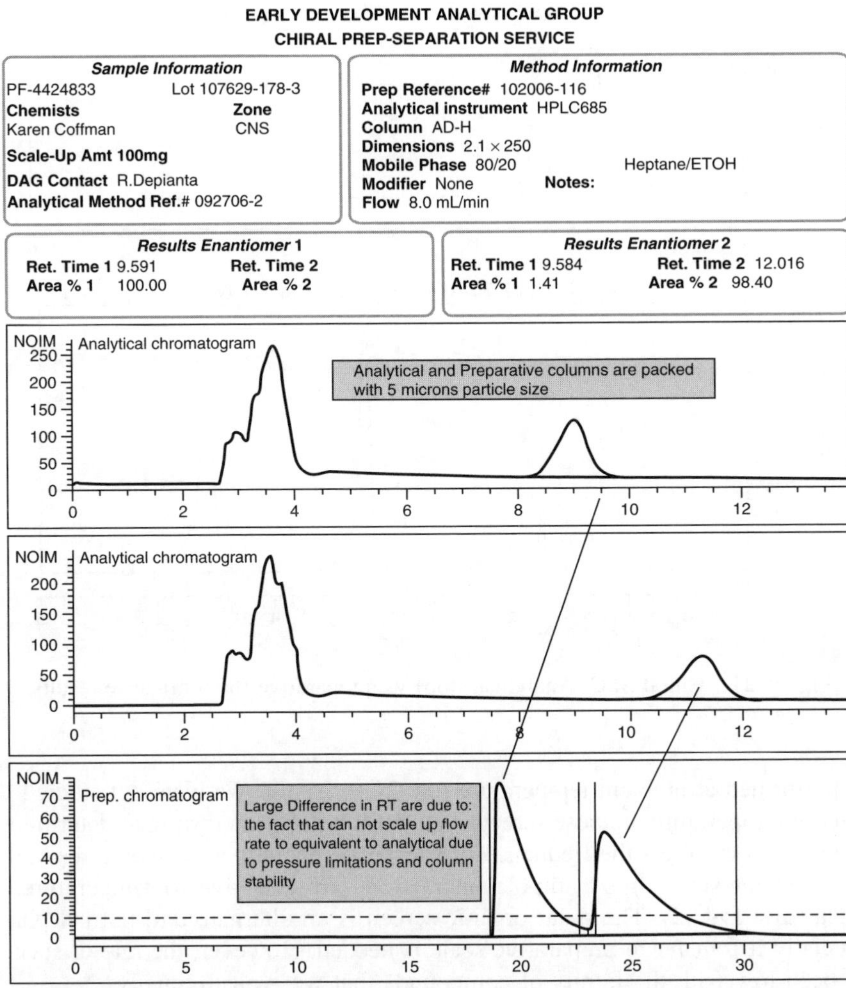

Figure 20 Chiral normal-phase HPLC. Analytical vs. preparative-scale results.

studies include spectroscopy, mass spectrometry, and computer modeling [48]. The current school of thought is related to the helical configuration of glucose. The glucose residues are regularly arranged along the helix. The helix has a grove where the inside is occupied by the polar carbamates and the hydrophobic groups are located outside. The incoming enantiomeric analytes will easily interact with the polar carbamate groups. The potential interactions for the polysaccharide CSPs are summarized in the type II classification shown in Table 1. The fact that the analytes can move within close proximity to the carbamate groups favors the formation of hydrogen bonding and dipole–dipole interactions.

Our screens are not temperature controlled because we do not have such capability in the preparative systems; therefore, to facilitate the transfer, all runs

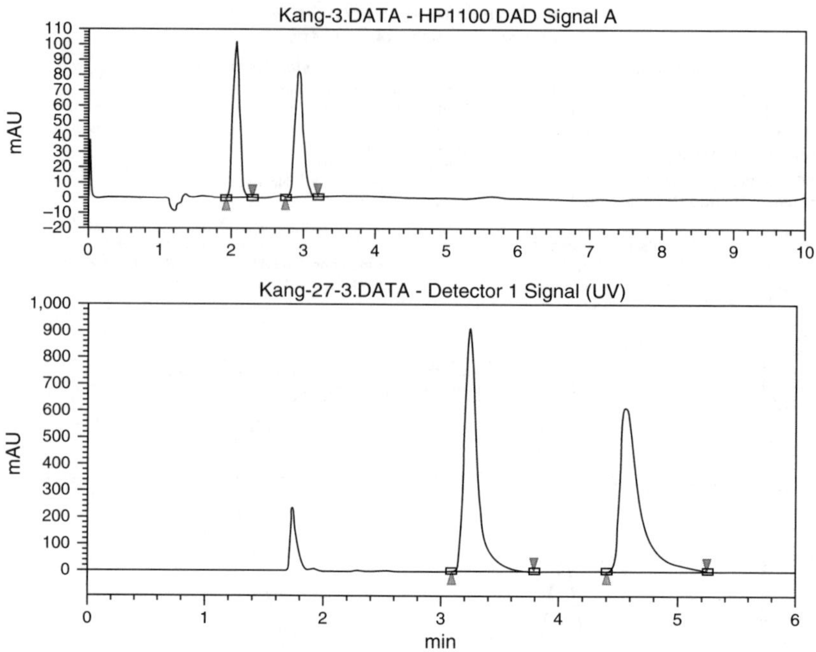

Figure 21 Chiral SFC. Analytical (top) vs. preparative (bottom)-scale results.

are performed at ambient temperature (ca. 22 to 25°C). We have not studied the effect of temperature on these screens yet, but it will be interesting to determine if the temperature alters the frequency of success, especially at temperatures higher than ambient since dipole–dipole interactions are sensitive to temperature, as indicated in Eq. (2). The polar organic screen is an alternative to normal phase, especially if transfer to preparative scale is needed. However, the rate of success has been low with the family of compounds that we typically analyze.

Acquisition of the Sepmatix system to perform parallel chiral NPLC allowed a significant increase in our productivity and efficiency. Perhaps the main drawback is in its poor sensitivity and the large injection volume required splitting the sample solution across the eight columns. This imposes limitations when the compound availability is low. We have, however, found these drawbacks to be minor overall. In cases where we face these situations, our large database allows us to compare structures and methods to identify three to four potential CSPs and to set up a customized screen for these compounds. A similar approach is taken if none of the screens provide a positive hit. In these cases, we either set up screens with phases not normally used or we proceed by developing the method in a traditional approach.

The chiral SFC screen has become our main screen used to develop methods to be transferred to preparative scale. An interesting observation is that the analyses of fractions are performed efficiently as SFC provides faster methods and less

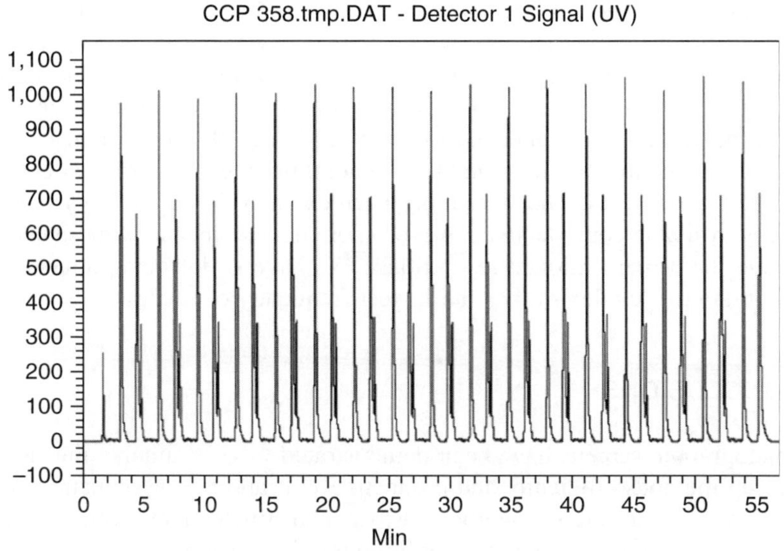

Figure 22 Stacking with chiral SFC.

solvent consumption. The lifetime of the columns exposed to SFC is noticeable longer than with normal-phase LC, thus reducing the cost of column replacement, especially for the case of preparative SFC. The frequency of success is shared by the Chiralpak AD and Chiralcel OJ. These two columns have been reported to have a higher level of success in separating racemates than that of the rest of the polysaccharide CSPs and other chiral phases [49]. Without a doubt, preparative chiral SFC has been demonstrated to be a very efficient way to isolate or purify enantiomers by increasing the yield and turnaround time of the process.

The chiral reversed-phase screen is not often used at the early stage of the pharmaceutical development in our laboratories because the yields are not favorable when transferred to preparative HPLC systems. We have, however, observed a good degree of success at the later stage of development: principally, for three reasons. First, it minimizes use of the environmentally unfriendly and costly solvents used in chiral NPLC; second, there is good solubility of the polar chiral compounds in the RPLC media; and third, the resulting methods are, in general, quality control friendly (e.g., rugged and robust). An interesting aspect of chiral RPLC is the use of inorganic anions such as PF_6^- and ClO_4^-. Both ions are chaotropic. Chaotropic ions disrupt the analyte solvation shell, and this increases the apparent hydrophobicity of the analyte; therefore, there is an observed increase in the analyte's retention time [50–52]. We have, indeed, observed this in our screens for compounds when the compounds are polar/ionic (strong ionic bases). The example in Figure 18 shows that the AD and AS chiral phases effectively separate both enantiomers, where the AD column has better resolution. There is no reversal of order, but the retention times are noticeable different, as are the gradients used.

The chiral CE screen has the same advantages as the Sepmatix system: namely, both are parallel systems. The advantage of the CE system is the capability, in theory, of performing 96 different compounds simultaneously. However, we found a small but important frequency of false negatives (ca. 15%), which forces us to perform the analysis in duplicate, as shown in Figure 19. The false negatives are attributed to injection problems and lack of conditioning of a particular column in the capillary array [46]. The system has been extensively utilized to screen commercially available compounds; however, more developmental work is needed to overcome the aforementioned restrictions.. We have so far not evaluated which highly sulfated cyclodextrin has the largest frequency of success.

9 CONCLUSIONS

Chromatographic screens have been demonstrated to be a suitable means to use to develop methods efficiently and productively. It allows a systematic approach to implementing different conditions which can affect many factors that contribute to separation of enantiomers. Minimum supervision of the screens during operation allows scientists to multitask, thus increasing the efficiency and productivity of the laboratory. The systematic approach allows evaluation of the impact of different conditions on the separations in a short period of time. It also establishes a database that can be used to study trends that will facilitate the development of compounds that have similar specific characteristics without using screens or traditional method development approaches.

We have covered a wide area with respect to the various modes of chiral separation (with the exception of gas chromatography). The use of different chiral modes allows the analyst efficiently to obtain the best possible method when samples are screened using all the different modes simultaneously. Finally, it is important to reiterate that these screens are not static but dynamic, and that feedback and optimization are necessary to improve them and to assure their alignment with the technical advances in the field.

REFERENCES

1. N. M. Maier, W. Lindner. In *Chirality in Drug Research*, E. Francotte, W. Lindner, Eds. pp. Wiley-VCH, Weinheim, Germany, 2006, pp. 189–260.
2. A. J. Hutt. In *Experimental Therapeutics*. K. Davies, Ed. Martin Dunitz, London, U.K, 2003, pp. 125–144.
3. T. D. Stephens, C. J. W. Bunde, B. J. Fillmore. *Biochem. Pharmacol.*, 59:1489–1499, 2000.
4. K. Dredge, J. B. Marriot, A. G. Dalgleish. *Crit. Rev. Immunol.*, 22:425–437, 2002.
5. H. J. Schmahl, H. Nau, D. Neubert. *Arch. Toxicol.*, 62:200–204, 1988.
6. T. Eriksson, S. Bjoerkman, B. Roth, A. Fyge, P. Hoglund. *Chirality*, 7:44–52, 1995.
7. T. Eriksson, S. Bjoerkman, B. Roth, A. Fyge, P. Hoglund. *Chirality*, 10:223–228, 1998.

8. T. Eriksson, S. Bjoerkman, P. Hoglund. *Eur. J. Clin. Pharmacol.*, 57:365–376, 2001.
9. S. K. Teo, D. I. Stirling, J. B. Zeldis. *Drug Discov. Today*, 10:107–114, 2005.
10. N. Matthijs, C. Perrin, M. Maftouh, D. L. Massart, Y. Vander Hayden. *J. Chromatogr. A*, 1004:119–133, 2004.
11. M. Maftouh, C. Granier-Loyaoux, E. Chavana, J. Marini, A. Pradines, Y. Vander Hayden, C. Picard, *J. Chromatogr. A*, 1008:67–81, 2005.
12. C. Perrin, M. Maftouh, Y. Vander Hayden, D. L. Massart. *Electrophoresis*, 22:3202, 2001.
13. C. Perrin, V. A. vu, N. Matthijs, M. Maftouh, D. L. Massart, Y. Vander Hayden. *J. Chromatogr. A*, 947:69, 2002.
14. C. Perrin, N. Matthijs, C. Granier-Loyaoux, M. Maftouh, D. L. Massart, Y. Vander Hayden. *J. Chromatogr. A*, 966:119, 2002.
15. W. H. Pirkle, T. C. Pochapsky. *Chem. Rev.*, 89:347–362, 1989.
16. R. Kaliszan. *Structure and Retention in Chromatography: A Chemometric Approach*. Hardwood Academic, New York, 1997, Chap. 3.
17. T. E. Beesley, R. P. W. Scott. *Chiral Chromatography*, Wiley, New York, 1998, Chap. 3.
18. J. M. Cintron, L. A. Colon. *Analyst*, 127:701–704, 2002.
19. M. E. Swartz. *J. Liq. Chromatogr. Relat. Technol.*, 28:1253, 2005.
20. M. E. Swartz. In *HPLC Method Development for Pharmaceuticals*. Separation Science and Technology Series, Vol. 8, S. Ahuja, Series Ed. Academic Press, San Diego, CA, 2007, Chap. 6.
21. J. Ringling, C. Wood, R. Borjas, C. Foti. *Am. Pharm. Rev.*, Nov.–Dec. 2008.
22. Y. Xiang, N. Wu, J. A. Lippert, M. L. Lee. *Chromatographia*, 55:399–403, 2002.
23. Y. Gong, Y. Xiang, B. Yue, G. Xue, J. S. Bradshaw, H. K. Lee, M. L. Lee. *J. Chromatogr. A*, 1002:63–70, 2003.
24. S. Ma, N. Grinberg. *Am. Pharm. Rev.*, Apr. 2009.
25. S. Ma, S. Shen, N. Haddad, W. Tang, J. Wang, H. Lee, N. Yee, C. Senanayake, N. Grinberg. *J. Chromatogr. A*, 1216:1232–1240, 2009.
26. M. C. Ringo, C. E. Evans. *Anal. Chem.*, 69:4964–4971 1997.
27. X. Sun, L. Wrisley. *Am. Drug Discov.*, June–July 2007.
28. J. P. C. Vissers. Capillary LC Columns: Packing Techniques and Applications. Dissertation, Eindhoven University of Technology, 1998.
29. R. A. Hartwick, R. F. Meyer, In *Microcolumn Separations: Columns, Instrumentation and Ancillary Techniques*. Journal of Chromatography Library, Vol. 30. Elsevier Science, North-Holland, Amsterdam, 1985, pp. 87–104.
30. L. R. Snyder, J. W. Dolan. High performance gradient elution. In *The Practical Application of the Linear-Solvent-Strength Model*. Wiley, Hoboken, NJ, 2007, Chap. 9.
31. R. Xu, C. Nemes, K. M. Jenkins, R. A. Rourick, D. B. Kassel, C. Z. C. Liu. *J. Am. Soc. Mass Spectrom.*, 13:155–165, 2002.
32. D. B. Kassel. In *HPLC for Pharmaceutical Scientists*, Y. Kazakevich, R. LoBrutto, Eds. Wiley, Hoboken, NJ, 2007, Chap. 11.
33. R. God, H. Gumm. In *Analysis and Purification Methods in Combinatorial Chemistry*, B. Yan, Ed. Wiley, Hoboken, NJ, 2004, Chap. 12.

34. L. Muller-Kuhrt, R. God, H. Gumm, J. Binkele. German patent DE 199 57 489 A1 (to SEPIAtec GmbH), May 31, 2000.
35. Hagen–Poiseuille equation. en.wikipedia.org/wiki/Poiseuille's_law.
36. C. J. Welch, P. Sajonz, G. Spencer, W. Leonard, D. Henderson, W. Schafer, F. Bernardoni. *Org. Process Res. Dev.*, 12:674–677, 2008.
37. R. D. Smith, H. R. Udseth. *Anal. Chem.*, 55:2266–2272, 1983.
38. T. A. Berger. In *Supercritical Fluid Technology for Drug Product Development*. Drug and the Pharmaceutical Sciences, Vol. 138, P. York, U. B. Kompella, B. Y. Shekunov, Eds. Marcel Dekker, New York, 2004, Chap. 12.
39. J. Zheng, T. Glass, L. T. Taylor, J. D. Pinkston. *J. Chromatogr. A*, 1090:155–164, 2005.
40. J. Zheng, L. T. Taylor, J. D. Pinkston, M. L. Mangels. *J. Chromatogr. A*, 1082:220–229, 2005.
41. R. Weinberger. *Practical Capillary Electrophoresis*, 2nd Ed. Academic Press, San Diego, CA, 2000, Chap. 2.
42. W. Kuhr. In *Capillary Electrophoresis: Theory and Practice*, 2nd Ed. CRC Press, Boca Raton, FL, 1998, Chap. 3.
43. B. Chankvetadze. *Capillary Electrophoresis in Chiral Analysis*. Wiley, New York, 1997, Chaps. 2 and 3.
44. H. Kubota, S. Tsuda, M. Murata. *Rev. Phys. Chem. Jpn.*, 49:59–69, 1979.
45. D. Corradine. Electro-osmotic flow. In *Encyclopedia of Chromatography*. Marcel Dekker, New York, 2002.
46. L. Saavedra, B. Nickerson, R. E. Borjas, F. Lynen, P. Sandra. *J. Chromatogr. B*, 875:248–253, 2008.
47. G. B. Cox. In *Preparative Enantioselective Chromatography*, G. B. Cox, Ed. Blackwell Publishing Ltd., Oxford, U.K., 2005, Chap. 2.
48. H. Y. Aboul-Enein, I. Ali. *Chiral Separations by Liquid Chromatography and Related Technologies*. Marcel Dekker, New York, 2003, Chap. 2.
49. R. M. Smith. In *Supercritical Fluid Chromatography with Packed Columns: Techniques and Applications*, K. Anton, C. Berger, Eds., Marcel Dekker, New York, 1998, Chap. 8.
50. J. M. Roberts, A. R. Diaz, D. T. Fortin, J. M. Friedle, S. D. Piper. *Anal. Chem.*, 74:4927, 2002.
51. Y. V. Kazakevich, R. LoBrutto, R. Vivilecchia. *J. Chromatogr. A*, 1064:9–18, 2005.
52. T. Cecchia, P. Passamonti. *J. Chromatogr. A*, 1216:1789–1797, 2009.
53. D. Haiger. *A Primer: High Performance Capillary Electrophoresis*. Agilent Technologies, Santa Clare CA, 2000, Intro. and Chap. 2.

CHAPTER 9

Separation of Enantiomers by Gas Chromatography on Chiral Stationary Phases

VOLKER SCHURIG

Institute of Organic Chemistry, University of Tübingen, Tübingen, Germany

1 INTRODUCTION

There exist two approaches for enantioseparations by gas chromatography. The *indirect* approach is based on the formation of diastereomeric derivatives via the reaction of enantiomers with an enantiomerically pure chiral auxiliary and the separation of stereoisomers on a conventional achiral stationary phase in the spirit of Pasteur's resolution principles [1]. This method requires the absence of kinetic resolution and the absence of racemization of both reaction partners during derivatization as well as the unbiased detection of diastereomers. In the *direct* approach, enantiomers are separated via the noncovalent diastereomeric interaction with a nonracemic chiral stationary phase (CSP). The rapid and reversible formation of diastereomeric association complexes of distinct stabilities will bring about differences in the partition coefficients of the enantiomers, resulting in gas chromatographic resolution [2]. In the present account, only the direct approach is treated. As the mechanisms of chirality recognition in enantioselective gas chromatography are not fully understood, molecular modeling studies are not mentioned in this account, nor is the patent literature surveyed.

High efficiency, sensitivity, and speed of separation are important advantages of enantioseparation by *high-resolution capillary gas chromatography* (HRC-GC). Because of the high separation power of HRC-GC, contaminants and impurities are separated from the chiral analytes, and the simultaneous analysis of multicomponent mixtures of enantiomers (e.g., derivatized proteinogenic α-amino

Chiral Separation Methods for Pharmaceutical and Biotechnological Products, Edited by Satinder Ahuja

acids) is straightforward. Ancillary techniques such as multidimensional gas chromatography (MDGC) (i.e., in series-coupled column operation) and interfacing and coupling methods such as gas chromatography–mass spectrometry (GC-MS), are important tools in chiral analysis. Employing the selected ion monitoring (SIM) mode, trace amounts of enantiomers can be detected by GC-MS(SIM). The universal flame ionization detector (FID) is linear over five orders of magnitude, and detection sensitivity can be further increased to the picogram level by electron-capture detection (ECD) and by element-specific detection, usually aided by special derivatization strategies. In contrast to liquid chromatography or electromigration methods, the delicate choice of solvents (buffers), modifiers, and gradient elution systems is absent in GC. Yet prerequisites for the use of GC are volatility, thermal stability, and resolvability of the chiral analyte, restricting the exclusive use of enantioselective GC. The state of the art of enantioselective GC has been reviewed previously in a number of accounts [3–12]. A fundamental review on developments and applications in chiral separations, including enantioselective GC from January 2006 to January 2008 of the English language literature, has been compiled by Ward and Baker [13]. Reviews are also included in general treatises on chiral analysis [14–17]. More specific topical reviews are mentioned in the text at the appropriate places.

2 CHIRAL STATIONARY PHASES BASED ON α-AMINO ACID DERIVATIVES FOR HYDROGEN-BONDING GC

The first direct enantioseparation on a chiral stationary phase (CSP) by gas chromatography was discovered in 1966 by Gil-Av, Feibush, and Charles-Sigler [18,19]. Interestingly, they employed an experimental setup that was not much different from today's use. An account of the fundamental work carried out at the Weizmann Institute of Science in Israel has recently been advanced [20]. A home-made 100 m × 0.25 mm ID glass capillary column [21] was coated with a 20% solution of the CSP *N*-trifluoroacetyl-L-isoleucine lauryl ester (see Scheme 1) in diethyl ether by the plug method. After conditioning overnight and raising the temperature to 80°C, the column exhibited 140,000 plates for 2-heptylacetate, and trials were performed to resolve racemic *N*-trifluoroacetyl (TFA) alkyl esters of natural α-amino acids. The 2-propanol, *n*-butanol, and cyclopentanol esters of *N*-trifluoroacetylalanine were partially resolved. More intriguingly, the diastereomeric *N*-TFA-α-amino acid esters of chiral 2-butanol, which were previously applied for the indirect method, gave rise to four peaks on the CSP, corresponding to two enantiomeric pairs D,D/L,L (*like*) and D,L/L,D (*unlike*), whereas only the two peaks were observed of the like and unlike diastereomers on an achiral phase. The design of this enantioselective selector–selectand system was based on the idea of biomimetically imitating the stereoselective peptide enzyme interaction employing simple α-amino acids entities as model substances. To prove that enantiomers were indeed separated, control experiments were carried out by Gil-Av et al. [2]. The resolution vanished on an achiral stationary phase [18]

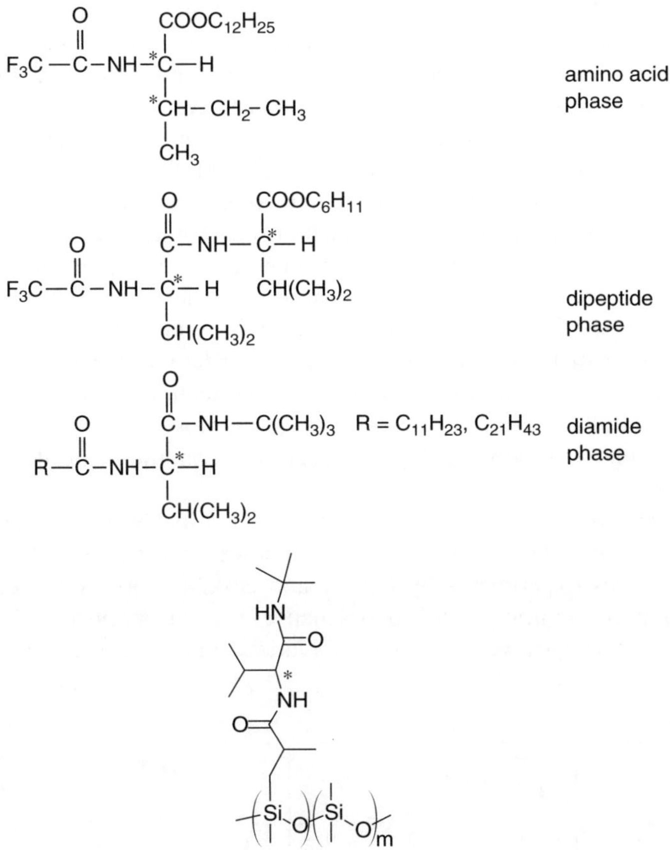

Scheme 1 Structure of α-amino acid CSPs (from [8]) and of Chirasil-Val (from [30]).

and mixtures of D- and L-amino acid derivatives in varying ratios were prepared. Upon injection onto the enantioselective column, it was found that the peak areas corresponded well to the proportions of the enantiomeric mixture investigated [2]. Also the configuration of the CSP was reversed from L- to D-isoleucine leading to a peak reversal (i.e., an inversion of the elution order of the samples enriched in one enantiomer). These experiments proved unequivocally that incongruent mirror-image isomers had been separated for the first time by gas chromatography [2]. The authors recognized the importance of their discovery [18]: *The achievement of resolution of optical isomers is one of the most striking demonstrations of the efficiency of gas–liquid partition chromatography. The columns described should have interesting analytical applications and might also serve as a tool for the study of stereospecific interactions*. In a follow-up paper, 18 pairs of enantiomers of *N*-trifluoroacetyl-α-amino acid esters were resolved using *N*-trifluoroacetyl-D-(or L-)isoleucine lauryl ester and *N*-trifluoroacetyl-L-phenylalanine cyclohexyl ester, respectively, as CSPs [22]. In all instances it was found that the derivatives of L-amino acids eluted after the corresponding D-enantiomers on columns coated with CSPs possessing the L-configuration (and vice versa for the D-configurated CSPs). Gil-Av et al. noted that the difference in the free energies of solvation $\Delta(\Delta G)$ of the diastereomeric associates amounted to only 0.006 to 0.03 kcal/mol at the temperature of enantioseparation. The enantioseparation was thought to be due to hydrogen bonding between NH · · · F and NH · · · O=C^{-} functions, whereby the latter contribution was considered to be more important. Whereas all previous enantioseparations were performed on an analytical scale, another pioneering paper appeared shortly afterward [23]. It can be regarded as the first approach to dealing with an enantioselective packed column for the semipreparative enantioseparation by GC [24]. Moreover, a chiroptical detector that produced opposite optical rotatory dispersion curves was employed for the first time for unequivocal identification of the separated enantiomers. In addition, a dipeptide derivative was employed as a second-generation CSP. The results are shown in Figure 1, and the conditions are detailed in the caption.

Further development of the direct enantioseparation by GC by dipeptide or α-amino acid diamide phases has been reviewed thoroughly [2,25–27]. By way of ingenious experiments by Gil-Av and collaborators (reviewed in [25]), the evolution to optimum CSP performance was accomplished later. Thus, peak reversal was observed with the derivatized mirror-image dipeptide CSPs

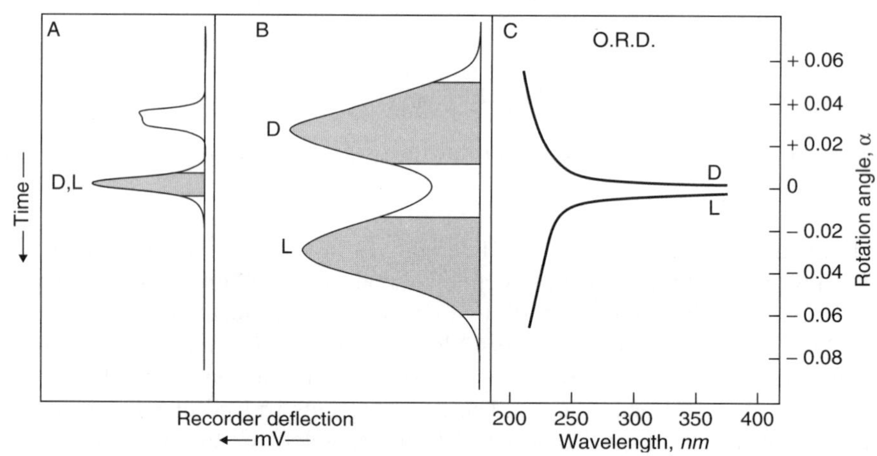

Figure 1 (A) Gas chromatogram of impure racemic D, L-*N*-TFA-alanine *tert*-butyl ester (shaded area; white area: impurity) on a 4 m × 6 mm (ID) column containing 20% achiral SE-30 on Chromosorb W at 125°C; no resolution occurs. (B) Gas chromatogram of the collected fraction (shaded area) corresponding to the second peak in (A) on a 2 m × 1 mm ID column containing 5% *N*-TFA-L-valyl-L-valine cyclohexyl ester (see Scheme 1) on Chromosorb W at 100°C; partial resolution occurs. (C) Optical rotatory dispersion diagram of the two collected fractions corresponding to the shaded areas in (B) showing opposite rotation angles. (From [23].)

L-Val-L-Val and D-Val-D-Val. Compared to D-Val-L-Val or L-Val-D-Val, peak reversal on the diastereomeric dipeptide CSP L-Val-L-Val occurred only on changing the configuration of the N-terminal amino acid. This was also confirmed by the finding that Gly-L-Val possessed significantly inferior enantioseparating properties than those of L-Val-Gly. Consequently, the C-terminal amino acid in the dipeptide phase was recognized as unimportant toward enantioselectivity, yet providing the second amide bond required for hydrogen bonding. The C-terminal amino acid was therefore substituted by a *tert*-butyl group to yield the versatile mono α-amino acid diamide selector *N*-lauroyl-L-valine-*tert*-butylamide (see Scheme 1) of Feibush [28]. The direct gas chromatographic enantioseparation on optically active mesophases has been achieved using mesomorphic chiral carbonyl-bis(amino acid esters) [29].

A breakthrough of enantioselective GC was achieved in 1977 when Frank et al. linked the valine diamide selector of Feibush to a polysiloxane backbone, thereby combining the inherent enantioselectivity of a CSP with the unique gas chromatographic properties of fluidic silicones [30]. A chiral polysiloxane containing valine diamide (see Scheme 1) termed *Chirasil-Val* was commercialized as the enantiomeric L- and D-form, allowing for peak switching scenarios in chiral analysis. The simultaneous enantiomeric separation of all proteinogenic α-amino acids as *N*(*O*, *S*)-trifluoroacetyl-*O*-*n*-propyl esters on Chirasil-Val is depicted in Figure 2 [31]. Similar enantioseparations were also obtained with *N*(*O*, *S*)-pentafluoropropionyl-*O*-isopropyl esters of α-amino acids. Compared to the rather archaic types of peaks in the first trials of Gil-Av et al., this high-resolution enantioseparation is truly remarkable.

The pretreatment of borosilicate glass columns prior to coating with Chirasil-Val has been described in detail [32] and the immobilization property

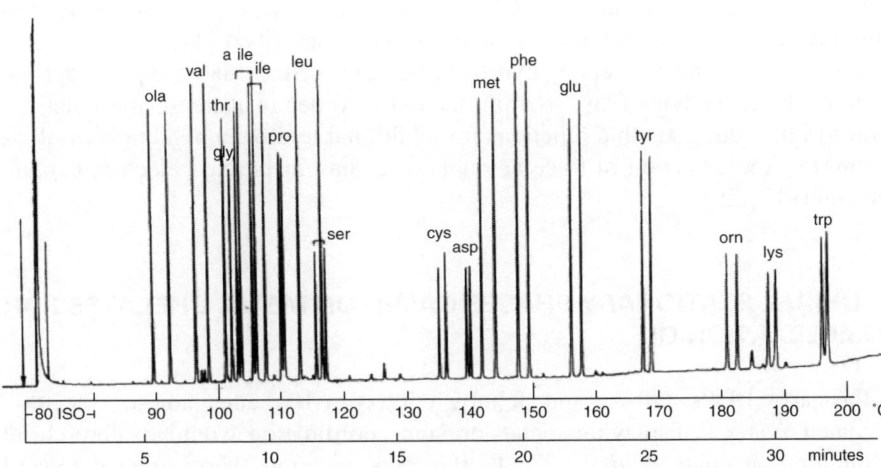

Figure 2 Enantioseparation of the *N*(*O*, *S*)-trifluoroacetyl *n*-propyl α-amino acids by HRC-GC on a glass capillary (20 m × 250 μm ID) coated with Chirasil-Val. The first eluted peak corresponds to the D-enantiomer. (From [31], with permission.)

of Chirasil-Val on the glass surfaces was also studied [33]. In later work, Chirasil-Val–coated borosilicate glass capillary columns were substituted by fused-silica capillary columns which are commercially available from Varian, Inc. in the United States (formerly from Chrompack, Middelburg, The Netherlands). A direct, straightforward approach to polymeric CSPs is based on the modification of cyanoalkyl-substituted polysiloxanes (XE-60, OV-225) [34,35]. Koppenhoefer et al. modified the chiral backbone in Chirasil-Val by variation of the loading and polarity of the CSP and by the introduction of rigid spacers [36–38]. In Chirasil-Val-C_{11}, a long undecamethylene spacer separates the valine diamide selector from the polymeric backbone [39]. In Chirasil-Val the chiral moieties are distributed statistically along the polymer chain. A more ordered Chirasil-type CSP has been obtained by block condensation of 1,5-bis(diethylamino)hexamethyltrisiloxane and 2′,2′,2′-trifluoroethyl-(3-dichloromethylsilyl)-2-methylpropionate followed by nucleophilic displacement of the functionalized polysiloxane with chiral amines and amino acids [40]. The immobilization of the CSPs by thermal [40] and radical-mediated cross-linking [41] has been studied, and the extent of radical-induced racemization was determined.

A highly ordered supramolecular structure has been prepared by linking chiral L-valine-*tert*-butylamide moieties to the eight hydroxyl groups of a resorcine[4]arene basked-type structure obtained from resorcinol and 1-undecanal. The calixarene was subsequently linked chemically via four spacer units to a poly(dimethylsiloxane) to give Chirasil-Calixval [42,43]. However, the ordered cyclic arrangement of the chiral selectors juxtaposed in close proximity to each other did not improve enantioselectivity compared to Chirasil-Val, and aromatic α-amino acid derivatives did not exhibit any beneficial effect due to supramolecular inclusion. The synthesis of thiacalix[4]arenes with pendant chiral amines and their application as CSP for the enantioseparation of derivatized amino acids, alcohols, and amines have also been described [44].

Separation of enantiomers by hydrogen-bonding CSPs usually requires derivatization of the analytes (e.g., α-amino acids) in order to increase the volatility and/or to introduce suitable functions for additional hydrogen bonding as well as to improve the detection of trace amounts of enantiomers (e.g., electron-capture detection) [25,26].

3 CHIRAL STATIONARY PHASES BASED ON METAL CHELATES FOR COMPLEXATION GC

In the early 1970s, Gil-Av and Schurig conjectured whether abiotic selector–selectand systems displaying metal–organic coordination would display chiral recognition in *complexation GC*. To this end, attempts were directed toward resolving a chiral olefin on an optically active metal coordination compound by GC. The π-complexation ability of silver(I) ions toward olefins, which was utilized for selective separations of unsaturated from saturated hydrocarbons by *argentation GC*, was already well established at the beginning of the 1960s [45,46], and terpene-derived β-diketonate anions were used as ligands for metal ions and the chelates were used in complexation GC for selective achiral separations by Feibush et al. and Schurig et al. [47,48]. Dicarbonylrhodium(I) 3-(trifluoroacetyl)-(1R)-camphorate (see Scheme 2) dissolved in squalane and coated on a long capillary column displayed isotopic selectivity toward deuteriated ethenes [49] and, as a first example, also enantioselectivity toward the chiral olefin 3-methylcyclopentene [50] by complexation GC.

Whereas the scope of enantiomeric separation of olefins on the rhodium (I)-containing selector was limited [51], the use of lanthanide(III) tris[3-(trifluoroacetyl)-(1R)-camphorates] [52] and particularly metal(II) bis[3-(trifluoroacetyl)-(1R)-camphorates] (see Scheme 2) and metal(II) bis[3-(heptafluorobutanoyl)-(1R)-camphorates] [53,54] (metal = manganese, cobalt, and nickel) as chiral selectors for the enantioseparation of the enantiomers of nitrogen-, oxygen- and sulfur-containing selectands soon emerged as a routine method for enantioselective GC. The classes of compounds resolved encompassed cyclic ethers, esters, acetals, underivatized alcohols, and ketones. Most of the compounds were not previously amenable to enantioseparation by

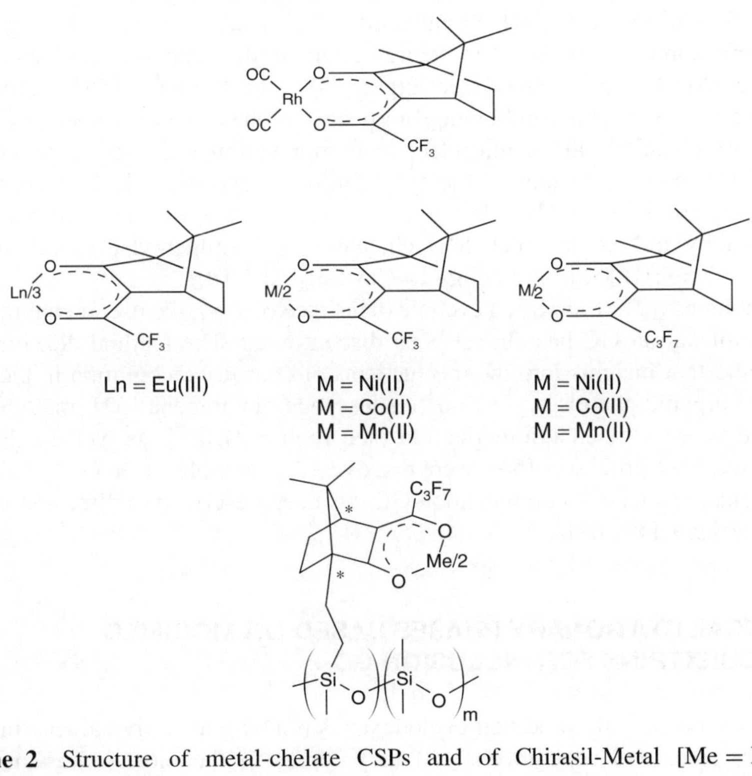

Scheme 2 Structure of metal-chelate CSPs and of Chirasil-Metal [Me = Ni(II)/2, Eu(III)/3]. (From [56].)

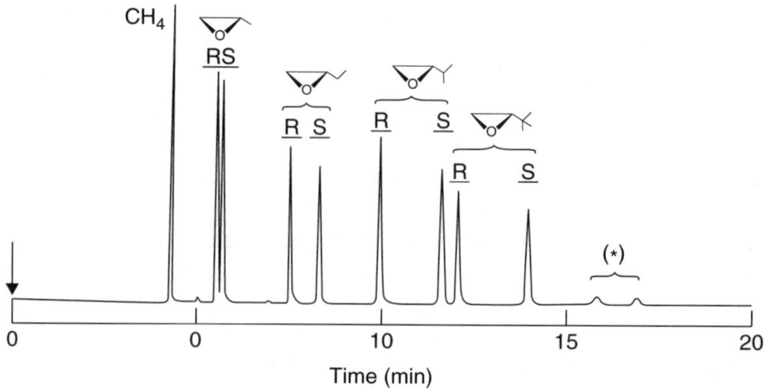

Figure 3 Enantioseparation of simple alkyloxiranes on a glass capillary column (42 m × 250 μm ID) coated with manganese(II) bis[3-(heptafluorobutanoyl)-(1*R*)-camphorate] in OV-101 at 40°C. (*) = unknown chiral impurity. (From [4], with permission.)

hydrogen-bonding CSPs. The application of high-resolution glass or fused-silica open tubular column technology in complexation GC improved the state of the art, and squalane was later replaced by poly(dimethylsiloxane) as a useful solvent for the selectors [54]. Significantly, the metal selectors were capable of separating some of the smallest chiral compounds: namely, alkyl-substituted aliphatic aziridines, oxiranes (see Figure 3), and thiiranes [53]. Altogether, 11 different 3-(heptafluorobutanoyl)terpene ketones were probed as chiral ligands of nickel(II) in enantioselective complexation GC, among them CSPs based on 3- and 4-pinanone, 4-methyl-3-thujone, carvone, pulegone, menthone, and isomenthone [55]. The low-temperature stability of the CSP was later improved by linking the nickel(II) chelate to poly(dimethylsiloxane) to yield Chirasil-Nickel(II) [56], in analogy to Chirasil-Val [30].

With the advent of modified cyclodextrins (see below), the use of enantioselective complexation GC has almost been discontinued. The method did, however, offer important insights into the mechanisms of chirality recognition in the realm of metal organic chemistry. Various coalescence phenomena [57], including the dynamic process of enantiomerization (see Figure 4) [58], as well as different enantioselective processes [56], were discovered in complexation GC. The evolution of enantioselective complexation GC has been described in three consecutive review articles [59–61].

4 CHIRAL STATIONARY PHASES BASED ON MODIFIED CYCLODEXTRINS FOR INCLUSION GC

Native and chemically modified cyclodextrins (CDs) play a pivotal role in many areas for analytical separations, and they represent the most important chiral stationary phases (CSPs) for the separation of enantiomers by chromatography

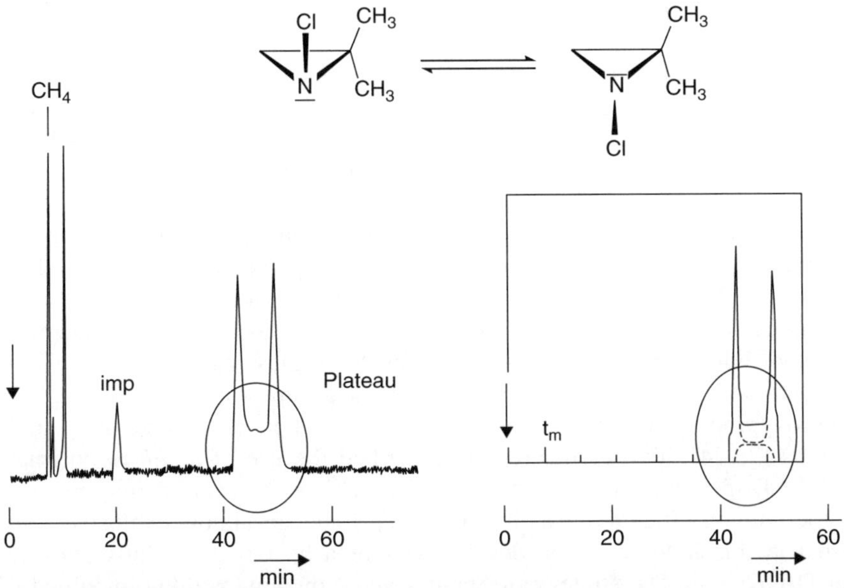

Figure 4 Plateau formation due to enantiomerization (distorted elution profile caused by inversion at the nitrogen atom) of 1-chloro-2,2-dimethylaziridine upon complexation GC on nickel(II)bis[3-(trifluoroacetyl)-(1*R*)-camphorate] in squalane at 60°C. Left: experimental trace; right: simulated trace. (From [58], with permission.)

and electromigration methods [62]. A number of review articles describe the use of modified cyclodextrins in enantioselective GC [62–65].

The first gas chromatographic enantioseparation of terpenoic hydrocarbons (α- and β-pinene, *cis*- and *trans*-pinane and carene) were observed by Kóscielski et al. on native α-cyclodextrin hydrate dissolved in formamide and coated on celite [66,67]. Despite large separation factors α, peak resolution on the packed column was low, the column temperature was limited to 70°C, and the lifetime of the chromatographic system was short due to extensive column bleeding and dehydration of the CD. The presence of water was subsequently identified as being essential for maintaining the high enantioselectivity of the setup [68]. Nevertheless, this work started an impressive development of enantioselective GC employing selectively derivatized CDs as CSPs. The high potential of CDs for enantioselective separations is (1) due to the presence of D-glucose building blocks that form cavities of different molecular dimensions, and (2) due to the different reactivity of the 2-, 3-, and 6-hydroxy groups of the glucose moiety, which can be modified by regioselective alkylation and acylation [62], thereby giving rise to a plethora of possible chemically modified CDs [69]. The only disadvantage is the fact that CDs are only available in the naturally occurring D-form of the glucose building blocks. It should be realized that absence of the L-form eliminates the possibility of reversing the sense of enantioselectivity by employing the CSP with opposite configuration, which is important for validation purposes.

Since 1987, modified CDs have been introduced in enantioselective capillary GC and they have begun to play an important role in chiral separations [62]. To coat capillary columns (Pyrex glass, replaced later by fused silica) efficiently, the CSP should be fluid. Two approaches have been developed independently. Schurig and Nowotny dissolved peralkylated CDs in a moderately polar polysiloxane (e.g., OV 1701), thereby combining enantioselectivity with the unique properties of silicones in GC [70,71], whereas König et al. [72] used low-melting-point CD derivatives containing *n*-pentyl groups (e.g., per-*O*-pentylated and 3-*O*-acylated-2,6-di-*O*-pentylated α-, β-, and γ-CD [64]) as undiluted fluid stationary phases coated on Pyrex glass capillary columns. This approach was later extended by Armstrong's group, who used permethylated 2-hydroxypropyl and pentylated/acylated CDs coated on fused-silica capillary columns [73,74]. Two earlier reports described the use of *undiluted* permethylated β-CD in a supercooled state below its melting point for enantioseparations, but the column efficiency was low [75–77]. Later an almost explosive activity ensued in this field, with the development of a host of differently substituted CDs [16,62–65]. The strategy of Schurig and Nowotny to dilute modified CDs such as permethylated β-CD and 2,6-di-*O*-methyl-3-*O*-trifluoro-β-CD [70,71] in semipolar polysiloxanes soon emerged as the most frequently used methodology. Although the impressive enantioseparations described by König [64] were originally obtained on *undiluted* CDs, this group extended their investigations to diluted systems employing fused-silica capillary columns [6]. The use of permethylated α-, β-, and γ-CD dissolved in polysiloxanes has a number of merits [78]:

1. The unique physicoselectivity of silicones is maintained and combined with the inherent chemoselectivity of cyclodextrins.
2. The universal coating properties of silicones for producing high-resolution, high-efficiency capillary columns by HRC-GC are maintained.
3. The high melting points or phase transitions of modified cyclodextrins are not detrimental to column performance and can thus be ignored.
4. Multicomponent (mixed) cyclodextrin-based stationary phases can be employed (vide infra).

In a comprehensive study involving more than 150 racemates, the advantages of diluted permethylated β-CD compared to the α- and γ-CD congeners was demonstrated [79]. A comparison between permethylated α-, β-, and γ-CD showed complementary behavior and in selected cases, even a reversal of the elution order of the enantiomers [80]. A useful collection of enantioseparation factors of racemates observed on octakis(6-*O*-methyl-2,3-di-*O*-pentyl)-γ-cyclodextrin (Lipodex G) [81], on hepakis(2,6-*O*-methyl-3-*O*-pentyl)-β-cyclodextrin vs. octakis(2,6-*O*-methyl-3-*O*-pentyl)-γ-cyclodextrin [82], and on octakis(3-*O*-butanoyl-2,6-di-*O*-pentyl)-γ-cyclodextrin (Lipodex E) [83] was compiled by König. Different types of CD-coated capillary columns available commercially under the trade name Lipodex (distributed by Macherey Nagel, Düren, Germany) are prepared according to procedures developed by König and collaborators. Schurig–Nowotny-type CD columns are produced by Chrompack Int. (now Varian, Inc., Middelburg, The Netherlands). Chiraldex is a trade mark for permethylated-2-hydroxypropyl- and trifluoroacetyldipentyl α-, β-, and γ-CDs developed by Armstrong et al., produced by Astec, Advanced Separation Technologies (Whippany, NJ). Additional commercial vendors of fused-silica capillary columns coated with modified CDs are listed in a review by He and Beesley [12] and a book by Schreier et al. [16]. Permethylated β-CD has also been dissolved in ionic liquids and then coated on capillary columns [84].

The bulky *tert*-butyldimethylsilyl (TBDMS) residue represents an important protecting group in various synthetic transformations involving CDs [84]. Yet per-TBDMS-β-cyclodextrin diluted in polysiloxane PS-086 and coated on glass capillary columns can be used in its own right as a versatile CSP for the gas chromatographic enantioseparation up to 250°C [85]. Whereas enantioseparation factors were small, two related CD derivatives carrying the TBDMS groups only in the 6-position, which emerged as very important CSPs in enantioselective GC, have been employed by Dietrich et al.: heptakis(2,3-di-*O*-acetyl-6-*O*-*tert*-butyldimethylsilyl)-β-cyclodextrin [86] and heptakis(2,3-di-*O*-methyl-6-*O*-*tert*-butyldimethylsilyl)-β-cyclodextrin [87]. The presence of a bulky substituent at the primary 6-hydroxy groups may influence the conformation of the CDs and may block entrance to the cavity from the smaller rim, which could have an impact on enantioselectivity. Compared to permethyl-β-CD, the TBDMS derivatives of β-CD also show higher solubility in semipolar polysiloxanes. Capillary columns coated with TBDMS-substituted CDs are commercially available from BGB Analytik KG (Adliswil, Switzerland) and from MEGA (Legnano, Italy).

The dissolution of modified CDs in semipolar polysiloxanes (e.g., OV-1701) [70,71] represents the most useful approach for enantioseparation by GC. In diluted systems the enantioseparation factor α is rendered concentration dependent due to the two different retention mechanisms arising (1) from the presence of the achiral solvent and (2) from the chiral CD selector, both comprising the total stationary phase [88–90]. A theoretical treatment has shown that α does not increase linearly with the CD concentration but often reaches an optimum at a low CD concentration [90]. Hence, no gain in enantioselectivity above an optimum value is obtained. Thus, the use of *undiluted* CDs [64] is closer and closer to being discontinued.

With the strategy to dissolve modified CDs in semipolar polysiloxanes, Schurig and Nowotny [70,71] had already noted that an obvious extension of the dilution approach consists in the fixation of the enantioselective CD selector to a poly(dimethylsiloxane) backbone by a permanent chemical linkage yielding the CSP Chirasil-Dex (a chiral polysiloxane containing cyclodextrin) in the spirit of the CSP Chirasil-Val [30] (the chiral polysiloxane containing valine) (Scheme 3). The synthesis of Chirasil-Dex has been realized independently by two groups [91–93]. The application of Chirasil-Dex in HRC-GC offers the following advantages:

- Hexakis(2,3,6-tri-*O*-pentyl)-α-cyclodextrin
- Hexakis(3-*O*-acetyl-2,6-di-*O*-pentyl)-α-cyclodextrin
- Heptakis(2,3,6-tri-*O*-pentyl)-β-cyclodextrin
- Heptakis(3-*O*-acetyl-2,6-di-*O*-pentyl)-β-cyclodextrin
- Octakis(2,3,6-tri-*O*-pentyl)-γ-cyclodextrin
- Octakis(3-*O*-butanoyl-2,6-di-*O*-pentyl)-γ-cyclodextrin (Lipodex E)

- Hexakis(per-*O*-2-hydroxypropyl)-per-*O*-methyl)-α-cyclodextrin (PMHP-α-CD)
- Heptakis(per-*O*-2-hydroxypropyl)-per-*O*-methyl)-β-cyclodextrin (PMHP-β-CD)
- Hexakis(2,6-di-*O*-pentyl)-α-cyclodextrin (dipentyl-α-CD)
- Heptakis(2,6-di-*O*-pentyl)-β-cyclodextrin (dipentyl-β-CD)
- Heptakis(3-*O*-trifluoroacetyl-2,6-di-*O*-pentyl)-β-cyclodextrin (DPTFA-β-CD)
- Octakis(3-*O*-trifluoroacetyl-2,6-di-*O*-pentyl)-γ-cyclodextrin (DPTFA-γ-CD)

- Heptakis(2,3-di-*O*-acetyl-6-*O*-*tert*-butyldimethylsilyl)-β-cyclodextrin
- Heptakis(2,3-di-*O*-methyl-6-*O*-*tert*-butyldimethylsilyl)-β-cyclodextrin

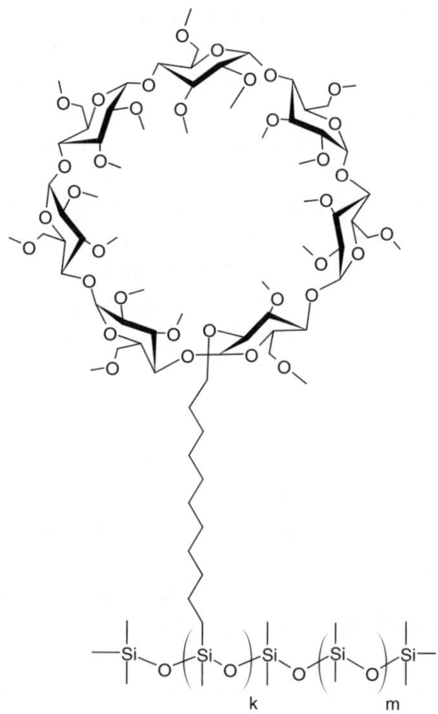

Scheme 3 Modified cyclodextrin CSPs and structure of Chirasil-Dex. (From [96].)

1. The use of apolar polysiloxanes in which CD derivatives cannot be physically diluted [e.g., permethylated CDs in poly(dimethylsiloxane)]
2. A high degree of inertness, allowing the fast analysis of polar racemates
3. The use of low-to-high CD concentrations
4. The immobilization of the CSP by thermal treatment, leading to solvent tolerance (in on-line injection or by rinsing of columns)

5. Compatibility with a very wide temperature range (−25 to 250°C)
6. The reduction of column bleeding

The thermal immobilization of Chirasil-Dex on all types of silica surfaces (e.g., glass, fused silica, silica particles) offers a universal applicability of this CSP in different chromatographic and electromigration modes, including an enantioselective unified approach (see below) [94,95]. In the synthesis of Chirasil-Dex, a monoalkenyl (e.g., allyl, 1-pentenyl, 1-octenyl) residue is introduced into one hydroxyl group of CDs, followed by permethylation and linking the selector to a poly(hydridopermethylsiloxane) via hydrosilylation with a platinum catalyst [91,93]. Statistical synthesis involving chromatographic purification furnishes preferentially the monosubstituted 2-alkenylated product and not, as had earlier been inferred, the 6-alkenylated regioisomer [91–93]. Via hydroxyl group protection chemistry, all three regioisomeric Chirasil-β-Dex CSPs (with 2-, 3-, and 6-octamethylene spacers) have been obtained and compared. They exhibit nearly identical gas chromatographic enantioselectivity for the racemates investigated [96]. The CSP Chirasil-γ-Dex refers to poly(dimethylsiloxane)-linked octakis(3-*O*-butanoyl-2,6-di-*O*-pentyl)-γ-cyclodextrin (Lipodex E) [97]. Another immobilization strategy to link β-CD to siloxanes has been advanced by Armstrong et al. [98] and Bradshaw et al. [99].

The option to mix different cyclodextrin (CD) selectors in one chiral stationary phase (CSP) has been proposed for enantioseparations by gas chromatography [63]. Subsequently, in a number of papers the use of mixed binary CD selector systems for the enantioseparation of different classes of chiral compounds has been described [79,100–105]. Recently, a comprehensive enantioselectivity spectrum toward racemic unfunctionalized alkanes and racemic derivatized α-amino acids as well as to other classes of racemic compounds, such as lactones, diols, secondary alcohols, ketones, and terpenes, has been realized by diluting the two versatile CD selectors octakis(6-*O*-methyl-2,3-di-*O*-pentyl)-γ-cyclodextrin (Lipodex G) [81] and heptakis(2,3-di-*O*-methyl-6-*O*-*tert*-butyldimethylsilyl)-β-cyclodextrin [87] in the polysiloxane PS-086 [106].

The gas chromatographic enantioseparation of a wide variety of racemic compounds of different classes of compounds on modified cyclodextrins usually displays small enantioseparation factors ($1.02 < \alpha < 1.20$), corresponding to a low enantioselectivity, $-\Delta_{D,L}(\Delta G) = RT \ln \alpha$, in the range 0.014 to 0.140 kcal/mol at 100°C. The low enantioselectivities elicited by cyclodextrins under gas chromatographic conditions may challenge the role of molecular inclusion. In principle, the chiral environment of the stationary phase may be sufficient to create a minute enantioselectivity. To this end, *cyclic* heptakis(2,3-di-*O*-acetyl-6-*O*-*tert*-butyldimethylsilyl)-β-cyclodextrin (CD 7) [86] was compared with *acyclic* heptakis[(2,3-di-*O*,4″-*O*)-acetyl-(1′-*O*,6-*O*)-*tert*-butyldimethylsilyl]maltoheptaose (G7) and its congeners (G8), (G4), and (G1) (G refers to the number of glucose units in the linear dextrins) for the enantioseparation of various racemic compounds. It was found that in selected cases the "acyclodextrins" showed even larger enantioseparation factors α than those shown by the cyclodextrins [107,108]. A change in the elution order has also been observed [108]. It was shown that the enantioselectivity of the linear dextrins was nearly independent of the number of glucose units. Even the parent building block glucose as the 2,3,4-tri-*O*-acetyl-1,6-di-*O*-*tert*-butyldimethylsilyl derivative was able to enantioseparate α-amino acids as *N*-TFA methyl esters [108]. One of the main advantages of the use of linear dextrin derivatives resides in the possibility of obtaining both D- and L-configurated (as well as epimeric) forms of the selectors, thereby widening the scope of enantioseparation, including tuning of the elution order of the enantiomers [108]. The results further underscore the difficulties in comprehending mechanisms of chirality recognition exhibited by dextrins.

5 THERMODYNAMIC PARAMETERS OF ENANTIOSEPARATION BY GC: THE ISOENANTIOSELECTIVE TEMPERATURE T_{ISO}

Gas chromatographic enantioseparation requires a *fast* and *reversible* 1 : 1 selector–selectand interaction and is therefore governed by thermodynamics and not by kinetics. Different stabilities of the transient diastereomeric associates AD and AL, formed between an enantiopure selector A and the selectand enantiomers D and L, is the prerequisite of gas chromatographic resolution. The equilibrium constants of association, $K_{\mathrm{D}}^{\mathrm{assoc}}$ and $K_{\mathrm{L}}^{\mathrm{assoc}}$, quantify the formation of energetically distinct diastereomeric association complexes. The true enantioseparation factor α_{assoc} is defined as the ratio of the association constants:

$$\alpha_{\mathrm{assoc}} = \frac{K_{\mathrm{D}}^{\mathrm{assoc}}}{K_{\mathrm{L}}^{\mathrm{assoc}}} \tag{1}$$

Thus, α_{assoc} quantifies the *true enantioselectivity* imparted by the chiral selector A on the selectand enantiomers D and L [109]. This applies for a pure (undiluted) selector [110]. In case of a diluted selector present in an achiral matrix, the apparent enantioseparation factor α_{app} is defined as the ratio of the retention factors of the two enantiomers, k_D and k_L. α_{app} is customarily used as a practical measure of the *apparent enantioselectivity*:

$$\alpha_{\mathrm{app}} = \frac{k_{\mathrm{D}}}{k_{\mathrm{L}}} \tag{2}$$

Since the retention factors of the enantiomers D and L also include (identical) nonenantioselective contributions to retention due to the presence of the achiral matrix (i.e., the solvent S containing the chiral selector A in the total CSP), α_{app} is always lower than α_{assoc}. By knowledge of the true enantioseparation factor α_{assoc}, the true Gibbs energy difference, $\Delta_{D,L}(\Delta G)$, between the diastereomeric association complexes can be obtained via the Gibbs–Helmholtz equation,

$$-\Delta_{\mathrm{D,L}}(\Delta G) = -\Delta_{\mathrm{D,L}}(\Delta H) + T\Delta_{\mathrm{D,L}}(\Delta S) = RT \ln \alpha_{\mathrm{assoc}} \tag{3}$$

where R is the gas constant and T is the absolute temperature. Using the van't Hoff relation

$$\ln\alpha_{\text{assoc}} = \frac{\Delta_{\text{D,L}}(\Delta S)}{R} - \frac{\Delta_{\text{D,L}}(\Delta H)}{R}\frac{1}{T} \tag{4}$$

the true enantioseparation factor α_{assoc} is related to the enthalpy and entropy differences, $\Delta_{\text{D,L}}(\Delta H)$ and $\Delta_{\text{D,L}}(\Delta S)$, between the diastereomeric selector–selectand associates in the solvent S. As $\Delta_{\text{D,L}}(\Delta H)$ and $\Delta_{\text{D,L}}(\Delta S)$ have the same sign in a 1 : 1 association process and are considered to be independent of the temperature, no enantioseparation can be achieved at the *isoenantioselective temperature* T_{iso}:

$$T_{\text{iso}} = \frac{\Delta_{\text{D,L}}(\Delta H)}{\Delta_{\text{D,L}}(\Delta S)} \tag{5}$$

At T_{iso} it follows that $K_D^{\text{assoc}} = K_{\text{L}}^{\text{assoc}}$ and $-\Delta_{\text{D,L}}(\Delta G) = RT \ln \alpha_{\text{assoc}} = 0$, due to *entropy/enthalpy compensation*. At T_{iso}, peak coalecence ensues and the elution order of the enantiomers above and below T_{iso} is opposite, causing a temperature-dependent peak inversion.

The existence of the isoenantioselective temperature T_{iso} in enantioselective gas chromatography was predicted by Koppenhoefer and Bayer [111] and by Schurig and Link [112]. In 1989, different examples of entropy/enthalpy compensations—most of them provided with the experimental proof of the inversion of the elution order—were observed in hydrogen-bonding GC systems [113,114] and in complexation GC [115–117]. A report of peak inversion for methyl lactate on a modified cyclodextrin selector (Lipodex E) [118] could not be ascertained [109], whereas temperature-induced inversions of the elution order have been observed for N-trifluoroacetyl-α-amino acid ethyl esters on the chiral selector Chirasil-Dex [109]. Thus, the derivatives of valine and leucine showed an inversion of the elution order on Chirasil-Dex below and above $T_{\text{iso}} = 70°\text{C}$. Only a single peak was observed at the coalescence temperature. For the isoleucine derivative, the isoenantioselective temperature was as low as $T_{\text{iso}} = 30°\text{C}$ [109]. A recent example of peak inversion due to enthalpy/entropy in a hydrogen-bonding system is depicted in Figure 5.

Enthalpy/entropy compensation must be considered for molecular modeling studies, and the importance of entropy changes should be taken into account. Most gas chromatographic enantioseparations on modified cyclodextrins are governed by the enthalpy term of the Gibbs–Helmholtz equation. Consequently, enantioselectivity increases by reducing the elution temperature. As involatile racemates usually require a high elution temperature, it is advisable to use short columns (1 to 5m × 0.25 mm ID). The loss of efficiency arising from the smaller theoretical plate number N of a short column is often compensated by the gain of enantioselectivity due to the increased enantioseparation factor α at the low elution temperature (see below).

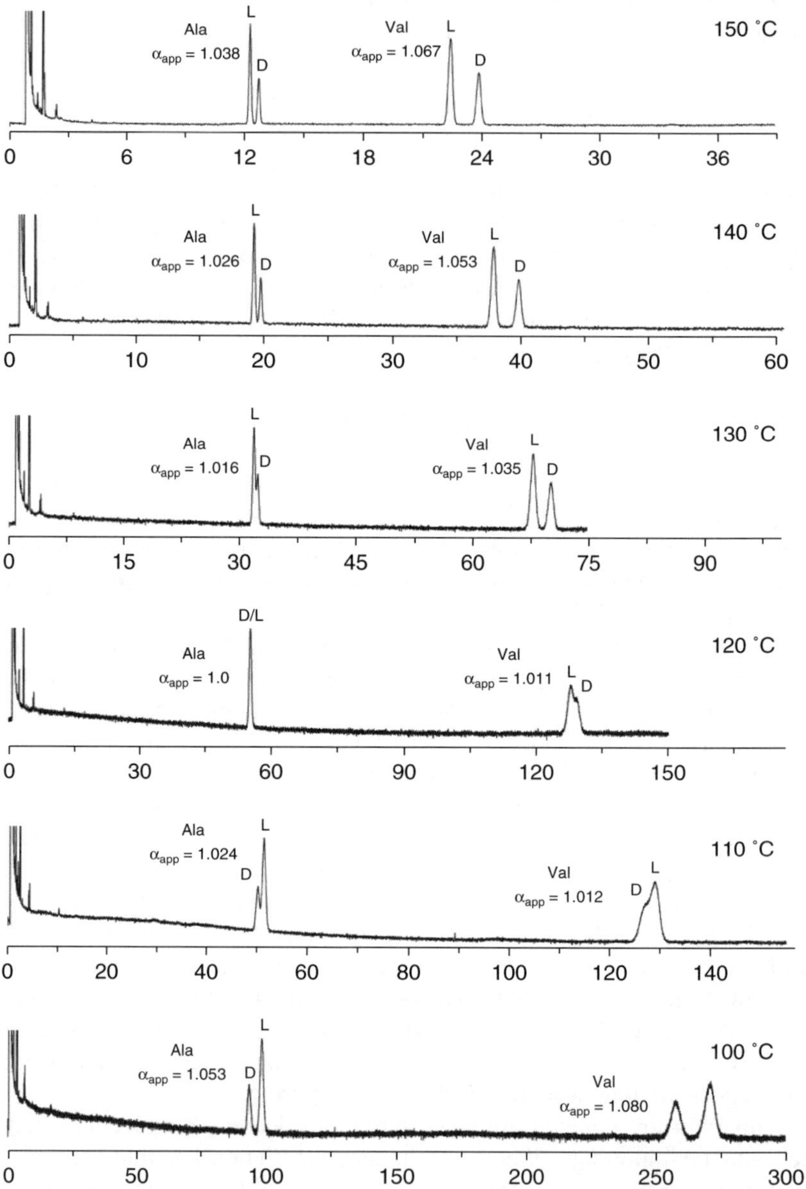

Figure 5 Temperature-dependent reversal of the elution order of the enantiomer of alanine-ECPA and valine-ECPA (ECPA = N-ethoxycarbonyl n-propylamide) on Chirasil-L-Val-C_{11}; α-amino acids are enriched with the L-enantiomer. Isoenantioselective temperatures, T_{iso}, for alanine-ECPA and valine-ECPA are 120 and 114°C, respectively. Column: fused silica, 20 m × 250 μm ID ×0.25 μm (film thickness); carrier gas: H_2; head pressure: 50 kPa (120 to 170°C) and 100 kPa (100 to 110°C); detector: FID. (From [109], with permission.)

Another thermodynamic treatment of enantioselectivity based on the enthalpy/entropy compensation formalism, introduced by Melander et al. for studies of hydrophobic interactions and separation mechanisms in reversed-phase HPLC [119], has also been advanced in cyclodextrin-mediated enantioselective GC [110,120]. A three-phase (pseudo) model for enantioselective GC, comprising a permethylated CD/polysiloxane CSP, was developed by Pino et al. [121].

6 APPLICATIONS

Chiral analyses are important in many fields (see Figure 6) [20]. Gas chromatographic capillary columns coated with Chirasil-Val have been used for many applications [122] involving amino acid analysis: the detection of D-amino acids in bacterial cell walls and in peptide antibiotics, the monitoring of amino acid enantiomeric purity in peptide synthesis, the determination of the degree of racemization during peptide hydrolysis (see below), the dating of paleontological and archeological artifacts by exploiting the time dependence of amino acid racemization, the search for biogenic amino acids in extraterrestrial material, and the amplification of optical activity under abiotic and prebiotic conditions (see references in [25]). Chirasil-Val has also been used for the stereochemical GC analysis of the nerve gas soman [123,124].

Gas chromatographic capillary columns coated with chiral metal chelates have been used in the field of chiral pheromones [125] and chiral flavors [126]. The possibility of determining the enantiomeric ratio of small chiral aliphatic oxiranes permitted study of the asymmetric epoxidation of simple unfunctionalized olefins and the kinetic resolution of racemic oxiranes by metal catalysts [116] and by microsomal enzymes [127,128].

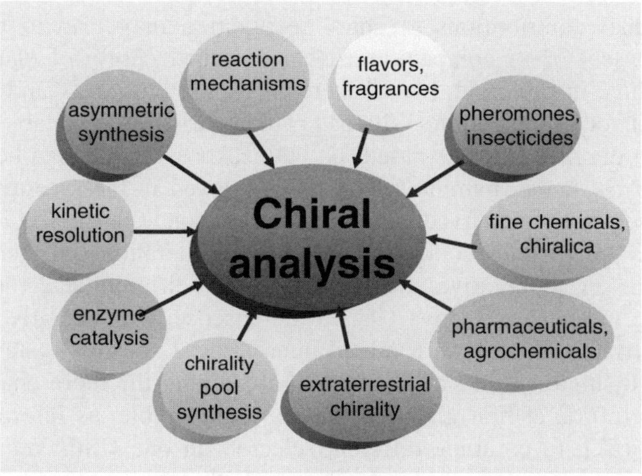

Figure 6 The importance of chiral analysis in many fields. (From [20], with permission.)

Gas chromatographic capillary columns coated with modified CDs were used in enzyme and catalyst screening [129–131] and for the enantiomeric analysis of different classes of chiral compounds, such as essential oils, flavors and fragrances [86,87,132–134], lactones [135], branched fatty acid esters [136,137], organochlorines [101,138], pollutants [139,140], silicon compounds [141], alkyl nitrates as atmospheric constitutents [110], inhalational anesthetics [142,143], compounds of clinical interest [144,145], α-amino acids for space exploration [146,147], as well as unfunctionalized 1,2-dialkylcyclohexanes [148] and aliphatic hydrocarbons [149–151] (see also references 97–105 in [10] and 270 references on pages 222–231 in the book of Schreier et al. [16]). Pharmaceutical enantioseparations on Chirasil-Dex, although limited in scope, have also been described [152]. On Chirasil-Val, enantioenriched amino acids have been determined in the Murchison meteorite [153].

For the determination of extraterrestrial homochirality, enantioselective GC columns coated with commercially available Chirasil-Val [30], Chirasil-Dex [91], and octakis(2,6-di-*O*-pentyl-3-*O*-trifluoroacetyl)-γ-cyclodextrin (G-TA) [12] are integrated in the COSAC experiment [149,154–156] as part of the payload of the Rosetta mission of ESA launched in 2004 and heading at present toward the comet 67P/Churyumov-Gerasimenko.

7 MIXED CHIRAL STATIONARY PHASES COMPRISING DIAMIDES AND MODIFIED CYCLODEXTRINS IN ENANTIOSELECTIVE GC

To combine the individual enantioselectivity entailed by diamide selectors with that of cyclodextrin selectors, the use of *mixed binary selectors systems* has been considered, either by combining two columns with different selectors (tandem-column arrangement) [157] or by mixing two selectors in one CSP and using a single-column format [158]. When two enantioselective selectors are employed, their individual contributions to chiral recognition may lead to enhancement ("matched case") or to compensation ("mismatched case") of enantioselectivity [159]. Thus, in principle, two diastereomeric systems (D,D′ and D,L′) should be employed for optimization. Whereas cyclodextrins are available only in the all-D form, a diamide selector based, on valine, for example, can be used in the D′ and L′ forms. If the diamide selector is employed in the racemic D′/L′ form, any enantioselectivity observed would then reside solely in the CD selector. This scenario may be of interest for mechanistic considerations. Hitherto no clear-cut indications of cooperative effects between two different selectors have been substantiated in enantioselective GC. In the absence of cooperative effects, the enantioselectivity obtained on a mixed binary chiral selector system is smaller than that of a single chiral selector system containing the more enantioselective selector. Therefore, at first glance it may be unfavorable, as inferred by Pirkle and Welch [157], to combine different selectors in one CSP. Yet for practical purposes, the combination of chiral selectors with complementary enantioselectivity toward enantiomers of very different classes of racemic compounds in one

CSP may result in a broader spectrum of enantioselectivities than those provided by either of the single-selector CSPs. A comprehensive quantitative analysis of mixed chiral selector systems, including the treatment of matched–mismatched enantioselectivities, has been advanced recently [159].

Three different approaches of mixed binary selectors systems in enantioselective GC have been described:

1. The two CSPs Chirasil-Calixval [42,43] and Chirasil-β-Dex [91] were bonded to poly(dimethylsiloxane) to furnish Chirasil-Calixval-Dex [160,161].
2. The two CSPs Chirasil-Val-C_{11} and Chirasil-β-Dex-C_{11} were bonded to poly(dimethylsiloxane) to furnish Chirasil-Dex-Val-C_{11} [39]. On the mixed CSP, both unfunctionalized 1,2-dialkylcycloalkanes and α-amino acids (as *N*-TFA ethyl esters) can be enantioseparated. This type of mixed selector system might be used advantageously for classes of compounds (e.g., derivatized β-amino acids [162]), for which homologous members are enantioseparated only on either of the CSPs Chirasil-Val *or* Chirasil-β-Dex.
3. Octakis(3-*O*-butanoyl-2,6-di-*O*-pentyl)-γ-cyclodextrin (Lipodex E) was dissolved in the CSP Chirasil-Val-C11 to furnish Chirasil-Val(γ-Dex) [163]. The mixed phase (one CSP doped into another CSP) was found to have a greatly improved enantioselectivity toward proline and aspartic acid (as *N*-TFA ethyl or methyl esters) compared to Chirasil-Val. Furthermore, the presence of Lipodex E extended the scope of enantioseparations achievable on Chirasil-Val toward underivatized alcohols, terpenes, and many other racemic compounds [83].

A viable strategy to combine the enantioselectivities of hydrogen-bonding and inclusion-type selectors as a single CSP consists of linking L-valine moieties directly to the permethylated β-cyclodextrin selector, similar to Chirasil-Calixval [42,43]. The selector Valdex (heptakis[6-*O*-(*N*-acetylyl-L-valine-*tert*-butylamide)-2,3-*O*-methyl]-β-cyclodextrin) represents a versatile chiral solvating agent (CSA) for the NMR spectroscopic discrimination of enantiomers which associate either with the permethylated cyclodextrin selector (e.g., for apolar trisubstituted allenes) or with the diamide selector (e.g., for polar 2-alkyl-α-amino acid derivatives) [164], but no remarkable enantioseparations of different classes of compounds were observed on Chirasil-Valdex in the GC mode. However, promising results have been obtained on a selector that carries only a single L-valine diamide moiety in the C6-position of permethylated β-cyclodextrin for enantioseparations by GC [165].

8 UNIFIED ENANTIOSELECTIVE CAPILLARY CHROMATOGRAPHY

The van Deemter equation (for packed columns) and the Golay equation (for open-tubular capillary columns), respectively, describe the efficiency of a column as a function of the mobile phase flow rate [166]. A theoretical assessment of the Golay equation predicts that the highest efficiency of a capillary column [i.e., the maximum plate number N (or the minimum plate height H)] is independent of the nature of the mobile phase: a gas in GC, a supercritical fluid in SFC, a liquid in LC, or a buffer system in CEC (capillary electrochromatograpy), whereas the optimum mobile-phase velocity depends on the diffusion coefficient of the analyte in the mobile phase, which differs by four orders of magnitude (GC: fast, LC and CEC: slow) [94]. If a coated open-tubular column (1m × 50 μm ID) is considered, in GC retention factors k are high because the analyte prefers to spend most of its time in the liquid stationary phase, whereas in LC and CEC retention factors k are low because the analyte resides mainly in the liquid mobile phase. Moreover, in GC the mobile-phase velocity is very high, with breakthrough times (void volumes) of a few seconds, whereas in LC and CEC the mobile-phase velocity is very slow, with breakthrough times of about half an hour. These conditions are attractive for a *unified enantioselective chromatographic approach*, in which a single enantioselective open-tubular column can be employed for the enantioseparation of the same chiral analyte by all the contemporary methods: o-GC, o-SFC, o-LC, and o-CEC. This has been demonstrated for the enantioseparation of racemic hexobarbital by four different methods with the same enantioselective column coated with Chirasil-Dex (Figure 7) [94,95].

9 MINIATURIZATION

The unified chromatographic experiment has also shown that enantiomers can be separated by o-GC employing a short miniaturized column (1m × 50 μm

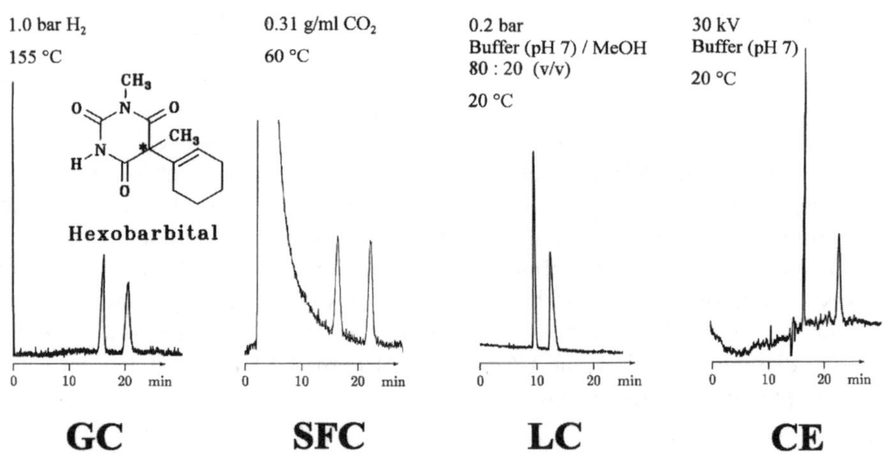

Figure 7 Unified enantioselective capillary chromatography. Enantioseparation of hexobarbital on 25-μm Chirasil-β-Dex coated on a 1 m × 50 μm ID fused-silica capillary by the four methods GC, SFC, LC, and CEC (the effective capillary length in LC and CEC is 80 cm for the online detection). Experimental conditions are inserted in the figure. (From [94], with permission.)

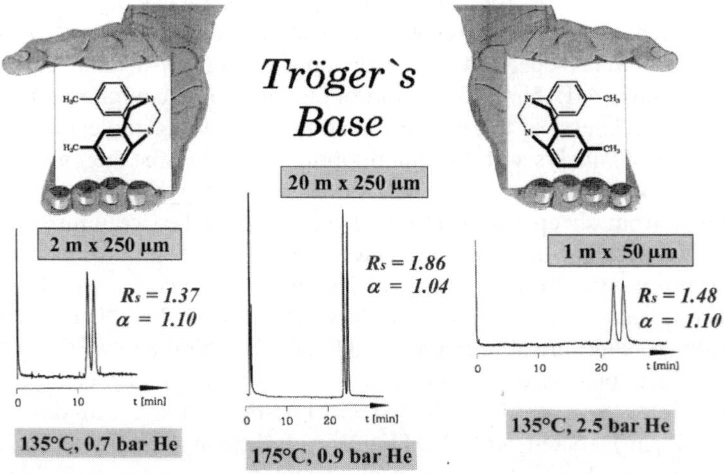

Figure 8 Gas chromatographic enantioseparation of Tröger's base on Chirasil-Dex at various conditions. (From [168].)

ID) coated with Chirasil-Dex (Figure 7, left). Indeed, unless enantiomers are analyzed in complex matrices requiring longer columns, the discrimination of a single enantiomeric pair represents a binary separation task for which the entire separation window of a long column (25 m × 250 µm ID) is actually not required. The loss of efficiency of a short column can be compensated by a gain in enantioselectivity at the reduced elution temperature applied for a shorter column, as first demonstrated by Lindström [167]. A comparison of different column dimensions used for the enantioseparation of Tröger's base on Chirasil-Dex is shown in Figure 8 [168].

Since small-inside-diameter capillary columns (50 µm ID) are difficult to coat with CSPs and are hence not available commercially, a good compromise is the use of short 2 to 5 m × 250 µm ID columns containing, for example, immobilized Chirasil-Dex. Comprehensive studies on column miniaturization in enantioselective GC have been advanced [168,169].

10 HYPHENATED APPROACHES IN ENANTIOSELECTIVE GC

10.1 Chiral Analysis by GC-MS(SIM)

When high sensitivity and analyte identification are required, the coupling of gas chromatography with mass spectrometry (GC-MS) is the method of choice [11]. GC-MS in the selected ion monitoring mode [GC-MS(SIM)] represents another refinement. In the MS(SIM) mode only selected ions are monitored, thus prolonging the detection time of these ions and thereby increasing the signal-to-noise ratio. The four stereoisomers of E,Z-chalcogran (2-ethyl-1,6-dioxaspiro[4.4]nonane), the aggregation pheromone of the bark beetle *Pityogenes chalcographus*, were separated at the 20-ppm level in n-hexane by complexation GC on nickel(II) bis[3-(heptafluorobutanoyl)-(1R)-camphorate] in SE-54 (0.2-μm film thickness) at the $m/z = 29$ and $m/z = 127.4$ molecular ions [60]. The GC-MS(SIM) analysis of 2,6-diaminopimelic acid (three stereoisomers) and proteinogenic D-amino acids as N-trifluoroacetyl/2-propyl esters in rumen bacteria isolated from sheep was performed on a fused-silica column coated with Chirasil-L-Val [170]. Similarly, the determination of D-amino acids as $N(O$-pentafluoropropionyl)/2-propyl esters in Rodentia, Carnivora, Artiodactyla, and Perissodactyla on Chirasil-Val by GC-MS(SIM) established that unnatural D-amino acids are widespread in mammals [171]. The chiral analysis of γ-lactones in milk products produced by spores of *Penicillium roquefort* catalyzed by an exogenous lipase was performed by GC-MS(SIM) on a fused-silica column (30 m × 250 μm) coated with PDTFA-γ-cyclodextrin (see Table 3) of 0.25-μm film thickness [172]. Glausch et al. detected an enantiomeric bias of the atropisomeric polychlorinated biphenyl PCB 132 in human milk samples in the GC-MS(SIM) mode [173]. The enantiomeric excess (ee) of R-limonene in citrus matrices (peel, juice), 99% for grapefruits and oranges but only 96% for lemons, was also determined in the GC-MS(SIM) mode at $m/z = 68$ on permethylated β-cyclodextrin [174]. A headspace GC-MS(SIM)-assisted study revealed that the racemic inhalational anesthetic isoflurane [2-chloro-2-(difluoromethoxy)-1,1,1-trifluoroethane] showed a deviation of the enantiomeric ratio of the racemic composition administered, from $S/R = 50:50$ to $S/R = 54:46$ during and after surgery in clinical patients. A multipurpose sampler was used for headspace GC-MS(SIM) ($m/z = 117$ and 149), combined with a cold injection system for trapping, enrichment, and focusing of the narcotic, whereby the enantiomers of isoflurane were separated on Lipodex E [175]. The quantitative stereoisomeric composition of the chiral hydrocarbons 1-methyltetralin and *cis*-1,2- and 1,3-dimethylindane as biomarkers in crude oil and coal samples were determined by GC-MS(SIM) at the parts per thousand level [176].

10.2 In Situ Deuteration and Chiral Analysis of α-Amino Acids by GC-MS(SIM)

The racemization of α-amino acids during the acid-catalyzed hydrolysis of peptides can falsify the true enantiomeric composition of the building blocks of a peptide [25,122]. This problem can be overcome by performing the hydrolysis in a fully deuterated medium (e.g., in 6N D_2O/DCl). In this case, racemization during hydrolysis is accompanied by substitution of the hydrogen atom attached to the stereogenic carbon atom by deuterium, and the hydrogenated and deuterated species can be diffentiated by mass spectrometry. After hydrolysis, the amino acids are derivatized and gas chromatographically enantioseparated on a suitable CSP (Chirasil-Val, Lipodex E, or Chirasil-γ-Dex) and on-line detected by GC-MS(SIM) [177] (see also Table 2 in [178]). For each amino acid, a characteristic ion containing the proton at the stereogenic carbon atom is monitored, whereas the deuterated species formed during the hydrolysis is disregarded. The reliable determination of enantiomeric purities of L-α-amino acids in peptides up to 99.9% is thus possible [177]. This method has also been used to determine the rate constants of configurational inversion at the stereogenic center of α-amino acids under acid hydrolysis conditions (110°C, 6*N* DCl) [177]. The time-dependent racemization of α-amino acids has been widely used for dating purposes of archeological artifacts containing α-amino acids [25]. An automated gas chromatographic chiral analysis system for α-amino acids has been described [179]. Freeze-dried body fluid, tissue, and food proteins were hydrolyzed directly and the hydrolysates were derivatized and analyzed automatically on Chirasil-Val or Chirasil-γ-Dex [179]. In situ deuteration (see above) and enantiomer labeling (see below) are further refinements of the method.

10.3 Chiral Analysis by GC-NMR

The off-line NMR-spectroscopic investigations of isomers separated by microscale–preparative capillary GC is straightforward [180]. The first direct on-line coupling of enantio-GC-NMR was demonstrated recently. Thus, 0.2 μL of 2,4-dimethylhexane (C*HMeEt*i*Bu) were enantioseparated at 60°C on a 30 m × 250 μm ID fused-silica capillary coated with 0.5-μm octakis(6-*O*-methyl-2,3-di-*O*-pentyl)-γ-cyclodextrin (Lipodex G). A two-dimensional plot of the GC-^{1}H-NMR (400 MHz) experiment showed two separate clustered signals of the alkyl resonance adsorptions [181]. Previously, the diastereomeric *cis*- and *trans*-1,2-dimethylcyclohexanes were separated by GC on a 25 m × 250 μm ID fused-silica capillary coated with 0.25 μm of Chirasil-β-Dex and differentiated by ^{1}H-NMR (400 MHz) [182]. The method, still in its infancy, must be improved in terms of sensitivity, heating of transfer lines, and minimizing dead volumes in capillary connections.

10.4 Multidimensional Approaches in Enantioselective GC

The presence of chiral compounds in multicomponent matrices will lead to the doubling of peaks when using an enantioselective column, thereby increasing the complexity of the elution pattern. Therefore, multidimensional approaches are used. In the commonly used two-dimensional approach, a second dimension of separation is employed. The conventional enantioselective technique involves the heart-cut GC-GC approach (enantio-MDGC) first demonstrated by Schomburg et al. [183]. Here the first nonenantioselective column, coated with an achiral polar stationary phase, is used to pre-separate components of interest (first dimension), whereas in the second enantioselective column, coated with a CSP, fractions of chiral analytes are enantioseparated after on-line transfer through a pneumatic or flow-controlled low-dead-volume heart-cut interface (second dimension). There are many applications in the field of terpenes and food chemistry [184–189]. The traditional one-column operation has been compared with enantio-MDGC for the enantioseparation of monoterpenes from essential oils [190]. A typical sequential tandem dual-column approach was also used for the separation of all eight stereoisomers of the sex pheromone of a pine beetle: derivatized dipropionol (2-hydroxy-3,7-dimethyl-heptadecane) containing three stereogenic centers. The achiral column pre-separated diastereomers, whereas the enantioselective column based on polysiloxane-linked XE-60-L-valine-(*S*)-α-phenylethylamide [26] separated the enantiomers [191]. The analysis of soman stereoisomers involved thermodesorption/cold trap injection and two-dimensional GC with an achiral precolumn followed by resolution of all four stereoisomers and an internal standard on a Chirasil-Val–coated capillary column [123,124]. From a complex mixture of atropisomeric polychlorinated biphenyls (PCBs) present in the formulation Clophen A 60, one single peak was heart-cut in the first dimension and resolved into three peaks in the second dimension, comprising an enantioselective fused-silica capillary column coated with Chirasil-β-Dex. The peaks were assigned to PCB 151 (chiral and containing *two* chlorine substituents in *ortho*-positions: unresolved) and to PCB 132 (chiral and containing *three* chlorine substituents in *ortho*-positions: resolved) [192] (see Figure 9).

In the realm of flavor chemistry, enantioselective two-dimensional GC is preferentially hyphenated with mass spectrometry (enantio-MDGC-MS)

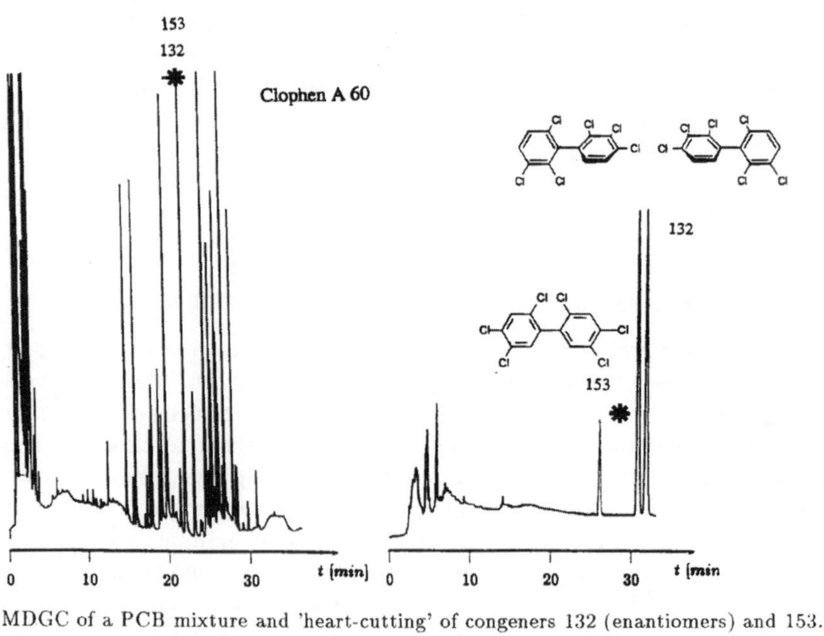

Figure 9 MDGC-ECD of PCBs. Left: separation of PCB congeners on an achiral first column (DB-5). Right: separation of the heart-cut fraction into racemic PCB 132 (resolved) and racemic PCB 153 (unresolved) on the second column coated with Chirasil-Dex [192]. (From [138], with permission.)

[133,134,193,194]. A recent example is provided by the stir-bar sorptive extraction and (SBSE)-enantio-MDGC-MS analysis of chiral flavor compounds in strawberries [195,196].

The development of *comprehensive* two-dimensional gas chromatography GC × GC provides an orthogonal separation of compounds on two columns using cryogenic modulation without a requirement for heart-cutting [134,197]. In case an achiral capillary column system is employed in the first dimension, a very fast enantioseparation is required in the second dimension (GC × enant-GC). Few enantioseparations are known today that fulfill this hard requirement (see Figure 3 in [97]). Retention times as short as 8 s were reported for racemic limonene with an adequate enantioseparation ($R_s \sim 1.0$) on a 1 m × 100 μm ID fused capillary column coated with heptakis(2,3-di-*O*-ethyl-6-*O*-*tert*-butyldimethylsilyl)-β-cyclodextrin (MEGA, Lignano, Italy) [198]. Sufficiently fast elution on the second column was achieved by using GC-MS in which the subambient pressure (vacuum outlet) conditions promoted the increased diffusion coefficients and higher component volatility. A fourfold reduction of second-dimension retention time was observed as compared with ambient pressure outlet conditions. The enantiomeric distribution of several monoterpene compounds in bergamot essential oil was reported as a demonstration of the method [198]. The column geometry can also be inverted [199]; that is, the first column represents an enantioselective separation system, whereas the second miniaturized achiral column is used to obtain a two-dimensional contour plot (enant-GC × GC).

For achiral diastereomers (e.g., *E*- and *Z*-configurations of oximes), the time-resolved cryogenic modulation in a dynamic DGC × DGC experiment allowed deconvolution of the interconversion profile (plateau) observed in the first dimension into individual isomeric profiles in the second dimension, thus representing a two-dimensional contour plot of molecular inversion [200].

10.5 Enantioselective Stopped-Flow Multidimensional Gas Chromatography

By this method the configurational lability of a chiral compound can be evaluated at elevated temperatures with only a minute amount of the racemic probe. Three columns are employed in series. The enantiomers are enantioseparated on-line in the first enantioselective column. Either one of the pure enantiomers is then transferred into the second uncoated reactor column. By stopping the flow for the time t, enantiomerization (inversion of configuration) is allowed to proceed at the elevated temperature T in the gas phase. After the time t, the enantiomers are focused by cooling and resuming the flow. The enantiomeric ratio *er* is then determined in the third enantioselective column (or by backflushing into the first column). From T, t, and *er*, the rate constant of enantiomerization can be calculated. This method has been used to determine the activation barriers of molecular inversion of atropisomeric polychlorinated biphenyls (chiral PCBs) [201,202] and of 1-chloro-2,2-dimethylaziridine [202–204].

10.6 Enantioselective Two-Dimensional LC-GC

In an off-line approach, the chiral hydrocarbons 1-methyltetralin, 1-methyl-, and *cis*-1,2- and 1,3-dimethylindane as biomarkers in crude oil and coal were separated by LC prior to chiral analysis by GC-MS(SIM) [176]. Enantioselective on-line LC-GC multidimensional approaches [205] are more straightforward. The enantiomeric ratios of linalool in orange and lemon oils were analyzed with a fully automated on-line LC-GC system to evaluate their genuineness, quality, and origin [206,207].

10.7 Enantioselective Multidimensional Gas Chromatography Coupled On-line with Isotope Ratio Mass Spectrometry

Often, the authenticity of genuine natural products is evaluated simply through determination of the enantiomeric excess (ee). However, fortification of genuine flavors with artificial synthetic enantiomers cannot be recognized by this means. Mosandl et al. therefore developed the enantio-MDGC-IRMS method based on the fact that natural and synthetic compounds show subtle differences in their isotopic ratios, due to kinetic isotopic effects operating in various synthetic pathways [208]. Thus, the enantiomers eluted from the GC column are burned in a pyrolysis interface and the $^{18}O/^{16}O$ and/or the $^{13}C/^{12}C$ ratio of the carbon dioxide formed is then determined by mass spectrometry [208–210]. Also biosynthetic routes or biogenetic pathways may be discerned by MDGC-IRMS [210,211]. GC-IRMS has also been used to determine the carbon isotopic ratio of extraterrestrial amino acids in the Murchison meteorite [153]

11 METHOD OF ENANTIOMER LABELING

If a quantitative enantioseparation on a CSP is available, the enantiomer of opposite configuration L represents an ideal standard for the quantification of the targed enantiomer D present in a mixture. This approach has been referred to as *enantiomer labeling* [212,213]. In the absence of self-recognition between enantiomers in (concentrated) nonracemic mixtures (the EE effect) [214], the measured enantiomeric ratio *er* after addition of the standard is not obscured by workup, sample manipulations, split injection, and detection. Also the amount of a racemate present in a complex matrix can be determined quantitatively via the enantiomer labeling method when a known amount of a single enantiomer is added to the mixture and the change in *er* is determined by enantioselective gas chromatography. As an example of the enantiomer labeling method, the amount of the racemic inhalation anesthetic isoflurane [2-chloro-2-(difluoromethoxy)-1,1,1-trifluoroethane] in blood samples during and after surgery has been quantified by employing *S*-(+)-isoflurane (obtained by preparative GC; see below) as internal standard and recording the *er* after the addition of the enantiomerically pure internal standard by enantioselective headspace GC-MS [215].

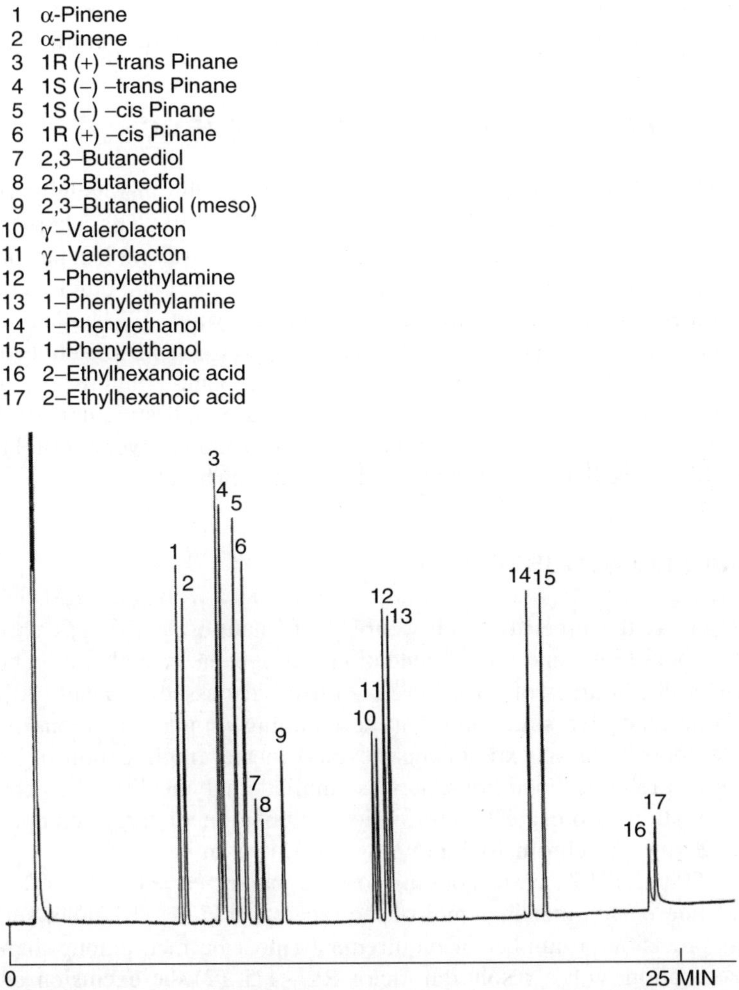

Figure 10 Simultaneous enantioseparation of various racemic compounds belonging to different classes of compounds ("Schurig test mixture") on permethylated β-cyclodextrin diluted in OV-1701 (70°C for 5 min followed by 3°C/min, 0.65 bar H_2, 25 m × 250 μm ID fused-silica capillary column). (From [8], with permission.)

12 CHIRAL TEST MIXTURES

A test mixture for the performance of enantioselective GC columns containing hydrogen-bonding diamide CSPs has been proposed [216]. For permethylated β-CD dissolved in OV-1701, another test mixture, covering the entire spectrum from apolar to highly polar, has been suggested (see Figure 10) [78], which was modified for another CSP [80]. Clearly, the choice of the constituents of those chiral test mixtures is highly arbitrary, limiting their general applicability.

Nevertheless, the wide range of different classes of chiral compounds amenable to enantioseparation on permethylated β-CD is clearly evident from Figure 10.

13 DATA RETRIEVAL FOR ENANTIOSELECTIVE GC

The wealth of information on enantioselective gas chromatography has been documented in the ChirBase/GC data bank [217–219]. This data bank has also been exploited for predictive searches of intrinsic molecular properties that influence enantiomer discrimination for chiral compounds containing three-membered rings [220] and four-membered rings [221]. The database ChirBase/GC contains method information for over 24,000 enantioseparations of more than 8000 chiral molecules. The data bank contains experimental conditions, as well as the structure, the substructure, and structural similarities. ChirBase/GC was developed by Koppenhoefer's group (University of Tübingen, Germany: www.acdlabs.com/products/adh/chrom/chirbase/).

14 MISCELLANEOUS TOPICS

Topics such as the precision and accuracy of enantioselective gas chromatography, practical hints, and recommendations have been treated comprehensively [4,8,26,64,138]. Sources of error have also been discussed in detail [60].

In a comprehensive study aimed at determining minute deviations from 1 : 1 racemic compositions in experiments devoted to the amplification of homochirality under prebiotic conditions, it was implied that the limit in determining minute deviations from the true racemic composition of target compounds by enantioselective gas chromatography (e.g., of D,L-α-amino acids) is generally in the range $50.0 \pm 0.1\%$, corresponding to an apparent ee = 0.2% (or 1.2×10^{21} excess of one enantiomer in 1 mol of the racemate) [222]. To attain such accuracy *and* precision, a number of requirements must be met, among them (1) an enantioseparation with a resolution factor $Rs > 1.5$, (2) the exclusion of coeluting impurities, (3) the absence of decomposition of the sample (or its derivative) during the gas chromatographic process, (4) an elution-time-independent detector response, and (5) a correct peak area integration. Measurement with a precision (reproducibility) equal or in excess of $\pm 0.05\%$ is feasible, but the accuracy (trueness) of the value determined must be validated carefully by employing oppositely configurated CSPs (e.g., in case of diamide phases), leading to reversal of the elution order of the enantiomers separated. The mean of the values obtained from different CSPs was considered to be a better estimate of the correct value of the enantiomeric fraction than measurement on one CSP only, provided that the degree of deactivation of different columns is comparable. Column deactivation is best assessed by the standard Grob test for capillaries [222]. Incidentally, the racemic composition of a chiral compound represents an ideal probe for the accuracy of integration devices because a strict 1 : 1 ratio should be encountered (statistical fluctuations of enantiomers are well beyond detectability).

For the development of a specific MS library used for the identification of 134 racemates belonging to flavors and fragrances, "interactive" linear retention index values I^{T} were determined on four modified β-cyclodextrin derivatives [223].

A chiral selector can be employed not only as a chiral stationary phase (CSP) but may also used as a chiral mobile-phase additive (CMPA) in liquid chromatography [224–226]. The successful use of a CMPA for enantioseparation by GC is still elusive. Adding gaseous *S*-(+)-isoflurane (obtained by preparative enantioselective GC; see below) to the carrier gas helium did not lead to the enantioseparation of any of the nine tested racemic analytes by GC [168].

15 SOME STATISTICS

In a 1995–1996 survey on the chromatographic techniques used in European laboratories, capillary GC scored the highest percentage of use [227]. As for analytical chiral separations surveyed between 1960 and 2000, enantioselective GC preceded the routine use but was then surpassed by enantioselective LC and enantioselective electromigration methods [228]. A survey of the analytical methods used in the determination of enantiomeric composition required by the journal *Tetrahedron Asymmetry* for 1995 (322 entries), 1999 (357 entries), and 2002 (290 entries) gave the following data: HPLC (118,146,129), NMR (104,79,60), chiral GC (66,85,66), polarimetry (35,30,24), and others (35,30,24) [229]. Today, analytical enantioselective GC is undoubtedly of widespread use as judged from the availability of commercial capillary columns containing chiral stationary phases by all leading suppliers of gas chromatographic items. Yet of 100 CSPs described in the literature until 1993, only 34% were available commercially according to the ChirBase data bank [217]. Up to the end of 1995, following the first enantioseparations described by Gil-Av et al. in 1966, Koppenhoefer estimated from the ChirBase data bank that 8000 enantioseparations of chiral analytes had been described (many of them repeatedly) in 3000 publications using 700 CSPs, of which 120 were commercially available. Altogether more than 20,000 enantioseparations by GC were known by the end of 1995 [219], of which 14,785 enantioseparations were performed with modified cyclodextrins, 7319 with diamide CSPs, 627 with metal complexes, and 269 with other CSPs (private communication from B. Koppenhoefer, Tübingen).

16 PREPARATIVE-SCALE GAS CHROMATOGRAPHIC ENANTIOSEPARATION

The state-of-the-art of enantioselective (semi)preparative gas chromatography has been reviewed [24]. This method is restricted to thermally stable and volatile compounds. Contrary to liquid chromatography (LC), the recovery of isolated enantiomers from the gaseous mobile phase (carrier gas) is straightforward when aerosol and mist formation is prevented by using specially designed collection vessels [24]. The GC approach does not match the overwhelming success of preparative enantioseparation achieved in the realm of LC [229,230]. Modern commercial instrumentation for preparative-scale GC is not available. In contrast to LC, enantioseparation factors α are usually small ($\alpha = 1.01$ to 1.20) in GC. Whereas this feature is beneficial for fast analytical enantioseparations, it is detrimental to preparative-scale enantioseparations. The method has, however, been very successful in cases where enantioseparation factors $\alpha > 1.5$ were involved [24]. Preparative chromatography relies on a compromise between the three variables in a magic triangle: (1) peak resolution (governed by selectivity, efficiency, and retention), (2) speed of separation, and (3) column sample capacity. Any of two desired goals may only be achieved at the expense of the third. If a large sample throughput is required in a short time, resolution must be high. However, if resolution is insufficient, either the column load is limited or the time of separation is long [24].

Interestingly, in the development of enantioselective gas chromatography, packed columns containing supported CSPs were employed initially for a hydrogen-bonding type of CSP [23], a coordination type of CSP [52], and an inclusion type of CSP [66]. In complexation gas chromatography, semipreparative enantiomer separations at the milligram scale have been reported for spiroketals (among them pheromones) [60,126,231]. The preparative invertomer separation of 1-chloro-2,2-dimethylaziridine permitted the determination of chiroptical data, the absolute configuration, and the inversion barrier [232]. Large enantioseparation factors α were observed for the saturated hydrocarbons *cis*- and *trans*-pinane on a mixture of α-cyclodextrin and formamide impregnated on celite [66]. The preparative separation of enantiomers of camphene was subsequently realized on a packed column [68].

With the advent of alkylated/acylated CDs as versatile CSPs for analytical enantioseparations by GC, their potential for preparative-scale enantioseparations of flavors, fragrances, and terpenoids was recognized. Micropreparative enantioseparations by GC on modified cyclodextrins have been achieved on thick-film wide-bore fused-silica capillary columns [233]. The enantiomers of 2 mg of racemic methyl jasmonate were enantioseparated within 80 min on a 1.8 m × 4mm i.d. stainless steel column packed with 5% (w/w) heptakis(2,6-di-*O*-methyl-3-*O*-pentyl)-β-cyclodextrin in polysiloxane OV-1701 (1 : 1, w/w) coated onto Chromosorb W-HP (100 to 120 mesh) at 120°C and 0.4 bar of helium [234]. The analytical and semipreparative enantioseparation of all-anti-*trans*-perhydrotriphenylene on heptakis(2,3-di-*O*-acetyl-6-*O*-*tert*-butyldimethylsilyl)-β-cyclodextrin has also been described [235], demonstrating that even a rather involatile racemate exhibiting a low enantioseparation factor α is amenable to preparative resolution by GC.

The human inhalational anesthetics enflurane, isoflurane, and desflurane exist as enantiomers, although they are administered as racemic mixtures. The chiral

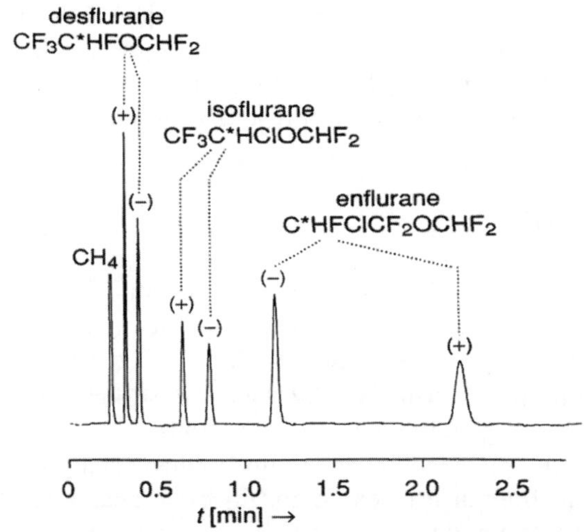

Figure 11 Gas chromatographic enantioseparation of racemic inhalation anesthetics (28°C, 10 m × 250 μm ID fused-silica capillary coated with 0.18 μm (film thickness) of immobilized Chirasil-γ-Dex). (From [97], with permission.)

haloethers exhibit unprecedented high enantioseparation factors α on octakis(3-*O*-butanoyl-2,6-di-*O*-pentyl)cyclodextrin (Lipodex E) (Figure 11). Thirty microliters of racemic enflurane was enantioseparated on a 4 m × 7 mm ID glass column containing 95 g of Lipodex E dissolved in the methylphenyl (5%) polysiloxane SE-54 (10% w/w) and coated onto Chromosorb P (AW, DMCS, 80 to 100 mesh) (20% w/w) [97]. A chemical purity of at least 99.9% and an enantiomeric purity of at least 99.8% were achieved. Upscaling toward the use of a packed 1 m × 6 mm ID stainless steel column required the synthesis of 500 g of stationary phase, consisting of 10.75 g of (unpurified) Lipodex E and 90.5 g of SE-54 coated onto 400 g of Chromosorb P. Three hundred milligrams of the single enantiomers of isoflurane (ee = 99.8%) was obtained per day with 130 automated repetitive injections isothermally at room temperature in a Hupe & Bush preparative gas chromatograph [236]. The enantiomers of isoflurane and enflurane were also preparatively separated on undiluted 3-*O*-trifluoroacetylated-2,6-di-*O*-pentylated γ-cyclodextrin (PDTFA-γ-cyclodextrin) [237,238]. The state of the art in this area has been reviewed [239].

In an attempt to further scale-up the batchwise discontinuous preparative-scale GC approach to a continuous enantioseparation of racemic enflurane, simulated moving bed (SMB) technology had been employed in enantioselective GC [240]. The SMB unit consisted of eight 80 cm × 15 mm ID stainless steel columns, each containing 20% of (unpurified) Lipodex E in polysiloxane SE-54 coated onto Chromosorb A (NAW, 20 to 30 mesh) (17% w/w). In enantioselective GC-SMB, the continuous countercurrent flow of the fluid and of the solid adsorbent (moving bed chromatography) is simulated by periodic switching of the different inlets and outlets in the eight-column array. GC-SMB was also used successfully for the enantioseparation of racemic α-ionone [241]. A key difference between the enantioselective liquid- and gas-phase SMB is the strong influence of the pressure drop on the flow rate in the latter approach and the limited solubility of the feed in the gas phase, which in turn results in limited productivities [242].

17 CONCLUSIONS

Analytical enantioselective GC represents a versatile and mature technique employed frequently in academia and industry. Advantages are speed, reproducibility, sensitivity, and hyphenation to MS. In contrast to LC and CEC, choice of the mobile phase, usually dihydrogen, is straightforward. The method is not well suited for preparative enantioseparations.

In contrast to the pharmaceutical environment, many analytical applications are found in flavor and fragrance research and in basic enantioselective chemistry. Less volatile chiral compounds can often be resolved by proper derivatization strategies and/or by column miniaturization. Chirasil-Val and Chirasil-Dex represent robust stationary phases over a wide temperature range. A universal GC chiral stationary phase is elusive, and a comprehensive understanding of mechanistic details of enantiorecognition is not yet available.

REFERENCES

1. E. Gil-Av, D. Nurok. Resolution of optical isomers by gas chromatography of diastereomers. In *Advances in Chromatography*, Vol. 10, G. C. Giddings, R. A. Heller, Eds. Marcel Decker, New York, 1974, pp. 99–172.
2. E. Gil-Av. Present status of enantiomeric analysis by gas chromatography. *J. Mol. Evol.*, 6:131–144, 1975.
3. C. H. Lochmüller, R. W. Souter. Chromatographic resolution of enantiomers. Selective review. *J. Chromatogr.*, 113:283–303, 1975.
4. V. Schurig. Gas chromatographic methods. In *Asymmetric Synthesis*, Vol. 1, *Analytical Methods*, J. D. Morrison, Ed. Academic Press, New York, 1983, pp. 59–86.
5. V. Schurig. Current methods for determination of enantiomeric compositions: 3. Gas chromatography on chiral stationary phases. *Kontakte* (*Darmstadt*), 1: 3–22, 1986.
6. W. A. König. Enantioselective gas chromatography. *Trends Anal. Chem.*, 12:130–137, 1993.
7. V. Schurig. Review. Enantiomer separation by gas chromatography on chiral stationary phases. *J. Chromatogr. A*, 666:111–129, 1994.
8. V. Schurig. Determination of enantiomeric purity. In *Houben-Weyl Methods of Organic Chemistry*, Vol. E21a, *Stereoselective Synthesis*. Thieme, New York, 1995, pp. 168–192.
9. Z. Juvancz, P. Petersson. Enantioselective gas chromatography. *J. Microcolumn Sep.*, 8:99–114, 1996.
10. V. Schurig. Separation of enantiomers by gas chromatography. *J. Chromatogr. A*, 906:275–299, 2001.

11. V. Schurig. Chiral separations using gas chromatography. *Trends Anal. Chem.*, 21:647–661, 2002.
12. L. He, T. E. Beesley. Applications of enantiomeric gas chromatography: a review. *J. Liq. Chromatogr. Relat. Technol.*, 28:1075–1114, 2005.
13. T. J. Ward, B. A. Baker. Chiral separations. *Anal. Chem.*, 80:4363–4372, 2008.
14. S. Allenmark, V. Schurig. Chromatography on chiral carriers. *J. Mater. Chem.*, 7:1955–1963, 1997.
15. S. G. Allenmark. *Chromatographic Enantioseparation: Methods and Applications*, 2nd rev. ed. Ellis Horwood Series in Analytical Chemistry. Ellis Horwood, New York, 1991.
16. P. Schreier, A. Bernreuther, M. Huffer. *Analysis of Chiral Organic Molecules*. Walter de Gruyter, New York, 1995, pp. 132–233.
17. T. E. Beesley, R. P. W. Scott. *Chiral Chromatography*. Wiley, New York, 1999.
18. E. Gil-Av, B. Feibush, R. Charles-Sigler. Separation of enantiomers by gas–liquid chromatography with an optically active stationary phase. *Tetrahedron Lett.*, 1009–1015, 1966.
19. B. Feibush. Separation of optical antipodes by the method of gas chromatography [in Hebrew]. Ph.D. dissertation, Feinberg Graduate School of the Weizmann Institute of Science, Rehovot, 1967.
20. V. Schurig. Emanuel Gil-Av and the separation of enantiomers on chiral stationary phases by chromatography. In *Milestones in Chromatography*, L. S. Ettre, Ed. *LCGC North Am*. Apr. 25(4):382–395, 2007.
21. D. H. Desty, J. N. Haresnape, B. H. F. Whyman. Construction of long lengths of coiled glass capillary. *Anal. Chem.*, 32:302–304, 1960.
22. E. Gil-Av, B. Feibush, R. Charles-Sigler. Separation of enantiomer by gas–liquid chromatography with an optically active stationary phase. In *Gas Chromatography 1996*, A. B. Littlewood, Ed. Institute of Petroleum, London, 1967, pp. 227–239 (discussion, pp. 254–257).
23. E. Gil-Av, B. Feibush. Resolution of enantiomers by gas–liquid chromatography with optically active stationary phases: separation on packed columns. *Tetrahedron Lett.*, 3345–3347, 1967.
24. V. Schurig. Preparative-scale separation of enantiomers on chiral stationary phases by gas chromatography. In *Enantiomer Separation: Fundamentals and Practical Methods*, F. Toda, Ed. Kluwer, Dordrecht, The Netherlands, 2004, pp. 267–300.
25. V. Schurig. Gas chromatographic separation of enantiomers on optically active metal-complex-free stationary phases. *Angew. Chem. Int. Ed. Engl.*, 23:747–765, 1984.
26. W. A. König. *The Practice of Enantiomer Separation by Capillary Gas Chromatography*. Hüthig, Heidelberg, Germany, 1987.
27. B. Feibush. Chiral separation of enantiomers via selector/selectand hydrogen bondings. *Chirality*, 10:382–395, 1998.
28. B. Feibush. Interaction between asymmetric solutes and solvents: *N*-lauroyl-L-valyl-*t*-butylamide as stationary phase in gas–liquid partition chromatography. *J. Chem. Soc. Chem. Commun.*, 544–545, 1971.
29. C. H. Lochmüller, R. W. Souter. Direct gas chromatographic resolution of enantiomers on optically active mesophases: II. Effects of stationary phase structure on selectivity. *J. Chromatogr.*, 88:41–54, 1974.

30. H. Frank, G. J. Nicholson, E. Bayer. Rapid gas chromatographic separation of amino acid enantiomers with a novel chiral stationary phase. *J. Chromatogr. Sci.*, 15:174–176, 1977.
31. H. Frank. Gas chromatography of enantiomers on chiral stationary phases. In *Chirality and Biological Activity*, B. Holmstedt, H. Frank, B. Testa, Eds. Alan R. Liss, New York, 1990, pp. 33–54.
32. G. J. Nicholson, H. Frank, E. Bayer. Glass capillary gas chromatography of amino acid enantiomers. *J. High Resolut. Chromatogr. Commun.*, 2:411–415, 1979.
33. G. Lai, G. Nicholson, E. Bayer. Immobilization of Chirasil-Val on glass capillaries. *Chromatographia*, 26:229–233, 1988.
34. T. Saeed, P. Sandra, M. Verzele. Synthesis and properties of a novel chiral stationary phase for the resolution of amino acid enantiomers. *J. Chromatogr.*, 186:611–618, 1979.
35. W. A. König, I. Benecke. Gas chromatographic separation of enantiomers of amines and amino alcohols on chiral stationary phases. *J. Chromatogr.*, 209:91–95, 1981.
36. B. Koppenhoefer, U. Mühleck, K. Lohmiller. Backbone modification of Chirasil-Val: I. Effect of loading on the separation of enantiomers by gas chromatography. *J. Chromatogr. A*, 699:215–221, 1995.
37. B. Koppenhoefer, U. Mühleck, M. Walser, K. Lohmiller. Backbone modification of Chirasil-Val: II. Introduction of a rigid cyclohexyl spacer. *J. Chromatogr. Sci.*, 33:217–222, 1995.
38. B. Koppenhoefer, U. Mühleck, K. Lohmiller. Backbone modification of Chirasil-Val: effect of nonpolar side chains on enantiomer separation in gas chromatography. *Chromatographia*, 40:718–723, 1995.
39. P. A. Levkin, A. Levkina, V. Schurig. Combining the enantioselectivities of L-valine diamide and permethylated β-cyclodextrin in one gas chromatographic chiral stationary phase. *Anal. Chem.*, 78:5143–5148, 2006.
40. H. Frank, I. Abe, G. Fabian. A versatile approach to the reproducible synthesis of functionalized polysiloxane stationary phases. *J. High Resolut. Chromatogr.*, 15:444–448, 1992.
41. I. Abe, K. Terada, T. Nakahara, H. Frank. New stereoselective GC phases: immobilized chiral polysiloxanes with (*S*)-(−)-*t*-leucine derivatives as selectors. *J. High Resolut. Chromatogr.*, 21:592–596, 1998.
42. J. Pfeiffer, V. Schurig. Enantiomer separation of amino acid derivatives on a new polymeric chiral resorc[4]arene stationary phase by capillary gas chromatography. *J. Chromatogr. A*, 840:145–150, 1999.
43. A. Ruderisch, J. Pfeiffer, V. Schurig. Synthesis of an enantiomerically pure resorcarene with pendant L-valine residues and its attachment to a polysiloxane (Chirasil-Calix). *Tetrahedron Asymm.*, 12:2025–2030, 2001.
44. F. Narumi, N. Iki, T. Suzuki, T. Onodera, S. Miyano. Syntheses of chirally modified thiacalix[4]arenas with enantiomeric amines and their application to chiral stationary phases for gas chromatography. *Enantiomer*, 5:83–93, 2000.
45. M. A. Muhs, F. T. Weiss. Determination of equilibrium constants of silver–olefin complexes using gas chromatography. *J. Am. Chem. Soc.*, 84:4697–4705, 1962.
46. J. Herling, J. Shabtai, E. Gil-Av. Gas chromatography with stationary phases containing silver nitrate: III. Isomeric C_8 and C_9 cyclohexenes and *p*-menthenes. *J. Chromatogr.*, 8:349–354, 1962.

47. B. Feibush, M. F. Richardson, R. E. Sievers, C. S. Springer, Jr. Complexes of nucleophiles with rare earth chelates: 1. Gas chromatographic studies of lanthanide nuclear magnetic resonance shift reagents. *J. Am. Chem. Soc.*, 94:6717–6724, 1972.
48. V. Schurig, R. C. Chang, A. Zlatkis, B. Feibush. Thermodynamics of molecular association by gas–liquid chromatography: σ-donor molecules and dimeric 3-trifluoroacetylcamphorates of Mn(II), Co(II) and Ni(II). *J. Chromatogr.*, 99:147–171, 1974.
49. V. Schurig. Separation of deuteriated ethylenes $C_2H_{4-n}D_n$ by complexation chromatography on a rhodium(I) complex. *Angew. Chem. Int. Ed.*, 15:304, 1976.
50. V. Schurig. Resolution of a chiral olefin by complexation chromatography on an optically active rhodium(I) complex. *Angew. Chem. Int. Ed.*, 16:110, 1977.
51. V. Schurig, E. Gil-Av. Chromatographic resolution of chiral olefins: specific rotation of 3-methylcyclopentene and related compounds. *Isr. J. Chem.*, 15:96–98, 1976/1977.
52. B. T. Golding, P. J. Sellars, A. K. Wong. Resolution of racemic epoxides on g.l.c. columns containing optically active lanthanoid complexes. *J. Chem. Soc. Chem. Commun.*, 570–571, 1977.
53. V. Schurig, W. Bürkle. Extending the scope of enantiomer resolution by complexation gas chromatography. *J. Am. Chem. Soc.*, 104:7573–7580, 1982.
54. V. Schurig, R. Weber. Manganese(II) bis(3-heptafluorobutyryl-1*R*-camphorate): a versatile agent for the resolution of racemic cyclic ethers by complexation gas chromatography. *J. Chromatogr.*, 217:51–70, 1981.
55. V. Schurig, W. Bürkle, K. Hintzer, R. Weber. Evaluation of nickel(II) bis[α-(heptafluorobutanoyl)-terpeneketonates] as chiral stationary phases for the enantiomer separation of alkyl-substituted cyclic ethers by complexation gas chromatography. *J. Chromatogr.*, 475:23–44, 1989.
56. V. Schurig. Contributions to the theory and practice of the chromatographic separation of enantiomers. *Chirality*, 17:S205–S226, 2005.
57. V. Schurig. Peak coalescence phenomena in enantioselective chromatography. *Chirality*, 10:140–146, 1998.
58. W. Bürkle, H. Karfunkel, V. Schurig. Dynamic phenomena during enantiomer resolution by complexation gas chromatography: a kinetic study of enantiomerization. *J. Chromatogr.*, 288:1–14, 1984.
59. V. Schurig. Resolution of enantiomers and isotopic compositions by selective complexation gas chromatography on metal complexes. *Chromatographia*, 13:263–270, 1980.
60. V. Schurig. Enantiomer analysis by complexation gas chromatography: scope, merits and limitations. *J. Chromatogr.*, 441:135–153, 1988.
61. V. Schurig. Review. Practice and theory of enantioselective complexation gas chromatography. *J. Chromatogr. A*, 965:315–356, 2002.
62. J. Snopek, E. Smolková-Keulemansová, T. Cserháti, K. H. Gahm, A. Stalcup. Cyclodextrins in analytical separation methods. In *Comprehensive Supramolecular Chemistry*, Vol. 3, *Cyclodextrins*, J. Szejtli, T. Osa, Eds. Pergamon Press, Elmsford, NY, 1996, pp. 516–571.
63. V. Schurig, H.-P. Nowotny. Gas chromatographic separation of enantiomers on cyclodextrin derivatives. *Angew. Chem. Int. Ed. Engl.*, 29:939–957, 1990.

64. W. A. König. *Gas Chromatographic Enantiomer Separation with Modified Cyclodextrins*. Hüthig, Heidelberg, Germany, 1992.

65. W. Li, T. M. Rossi. Derivatized cyclodextrins as chiral gas chromatographic stationary phases and their potential applications in the pharmaceutical industry. In *The Impact of Stereochemistry on Drug Development and Use*, H. Y. Aboul-Enein, I. W. Wainer, Eds., Vol. 142 in *Chemical Analysis*, J. D. Winefordner, Ed. Wiley, New York, 1997, pp. 415–436.

66. T. Kóscielski, D. Sybilska, J. Jurczak. Separation of α- and β-pinene into enantiomers in gas–liquid chromatography systems via α-cyclodextrin inclusion complexes. *J. Chromatogr*., 280:131–134, 1983.

67. T. Kóscielski, D. Sybilska, J. Jurczak. New chromatographic method for the determination of the enantiomeric purity of terpenoic hydrocarbons. *J. Chromatogr*., 364:299–303, 1986.

68. M. Lindström, T. Norin, J. Roeraade. Gas chromatographic separation of monoterpene hydrocarbon enantiomers on α-cyclodextrin. *J. Chromatogr*., 513:315–320, 1990.

69. P. Sandra. Editorial: the enantioselectivity of derivatized cyclodextrins: a plea for more systematic studies. *J. High Resolut. Chromatogr*., 13:665, 1990.

70. V. Schurig, H.-P. Nowotny. Separation of enantiomers on diluted permethylated β-cyclodextrin by high-resolution gas chromatography. In *Proceedings of Advances in Chromatography 1987*, Berlin, Sept. 8–10, 1987, A. Zlatkis, Ed.; *J. Chromatogr*., 441:155–163, 1988.

71. H.-P. Nowotny, D. Schmalzing, D. Wistuba, V. Schurig. Extending the scope of enantiomer separation on diluted methylated β-cyclodextrin derivatives by high-resolution gas chromatography. *High Resolut. Chromatogr*., 12:383–393, 1989.

72. W. A. König, S. Lutz, P. Mischnick-Lübbecke, B. Brassat, G. Wenz. Cyclodextrins as chiral stationary phases in capillary gas chromatography: I. Pentylated α-cyclodextrin. *J. Chromatogr*., 447:193–197, 1988.

73. D. W. Armstrong, W. Li, C.-D. Chang. Polar-liquid, derivatized cyclodextrin stationary phases for the capillary gas chromatography separation of enantiomers. *Anal. Chem*., 62:914–923, 1990.

74. W.-Y. Li, H. L. Jin, D. W. Armstrong. 2,6-Di-*O*-pentyl–3-*O*-trifluoroacetyl cyclodextrin liquid stationary phases for capillary gas chromatographic separation of enantiomers. *J. Chromatogr*., 509:303–324, 1990.

75. Z. Juvancz, G. Alexander, J. Szejtli. Permethylated β-cyclodextrin as stationary phase in capillary gas chromatography. *J. High Resolut. Chromatogr*., 10:105–107, 1987.

76. G. Alexander, Z. Juvancz, J. Szejtli. Cyclodextrins and their derivatives as stationary phases in GC capillary columns. *J. High Resolut. Chromatogr*., 11:110–113, 1988.

77. A. Venema, P. J. A. Tolsma. Enantiomer separation with capillary gas chromatography columns coated with cyclodextrins: I. Separation of enantiomeric 2-substituted propionic acid esters and some lower alcohols with permethylated β–cyclodextrin. *J. High Resolut. Chromatogr*., 12:32–34, 1989.

78. S. Mayer, D. Schmalzing, M. Jung, M. Schleimer. A chiral test mixture for permethylated β-cyclodextrin-polysiloxane gas–liquid chromatography phases: the Schurig test mixture. *LCGC Int*., 5(4): 58–59, 1992.

79. W. Keim, A. Köhnes, W. Meltzow, H. Römer. Enantiomer separation by gas chromatography on cyclodextrin chiral stationary phases. *J. High Resolut. Chromatogr.*, 14:507–529, 1991.

80. C. Bicchi, G. Artuffo, A. D'Amato, G. M. Nano, A. Galli, M. Galli. Permethylated cyclodextrins in the GC separation of racemic mixtures of volatiles: Part 1. *J. High Resolut. Chromatogr.*, 14:301–305, 1991.

81. Forum: Collection of enantiomer separation factors obtained by capillary gas chromatography on chiral stationary phases. *J. High Resolut. Chromatogr.*, 16:312–323, 1993.

82. W. A. König. Forum: Collection of enantiomer separation factors obtained by capillary gas chromatography on chiral stationary phases. *J. High Resolut. Chromatogr.*, 16:338–352, 1993.

83. W. A. König. Forum: Collection of enantiomer separation factors obtained by capillary gas chromatography on chiral stationary phases. *J. High Resolut. Chromatogr.*, 16:569–586, 1993.

84. A. Berthod, L. He, D. W. Armstrong. Ionic liquids as stationary phase solvents for methylated cyclodextrins in gas chromatography. *Chromatographia*, 53:63–68, 2001.

85. W. Blum, R. Aichholz. Gas chromatographic enantiomer separation on *tert*-butylsilylated β-cyclodextrin diluted in PS–086: a simple method to prepare enantioselective glass capillary columns. *J. High Resolut. Chromatogr.*, 13:515–518, 1990.

86. A. Dietrich, B. Maas, V. Karl, P. Kreis, D. Lehmann, B. Weber, A. Mosandl. Stereoisomeric flavor compounds: LV. Stereodifferentiation of some chiral volatiles on heptakis(2,3-di-*O*-acetyl–6-*O*-*tert*-butyldimethylsilyl)-β-cyclodextrin. *J. High Resolut. Chromatogr.*, 15:176–179, 1992.

87. A. Dietrich, B. Maas, W. Messer, G. Bruche, V. Karl, A. Kaunzinger, A. Mosandl. Stereoisomeric flavor compounds: LVIII. The use of heptakis(2,3-di-*O*-methyl–6-*O*-*tert*-butyldimethylsilyl)-β-cyclodextrin as a chiral stationary phase in flavor analysis. *J. High Resolut. Chromatogr.*, 15:590–593, 1992.

88. V. Schurig, M. Juza. Approach to the thermodynamics of enantiomer separation by gas chromatography: enantioselectivity between the chiral inhalation anesthetics enflurane, isoflurane and desflurane and a diluted γ-cyclodextrin derivative. *J. Chromatogr. A*, 757:119–135, 1997.

89. I. Špánic, J. Krupčik, V. Schurig. Comparison of two methods for the gas chromatographic determination of thermodynamic parameters of enantioselectivity. *J. Chromatogr. A*, 843:123–128, 1999.

90. M. Jung, D. Schmalzing, V. Schurig. Theoretical approach to the gas chromatographic separation of enantiomers on dissolved cyclodextrin derivatives. *J. Chromatogr.*, 552:43–57, 1991.

91. V. Schurig, D. Schmalzing, U. Mühleck, M. Jung, M. Schleimer, P. Mussche, C. Duvekot, J. C. Buyten. Gas chromatographic enantiomer separation on polysiloxane-anchored permethyl-β-cyclodextrin (Chirasil-Dex). *J. High Resolut. Chromatogr.*, 13:713–717, 1990.

92. P. Fischer, R. Aichholz, U. Bölz, M. Juza, S. Krimmer. Permethyl–β–cyclodextrin, chemically bonded to polysiloxane: a chiral stationary phase with wider application range for enantiomer separation by capillary gas chromatography. *Angew. Chem. Int. Ed. Engl.*, 29:427–429, 1990.

93. V. Schurig, D. Schmalzing, M. Schleimer. Enantiomer separation on immobilized Chirasil-Metal and Chirasil-Dex by gas chromatography and supercritical fluid chromatography. *Angew. Chem. Int. Ed. Engl.*, 30:987–989, 1991.

94. V. Schurig, M. Jung, S. Mayer, M. Fluck, S. Negura, H. Jakubetz. Unified enantioselective capillary chromatography on a Chirasil-DEX stationary phase: advantages of column miniaturization. *J. Chromatogr. A*, 694:119–128, 1995.

95. V. Schurig, S. Mayer, M. Jung, M. Fluck, H. Jakubetz, A. Glausch, S. Negura. Unified enantioselective chromatography involving Chirasil-Dex. In *The Impact of Stereochemistry on Drug Development and Use*, H. Y. Aboul-Enein, I. W. Wainer, Eds., Vol. 142 in *Chemical Analysis*, J. D. Winefordner, Ed. Wiley, New York, 1997, pp. 401–413.

96. H. Cousin, O. Trapp, V. Peulon-Agasse, X. Pannecoucke, L. Banspach, G. Trapp, Z. Jiang, J. C. Combret, V. Schurig. Synthesis, NMR spectroscopic characterization and polysiloxane-based immobilization of the three regioisomeric monooctenylpermethyl-β-cyclodextrins and their application in enantioselective GC. *Eur. J. Org. Chem.*, 2003; 3273–3287.

97. H. Grosenick, V. Schurig. Enantioselective capillary gas chromatography and capillary supercritical fluid chromatography on an immobilized γ-cyclodextrin derivative. *J. Chromatogr.*, 761:181–193, 1997.

98. D. W. Armstrong, Y. Tang, T. Ward, M. Nichols. Derivatized cyclodextrins immobilized on fused-silica capillaries for enantiomeric separations via capillary electrophoresis, gas chromatography, or supercritical fluid chromatography. *Anal. Chem.*, 65:1114–1117, 1993.

99. J. S. Bradshaw, Z. Chen, G. L. Yi, B. E. Rossiter, A. Malik, D. Pyo, H. Yun, D. R. Black, S. S. Zimmermann, M. L. Lee, W. D. Tong, V. T. d'Souza. 6A,6B-β-Cyclodextrin-hexasiloxane copolymers: enantiomeric separations by a β-cyclodextrin-containing rotaxane copolymer. *Anal. Chem.*, 67:4437–4439, 1995.

100. M. Y. Nie, L. M. Zhou, Q. H. Wang, D. Q. Zhu, Gas chromatographic enantiomer separation on single and mixed cyclodextrin derivative chiral stationary phases. *Chromatographia*, 51:736–740, 2000.

101. J. Magnusson, L. G. Blomberg, S. Claude, R. Tabacchi, A. Saxer, S. Schürch. Gas chromatographic enantiomer separation of atropisomeric PCBs using modified cyclodextrins as chiral phases. *J. High Resolut. Chromatogr.*, 23:619–627, 2000.

102. S. Tamogami, K.-I. Awano, M. Amaike, Y. Takagi, T. Kitahara. Development of an efficient GLC system with a mixed chiral stationary phase and its application to the separation of optical isomers. *Flavour Fragr. J.*, 16:349–352, 2001.

103. M. Bayer, A. Mosandl. Improved gas chromatographic stereodifferentiation of chiral main constituents from different essential oils using a mixture of chiral stationary phases. *Flavour Fragr. J.*, 19:515–517, 2004.

104. S. Qi, P. Ai, C. Wang, L. Yuan, G. Zhang. The characteristics of a mixed stationary phase containing permethylated-β-CD and perpentylated-β-CD in gas chromatography. *Sep. Purif. Technol.*, 48:310–313, 2006.

105. W. Vetter, K. Lehnert, G. Hottinger. Enantioseparation of chiral organochlorines on permethylated β- and γ-cyclodextrin, as well as 1 : 1 mixtures of them. *J. Chromatogr. Sci.*, 44:596–601, 2006.

106. D. Kreidler, H. Czesla, V. Schurig. A mixed stationary phase containing two versatile cyclodextrin-based selectors for the simultaneous gas chromatographic enantioseparation of racemic alkanes and racemic α-amino-acid derivatives. *J. Chromatogr. B*, 875:208–216, 2008.

107. G. Sicoli, Z. Jiang, L. Jicsinsky, V. Schurig. Modified linear dextrins ("acyclodextrins") as new chiral selectors for the gas-chromatographic separation of enantiomers. *Angew. Chem. Int. Ed*., 44:4092–4095, 2005.

108. G. Sicoli, F. Pertici, Z. Jiang, L. Jicsinszky, V. Schurig. Gas-chromatographic approach to probe the absence of molecular inclusion in enantioseparations by carbohydrates: investigation of linear dextrins ("acyclodextrins") as novel chiral stationary phases. *Chirality*, 19:391–400, 2007.

109. P. A. Levkin, A. Levkina, H. Czesla, V. Schurig. Temperature-induced inversion of the elution order of enantiomers in gas chromatography: *N*-ethoxycarbonyl propylamides and *N*-trifluoroacetyl ethyl esters of α-amino acids on Chirasil-Val-C_{11} and Chirasil-Dex stationary phases. *Anal. Chem*., 79:4401–4409, 2007.

110. M. Schneider, K. Ballschmiter. Alkyl nitrates as achiral and chiral solute probes in gas chromatography: novel properties of a β-cyclodextrin derivative and characterization of its enantioselective forces. *J. Chromatogr. A*, 852:525–534, 1999.

111. B. Koppenhoefer, E. Bayer. Chiral recognition in the resolution of enantiomers by GLC. *Chromatographia*, 19:123–130, 1984.

112. V. Schurig, R. Link. Recent developments in enantiomer separation by complexation gas chromatography. In: *Chiral Separations*, D. Stevenson, I. D. Wilson, Eds. Plenum Press, New York, 1988, pp. 91–114.

113. K. Watabe, R. Charles, E. Gil-Av. Temperature dependent inversion of elution sequence in the resolution of α-amino acid enantiomers on chiral diamide selectors. *Angew. Chem. Int. Ed*., 28:192–194, 1989.

114. B. Koppenhoefer, B. Lin. Thermodynamic properties of enantiomers of underivatized diols versus the cyclic carbonates in gas chromatography on Chirasil-Val. *J. Chromatogr*., 481:17–26, 1989.

115. V. Schurig, J. Ossig, R. Link. Evidence for a temperature dependent reversal of the enantioselectivity in complexation gas chromatography on chiral phases. *Angew. Chem. Int. Ed*., 28:194–196, 1989.

116. V. Schurig, F. Betschinger. Metal-mediated enantioselective access to unfunctionalized aliphatic oxiranes: prochiral and chiral recognition. *Chem. Rev*., 92:873–888, 1992.

117. Z. Jiang, V. Schurig. Existence of a low isoenantioselective temperature in complexation gas chromatography: profound change of enantioselectivity of a nickel(II) chiral selector either bonded to, or dissolved in, poly(dimethylsiloxane). *J. Chromatogr. A*, 1186:262–270, 2008.

118. W. A. König, D. Icheln, I. Hardt. Unusual retention behaviour of methyl lactate and methyl 2-hydroxybutyrate enantiomers on a modified cyclodextrin. *J. High Resolut. Chromatogr*., 14:694–695, 1991.

119. W. Melander, D. E. Campbell, C. Horváth. Enthalpy–entropy compensation in reversed-phase chromatography. *J. Chromatogr*., 158:215–225, 1978.

120. A. Berthod, W. Li, W. A. Armstrong. Multiple enantioselective retention mechanisms on derivatized cyclodextrin gas chromatographic chiral stationary phases. *Anal. Chem*., 64:873–879, 1992.

121. V. Pino, A. W. Lantz, J. L. Anderson, A. Berthod, D. W. Armstrong. Theory and use of the pseudophase model in gas–liquid chromatographic enantiomeric separations. *Anal. Chem.*, 78:113–119, 2006.

122. E. Bayer. Chirale Erkennung von Naturstoffen an optisch aktiven Polysiloxanen (Chiral recognition of natural products on optically active polysiloxanes). *Z. Naturforsch.*, 38b:1281–1291, 1983.

123. H. P. Benschop, C. A. G. Konings, L. P. A. De Jong. Gas chromatographic separation and identification of the four stereoisomers of 1,2,2-trimethylpropyl methylphosphonofluoridate (soman): stereospecificity of in vitro "detoxification" reactions. *J. Amer. Chem. Soc.*, 103:4260–4262, 1981.

124. H. P. Benschop, L. P. A. De Jong, Toxicokinetics of soman: species variation and stereospecificity in elimination pathways. *Neurosci. Biobehav. Rev.*, 15:73–77, 1991.

125. R. Weber, V. Schurig. Complexation gas chromatography: a valuable tool for the stereochemical analysis of pheromones. *Naturwissenschaften*, 71:408–413, 1984.

126. V. Schurig. Enantiomer separation by complexation gas chromatography: applications in chiral analysis of pheromones and flavours. In *Bioflavour '87*, P. Schreier, Ed. Walter de Gruyter, New York, 1988, pp. 35–54.

127. D. Wistuba, H.-P. Nowotny, O. Träger, V. Schurig. Cytochrome P–450-catalyzed asymmetric epoxidation of simple prochiral and chiral aliphatic alkenes: species dependence and effect of enzyme induction on enantioselective oxirane formation. *Chirality*, 1:127–136, 1989.

128. D. Wistuba, O. Träger, V. Schurig. Enantio- and regioselectivity in the epoxide-hydrolase-catalyzed ring opening of simple aliphatic oxiranes. *Chirality*, 4:178–184, 185–192, 1992.

129. A. Ghanem, V. Schurig. Lipase-catalyzed irreversible transesterification of 1-(2-furyl)ethanol using isopropenyl acetate. *Chirality*, 13:118–123, 2001.

130. C. Petucci, L. Di, O. McConnell. Rapid screening of enzymes for the enzymatic hydrolysis of chiral esters in drug discovery. *Chirality*, 19:701–705, 2007.

131. M. T. Reetz, K. M. Kuhling, S. Wilensek, H. Husmann, U. W. Hausig, M. Hermes. A GC-based method for high-throughput screening of enantioselective catalysts. *Catal. Today*, 67:389–396, 2001.

132. A. Mosandl. Enantioselective capillary gas chromatography and stable isotope ratio mass spectrometry in the authenticity control of flavours and essential oils. *Food Rev. Int.*, 11:597–664, 1995.

133. C. Bicchi, A. D'Amato, P. Rubiolo. Review: Cyclodextrin derivatives as chiral selectors for direct gas chromatographic separation of enantiomers in the essential oil, aroma and flavour field. *J. Chromatogr. A*, 843:99–121, 1999.

134. P. J. Marriott, R. Shellie, C. Cornwell. Review: Gas chromatographic technologies for the analysis of essential oils. *J. Chromatogr. A*, 936:1–22, 2001.

135. R. Fellous, L. Lizzani-Cuvelier, A.-M. Loiseau. Permethylated maltosyl cyclodextrins as chiral stationary phases in capillary gas chromatography: enantiomer separation of chiral lactones. *J. High Resolut. Chromatogr.*, 13:785–789, 1990.

136. R. Stritzel, B. Dobner, F. Bringezu, P. Nuhn. Separation of racemic 2-alkyl-branched fatty acid methyl esters by gas chromatography on a commercially available chiral phase. *J. High Resolut. Chromatogr.*, 19:121–123, 1996.

137. S. Thurnhofer, G. Hottinger, W. Vetter. Enantioselective determination of anteiso fatty acids in food samples. *Anal. Chem.*, 79:4696–4701, 2007.

138. W. Vetter, V. Schurig. Review: Enantioselective determination of chiral organochlorine compounds in biota by gas chromatography on modified cyclodextrins. *J. Chromatogr. A*, 774:143–175, 1997.

139. A. Imran, H. Y. Aboul-Enein. Chiral pullutants: distribution, toxicity and analysis by chromatography and capillary electrophoresis. In *The Analysis of Chiral Pollutants by Gas Chromatography*. Wiley-VCH, New York, 2004.

140. W. Vetter, K. Bester. Gas chromatographic enantioseparation of chiral pollutants: techniques and results. In *Chiral Analysis*, K. Busch, and M. Busch, Eds. Elsevier Science, Amsterdam, 2006, Chap. 6.

141. B. Feibush, C. L. Woolley, V. Mani. Separation of chiral silicon compounds using permethylated α-, β-, and γ-cyclodextrin capillary GC columns. *Anal. Chem.*, 65:1130–1133, 1993.

142. V. Schurig, H. Grosenick, M. Juza. The chiral inhalation anesthetics enflurane, isoflurane and desflurane. *Chimica Oggi (Chemistry Today)*, 16:40–44, June 1998.

143. J. Meinwald, W. R. Thompson, D. L. Pearson, W. A. König, T. Runge, W. Francke. Inhalational anesthetics stereochemistry: optical resolution of halothane, enflurane, and isoflurane. *Science*, 251:560–561, 1991.

144. A. Kaunzinger, A. Rechner, T. Beck, A. Mosandl, A. C. Sewell, H. Böhles. Chiral compounds as indicators of inherited metabolic disease: simultaneous stereodifferentiation of lactic-, 2-hydroxyglutaric- and glyceric acid by enantioselective cGC. *Enantiomer*, 1:177–182, 1996.

145. F. Podebrad, M. Heil, A. Scharrer, S. Feldmer, O. Schulte-Mäter, A. Mosandl, A. C. Sewell, H. J. Böhles. Analysis of methylcitric acid by enantioselective multidimensional gas chromatography–mass spectrometry. *J. High Resolut. Chromatogr.*, 22:604–608, 1999.

146. M. Zampolli, D. Meunier, R. Sternberg, F. Raulin, C. Szopa, M. C. Pietrogrande, F. Dondi. GC-MS analysis of amino acid enantiomers as their N(O,S)-perfluoroacyl perfluoroalkyl esters: application to space analysis. Chirality, 18:279–295, 2006.

147. M. Zampolli, G. Basaglia, F. Dondi, R. Sternberg, C. Szopa, M. C. Pietrogrande. Gas chromatography–mass spectrometry analysis of amino acid enantiomers as methyl chloroformate derivatives: application to space analysis. *J. Chromatogr. A*, 1150:162–172, 2007.

148. V. Schurig, H.-P. Nowotny, D. Schmalzing. Gas-chromatographic enantiomer separation of unfunctionalized cycloalkanes on permethylated-β-cyclodextrin. *Angew. Chem. Int. Ed. Engl.*, 28:736–737, 1989.

149. U. J. Meierhenrich, W. H.-P. Thiemann, F. Goesmann, R. Roll, H. Rosenbauer. Enantiomer separation of hydrocarbons in preparation for *Rosetta*'s "chirality-experiment." *Chirality*, 13:454–457, 2001.

150. U. J. Meierhenrich, M.-J. Nguyen, B. Barbier, A. Brack, W. H.-P. Thiemann. Gas chromatographic separation of saturated aliphatic hydrocarbon enantiomers on permethylated β-cyclodextrin. *Chirality*, 15:S13–S16, 2003.

151. G. Sicoli, D. Kreidler, H. Czesla, H. Hopf, V. Schurig. Gas chromatographic enantioseparation of unfunctionalized chiral alkanes: a challenge in separation science (overview, state of the art, and perspectives). *Chirality*, 21:183–198, 2009.

152. Z. Juvancz, K. Grolimund, V. Schurig. Pharmaceutical applications of a bonded cyclodextrin stationary phase. *J. Microcolumn Sep.*, 5:459–468, 1993.

153. S. Pizzarello, Y. Huang, M. Fuller. The carbon isotopic distribution of Murchison amino acids. *Geochim. Cosmochim. Acta*, 23:4963–4969, 2004.

154. F. Goesmann, H. Rosenbauer, R. Roll, H. Böhnhardt. COSAC on board *Rosetta*: a bioastronomy experiment for the short-period comet 67P/Churyumov-Gerasimenko. *Astrobiology*, 5:622–631, 2005.

155. C. Szopa, F. Goesmann, H. Rosenbauer, R. Sternberg and the COSAC team. The COSAC experiment of the *Rosetta* mission: performance under representative conditions and expected scientific return. *Adv. Space Res.*, 40:180–186, 2007.

156. U. J. Meierhenrich. *Amino Acids and the Asymmetry of Life* (with a foreword by Henri B. Kagan). Springer-Verlag, Heidelberg, Germany, 2008.

157. W. H. Pirkle, C. J. Welch. Some thoughts on the coupling of dissimilar chiral columns or the mixing of chiral stationary phases for the separation of enantiomers. *J. Chromatogr. A*, 731:322–326, 1996.

158. V. Schurig, M. Juza, M. Preschel, G. J. Nicholson, E. Bayer. Gas-chromatographic enantiomer separation of proteinogenic amino acid derivatives: comparison of Chirasil-Val and Chirasil-γ-Dex used as chiral stationary phases. *Enantiomer*, 4:297–303, 1999.

159. P. A. Levkin, V. Schurig. Apparent and true enantioselectivity of single- and binary-selector chiral stationary phases in gas chromatography. *J. Chromatogr. A*, 1184:309–322, 2008.

160. A. Ruderisch, J. Pfeiffer, V. Schurig. Mixed chiral stationary phase containing modified resorcinarene and β-cyclodextrin selectors bonded to a polysiloxane for enantioselective gas chromatography. *J. Chromatogr. A*, 994:127–135, 2003.

161. P. A. Levkin, A. Ruderisch, V. Schurig. Combining the enantioselectivity of a cyclodextrin and a diamide selector in a mixed binary gas–chromatographic chiral stationary phase. *Chirality*, 18:49–63, 2006.

162. E. Forró. New gas chromatographic method for the enantioseparation of β-amino acids by a rapid double derivatization technique. *J. Chromatogr. A*, 1216:1025–1029, 2009.

163. P. A. Levkin, A. Levkina, H. Czesla, S. Nazzi, V. Schurig. Expanding the enantioselectivity of the gas-chromatographic chiral stationary phase Chirasil-Val-C–11 by doping it with octakis(3-*O*-butanoyl–2,6-di-*O*-*n*-pentyl)-γ-cyclodextrin. *J. Sep. Sci.*, 30:98–103, 2007.

164. G. Uccello-Barretta, S. Nazzi, S. Balzano, P. A. Levkin, V. Schurig, P. Salvadori. Heptakis[2,3-di-*O*-methyl-6-*O*-(L-valine-*tert*-butylamide-*N*-alphaylcarbonylmethyl)]-β-cyclodextrin: a new multifunctional cyclodextrin CSA for the NMR enantiodiscrimination of polar and apolar substrates. *Eur. Org. Chem.* 3219–3226, 2007.

165. O. Stephany, F. Dron, S. Tisse, A. Martinez, J.-M. Nuzillard, V. Peulon-Agasse, P. Cardinaël, J.-P. Bouillon. (L)- or (D)-valine *tert*-butylamide grafted on permethylated β-cyclodextrin derivatives as new mixed binary chiral selectors: versatile tools for capillary gas chromatographic enantioseparation. *J. Chromatogr. A*, 1216:4051–4062, 2009.

166. C. F. Poole. Rate theories. In *The Essence of Chromatography*. Elsevier, Amsterdam, 2003, pp. 29–38.

167. M. Lindström. Improved enantiomer separation using very short capillary columns coated with permethylated β–cyclodextrin. *J. High Resolut. Chromatogr.*, 14:765–767, 1991.

168. V. Schurig, H. Czesla. Miniaturization of enantioselective gas chromatography. *Enantiomer*, 6:107–128, 2001.

169. C. Bicchi, E. Liberto, C. Cagliero, C. Cordero, B. Sgorbini, P. Rubiolo. Conventional and narrow bore short capillary columns with cyclodextrin derivatives as chiral selectors to speed-up enantioselective gas chromatography and enantioselective gas chromatography–mass spectrometry analyses. *J. Chromatogr. A*, 1212:114–123, 2008.

170. A. Schieber, H. Brückner, J. R. Ling. GC-MS analysis of diaminopimelic acid stereoisomers and amino acid enantiomers in rumen bacteria. *Biomed. Chromatogr.*, 13:46–50, 1999.

171. H. Brückner, A. Schieber. Determination of free D-amino acids in mammalia by chiral gas chromatography–mass spectrometry. *J. High Resolut. Chromatogr.*, 23:576–582, 2000.

172. P. Chalier, J. Crouzet. Enantiodifferentiation of four γ-lactones produced by *Penicillium roqueforti*. *Chirality*, 10:786–790, 1998.

173. A. Glausch, J. Hahn, V. Schurig. Enantioselective determination of chiral 2,2′,3,3′,4,6′-hexachloro-biphenyl (PCB 132) in human milk samples by multidimensional gas chromatography/electron capture detection and by mass spectrometry. *Chemosphere*, 30:2079–2085, 1995.

174. V. Schurig, M. Schleimer, M. Jung, S. Mayer, A. Glausch. In *Progress in Flavour Precursor Studies*: Proceeedings of the International Conference, Würzburg, Germany, 1992, P. Schreier, P. Winterhalter, Eds. Allured Publishing, Carol Stream, IL, 1993, pp. 63–75.

175. R. Schmidt, H. G. Wahl, H. Häberle, H.-J. Dieterich, V. Schurig. Headspace gas chromatography–mass spectrometry analysis of isoflurane enantiomers in blood samples after anesthesia with the racemic mixture. *Chirality*, 11:206–211, 1999.

176. A. Berthod, X. Wang, K. H. Gahm, D. W. Armstrong. Quantitative and stereoisomeric determination of light biomarkers in crude oil and coal samples. *Geochim. Cosmochim. Acta*, 62:1619–1630, 1998.

177. R. Liardon, S. Ledermann, U. Ott. Determination of D-amino acids by deuterium labelling and selected ion monitoring. *J. Chromatogr.*, 203:385–395, 1981.

178. M. Scholz, L. Bachmann, G. J. Nicholson, J. Bachmann, I. Giddings, B. Rüschoff-Thale, A. Czarnetzki, C. M. Pusch. Genomic differentiation of Neanderthals and anatomically modern man allows a fossil-DNA-based classification of morphologically indistinguishable hominid bones. *Am. J. Hum. Genet.*, 66:1927–1932, 2000.

179. K. Nokihara, J. Gerhardt. Development of an improved automated gas-chromatographic chiral analysis system: application to non-natural amino acids and natural protein hydrolysates. *Chirality*, 13:431–434, 2001.

180. G. T. Eyres, S. Urban, P. D. Morrison, P. J. Marriott. Application of microscale-preparative multidimensional gas chromatography with nuclear magnetic resonance spectroscopy for identification of pure methylnaphthalenes from crude oil. *J. Chromatogr. A*, 1215:168–176, 2008.

181. D. Kreidler. Doctoral dissertation, University of Tübingen, 2008, pp. 109–111.

182. M. Kühnle, D. Kreidler, H. Czesla, K. Holtin, P. Schuler, W. Schaal, V. Schurig, K. Albert. On-line coupling of gas chromatography to nuclear magnetic resonance spectroscopy: a method for the analysis of volatile stereoisomers. *Anal. Chem.*, 80:5481–5486, 2008.

183. G. Schomburg, H. Husmann, E. Hübinger, W. A. König. Multidimensional capillary gas chromatography–enantiomeric separations of selected cuts using a chiral second column. *J. High Resolut. Chromatogr.*, 7:404–410, 1984.

184. L. Mondello, M. Catalfamo, A. R. Proteggente, I. Bonaccorsi, G. Dugo. Multidimensional capillary GC-GC for the analysis of real samples: 3. Enantiomeric distribution of monoterpene hydrocarbons and monoterpene alcohols of mandarin oils. *J. Agric. Food Chem.*, 46:54–61, 1998.

185. L. Mondello, A. Verzera, P. Previti, F. Crispo, G. Dugo. Multidimensional capillary GC-GC for the analysis of complex samples: 5. Enantiomeric distribution of monoterpene hydrocarbons and monoterpene alcohols, and linalyl acetate of bergamot (*Citrus bergamia* Risso et Poiteau) oils. *J. Agric. Food Chem.*, 46:4275–4282, 1998.

186. L. Mondello, M. Catalfamo, A. Cotroneo, G. Dugo, G. Dugo, H. McNair. Multidimensional capillary GC-GC for the analysis of real complex samples: IV. Enantiomeric distribution of monoterpene hydrocarbons and monoterpene alcohols of lemon oils. *J. High Resolut. Chromatogr.*, 22:350–356, 1999.

187. D. Lorenzo, D. Paz, P. Davies, J. Villamil, R. Vila, S. Cañigueral, E. Dellacassa. Application of multidimensional gas chromatography to the enantioselective characterisation of the essential oil of *Eupatorium buniifolium* Hooker et Arnott. *Phytochem. Anal.*, 16:39–44, 1999.

188. Y. Saritas, N. Bülow, C. Fricke, W. A. König, H. Muhle. Sesquiterpene hydrocarbons in the liverwort *Dumortiera hirsuta*. *Phytochemistry*, 48:1019–1023, 1998.

189. G. P. Blanch, J. Jauch. Enantiomeric composition of filbertone in hazelnuts in relation to extraction conditions: multidimensional gas chromatography and gas chromatography/mass spectrometry in the single ion monitoring mode of a natural sample. *J. Agric. Food Chem.*, 46:4283–4286, 1998.

190. R. Shellie, L. Mondello, G. Dugo, P. Marriott. Enantioselective gas chromatographic analysis of monoterpenes in essential oils of the family Myrtaceae. *Flavour Fragr. J.*, 19:582–585, 2004.

191. H.-E. Högberg, E. Hedenström, A.-B. Wassgren, M. Hjalmarsson, G. Bergström, J. Löfqvist, T. Norin. Synthesis and gas chromatographic separation of the eight stereoisomers of diprionol and their acetates, components of the sex pheromone of pine sawflies. *Tetrahedron*, 46:3007–3018, 1990.

192. A. Glausch, G. J. Nicholson, M. Fluck, V. Schurig. Separation of the enantiomers of stable atropisomeric polychlorinated biphenyls (PCBs) by multidimensional gas chromatography on Chirasil-Dex. *J. High Resolut. Chromatogr.*, 17:347–349, 1994.

193. F. Podebrad, M. Heil, S. Leib, B. Geier, T. Beck, A. Mosandl, A. C. Sewell, H. Böhles. Analytical approach in diagnosis of inherited metabolic diseases: maple syrup disease (MSUD)—simultaneous analysis of metabolites in urine by enantioselective multidimensional capillary gas chromatography–mass spectrometry (enantio-MDGC-MS). *J. High Resolut. Chromatogr.*, 20:355–362, 1997.

194. M. Heil, F. Podebrad, T. Beck, A. Mosandl, A. C. Sewell, H. Böhles. Enantioselective multidimensional gas chromatography–mass spectrometry in analysis of urinary organic acids. *J. Chromatogr. B*, 714:119–126, 1998.

195. M. Kreck, A. Scharrer, S. Bilke, A. Mosandl. Stir bar sorptive extraction (SBSE)-enantio-MDGC-MS: a rapid method for the enantioselective analysis of chiral flavour compounds in strawberries. *Eur. Food Res. Technol.*, 213:389–394, 2001.

196. M. Kreck, A. Scherrer, S. Bilke, A. Mosandl. Enantioselective analysis of monoterpene compounds in essential oils by stir bar sorptive extraction (SBSE)-enantio-MDGC-MS. *Flavour Fragr. J.*, 17:32–40, 2002.

197. P. Marriott, R. Shellie. Principles and applications of comprehensive two-dimensional gas chromatography. *Trends Anal. Chem.*, 21:573–583, 2002.

198. R. Shellie, P. J. Marriott. Comprehensive two-dimensional gas chromatography with fast enantioseparation. *Anal. Chem.*, 74:5426–5430, 2002.

199. P. Shellie, P. Marriott, C. Cornwell. Application of comprehensive two-dimensional gas chromatography (GC×GC) to the enantioselective analysis of essential oils. *J. Sep. Sci.*, 24:823–830, 2001.

200. P. Marriott, O. Trapp, R. Shellie, V. Schurig. Time-resolved cryogenic modulation reveals isomer interconversion profiles in dynamic chromatography. *J. Chromatogr. A*, 919:115–126, 2001.

201. V. Schurig, S. Reich. Determination of the rotational barriers of atropisomeric polychlorinated biphenyls (PCBs) by a novel stopped-flow multidimensional gas chromatographic technique. *Chirality*, 10:316–320, 1998.

202. S. Reich, V. Schurig. Stopped-flow multidimensional gas chromatography: a new method for the determination of enantiomerization barriers. *J. Microcolumn Sep.*, 11:475–479, 1999.

203. S. Reich, O. Trapp, V. Schurig. Enantioselective stopped-flow multidimensional gas chromatography: determination of the inversion barrier of 1-chloro–2,2-dimethylaziridine. *J. Chromatogr. A*, 892:487–498, 2000.

204. J. Krupčik, J. Mydlová, P. Májek, P. Šimon, D. W. Armstrong. Methods for studying reaction kinetics in gas chromatography, exemplified by using the 1-chloro–2,2-dimethylaziridine interconversion reaction. *J. Chromatogr. A*, 1186:144–160, 2008.

205. L. Mondello, G. Dugo, K. D. Bartle. On-line microbore high performance liquid chromatography–capillary gas chromatography for food and water analyses: a review. *J. Microcolumn Sep.*, 8:275–310, 1996.

206. G. Dugo, A. Verzera, A. Cotroneo, I. S. d'Alcontres, L. Mondello, K. D. Bartle. Automated HPLC–HRGC: a powerful method for essential oil analysis: II. Determination of the enantiomeric distribution of linalol in sweet orange, bitter orange and mandarin essential oils. *Flavour Fragr. J.*, 9:99–104, 1994.

207. L. Mondello, G. Dugo, P. Dugo, K. D. Bartle. On-line HPLC-HRGC in the analytical chemistry of citrus essential oils. *Perfume Flavor.*, 21:25–49, 1996.

208. A. Mosandl. Enantioselective capillary gas chromatography and stable isotope ratio mass spectrometry in the authenticity control of flavors and essential oils. *Food Rev. Int.*, 11:597–664, 1995.

209. D. Juchelka, T. Beck, U. Hener, F. Dettmar, A. Mosandl. Multidimensional gas chromatography coupled on-line with isotope ratio mass spectrometry (MDGC-IRMS): progress in the analytical authentication of genuine flavor components. *J. High Resolut. Chromatogr.*, 21:145–151, 1998.

210. S. Reichert, D. Fischer, S. Asche, A. Mosandl. Stable isotope labelling in biosynthetic studies of dill ether, using enantioselective multidimensional gas chromatography, on-line coupled with isotope ratio mass spectrometry. *Flavour Fragr. J.*, 15:303–308, 2000.

211. K. Schumacher, U. Hener, C. Patz, H. Dietrich, A. Mosandl. Authenticity assessment of 2- and 3-methylbutanol using enantioselective and/or $^{13}C/^{12}C$ isotope ratio analysis. *Eur. Food Res. Techn.*, 209:12–15, 1999.

212. H. Frank, G. J. Nicholson, E. Bayer. Enantiomer labelling, a method for the quantitative analysis of amino acids. *J. Chromatogr.*, 167:187–196, 1978.

213. N. E. Blair, W. A. Bonner. Quantitative determination of D $\neq$ L mixtures of optical enantiomers by gas chromatography. *J. Chromatogr.*, 198:185–187, 1980.

214. W.-L. Tsai, K. Hermann, E. Hug, B. Rohde, A. S. Dreiding. Enantiomer-differentiation induced by an enantiomeric excess during chromatography with achiral phases. *Helv. Chim. Acta*, 68:2238–2243, 1985.

215. M. Juza, H. Jakubetz, H. Hettesheimer, V. Schurig. Quantitative determination of isoflurane enantiomers in blood samples during and after surgery via headspace gas chromatography–mass spectrometry. *J. Chromatogr. B*, 735:93–102, 1999.

216. R. Aichholz, U. Bölz, P. Fischer. A standard test mixture for assessing enantioselectivity of chiral phase capillary columns: Chiral-test I for amide phases. *J. High Resolut. Chromatogr.*, 13:234–238, 1990.

217. B. Koppenhoefer, A. Nothdurft, J. Pierrot-Sanders, P. Piras, C. Popescu, C. Roussel, M. Stiebler, U. Trettin. Chirbase, a graphical molecular database on the separation of enantiomers by liquid-, supercritical fluid-, and gas chromatography. *Chirality*, 5:213–219, 1993.

218. B. Koppenhoefer, R. Graf, H. Holzschuh, A. Nothdurft, U. Trettin, P. Piras, C. Roussel. Chirbase, a molecular database for the separation of enantiomers by chromatography. *J. Chromatogr. A*, 666:557–563, 1994.

219. B. Koppenhoefer. Eine graphische Moleküldatenbank zur Enantiomerentrennung. *GIT Laborfachzeitschr.*, 42:716, 1998.

220. B. Koppenhoefer, U. Epperlein. Overview of the gas chromatographic separation of enantiomers containing a three-membered ring. *J. Chromatogr. Sci.*, 33:244–255, 1995.

221. B. Koppenhoefer, U. Epperlein, M. Schwierskott. Enantioseparation of four-membered rings by gas chromatography: on overview. *Fresenius J. Anal. Chem.*, 359:107–114, 1997.

222. C. Reiner, G. J. Nicholson, U. Nagel, V. Schurig. Evaluation of enantioselective gas chromatography for the determination of minute deviations from racemic composition of α-amino acids with emphasis on tyrosine: accuracy and precision of the method. *Chirality*, 19:401–414, 2007.

223. E. Liberto, C. Cagliero, B. Sgorbini, C. Bicchi, D. Sciarrone, B. D'Acampora Zellner, L. Mondello, P. Rubiolo. Enantiomer identification in the flavour and fragrance fields by "interactive" combination of linear retention indices from enantioselective gas chromatography and mass spectrometry. *J. Chromatogr. A*, 1195:117–126, 2008.

224. H. Nakazawa, H. Yoeneda. Chromatographic study of optical resolution: II. Separation of optically active cobalt(III) complexes using potassium antimony *d*-tartrate as eluent. *J. Chromatogr.*, 160:89–99, 1978.

225. P. E. Hare, E. Gil-Av. Separation of D and L amino acids by liquid chromatography. *Science*, 204:1226–1228, 1979.

226. W. Lindner, J. N. LaPage, G. Davies, D. E. Seitz, B. L. Karger. Reversed-phase separation of optical isomers of Dns-amino acids and peptides using chiral metal chelate additives. *J. Chromatogr.*, 185:323–343, 1979.

227. Chromatographic techniques used in European laboratories in 1995 and 1996. *LCGC Int*., 10(4): 252, Apr. 1997.

228. T. Arai. Review: Chiral separation of pharmaceuticals possessing a carboxy moiety. *J. Chromatogr. B*, 717:295–311, 1998.

229. C. Yamamoto, Y. Okamoto. Practical resolution of enantiomers by high-performance liquid chromatography. In *Enantiomer Separation: Fundamentals and Practical Methods*, F. Toda, Ed. Kluwer, Dordrecht, The Netherlands, 2004, pp. 301–322.

230. E. R. Francotte. Enantioselective chromatography as a powerful alternative for the preparation of drug enantiomers. *J. Chromatogr. A*, 906:379–397, 2001.

231. V. Schurig. Semi-preparative enantiomer separation of 1,6-dioxaspiro[4.4]nonanes by complexation gas chromatography. *Naturwissenschaften*, 74:190–191, 1987.

232. V. Schurig, U. Leyrer. Semi-preparative enantiomer separation of 1-chloro–2,2-dimethylaziridine by complexation gas chromatography: absolute configuration and barrier of inversion. *Tetrahedron Asymm*., 1:865–868, 1990.

233. C. Bicchi, C. Balbo, A. D'Amato, V. Manzin, P. Schreier, A. Rozenblum, P. Brunerie. Cyclodextrin derivatives in GC separation of racemic mixtures of volatiles: XIV. Some applications of thick-film wide-bore columns for enantiomer GC micro-preparation. *J. High Resolut. Chromatogr*., 21:103–106, 1998.

234. I. Hardt, W. A. König. Preparative enantiomer separation with modified cyclodextrins as chiral stationary phases. *J. Chromatogr*., 666:611–615, 1994.

235. S. Schürch, A. Saxer, S. Claude, R. Tabacchi, B. Trusch, J. Hulliger. Semi-preparative gas chromatographic separation of all-*trans*-perhydrotriphenylene enantiomers on a chiral cyclodextrin stationary phase. *J. Chromatogr A*., 905:175–182, 2001.

236. M. Juza, E. Braun, V. Schurig. Preparative enantiomer separation of the inhalation anesthetics enflurane, isoflurane and desflurane by gas chromatography on a derivatized γ-cyclodextrin stationary phase. *J. Chromatogr. A*, 769:119–127, 1997.

237. D. U. Staerk, A. Shitangkoon, G. Vigh. Gas-chromatographic separation of the enantiomers of volatile fluoroether anesthetics by derivatized cyclodextrins: II. Preparative-scale separations for isoflurane. *J. Chromatogr. A*, 663:79–85, 1994.

238. D. U. Staerk, A. Shitangkoon, G. Vigh. Gas-chromatographic separation of the enantiomers of volatile fluoroether anesthetics by derivatized cyclodextrins: III. Preparative-scale separations for enflurane. *J. Chromatogr. A*, 677:133–140, 1994.

239. H. Y. Aboul-Enein, J. Bojarski, J. Szymura-Oleksiak. The impact of chirality of the fluorinated volatile inhalation anaesthetics on their clinical applications. *Biomed. Chromatogr*., 14:213–218, 2000.

240. M. Juza, O. Di Giovanni, G. Biressi, V. Schurig, M. Mazzotti, M. Morbidelli. Continuous enantiomer separation of the volatile inhalation anesthetic enflurane with a gas chromatographic simulated moving bed unit. *J. Chromatogr. A*, 813:333–347, 1998.

241. F. Quattrini, G. Biressi, M. Juza, M. Mazzotti, C. Fuganti, M. Morbidelli. Enantiomer separation of α-ionone using gas chromatography with cyclodextrin derivatives as chiral stationary phases. *J. Chromatogr. A*, 865:201–210, 1999.

242. A. Rajendran, G. Paredes, M. Mazzotti. Simulated moving bed chromatography for the separation of enantiomers. *J. Chromatogr. A*, 1216:709–738, 2009.

CHAPTER 10

Separations of Chiral Compounds by SFC

ZIQIANG WANG

TharSFC, A Waters Company, Pittsburgh, Pennsylvania

1 THEORIES AND PROPERTIES OF SUPERCRITICAL FLUIDS

A supercritical fluid is a substance (mostly liquids and gases) whose temperature and pressure are at or above its thermodynamic critical point, as shown in Figure 1 [1]. A supercritical fluid possesses unique characteristics compared to those of either the liquid or gaseous states under standard conditions. In supercritical fluids, molecules do not have the strong intermolecular forces that exist in normal liquids (i.e., they possess weak van der Waals forces, as in the gaseous phase). However, when they are heated and "squeezed," they look like a highly "condensed" liquid. These unique characteristics contribute to helping the supercritical fluid have some of the useful properties of both gases and liquids. For example, a supercritical fluid has very low viscosity and higher solute diffusivity, as in a gas. Also, its close molecular arrangement makes it have higher dissolving power, close to that of a normal liquid. Table 1 [2] lists some of these properties for a gas, a liquid, and a supercritical fluid. Carbon dioxide, water, nitric oxide, methane, ethane, and propane have all been explored under supercritical conditions for their unique advantages based on their thermodynamic properties [3].

2 FUNDAMENTALS OF SUPERCRITICAL FLUID CHROMATOGRAPHY

Supercritical fluid chromatography (SFC) was developed to take advantage of the aforementioned properties of supercritical fluids. The liquidlike density of a

Chiral Separation Methods for Pharmaceutical and Biotechnological Products, Edited by Satinder Ahuja

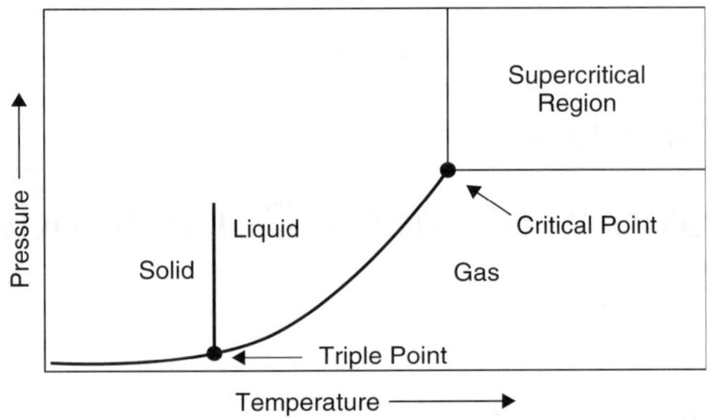

Figure 1 Phase diagram of a substance.

TABLE 1 Comparison of Some Physical Properties Among Gas, Liquid, and Supercritical Fluids

Fluids	Density (g/cm^2)	Diffusion (cm^2/s)	Viscosity (g/cm-s)
Gas	10^{-3}	10^{-1}	10^{-4}
Supercritical fluid	10^{-1} to 1	10^{-4} to 10^{-3}	10^{-4} to 10^{-3}
Liquid	1	$< 10^{-5}$	10^{-2}

Source: Adapted from [2].

supercritical liquid makes it possible to dissolve a wide variety of compounds, which in turn makes it possible to perform chromatography not only for volatile compounds but also for generally nonvolatile and thermally labile compounds such as pharmaceuticals.

The low viscosity of supercritical fluids makes it possible to perform separations on packed columns at higher speeds and/or to use longer columns because of lower pressure buildup across the column. Also, the higher flow rate increases the throughput in unit time because the run time is shorter. Throughput is increased manyfold by faster turnaround cycles, meaning higher productivity, which is the inherent advantage of this technique. As a result, higher resolving power and loadability can be realized through an increased plate count compared with that of conventional chromatography.

The higher diffusivity of a solute in supercritical fluid is another significant advantage that SFC offers over conventional chromatography. According to the van Deemter equation, under normal conditions the factor contributing most to minimal plate height (or peak broadening in chromatographic terms) in the column is the mass transfer rate between the mobile and stationary phases. This variable is known as the *C-term*. The van Deemter equation is written

$$H = A + \frac{B}{\mu} + C\mu,$$

where A is the coefficient for eddy diffusion, B the coefficient for longitudinal diffusion, C the coefficient for mass transfer kinetics of the analyte between the mobile and stationary phases, and μ the linear velocity of the mobile phase. In chromatography, the A-term is equal to the multiple paths taken by the chemical compound. The multiple paths occur in packed columns where there are several routes through the column packing, which results in band spreading. The term B/μ is equal to the longitudinal diffusion of the particles of the compound. The term $C\mu$ is equal to the equilibration point between phases. In a column, molecules interact between the mobile and stationary phases, and they are at constant equilibrating states along the flow path; $C\mu$ accounts for this.

Figure 2a [4] depicts the contributions of plate height from all three terms along the flow rate (x-axis). The rising part from the right of the curve is the

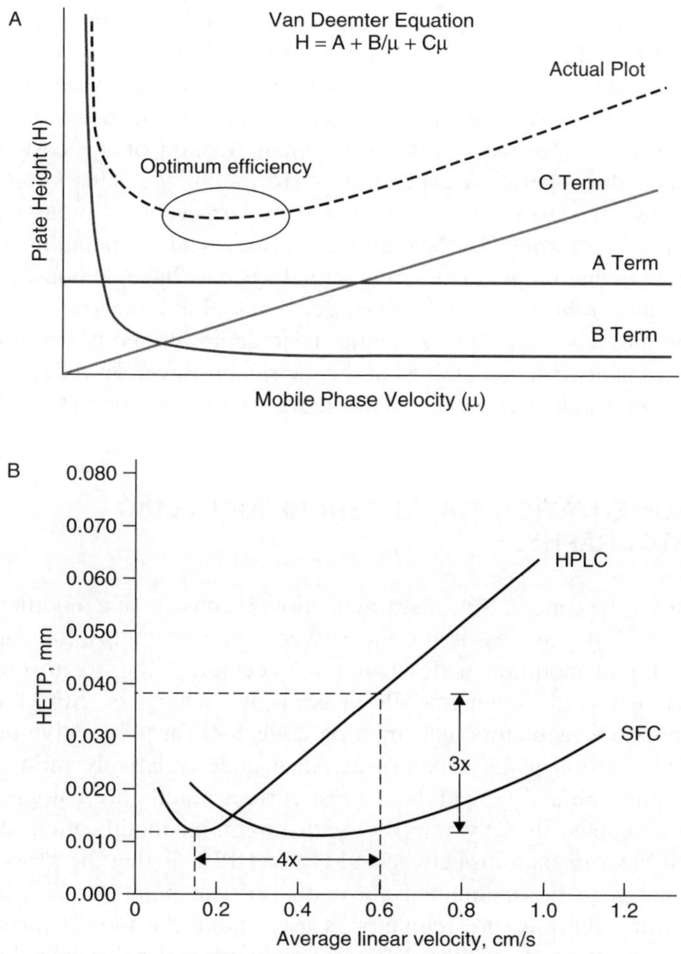

Figure 2 (A) Van Deemter curve; (B) HPLC vs. SFC curves.

contribution mostly from the C-term. Figure 2b [5] compares high-performance liquid chromatography (HPLC) with SFC on the effect of flow rate on plate height. As the flow rate increases, the HPLC curve rises much faster than the SFC curve. The higher mass transfer rate from higher solute diffusivity in supercritical fluid significantly reduces the C-term, making the SFC curve much flatter at the higher-flow region. What this implies in a practical sense is that in SFC, the separation can be run at a faster flow rate without compromising the efficiency; the peak is sharper, which means higher resolution, higher peak capacity, and higher loadability. Overall, in terms of chromatographic productivity, SFC is the technique preferred over traditional HPLC, as demonstrated by the van Deemter equation.

Today, CO_2-based SFC is the most widely used technology in industry because of its mild nature, the minimal requirements for forming CO_2 supercritical fluid (critical point $T = 31^\circ C$ and $P = 1070$ psi), and its green technology. CO_2 is inert and safe to use. By replacing the toxic organic solvents of the mobile phase with CO_2, the use of hydrocarbon solvents is reduced significantly; at the same time, there is a reduction of waste generation and waste disposal that is beneficial to the environment. In addition, CO_2 is inexpensive—industrial high-purity grade CO_2 costs only a very small fraction of the cost of organic solvents—and thus operation expenditures are minimized. This is an attractive feature for cost-stressed industries such as the pharmaceutical industry.

SFC can be performed successfully at various scales, including analytical, semipreparative, and preparative. Commercial systems have demonstrated excellent performance, robustness, and cost-effectiveness. For industrial purposes, SFC at simulated moving-bed (SMB) scale has been demonstrated in the lab as a prototype. The capacity can be obtained at the metric tons level. Excellent economies have been demonstrated compared with liquid-based SMB operations [6].

3 INSTRUMENTATION FOR SUPERCRITICAL FLUID CHROMATOGRAPHY

Figure 3 shows the general SFC instrumentation. It consists of a modifier reservoir and a CO_2 tank/dewar vessel for the solvent supply, pumps for supercritical CO_2 and a liquid modifier, a flowmeter for accurate CO_2 stream delivery, an autosampler, a column oven, mobile-phase heat exchangers, SFC- compatible detectors, pressure regulators, and fraction collectors for preparative purposes.

Since CO_2 can reach its supercritical state under relatively mild conditions in terms of pressure and temperature, most current liquid chromatography hardware can be adapted directly for SFC with minimum modification. The major difference in instrumentation between SFC and HPLC is that the pressure of the SFC system has to be maintained above the critical point of the supercritical fluid. Therefore, the pressure regulator is used along the mobile-phase streamline. For development of capillary SFC, the regulator is usually placed in front of the capillary column since the capillary itself can work as an effective pressure

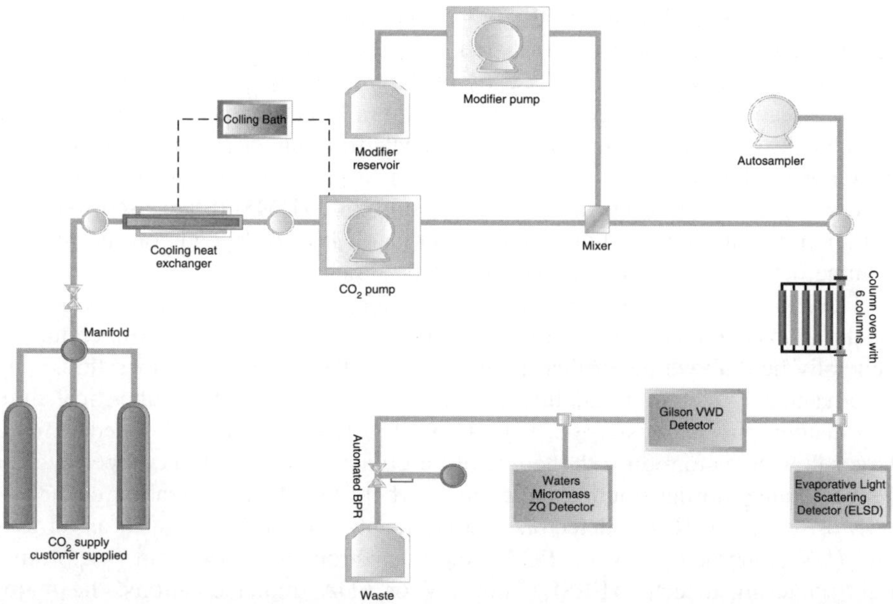

Figure 3 Instrumentation of SFC. (*See insert for color representation of the figure*).

restrictor. For most instruments, which are packed-column SFC systems with general HPLC format columns, the pressure regulator is put downstream (after the column) to ensure stable pressure across the column.

The supply of CO_2 is usually achieved by pumping high-pressure liquid CO_2 at room temperature, by either siphoning CO_2 from the bottom of the cylinder or by pressurizing CO_2 vapor into liquid form. The CO_2 pump head is usually chilled to prevent liquid CO_2 vaporization due to frictional heat generated from movement of the pistons.

There are two types of sample injection modes in SFC: direct modifier-stream injection, in which case the normal HPLC injectors can be adapted, or injection onto a mixed CO_2–modifier stream, in which case the HPLC injectors need to be modified with a pressure-release mechanism after each injection. This is particularly true for preparative-scale use because of the large amount of pressurized gas observed after injections.

Each injection mode has advantages and disadvantages. In the modifier–stream injection mode, the sample is kept in the liquid phase until it hits the column, thus avoiding the pressure-release issue from the injection valve and system pressure fluctuations; however, it may incur peak broadening and possible at-column precipitation, since the modifier line flows only at a percentage of the total flow rate. The immediate shock brought about by mixing with the lower solvent strength of CO_2 could push the compound out of the liquid stream at the end of the solvent line, perhaps causing line-clogging and an immediate modifier–stream pressure increase and bringing about a system shutdown by the safety function design. On the other hand, the mixed-stream injection mode relieves most of the sample crash risk and the sudden pressure hike (and the triggered system shutdown) caused by pushing a sample plug all at once with a mixed stream onto the larger area of the column front end. But stronger solvents used in the sample solution may cause peak fronting in some cases and also decrease loading capacity. This phenomenon occurs only when much stronger solvents, such as chloroform or dimethyl sulfoxide (DMSO), are used, but it is not so apparent with alcohols such as methanol. It also depends on the relative volume of diluent injected vs. the overall flow rate.

The heat exchanger and column oven (and/or column switchers) are usually easily adapted directly from HPLC configurations. The column temperature is generally held above the critical point (31°C), although lower temperatures may be advantageous in some applications; these are categorized as subcritical fluid chromatography in most cases [7]. An in-line heat exchanger is needed for a larger flow so as to reduce the temperature-gradient effect and to compensate for heat exchange during fraction collections, where fast depressurization occurs.

Most common HPLC detectors can be adapted for SFC use, such as ultraviolet (UV), photodiode array (PDA), mass spectrometry (MS), and evaporative light-scattering detector (ELSD). For UV or PDA implementations, the major modification necessary for safety reasons is a higher pressure-rating flow cell. Sensitivity and linearity range similar to HPLC applications have been demonstrated with SFC. The same is true for optically based chiral detectors such as a circular dichroism (CD) detector. Good sensitivity for enantiomers has been obtained with SFC [8–10] MS detectors have gained popularity in HPLC for their ultimate identification capability and high sensitivity; the same features are desirable in SFC as well. More and more uses of the SFC/MS configuration are being developed in analytical applications, and a mass-directed preparative SFC system is presently the focus of major development and shows great promise. Other types of detectors from liquid chromatography systems, such as corona-charged aerosol (CAD), are used with SFC and show good results [11,12].

4 ENANTIOMERIC SEPARATION IN SFC

4.1 Chirality and Chromatographic Separation Mechanisms

Compounds that have identical molecular formulas and the same sequence of atom connections but differ in spatial arrangement are called *stereoisomers* (Figure 4) [13]. Stereoisomers have similar physical and chemical properties, so they cannot easily be separated by regular chromatographic phases such as C_{18}-bonded or pure silica. They need specially designed chiral phases to achieve the desired separation.

In chiral chromatography, chirally pure functional groups are applied or bonded to surfaces of inert matrix material such as silica gel particles. The chiral phase forms *transient* diastereomers with the sample chiral molecules injected onto the column, based on various interactions, such as hydrogen bonding,

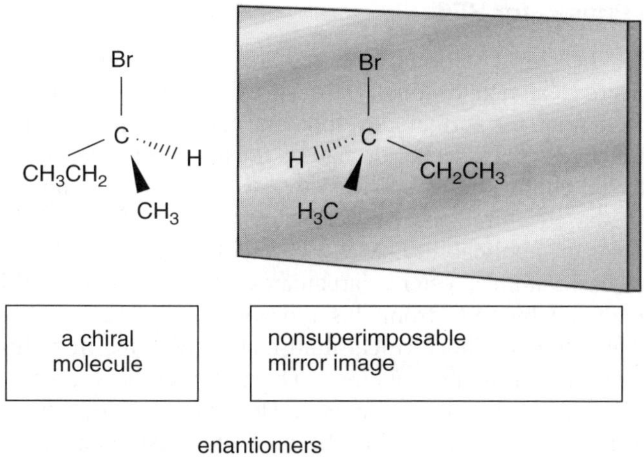

Figure 4 Stereoisomers/enantiomers.

dipole–dipole, $\pi-\pi$ coupling, or spatial inclusion. Diastereomers with different chiral isomer solutes in the compound mixtures have distinctive properties in chromatographic terms and will elute at different times; thus, chiral separation is achieved.

4.2 Chiral Drugs

Each of the stereoisomers in drugs that possess chiral structures can exhibit different biological responses, pharmacokinetic properties, and toxicologic effects. Since 1992 [14], the U.S. Food and Drug Administration (FDA) has stipulated that for chiral drugs, the enantiomers need to be fully identified with respect to pharmacologic, toxicologic, and clinical information. This is possible because modern chromatographic techniques are capable of differentiating and separating racemates into a pure form of the enantiomer. Since the 1990s, chiral chromatography applications such as chiral HPLC have seen tremendously increased use in the pharmaceutical industry.

In recent years, with the maturity of its instrumentation and from its green aspect and economical benefits, SFC has seen increased interest and applications in the pharmaceutical industry as an alternative or complementary tool for HPLC, especially for the separation of enantiomers. Today, more and more pharmaceutical labs implement this technology. In some cases it is used as a replacement or method of choice for chiral separations. Growing interest and accompanying investments by industry have also naturally stimulated faster progress of the technology. With improvements in method development for efficiency and for hardware refinement for system robustness, more explorations are under way, and SFC is being used more frequently in various chemical, biological, and clinical fields.

4.3 Chiral Phases for SFC

In chiral SFC, the chiral stationary phases are adopted directly from the same phases that were designed originally for HPLC applications. This is one area that has the potential for further development and growth. For the majority of chiral SFC practices, polysaccharide-based, brush-type Pirkle, and macrocyclic antibiotics–based chiral phases are used overwhelmingly for most applications. Polysaccharide-based chiral phases are the most successful in SFC because of their wide-spectrum applicability. Many types of drug molecules, such as carboxylic acids, basic amines, esters, carbamates, alcohols, diols, and ketones, are frequently separated by SFC from this type of phase. The excellent resolving power of the phases with supercritical fluid makes them the first choice for analytical purposes in almost all situations. These phases have also demonstrated good loadability in preparative processes. However, one disadvantage of this type of phase is solvent compatibility, because the most versatile phases in this category are all coated and not bonded to the silica particle. A limited number of solvents, such as an alcohol–hexane combination, can be used as the mobile phase without risking phase wash-off. This means that they are not strong in mechanical terms and have limited usefulness for many drug intermediates, for which compounds often have little or no solubility in alcohols. The lifetime of the column is short for the same reason. In recent years, immobilized phases with similar chiral function groups have been developed to extend applicability with more solvents, but the overall efficiency has not been demonstrated fully [15].

Brush-type Pirkle phases comprise another excellent phase for chiral SFC [16]. The phases in this category include rational-designed chiral functional groups derivatized and immobilized directly onto the silica surface. They demonstrate excellent physical stability against many extreme conditions, such as the use of acids and bases, as well as exotic solvents that otherwise cannot be used with polysaccharide phases. The Pirkle phases are logically designed to target certain chiral functional interactions in the drug molecules, with a proven mechanism in place. Their applicability is as wide as polysaccharide phases. However, since the functional group is bonded directly to the silica surface, the strong residual silanol activity from the silica particle is not eliminated or shielded effectively. With SFC, the pH of a supercritical CO_2–modifier combination is not 100% neutral, as it is closer to neutral conditions in normal-phase LC with alcohol–hexane systems. This inherent acidity will sometimes make silanol activity interfere with the chiral mechanism in SFC [17], so the overall success rate is somewhat decreased. At preparative scales, the phases show excellent loadability and robustness with many solvents because the phases are directly bonded chemically to silica particles.

Another widely used phase is the macrocyclic antibiotics–based phase: for example, vancomycin (V), tecoplanin (T), or ristocetin (R). These phases are very versatile, with multiple active chiral interaction sites from within their macromolecular structures. Successful applications have been reported routinely [18]. Many neutral and acidic compounds are well separated in SFC; for basic amines, certain amine additives in the modifier are necessary to prevent peak tailing and to ensure satisfactory separation. Preparative-scale purification is also reported. The overall larger size of the macromolecular structure on the phase makes the phase ratio relatively low; this limited column load capacity and throughput are restricted to some extent.

The use of other types of phases in SFC have also been reported [19]. Examples include networked polymer phases Kromasil CHI-TBB and CHI-DMB. Continuous developments are ongoing for product improvement, and new phases are introduced continually, targeting specific applications.

5 ANALYTICAL DEVELOPMENT OF CHIRAL SEPARATIONS

5.1 Chiral Screening and Method Optimization

In the modern practice of chiral LC analytical development, a racemate sample is usually injected onto a series of different chiral phases, combined with different solvents to pursue an applicable resolution. This process, called *chiral screening*, is widely used as the first priority for several methodologies. It is recognized as the quickest and most effective way for chiral method development because of specificity limitations of chiral phases and the complexity of chiral molecules. The same screening procedure is still the standard approach in analytical SFC for chiral analysis.

Table 2 shows major types of phases discussed previously. All column sizes are 4.6 × 250 mm, 5-μm particle size, and a flow rate of 4 mL/min. The gradient is set with modifier content from 5 to 40% in 5 min.

Figure 5 shows the resolution of a chiral standard *trans*-stilbene oxide (TSO) using this screening procedure, with methanol as the modifier. What is now fundamentally different from chiral LC is the higher efficiency of SFC. The usual long, slow process in LC, which may require many hours or even days for compound screening, takes only minutes or a few hours in SFC, with more chiral phases included in the screening procedure. This is possible because gradient elution has been adapted in chiral SFC as opposed to isocratic elution schemes used in LC screening. The low efficiency of chiral LC makes it impractical to run gradient screening in most cases, especially in the normal-phase mode. Isocratic

TABLE 2 Chiral Gradient Screening

Cosolvent/Modifier	Chiral Columns
Methanol	Chiralpak AD
Ethanol	Chiralpak AS
Isopropanol	Chiralcel OD
Acetonitrile	Chiralcel OJ
Methanol + 0.1% trifluoroacetic acid (optional)	Whelk-O
Methanol + 0.2% triethylamine (optional)	Chirobiotic R + V + T

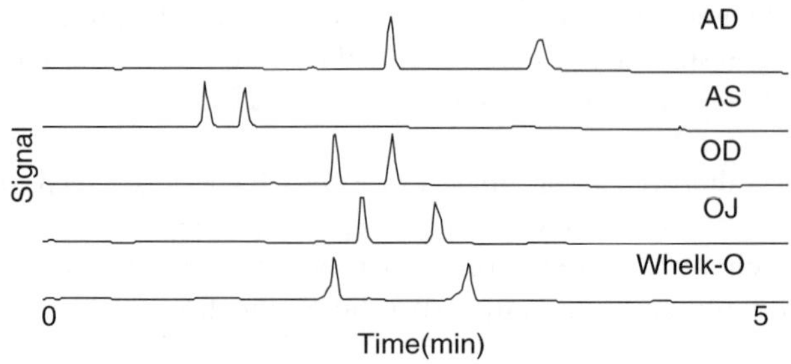

Figure 5 Chiral screening of *trans*-stilbene oxide (TSO).

elution is the only effective approach that can give affirmative and reproducible results in chiral LC.

In SFC, with a much faster process, the gradient approach is just as effective for chiral resolution and it is reproducible from practical implementations. The gradient shortens the overall run time, and more phases can be included in the procedure without prolonged duration. Each injection now takes only 3 to 5 min, and complete screening may take as little as 30 min. The entire process can be fully automated. The overnight run that is usually required when using LC for just a few samples is no longer necessary. In SFC, many more samples (10 or more times as many) can be screened with a high success rate in the same amount of time.

6 PREPARATIVE PROCESS/PURIFICATIONS OF CHIRAL COMPOUNDS BY SFC

6.1 Purification Process

SFC offers some significant advantages for preparative processes.

1. Difficult separations in LC become much easier using SFC, and the higher efficiency can improve the overall success rate for separation.
2. CO_2 is inert and can help relieve safety concerns because of the elimination of the large amount of toxic solvents used in preparative LC.
3. The faster process of SFC can be realized by higher throughput. The average improvement in terms of capacity of SFC is at least five or more times higher than that of LC.
4. The most profound effects that SFC brings about for preparative processes are economical and environmental improvements. CO_2 is very cheap compared to the expensive liquid solvents used in LC. The operational cost of preparative SFC is only a few percent (ca. 3 to 5%) of that used in running LC. SFC also eliminates significant waste disposal costs compared to the toxic organic solvents of LC, and it poses no threat to the environment.

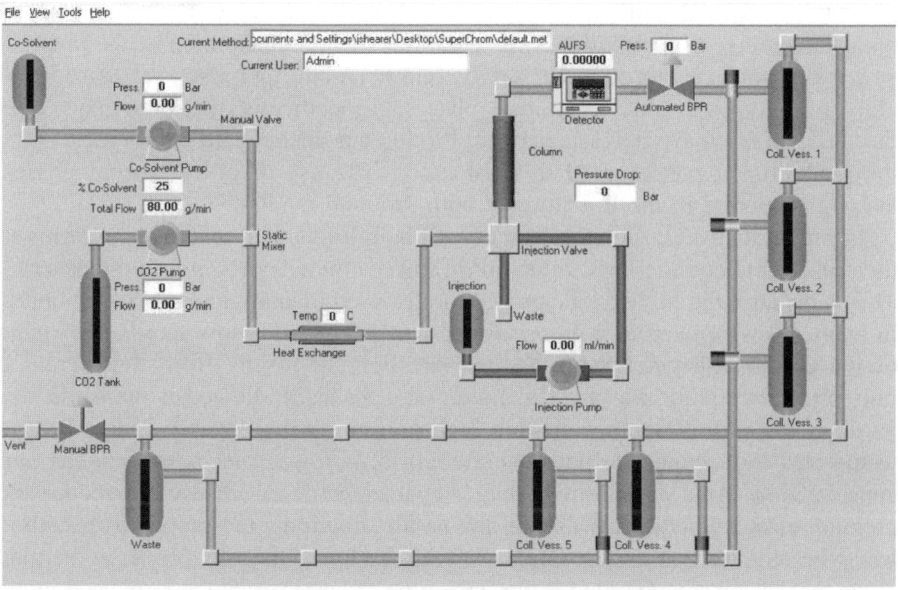

Figure 6 Preparative SFC instrumentation.

The instrumentation of preparative SFC is, in general, similar to that of LC because of the gentle nature of supercritical CO_2. Most hardware is adapted directly from preparative LC instrumentation with minimal modifications. Figure 6 shows a typical setup. The difference in the setup is the addition of a pressure-release mechanism before the flow stream goes into the fraction-collection module. The added benefit of this technology is that the greater part of the mobile phase is supercritical CO_2. The setup vents itself into the atmosphere when it is released from the column, leaving only a small fraction of liquid modifier with the higher concentration fractions in the collection bottles, making its solvent removal much easier by rotovaping afterward and saving energy.

6.2 Scale-up Study

Preparative chiral chromatographic processes in pharmaceutical labs are for the purpose of purification of enantiomers for toxicologic and potency studies or for further chemical developments. The compound amounts range from a few milligrams to grams, hundreds of grams, or even kilograms. Purification at the metric tons level falls into a bulk manufacturing category that utilizes very different instrumentation technologies, and is not within the scope of this chapter.

A routine preparative process is derived from the analytical method development with a scale-up adjustment for instrumentation factors, such as the flow rate, injection volume, or solvent use. In addition to these simple calculations, there are several things that need to be studied before a purification process can be performed successfully.

At the analytical scale, a variety of chiral phases are applied in an attempt to resolve isomers; from practical considerations of instrumentation at the preparative scale, however, throughput, safety, and economic purposes are considered. Only a few types of chiral stationary phases,—specifically, the three major ones discussed previously (polysaccharides, Pirkle, and macrocyclic antibiotics)—are the predominant phases used in most cases. The use of other type of phases, although reported in the literature, is both rare and random.

In preparative LC, the chiral phases, although identical in chemical terms to their analytical counterparts, often are of lower efficiency (bigger particles, etc.). This is because the higher-efficiency particles used in analytical processes cannot in many cases be used in a preparative process. First, the flow rate is the square of the column diameter ratio, so pressure buildup results from the increased rate, often exceeding the physical limits of the phases allowed if the same size particles are used. Running at such high pressure will probably cause damage to the phases. Second, the physical strength of hardware needs to be taken into consideration. As a compromise, larger coarser particles are used to reduce the pressure effect. This means that in almost all situations, the preparative method has to be redeveloped to be compatible with the low-efficiency phase. Some may argue that this can be avoided if the same type of larger particles were used in the analytical process. However, this approach may not always work as anticipated. Some other factors also contribute to the process.

Column packing, for example, is a factor that cannot be neglected in practice. It is harder to obtain uniform packing in big columns and easier to achieve in analytical-size columns; this creates nonuniformity across the column in all dimensions. In a slower process in LC, the variations across the column with this nonuniformity may cause significant peak broadening and degrade the separation. For throughput purposes, in preparative LC the sample is always overloaded, to speed up the process. The interactions between solutes and the stationary phase at a higher concentration level and at greater dimensions are not the same as from the analytical scale; as a result, the separation mechanism may change.

These considerations make it necessary to perform a full scale-up study for the preparative process in LC. In other words, the process from analytical to preparative LC is not straightforward and is less efficient since the highly automated feature in the analytical process is now interrupted by more human intervention. This is part of the reason that the preparative chiral LC process is still laborious, especially at the laboratory scale, for processing tens of grams or less, because the amount of time used for extra development is a significant portion of the overall process. By contrast, the scale-up study in preparative SFC becomes much more simple and straightforward because of the advantages offered by SFC.

The lower viscosity of supercritical fluid makes it practical to run the same separation at a fivefold higher flow rate without the pressure-buildup issue. This means that the same type of high-efficiency particles can be used in both analytical and preparative processes. At a faster flow rate, the molecular diffusion of solute and mass transfer between the phases is much faster, meaning that the solute will have a more efficient resolution and less chance to diffuse in other dimensions, as is the case in LC. At this flow rate, the peak shape and retention time are more uniform in SFC.

The faster process makes the separation cycle time significantly shorter, and it is now practical to run purification by "stacking" small injections in short time frames without compromising the throughput. The utilization rate of column material is much higher. The usual overloading mode with a long run time in LC is not necessarily advantageous, and the interactions between solutes on the column at lower concentrations are more linear and predictable.

In SFC, the processes are generally more uniform and are directly scalable; there are fewer, more limited, variations. All of these factors contribute to a more straightforward process with higher accuracy in a scale-up process. In most cases at the laboratory scale, it is no longer required that a scale-up study be performed. Most purifications can go forward directly from the general analytical screening results, thus shortening the overall turnaround time.

This is a revolutionary change from the perspective of process efficiency and automation. With sophisticated software development and proper hardware robustness improvements, preparative chiral SFC can be run in a highly automated fashion. In preparative SFC, the entire process can be fully automated, just as it is with fully automated reversed-phase LC purification development today. Realization of this automation can initiate all sorts of possibilities, such as open-access/autopurify for medicinal chemists to improve productivity and overall process optimization. It is estimated that within just a few years, the possible bottleneck at the purification step in a drug discovery process can be greatly improved or eliminated altogether with the implementation and development of SFC technology at full scale.

7 APPLICATIONS OF SFC IN THE PHARMACEUTICAL INDUSTRY

7.1 Developments of Chiral SFC for Drug Analysis

In 1986, Mourier et al. [20] conducted a study of the influence of various parameters on solute retention in SFC with CO_2, using columns packed with small particles for both chiral and achiral separations. The influence of the physical state of CO_2 on solute retention on bare silica has been studied thoroughly. The influence of a polar modifier addition to CO_2 has been investigated on three stationary phases—bare silica, Pirkle, and octadecyl-bonded silica—for enantiomeric separation. It was found that the retention mechanism in SFC with CO_2 is very similar to that observed in HPLC with an apolar solvent such as hexane. For enantiomeric separation on a chiral-bonded stationary phase, some phosphine oxides have been resolved, with retention times below 90 s.

In the1990s, the early adopters of packed-column SFC for chiral separations reported success with applications of three major types of chiral phases: brush-type, polysaccharide, and cyclodextrin-based CSPs in SFC. Wide ranges of pharmaceutical compounds were used to probe the advantages and limitations.

They demonstrated that faster column equilibrium and optimization were accomplished using SFC; better resolution was generally achieved compared to LC, along with a higher success rate [21–23].

Petersson and Markides [24] reported analytical development on several types of chiral phases including polysaccharides, brush-type Pirkle phases, cyclodextrins, and some other networked polymer phases. The packed columns usually were 5- to 10-μm particles with a 4.6-mm column ID and 150- to 250-mm length; the flow rate was 2 to 4 mL/min. The modifiers used were alcohols such as methanol, ethanol, and isopropanol, in decreased order of polarity. Wide ranges of compounds have been separated on these phases, including aromatic amides, esters, amino alcohols, aromatic oxides, and carboxyl acids.

Liu et al. [18] investigated separations of pharmaceutical compounds on macrocyclic antibiotics–based phases such as chirobiotic R, T, and TAG. The types of compounds investigated include heterocyclic compounds, acids, β-blockers, and amino acids. The more polar TAG phase seems to have the highest hit rate among all phases tested. Acidic and basic additives are also helpful in the majority of these separations.

A porous graphitic carbon stationary phase was used in chiral SFC [25] with a combination of dimethylated-β-cyclodextrin mixtures (MeCD) as chiral selector additives in a CO_2-polar modifier mobile phase. Several common drugs, such as tofizopam, warfarin, benzoxazine derivative, lorazepam, flurbiprofen, temazepam, chlorthalidone, and methylphenyl hydantoin were separated using this methodology. The quantity of MeCD adsorbed onto the porous graphitic carbon (Hypercarb) was measured for various chiral selector concentrations, using the breakthrough method with an ELSD. The effects of MeCD concentration in the mobile phase, the nature of the polar modifier, the outlet pressure, the column temperature, the nature of the commercial MeCD mixture on the retention, and the enantioselectivities were studied. For a given solute, the enantioselectivity is greatly dependent on the commercial MeCD mixture used. It was determined that the dominant mechanism for chiral discrimination is the diastereoisomeric complexation in the mobile phase.

New chiral phases have been continually developed for applications. For example, recently the synthesis of mono-6-(3-methylimidazolium)–6-deoxyperphenylcarbamoyl-β-cyclodextrin chloride as a chiral stationary phase for SFC has been reported [26]. This phase is coated onto silica gel, and good resolution of *p*-fluorophenylethanol, *p*-chlorophenylethanol, *p*-bromophenylethanol, *p*-iodophenylethanol, and *p*-fluorophenyl-3-buten-1-ol have been obtained under SFC separation conditions.

7.2 Chiral Screening Strategies for Practical SFC in the Lab

In many pharmaceutical labs today, SFC is chosen as the technique of choice for chiral chromatography because of its overall high success rate. In view of actual pharmaceutical developmental process environments, integration of chromatography-based analytical processes in drug research pipelines will undoubtedly expedite the project's overall pace and bring about productivity improvement. An efficient chiral screening strategy is critical to this type of implementation, not only in early-stage discovery but also in late-stage process development. A high degree of automation, wide applicability, high performance in terms of success rate and throughput, and system robustness of such a strategy need to be demonstrated throughout the screening, as well as the method optimization processes.

Francotte [27] has summarized the methodologies for chiral chromatography in the pharmaceutical industry and has concluded that over 90% of separations can be achieved with a set of only four columns (AD, OD, AS, OJ), and that the remainder of the cases can be resolved with a few other phases. This proposed column matrix is adopted by many pharmaceutical labs around the world and has significantly improved the efficiency of chiral chromatography in terms of overall workflow and sample turnaround times.

Beginning in the late 1990s, such fast, complete automated screening protocols by SFC have emerged from pharmaceutical labs to expedite chiral separation. In one report [28], modification of a commercial instrument to incorporate a six-way column selection valve, multiple chiral columns based on derivatized cellulose or amylose, and a four-way modifier selection valve provides a powerful combination for the rapid development of chiral separations. A wide set of columns and conditions can be tested sequentially, including unattended operation overnight. It shows that similar racemic compounds, even those from the same molecular class, are separated using different column and modifier combinations. Therefore, the use of program-controlled column and modifier selectors has great advantages. Using the fully automated system, the optimal chiral separation of several compounds can be obtained unattended within 24 hours.

In drug discovery, high-throughput and performance analyses, along with purification of enantiomers, are important components and provide high-quality compounds for pharmacological testing. To cope with increasing numbers of new compounds generated by chemistry programs, the development and implementation of a faster screen, using supercritical fluid chromatography (SFC) to maximize the chance of achieving rapid enantiomer resolution of a large number of compounds in a minimum of time, has been reported [29]. The SFC screen utilizes a narrow combination of only four columns (Chirlapak AD and AS and Chiralcel OD and OJ) and two solvent modifiers (methanol and isopropanol). A modifier and column-switching setup was employed to allow the entire screening process to be run serially in the order AD > OD > OJ > AS and methanol > isopropanol, so that the screening for a given molecule could be stopped when separation was achieved. The switching system was fully automated for unattended operation of multiple compounds. An optimization procedure was also defined, which can be performed if needed for unsuccessful separations in the screening step. The chiral SFC strategy proved its performance and robustness in the resolution of hundreds of proprietary chiral molecules generated by drug discovery programs, with a success rate exceeding 95%. In addition, the generic capability of the strategy was evaluated by applying the screen and optimization methodology to a test set comprising 40 marketed drugs differing from proprietary compounds in terms of chemical diversity, revealing a similar high success rate of 98%. Chiral separations developed at the analytical scale work easily, and they perform equally well at the semipreparative level. Also, the SFC screen allows resolution of compounds that were partially separated by HPLC or not separated at all by RPLC, demonstrating the utility of implementing complementary chromatographic techniques. The SFC screen is currently considered the first choice for chiral separations of new compounds because it offers a higher success rate, performance, and throughput.

In another report [30], a rapid chiral screening has been implemented using short columns (10-cm length), high flow rates, and fast gradients. A primary screen delivers a separation assessment using just one solvent modifier (methanol) and four columns (Chiralpak AD-H and AS-H and Chiralcel OD-H and OJ-H) run serially in a total of 24 min. A secondary screen then uses ethanol and isopropanol (IPA) modifiers across the same columns. The screens can be combined to run a sequence of samples overnight, where each racemate is analyzed over 80 min. The fast analytical screening and optimization process enables rapid purity assessment separation, method development, and identification of the purification method.

7.3 Development of Hyphenated Techniques

Development of hyphenated techniques has also been introduced to enhance productivity for practical purposes. (In reality, pure chiral compounds are generally impossible for chromatographers in pharmaceutical labs.) First, there is no optically pure national standard commercially available for reference purposes; second, the nature of pharmaceutical research determines that most chiral molecules will be new entities in a molecular sense, in that no pure chiral standards would exist. On the contrary, a chromatographer's sample may possibly contain many impurities, as well as residual solvents from its development processes. In most cases the portion of these types of impurities accounts significantly in terms of the overall composition. This brings about challenges to regular UV-based detectors, since they do not offer clear identification of molecules and the signals are not uniform. SFC-MS, with MS having powerful identification capability, has therefore been considered a useful tool to complement the real issues.

In one report [31], SFC-MS on four chiral stationary phases and eight different modifier concentrations (5 to 40% methanol–0.2% isopropylamine) was implemented. The screening was performed with an outlet pressure of 110 bar at 35°C, and a flow rate of 2.5 mL/min for the initial 20 min, then ramped up to 4 mL/min and held for 4.5 min to elute all the solutes from the column. The entire process is fully automated from injection to data processing and operates unattended for 15 h overnight to obtain optimal chiral separation for multiple compounds. The unique feature of using SFC-MS to monitor chiral synthesis with negligible interferences from achiral impurities has been demonstrated. In addition, with SFC-MS, enantiomeric excess can be determined with much lower detection limits than when using UV, and SFC-MS also provides a much shorter analysis time than that of normal-phase/reversed-phase LC.

A hyphenated SFC technique has also been developed for compound screening and structure identification [32]. SFC coupled to a hybrid mass spectrometer (Q-Tof2) equipped with an electrospray ion source has been used to separate and characterize a wide range of pharmaceutical racemates. A compound matrix with diverse molecular structures has been chosen to demonstrate the potential of such an experimental arrangement for high-throughput analyses. These drug molecules include alprenolol, homatropine, lercanidipine, nicardipine, disopyramide, sulpiride, propafenone, verapamil, tropicamide, pindolol, atenolol, flecainide, ofloxacin, atropine, salbutamol, clenbuterol, econazole, sulconazole, miconazole, and ketoconazole. Three different chiral stationary phases (Chiralpak AD and Chiralcel OD and OJ) were used. The use of mass spectrometry was found to be essential for the unambiguous assignment of the eluting components, particularly in the case of complex mixtures. Direct coupling of both systems without the need for a special interface resulted in similar peak shapes and peak widths in UV and total ion current (TIC) chromatograms (Figure 7).

SFC-UV-MS-ELSD instrumentation has been developed for discovery-stage library compound screening, purification, and quantification purposes [33]. The analytical SFC system equipped with a standard UV detector is coupled to the mass spectrometer for molecular identifications, and is coupled to ELSD for quantification of libraries before and after purifications. The system configuration has proved to be highly efficient, robust, easy to use, and has a good linearity response range.

7.4 Chiral SFC on Drug Compounds and Intermediates

A wide range of drug compounds has shown successful chiral separations with SFC. For applications published before the year 2000, several reviews can be of comprehensive use to readers [34–37]. Most of the citations given here discuss developments after 2000.

Racemic amino acid derivatives have been analyzed by SFC with rapid resolution results [38]. Racemic *N*–4-nitrobenzoyl amino acid isopropyl esters were separated on chiral valine–diamide phases with methanol as the modifier. Total separation time is less than 5 min, and the resolution achieved is between 1.25 and 10.8. When isopropanol replaces methanol as the modifier, the resolution can be improved further.

SFC has been used successfully to separate several stereoisomers (enantiomers and geometrical isomers) of furan derivatives [39], important intermediates for the synthesis of physiologically active natural products with tris(3,5-dimethylphenylcarbamate) of cellulose or amylose for the chiral stationary phases (Chiralcel OD and Chiralpak AD) respectively. The separation of thermally unstable furan derivatives was proved better with SFC.

Nonsteroidal anti-inflammatory drugs (NSAIDs) such as ibuprofen have been separated successfully by SFC, with excellent resolution and efficiency [40,41]. At an analytical scale, 11 different stationary phases have been tested; among them, the Kromasil CHI-TBB phase showed the best separation properties. It was

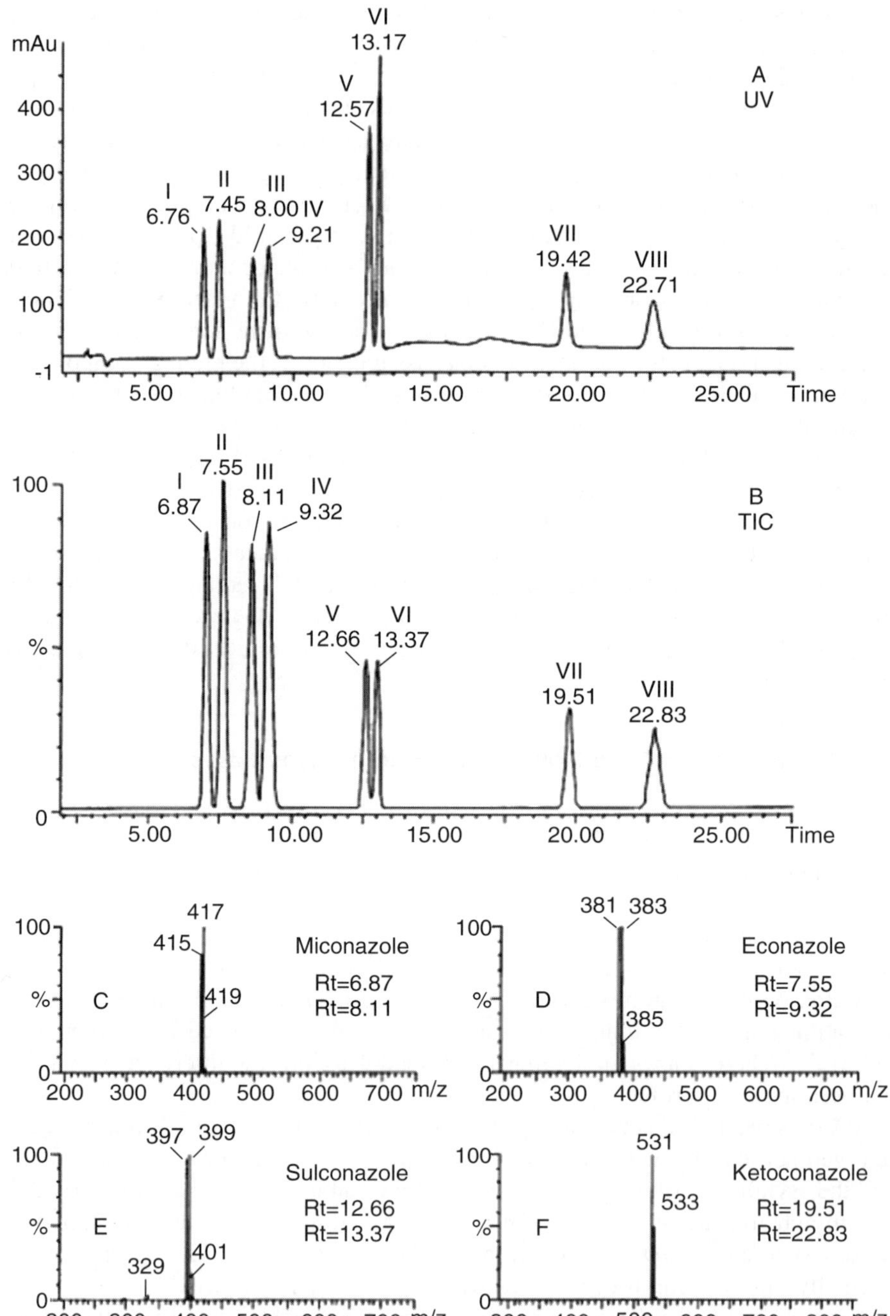

Figure 7 UV (A) and TIC (B) chromatograms of a solution containing four antimycotic drugs. (C–F) Mass spectra associated with the peaks present in the TIC chromatogram. (From [32]).

found that the separation behavior depends strongly on the type of modifier and the modifier content. Temperature and pressure are of less influence. Industrial-scale purifications such as SFC-SMB have also been developed based on this analytical method.

For the separation of naproxen enantiomers, an SFC method on the Kromasil CHI-TBB column has been reported [42]. The mobile phase was composed of supercritical CO_2 with 2-propanol as the modifier. The experimental conditions were a temperature of 293 to 323 K, pressure of 9.4 to 21.3 MPa, and a 2-propanol concentration of 6 to 15% (by mass), respectively. The enthalpic contribution to the overall enantiomer transfer energy was more important than the entropic contribution in the temperature range examined. The preferred operation conditions were 293 K, 9.4 MPa, and a concentration of 11% (by mass) 2-propanol in the mobile phase.

Triadimenol and triadimefon have been separated on a Chiralpak AD column using SFC (Figure 8) [43]. The effect of different modifiers (methanol, ethanol, and 2-propanol) was tested, with methanol and ethanol providing the best results for the enantiomeric separation of the two compounds. The enantioseparation of a mixture of triadimenol and triadimefon (six stereoisomers) was achieved in only 15 min using a gradient of ethanol, 200 bar, 35°C, and a flow rate of 2 mL/min. The separation of triadimenol diastereoisomers on different

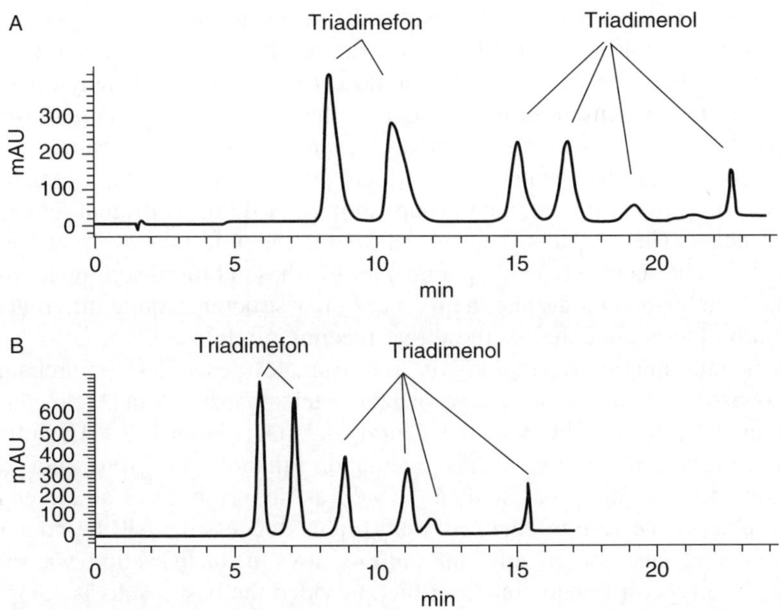

Figure 8 Simultaneous separation of triadimefon and triadimenol enantiomers. (A) Methanol programmed from 2% (1 min) to 5% at 0.2%/min, held for 3 min, and then to 25% at 20%/min. (B) Ethanol programmed from 5% (2 min) to 25% at 1.8%/min. (From [43].)

achiral columns (diol, silica, and ODS) was also investigated, in which case the type of organic modifier to be used depended on the stationary phase, the diol being the column that gave the best separation. Using this column, resolutions higher than 3 were obtained in analysis times of 5 min with any of the modifiers checked.

Brush-type Pirkle phases have often been used in SFC. The most successful is the Whelk-O 1 phase [44]. A broad range of druglike enantiomers have been resolved on Whelk-O phases under SFC conditions. The phases showed excellent resolution and high loadability; they exhibited robust stability in a wide range of temperatures, solvents, and additives. The now-obsolete PolyWhelk-O, which showed better enantioselectivity than the original Whelk-O, may be worth rediscovery in SFC.

Chiral stationary phases containing derivatives of *trans*−1, 2-diamino-cyclohexane (DACH-DNB and DACH-ACRYL) as selectors have been applied to SFC. Aryloxypropionic acid methyl esters, binaphthyl derivatives, sulfoxides, and β-receptor blockers have been separated with SFC, and the resolutions was superior to that obtained with HPLC for the same types of columns [45].

A number of chiral phenylpropanols and related compounds have been separated by using SFC with Chiralcel OD and Chiralcel OB cellulose-based columns [46]. A discrimination mechanism similar to chiral HPLC has been proposed. Lowering the temperature as well as proportions of modifier in the carbon dioxide eluent improved the resolution.

A series of potassium channel activator (KCA) analog-related compounds were separated by chiral SFC [47]. The temperature effect was examined at several temperatures between 0 and 52°C. Cellulose tris(3,5-dimethylphenylcarbamate) (Chiralcel-OD) was used as the stationary phase. The term *isoenantioselective temperature* (T_{iso}), at which separation of enantiomers is not possible, was proposed. Two of the test compounds, which differed only by replacement of a benzoyl group by an *n*-pentanoyl group, showed strikingly different temperature dependencies. These results indicate that one compound is above and the other below T_{iso}. The thermodynamic parameters for these chiral discriminations support the conclusion that despite their very similar structures, quite different chiral recognition factors operate for these two racemic mixtures.

The enantiomeric separation of six triazole pesticides (cyproconazole, propiconazole, diniconazole, hexaconazole, tebuconazole, and tetraconazole) is obtained by using SFC and the Chiralpak AD column [48]. The effect of different organic modifiers, such as methanol, ethanol, and 2-propanol, on the retention and resolution was studied, as well as the presence of additives in the mobile phase. The results obtained were highly successful: All the compounds were separated enantiomerically, and in most cases the analysis time was close to 10 min. The type of organic modifier that provided the best results is compound-dependent.

Four antifungal compounds—three imidazoles (miconazole, econazole, and sulconazole) and one triazole (itraconazole)—can be chirally separated by using SFC on the amylose-based chiral stationary phase Chiralpak AD [49]. The influence of pressure, type and percentage of organic modifier, and temperature on retention times and resolution was studied. Enantiomeric separation of the three imidazoles was achieved with resolutions higher than 2; and analysis times lower than 10 min, obtaining the best results using methanol as the modifier. However, the analysis time of the triazole was greater than 80 min because of the existence of a high number of functional groups that were able to interact with the chiral stationary phase. In this case, resolution of the four stereoisomers was achieved only partially with mixtures of ethanol and 2-propanol as modifier. The isoenantioselective temperatures were obtained from a study of the influence of the temperature; they were above the range of temperatures assayed, except for sulconazole, using 2-propanol.

Enantiomeric separation of cetirizine and oxfendazole on a Chiralpak AD column was achieved with SFC [50]. The enantioseparation of cetirizine was possible only when 2-propanol was used as a modifier, obtaining better results in the presence of the additives triethylamine (TEA) and trifluoroacetic acid (TFA). On the contrary, 2-propanol provided the lowest enantioresolutions for oxfendazole; in this case, the best results in terms of high resolution and short analysis time were obtained with ethanol. The study of the temperature effect revealed that in the case of cetirizine using 2-propanol, and oxfendazole using methanol, the separation was enthalpy driven, and the isoelution temperature was above the working range. Using ethanol or 2-propanol, the results showed that the oxfendazole enantioseparation was entropically driven and the isoelution temperatures were below the range studied.

Chiral separation of several dioxolane compounds has been achieved by two chiral columns (Chiralpak AD and Chiralcel OD) using SFC [51]. The compounds studied included ketoconazole and some of its precursors. The effects of the different modifiers and of the pressure on the chromatographic parameters were also evaluated. In general, the alcohol modifiers provided better results than did acetonitrile; and all the compounds could be separated with these two columns, but the selection of the column depends on the type of compound.

The separation of diltiazem (a calcium channel blocker) optical isomers is carried out using packed column SFC (Figure 9) [52,53]. A cellulose tris(4-chlorophenylcarbamate) coated on silica was used as a chiral stationary phase. The effect of temperature on the separation of the *cis*- and *trans*-enantiomers was studied in detail. The chiral separation of *cis*-enantiomers was improved with a decrease in temperature, whereas that of *trans*-enantiomers was improved with an increase in temperature. The isothermodynamic parameters revealed that the separation of *trans*-enantiomers was entropy-controlled in the range of temperatures examined, whereas enthalpy-controlled separation was observed for *cis*-enantiomers. The separations of the related compounds, 3-hydroxy isomers, were compared with those of diltiazem isomers. In addition, the differences in separation mechanisms between *cis*- and *trans*-enantiomers of diltiazem optical isomers were demonstrated by means of three-dimensional structures.

Chiral separation of some proline derivatives was achieved by SFC [54] using a Chiralpak AD column; the mobile phase was CO_2/ethanol. With SFC,

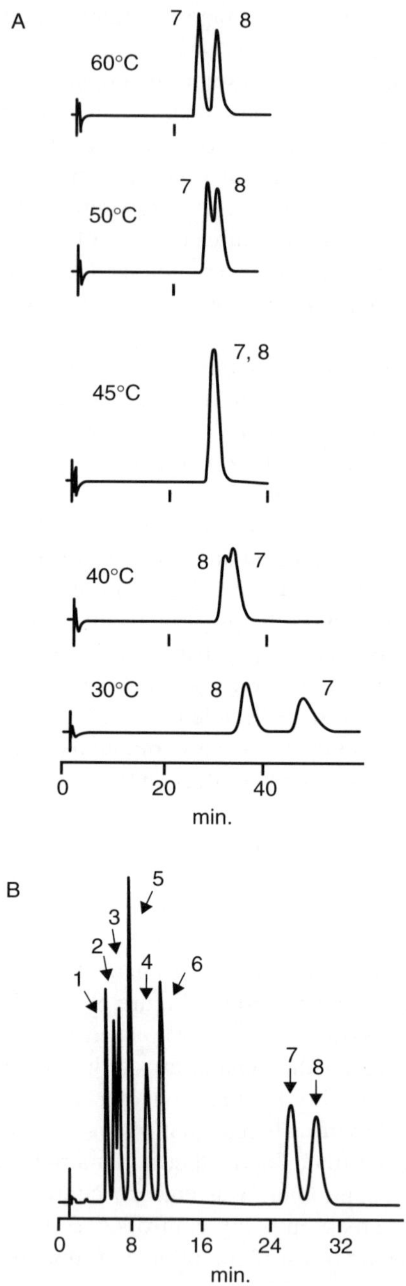

Figure 9 Chromatograms of diltiazem and 3-hydroxy isomers by packed column SFC. Conditions: column, Chiralcel OF; mobile phase, CO 22.5% (v/v) 2-propanol containing 0.1% (v/v) diethylamine; flow rate of CO, 2 mL/min; outlet pressure, 180 kg/cm; temperature, (A) 30, 40, 45, 50, and 60°C, (B) 60°C; detection, 254 nm. Peaks: 1, (2*S*, 3*R*)-dialtiazem; 2, (2*R*, 3*S*)-diltiazem; 3, (2*R*, 3*R*)-diltiazem; 4, (2*S*, 3*S*)-diltiazem; 5, (2*R*, 3*R*)-3-OH; 6, (2*S*, 3*S*)-3-OH; 7, (2*S*, 3*R*)-3-OH; 8, (2*R*, 3*S*)-3-OH. (From [52].)

the contribution from hydrogen bonding is most likely a dominant interaction for retention rather than chiral recognition.

An SFC method has been developed for amphetamine and methamphetamine enantiomers [55]. Addition of cyclic amines to the mobile phase showed the differences in separation with different modifiers.

Chiral separation of several antiulcer drugs, such as omeprazole, lansoprazole, rabeprazole, and pantoprazole is accomplished by using SFC on a Chiralpak AD column [56]. The results show that SFC allowed the enantiomeric separation of all the compounds studied with higher resolutions and lower analysis times than those for normal-phase HPLC, with a higher success rate.

In addition to regular chiral purity assessment and method screening, chiral SFC has been used in the assessment of pharmacokinetics of drug molecules and drug interactions. Chiral SFC has been incorporated into an automated screening method monitoring enzymatic transformations [57]. The SFC method developed has a run time of only 1.5 min. The method accelerates the enzyme–Amano hydrolases selection process for screening biocatalysts, where a large number of enzymes are evaluated for activity and enantioselectivity. Kinetic resolution of secondary alcohols by enzymatic transesterification was used as a prototype for method development. The rapid automated method can be used effectively for screening enzymes and optimizing reaction conditions in biocatalysis.

SFC has been used to study the interconversion of some 3-hydroxy-1,4-benzodiazepine enantiomers such as lorazepam, oxazepam, and temazepam [58]. Benzodiazepines have become the most widely prescribed drugs for the treatment of anxiety, insomnia, and related disorders. In this study, these drugs were separated by SFC on a chiral (*R*, *R*)-Whelk-O 1 column with supercritical carbon dioxide containing 12.5% methanol and 0.5% diethylamine as a mobile phase. The separation data, such as peak height, have been used to model the interconversion energy barrier of these drugs, and a kinetic relation equation has been proposed.

In one report [59], SFC is compared with normal-phase HPLC and steady-state recycling (SSR) for the enantiomeric resolution of three pharmaceutical intermediates at various sample scales. The separation conditions, the sample purities and yields, the productivities, and the solvent consumptions were discussed in three case studies. In case I, the SSR process was used for a low-selectivity resolution of 2.0 kg of a pharmaceutical intermediate. By using this separation process, a productivity of 750 g of racemate per kilogram stationary phase per day was achieved, while solvent usage was minimized (200 L/kg racemate). Case II pertained to the effectiveness of the SSR process. Productivity using SSR techniques increased by a factor of 4.5, while solvent usage decreased by a factor of 4.1 compared to the productivity and solvent usage of batch HPLC. Case III compared SFC purification to HPLC purification. The SFC process was more effective in terms of an increase in productivity and a reduction in solvent usage. Based on these results, it appears that SSR and SFC are very useful choices at the early stage of drug development for high throughput and rapid turnaround of samples.

The semipreparative purification of omeprazole drugs was carried out on a 250 × 10 mm column after successful analytical development [60]. The effect of two organic modifiers (ethanol and 2-propanol) was studied, and different injection volumes and concentrations of the omeprazole racemic mixture were evaluated to obtain high enantiomeric purities and production rates. Better results were achieved using concentration overloading rather than volume overloading. The recoveries decreased when the requirements of enantiomeric purity or the load increased, and it was possible to recover 100% of both enantiomers at an enantiomeric purity level higher than 99.9% under some loading conditions.

The semipreparative purification of albendazole sulfoxide enantiomers was also carried out in the similar fashion on SFC with a 250 × 10 mm Chiralpak AD column [61]. The compound has a low solubility (3 g/L in methanol). The throughput with maximum load yields a productivity of 37 mg/h of the first eluted enantiomer, with a purity of 99.9%, and 36.5 mg/h of the second eluted enantiomer, with a purity of 95%.

The optimization of large-scale SFC purifications has been studied with a modeling compound at analytical SFC scale [62]. Racemic 1-phenyl-1-propanol has been used as such a modeling compound, and separation was achieved on a Chiralcel OD column. The study was performed at 30°C, using supercritical CO_2 modified with methanol as the mobile phase. The parameters of the binary Langmuir adsorption isotherm were determined by the inverse method, comparing experimental and simulated peak responses. Isotherm parameters were derived for modifier concentrations between 1 and 5% (w/w) and operating pressures between 125 and 185 bar; the dependence of the isotherm parameters (i.e., the Henry constant and the saturation capacity) on operating conditions was investigated.

Chiral separation and purification of Nutlin-3, a small-molecule antagonist of mouse double minute 2 (*MDM2*) gene, using an SFC technique [63] has been reported. One enantiomer from the racemate is 100 to 200 times more active than the other. The chiral SFC method based on a Chiralcel OD column showed superior separation in terms of selectivity and efficiency. Optimization of the chiral separation method enabled high-throughput preparative-scale purification. Ultimately, 5 g of racemic mixture was purified on preparative SFC in 75 min, with a recovery rate above 92%.

SFC technique, not only at the analytical scale with preparative applications, has been evaluated with other chromatographic techniques in terms of efficiency, productivity, solvent usage, and so on. SFC has proven its advantages from all aspects, especially in production, cost reduction, and environmental friendliness.

8 PRACTICAL ISSUES FOR PROCESS OPTIMIZATION

8.1 Peak Tailing in SFC for Basic Compounds

Many chiral amines show severe peak tailing in SFC, which hampers resolution and efficiency. The carbobenzyloxy (cbz) protecting group is known for its potential to enhance the resolution of chiral amine enantiomers [64]. A series

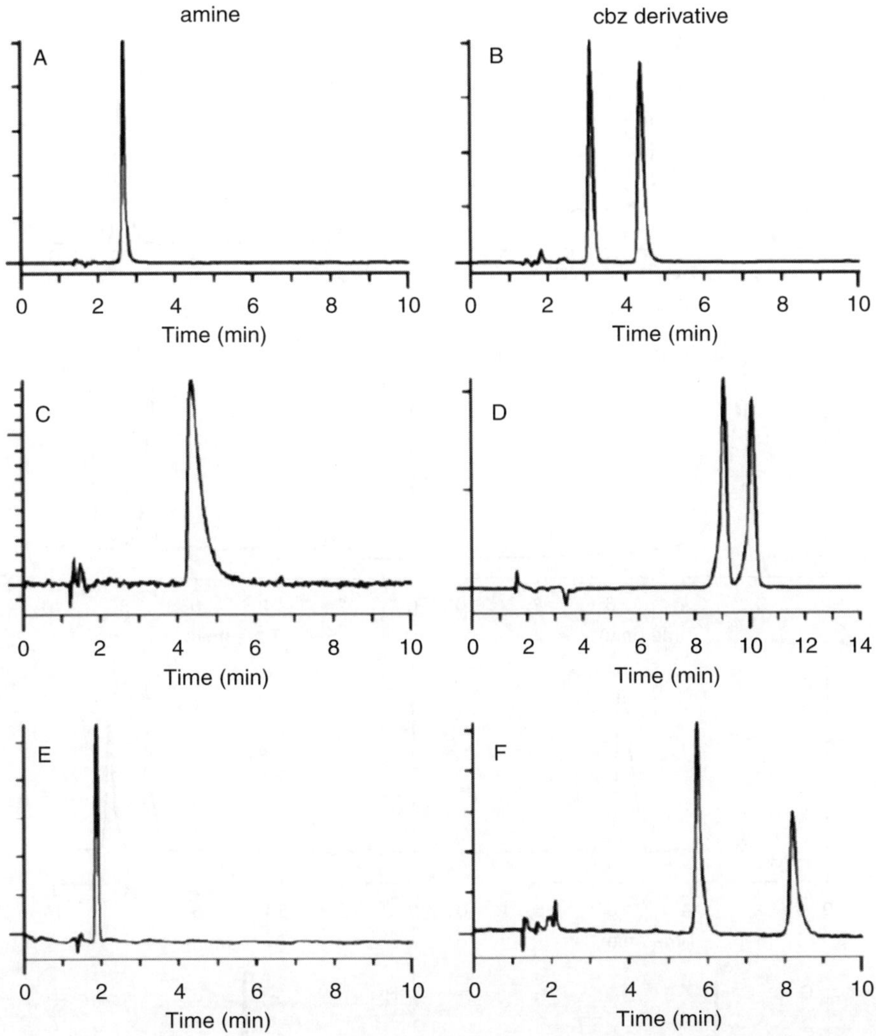

Figure 10 SFC separations of the cbz derivatives of 2-(diphenylhydroxymethyl)-pyrrolidine, fluoxetine, and *N*-benzyl-α-methylbenzylamine: (A) 2-(diphenylhydroxy-methyl)pyrrolidine, no separation; (B) cbz-2-(diphenylhydroxymethyl)pyrrolidine, OD, 40% 2-propanol (DEA)/CO_2; (C) fluoxetine, no separation; (D) cbz-fluoxetine, AD-H, 25% 2-propanol (DEA)/CO_2; (E) *N*-benzyl-α-methylbenzylamine, no separation; (F) cbz-*N*-benzyl-α-methylbenzylamine, Whelk-O 1, 20% 2-propanol (DEA)/CO_2. (From [64].)

of cbz derivatives of commercially available racemates was prepared and analyzed by enantioselective chromatography using a variety of mobile phases and polysaccharide and Pirkle-type chiral stationary phases (Figures 10 and 11). The cbz-derivatized product consistently demonstrated enhanced chiral resolution under SFC conditions. It is therefore concluded that for overall efficiency

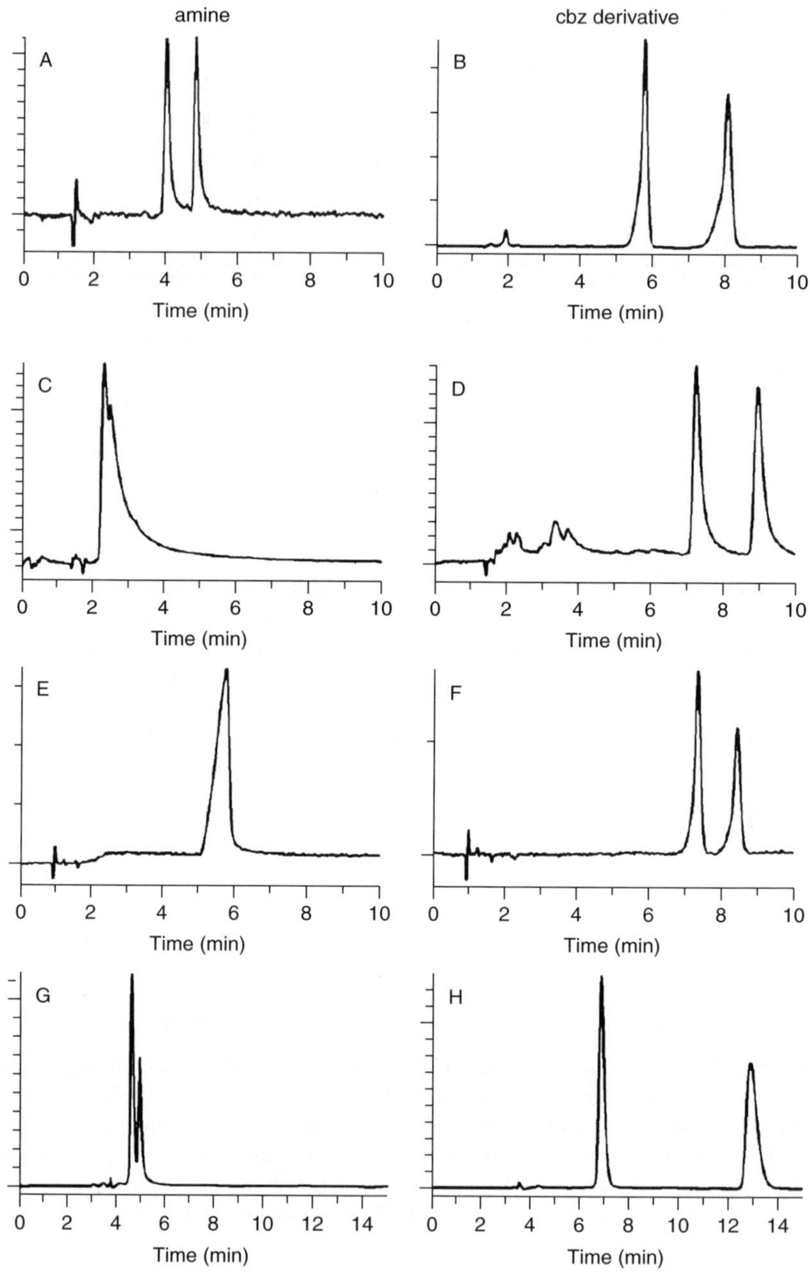

Figure 11 SFC and HPLC separations of the cbz derivatives of norephedrine, isoproterenol, phenylalaninol, and 2-(diphenylhydroxymethyl)pyrrolidine: (A) norephedrine, AD-H, 20% 2-propanol/CO_2; (B) cbz-norephedrine, AD-H, 20% methanol (DEA)/CO_2; (C) isoproterenol, AD-H, 20% methanol (DEA)/CO_2; (D) cbz-isoproterenol, AD-H, 20% methanol (DEA)/CO_2; (E) phenylalaninol, no separation; (F) cbz-phenylalaninol, AD-H, 20% 2-propanol (DEA)/CO_2; (G) 2-(diphenylhydroxymethyl)pyrrolidine, OJ, 1:1 hexane/ethanol (DEA); (H) cbz-2-(diphenylhydroxymethyl)pyrrolidine, OD, 8:2 hexane/2-propanol (DEA). (From [64].)

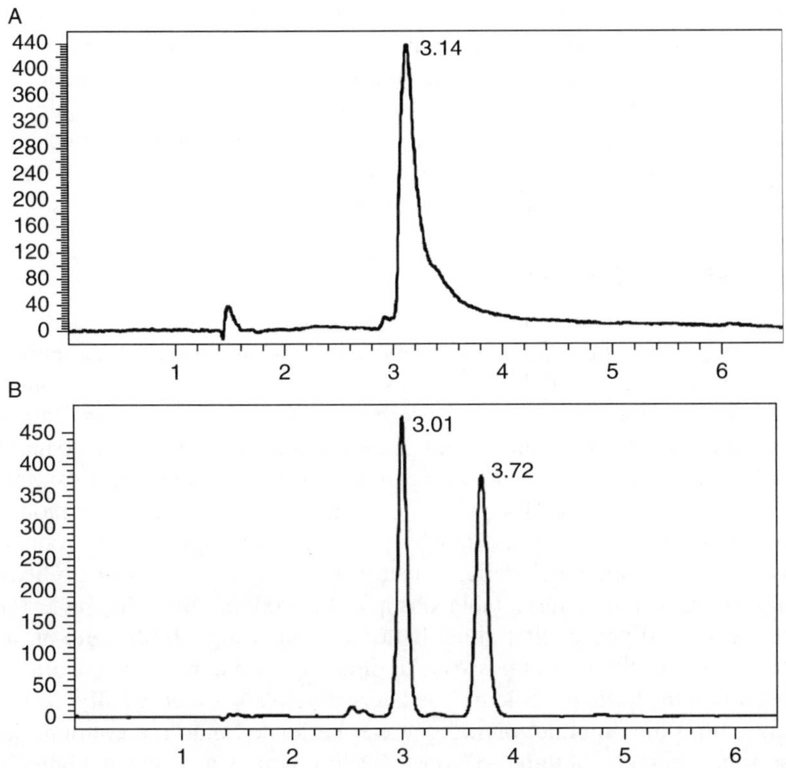

Figure 12 SFC chromatogram of chloramphetamine hydrochloride salt on a Chiralpak AD-H column using (A) 20% ethanol and (B) 20% ethanol containing 0.1% ESA modifier. Sample dissolved in modifier. (From [65].)

and productivity, it is worth the effort to derivatize the molecule if the chemistry is straightforward and there are no side reactions.

In another report, the resolution issue of basic amines under SFC conditions is tackled from a different approach [65]. Incorporation of a strong acid, ethanesulfonic acid (ESA), into the sample diluent and mobile-phase modifier provides dramatic improvement in these separations (Figure 12). Screening with ethanol containing 0.1% ESA on Chiralpak AD-H gave a separation of 36 of 45 basic compounds previously not separated in SFC. The mechanism appears to involve the separation of an intact salt pair formed between the basic compound and ESA. Other modifiers, other acids, and one additional stationary phase were examined and found to yield additional separations.

In a similar approach, ion pairing has been used in SFC to separate ionizable organic compounds [66]. The mobile phase was carbon dioxide mixed with an acetonitrile solution containing a counterion and a competing ion. The retention and separating efficiency was controlled by pressure and temperature as well as by the concentration and nature of the ionic mobile-phase component. Resolutions of enantiomeric 1,2-amino alcohols were obtained on SFC by the added chiral counterion of an antipode of a dipeptide, *N*-benzoxycarbonylglycyl(+)-proline (ZGP). A cyano phase was used as the stationary phase, and the separation mechanism is based on the formation of diastereomeric ion pairs with cationic solutes, which are in turn retained by the achiral solid phase. The additives' concentration, temperature, and pressure effects on resolutions were all studied. The resolutions obtained from SFC at higher flow rates were superior to those from HPLC.

8.2 Solvent Compatibility

There are some other significant contributions by SFC to bring the chiral separation process to a new stage. In many chiral LC processes, especially those that employ polysaccharide-based phases, there are limited types of compatible solvents used on the phases: namely, certain alcohols for most of the situations. This is very unfortunate, since many pharmaceutical compounds do not have good solubility in alcohols, and these chiral phases tend to be the most versatile and highest-efficiency ones for problem solving in enantiomer resolutions. Solvent compatibility is the biggest hurdle that limits throughput and productivity. Because of the compounds' delicate nature by design, this compatibility issue becomes an almost insurmountable obstacle to productivity. The same type of recently developed phases that allow better solvent compatibility with an immobilization functionality do not possess the same high loadability. There are several practical solutions to this problem: Use alternative phases that usually have lower efficiency, limit the amount/volume that can be loaded onto the column, or simply use much bigger columns to boost the throughput at a much higher price. The overall effect is a decrease in efficiency and low productivity with significant economic disasters.

It has been found that in SFC these same types of chiral phases demonstrate a much better tolerance to a wider range of solvents while maintaining the same efficiency. Some types of these solvents would be detrimental to phases if used in LC. For example, halogenated solvents, methylene chloride, and chloroform are not suitable, whereas they are quite acceptable for use in SFC. No negative effects, such as decreases of efficiency or loss of resolution that would normally be seen in LC, are observed over long-term use with these solvents. Even DMSO has been used successfully as a sample diluent for chiral SFC. Figure 13 shows one such example in preparative work [67]. The actual task took over 500 injections with DMSO as the sample solvent, with 5 mL of DMSO/500 mg per injection onto a 5-cm i.d. chiral column. Note that the high trace up front is the DMSO peak. The column showed excellent robustness and reproducibility during the entire period.

This success means two things. First, more choices are now possible, there is less restriction of solvents, and the success rate for chiral resolution will be higher. Second, with these "exotic" solvents available for preparative SFC, drug compounds can be prepared at higher concentrations than with alcohols, and the throughput/productivity can be improved significantly. This extends the SFC applicability to compounds that otherwise would not be attempted because

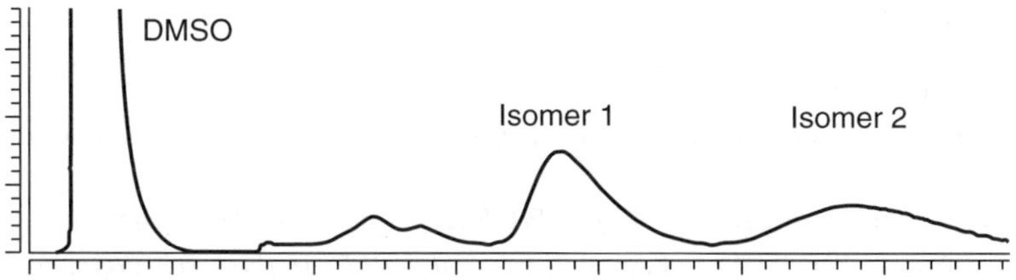

Figure 13 Preparative chiral SFC injection, Chiralpak AD, 5 × 25 cm, flow = 350 g/min, UV at 214 nm.

many analysts believe that SFC is only suitable for polar compounds that can be dissolved only in alcohols or in less potent solvents. In the author's practical experiences over many years, all compounds handled by SFC with various solubility levels can be resolved with appropriate solvents available in the laboratory, with no chromatographic issues and no restrictions. In this respect, it is important for medicinal chemists to optimize their efficiency. They do not have to modify the process to accommodate solubility conditions when it comes to the preparative process step. This can be translated into overall efficiency/productivity enhancement.

8.3 Column Lifetime in SFC

It is worth noting that chiral columns showed longer lifetimes in SFC use. The typical lifetime for a preparative chiral column in HPLC is about one year, at which time its efficiency starts to decrease, primarily from pressure-related factors. In SFC, the same type of column can last for three years with routine daily use and it will show no signs of efficiency change. This characteristic can contribute to very significant cost savings, since chiral columns are extremely expensive.

9 CONCLUSIONS

Since the new millennium, SFC has seen significant development and expansion into pharmaceutical laboratories worldwide because of its higher efficiency and higher productivity, especially in chiral analysis. Although these advantages were demonstrated more than two decades ago when SFC was first developed and commercialized, there are now other factors contributing to the resurgence of the technology. The robustness of instrumentation, more SFC-appropriate phases, and more published applications have led to wider acceptance into new territories. Also, the green features of SFC regarding cost reduction and decreased waste just happened to concur with the worldwide trend toward environmentally sustainable economic development. It is unarguably true that SFC will take off in the chemical analysis field and will be a real competitor for the well-developed liquid-based chromatographic techniques that are the mainstream in laboratories today.

REFERENCES

1. Thar Instruments, Inc., Pittsburgh, PA.
2. T. A. Berger. *Packed Column SFC*. Royal Society of Chemistry, Cambridge, UK, 1995.
3. G. M. Schneider. *Angew. Chem. Int. Ed. Engl.*, 17:716L–727L, 1978.
4. L. Snyder, J. Kirkland, J. Glajch. *Practical HPLC Method Development*. Wiley-Interscience, New York, 1997.
5. D. R. Gere et al. *Anal. Chem.*, 54:736–740, 1982.
6. A. Rajendran, S. Pepera, M. Johannsena, M. Mazzottib, M. Morbidellic, G. Brunner. *J. Chromatogr. A*, 1092:55–64, 2005.
7. B. J. Hoffman et al. *J. Chromatogr. A*, 1034:207–212, 2004.
8. L. Kott, W. B. Holzheuer, M. M. Wong, G. K. Webster. *J. Pharm. Biomed. Anal.*, July 31, 2006.
9. Y. Zhang, W. Watts, L. Nogle, O. McConnell. *J. Chromatogr. A*, 1049(1–2):75–84, Sept. 17, 2004.
10. G. Yanik. Chiral Separation Techniques, 3rd ed. Wiley-VCH, Weinheim, Germany, 2007.
11. B. Herbreteau, A. Salvador, M. Lafosse, M. Dreux. *Analusis*, 27:706–712, 1999.
12. K. Takahashi, S. Kinugasa, M. Senda, K. Kimizuka, K. Fukushima, T. Matsumoto, Y. Shibata, J. Christensen. *J. Chromatogr. A.*, 1193(1–2):151–155, June 6, 2008.
13. wps.prenhall.com/wps/media/objects/724/741576/Instructor_Resources/Chapter_05/Text_Images/FG05_01-10UN.JPG.
14. www.fda.gov/cder/guidance/stereo.htm.
15. Application book from Chiral Technologies, Inc.
16. Application book from Regis Technologies.
17. Z. Chen, T. Fuyumuro, K. Watabe, T. Hobo. *Anal. Chim. Acta*, 518(1–2):181–189, Aug. 2, 2004.
18. Y. Liu, A. Berthod, C. Mitchell, T. Xiao, B. Zhang, D. Armstrong. *J. Chromatogr. A*, 978:185, 2002.
19. A. Depta, T. Giese, M. Johannsen, G. Brunner. *J. Chromatogr. A*, 865:175, 1999.
20. P. Mourier et al. *J. Chromatogr.*, 353:61–75, 1986.
21. K. Williams, L. Sander, S. Wise. *J. Chromatogr. A*, 746:91, 1996.
22. K. Williams et al. *J. Pharm. Biomed. Anal.*, 15:1789–1799, 1997.
23. K. Williams, L. Sanders. *J. Chromatogr. A*, 785:149–158, 1997.
24. P. Petersson, K. Markides. *J. Chromatgr. A*, 666:381–394, 1994.
25. A. Salvador, B. Herbreteau, M. Dreux, A. Karlsson, O. Gyllenhaal. *J. Chromatogr. A*, 929:101–112, 2001.
26. T.-T. Ong, R.-Q. Wang, I. Wayan Muderawan, S.-C. Ng. *J. Chromatogr. A*, 1182:136–140, 2008.
27. E. Francotte. *J. Chromatogr. A*, 906(1–2):379–397, Jan. 12, 2001.
28. M. S. Villeneuve, R. J. Anderegg. *J. Chromatogr. A*, 826:217–225, 1998.
29. M. Maftough et al. *J. Chromatogr. A*, 1088:67–81, 2005.
30. C. White. *J. Chromatogr. A*, 1074:163–173, 2005.

31. Y. Zhao et al. *J. Chromatogr. A*, 1003:157–166, 2003.
32. M. Garzotti, M. Hamdan. *J. Chromatogr. B*, 770:53–61, 2002.
33. Z. Wang. *Int. Labmate*, Jan. 2007.
34. G. Terfloth. *J. Chromatgr. A*, 906:301, 2001.
35. J. Pinkston, T. Chester. *Anal. Chem.*, 650A, 1996.
36. K. Phinney. *Anal. Chem.*, 204A, 2000.
37. M. T. Combs, M. Ashraf-Khorassani, L. T. Taylor. *J. Chromatogr. A*, 785(1–2):85–100, Oct. 17, 1997.
38. A. Dobashi et al. *J. Chromatogr.*, 461:121–127, 1989.
39. H. Kasai et al. *J. Chromatogr. A*, 977:125–134, 2002.
40. M. Johannsen. *J. Chromatogr. A*, 937:135–138, 2001.
41. S. Peper, M. Johannsen, G. Brunner. *J. Chromatogr. A*, 1176(1–2):246–253, Dec. 28, 2007.
42. Y. Yang et al. *J. Pharm. Biomed. Anal*. 39:815–818, 2005.
43. M. J. del Nozal, L. Toribio, J. L. Bernal, N. Casta no. *J. Chromatogr. A*, 986:135–141, 2003.
44. W. H. Pirkle, C. J. Welch, B. Lamm. *J. Org. Chem.*, 57:3854, 1992.
45. F. Gasparrini, D. Misiti, C. Villani. *Trends Anal. Chem.*, 12(4), 1993.
46. R. M. Smith, L. Ma. *J. Chromatogr. A*, 785:179–184, 1997.
47. R. J. Smith, D. R. Taylor, S. M. Wilkins. *J. Chromatogr. A*, 697:591–596, 1995.
48. L. Toribio, M. J. del Nozal, J. L. Bernal, J. J. Jiménez, C. Alonso. *J. Chromatogr. A*, 1046:249–253, 2004.
49. L. Toribio et al. *J. Chromatogr. A*, 1144:255–261, 2007.
50. L. Toribio et al. *J. Chromatogr. A*, 1121:268–273, 2006.
51. L. Toribio et al. *J. Chromatogr. A*, 921:305–313, 2001.
52. K. Yaku et al. *J. Chromatogr. A*, 848:337–345, 1999.
53. K. Yaku et al. *J. Chromatogr. A*, 785:185–193, 1997.
54. Y. Zhao, W. A. Pritts, S. Zhang. *J. Chromatogr. A*, 1189:245–253, 2008.
55. Y. K. Ye, K. G. Lynam, R. W. Stringham. *J. Chromatogr. A*, 1041:211–217, 2004.
56. L. Toribio et al. *J. Chromatogr. A*, 1091:118–123, 2005.
57. L. Di et al. *J. Chromatogr. B*, 809:231–235, 2004.
58. P. Oswald et al. *J. Chromatogr. B*, 779:283–295, 2002.
59. T. Q. Yan, C. Orihuela. *J. Chromatogr. A*, 1156:220–227, 2007.
60. L. Toribio et al. *J. Chromatogr. A*, 1137:30–35, 2006.
61. L. Toribio, M. J. Nozal, J. L. Bernal, E. M. Nieto. *J. Chromatogr. A*, 1011:155–161, 2003.
62. Stefan Ottiger, Johannes Kluge, Arvind Rajendran, Marco Mazzotti. *J. Chromatogr. A*, 1162:74–82, 2007.
63. Z. Wang et al. *J. Pharm. Biomed. Anal*. 45:720–729, 2007.
64. C. Kraml et al. *J. Chromatogr. A*, 1100:108–115, 2005.
65. R. W. Stringham. *J. Chromatogr. A*, 1070:163–170, 2005.
66. W. Steuer et al. *J. Chromatogr.*, 441:287–296, 1988.
67. Z. Wang. Presentation Pittcon 2007, Chicago, Mar. 2007.

CHAPTER 11

Chiral Separations by Capillary Electrophoresis

DEBBY MANGELINGS and YVAN VANDER HEYDEN

Department of Analytical Chemistry and Pharmaceutical Technology, Vrije Universiteit Brussel–VUB, Brussels, Belgium

1 INTRODUCTION

The separation of chiral compounds has now been studied for half a century. The need to separate enantiomers originates from the fact that they display different behavior or interaction in a chiral environment such as the human body. Chirality has implications in the pharmaceutical industry primarily on the level of drug development, but we also observe chiral separation applications in the environmental [1] and in the food analysis fields. For chiral drugs, an eutomeric and a distomeric form can be distinguished, of which the former displays therapeutic activity and the latter exhibits no activity, another activity, an antagonistic effect, an adverse effect, or even a toxic effect. For these reasons, regulatory authorities have defined some guidelines to be followed when developing a chiral drug. The development of single-enantiomer drugs is preferred because of the reduced risk of side effects, but the development of racemates can be approved when the drug is, for example, converted into both enantiomers in the human body (racemization). In both cases, analytical methods must be presented that are able to quantify either both enantiomers in case of a racemic drug, or one enantiomer as an impurity of the active enantiomer in case of a single-enantiomer drug.

Many analytical techniques can be used to perform chiral separations: for example, liquid chromatography (LC), gas chromatography (GC), capillary electrochromatography (CEC), sub- or supercritical fluid chromatography (SFC), and capillary electrophoresis (CE). The latter technique is the subject of this chapter.

CE is an electrophoretic analytical technique that is conducted in fused-silica capillaries with an internal diameter between 25 and 100 μm, of which each

Chiral Separation Methods for Pharmaceutical and Biotechnological Products, Edited by Satinder Ahuja

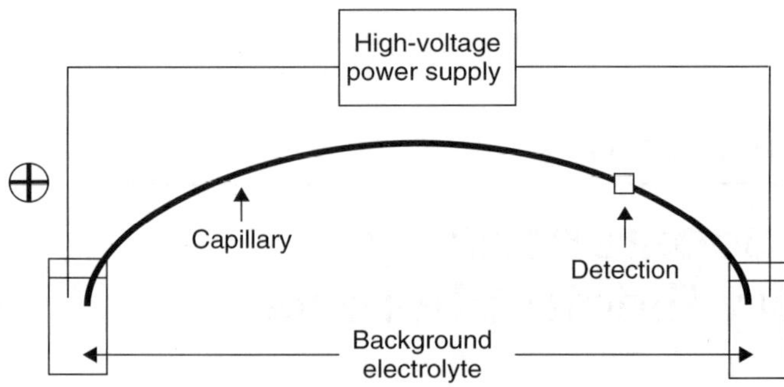

Figure 1 Capillary electrophoresis setup.

end is dipped in a vial that contains a buffer solution, also called the background electrolyte (BGE) (Figure 1). Electrodes are present in the vials so that an electrical field can be applied over the capillary, to perform separations. In a normal setup, an anode is placed at the capillary inlet and a cathode is located at the capillary outlet. To introduce the sample in the capillary, a sample vial is placed at the capillary inlet for a given time. Two types of injections can be performed in CE: hydrodynamic and electrokinetic injections. The first type uses either pressure, which is set at the sample vial, a vacuum set at the capillary outlet, or the siphoning effect when the sample vial is placed at a higher position than the capillary outlet. These are nonselective injections, as a given amount of sample is injected, independent of the charges of the analytes. With electrokinetic injections, an electrical field is applied for a few seconds over the capillary dipped into sample at the inlet and BGE at the outlet. This type of injection is selective, because positively charged ions will be injected preferentially. The most used injection type is the hydrodynamic type, which uses pressure at the sample inlet. Detection occurs on-capillary, usually at the cathodic side of the capillary. Figure 1 displays a practical setup for CE separation.

Due to the presence of acidic silanol (–Si–OH) groups at the capillary wall, negative charges ($-Si-O^-$) originate at pH values above 4 when an electrical field is applied over the capillary. A positive electrical double layer is formed by the ions of the BGE in the capillary to maintain electrical neutrality (Figure 2). This double layer is dense and immobile at the wall, but is more diffuse farther away. When an electrical field is applied, the diffuse layer starts to move from anode to cathode, which results in a displacement of the fluid inside the capillary. The resulting flow, called *electroosmotic flow* (EOF), is the driving force of this technique (Figure 2). The flow is characterized by a flat flow profile, which results in highly efficient separations.

Besides the EOF, analytes have their own electrophoretic mobility, which depends on the charge they bear and their ionic radius, and on the viscosity of the surrounding medium. Hence, positively charged species will migrate faster than the EOF and elute before the EOF, while negatively charged species will display counter-EOF mobility. This implies that the latter species cannot be detected

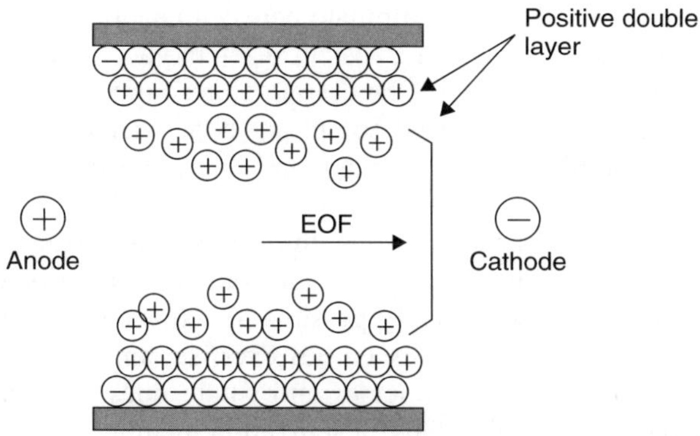

Figure 2 Generation of EOF inside a fused-silica capillary.

when the EOF is not high enough. The net mobility (i.e., the difference between own mobility and the EOF) should be in the direction of the cathode. Neutral species coelute with the EOF, as they have no own electrophoretic mobility. This implies that neutral species cannot be separated using this technique.

Chiral separations using CE can be conducted when a chiral selector is added to the BGE. The chiral selector forms transient diastereomeric complexes with both enantiomers, but with different binding constants, which results in a separation. Different types of selectors can be used to perform chiral separations, of which the most important are cyclodextrins, macrocyclic antibiotics, crown ethers, and proteins. Enantioseparations by means of CE have been the subject of many reviews [2–7].

In this chapter we begin with the basic principles of chiral separations with CE, and describe which models have been proposed. Further, the most commonly used chiral selectors in aqueous CE (e.g., cyclodextrins, crown ethers, macrocyclic antibiotics, and proteins) are discussed, and some applications of the last five years are highlighted for each of these selector types. No subclassification is made according to the type of application. The applications in ligand-exchange CE are also considered. Finally, we discuss the applications in a relatively new mode of CE, nonaqueous capillary electrophoresis (NACE). We evaluate the latest trends in the chiral separation field using CE as separation technique. Separations using micellar electrokinetic chromatography (MEKC) are not included. This technique, which uses combined electrophoretic and chromatographic separation principles, is discussed briefly toward the end of the chapter.

2 PRINCIPLES OF CHIRAL SEPARATIONS BY MEANS OF CAPILLARY ELECTROPHORESIS

As enantiomers do not differ in physical and chemical properties, they exhibit no difference in electrophoretic mobility when exposed to an electrical field. Therefore, a chiral separation of enantiomers cannot be achieved when a nonchiral separation medium is used. To discriminate between enantiomers, a chiral selector is needed which is able to form transient diastereomeric complexes between the selector and the enantiomers of the chiral solute. Forces involving the complex formation between selector and solute are noncovalent (i.e., hydrophobic, dipole–dipole, coulombic, van der Waals), and intermolecular interactions such as hydrogen bonding can occur.

One principal requirement to enable a chiral separation using capillary electrophoresis is that the binding constants for the complex formation differ for each enantiomer, although it is theoretically possible to achieve a separation when the binding constants are the same. This is, however, seldom seen in practice [8]; a high binding constant between selector and solute does not guarantee a successful chiral separation. Additionally, a separation medium is needed where the mobilities of free and complexed analyte are different. If both requirements are fulfilled, it is theoretically possible to obtain a chiral separation. The mobility difference requirement is not valid when working with neutral enantiomers and neutral selectors. This implies that it is impossible to separate neutral enantiomers with neutral selectors, no matter how high their binding constants are.

Two types of mathematical models were described to explain the separation process: the mobility difference model proposed by Wren and Rowe [9–11] and the separation model proposed by Vigh's group [12–16].

2.1 Mobility Difference Model

The mobility difference model starts from a freely soluble analyte, composed of enantiomers A and B, which both form a complex with a given chiral selector C according to

$$\mathrm{A} + \mathrm{C} \rightleftharpoons \mathrm{AC}$$

$$\mathrm{B} + \mathrm{C} \rightleftharpoons \mathrm{BC}$$

with K_1 and K_2 the equilibrium constants of the respective complex formations of A and B. Because free A and B are enantiomers, their electrophoretic mobilities in solution are the same. This mobility is designated as μ_1. The electrophoretic mobilities of the complexes AC and BC are assumed to be the same and are represented by μ_2. When the two enantiomers display different affinities for the selector, K_1 and K_2 will be different, which results in different electrophoretic mobilities of free and complexed enantiomer. In this case, chiral resolution is possible.

The difference in apparent electrophoretic mobilities of A and B can be calculated using the following equation [9]:

$$\Delta\mu = \frac{\mu_1 + \mu_2 K_1[\mathrm{C}]}{1 + K_1[\mathrm{C}]} - \frac{\mu_1 + \mu_2 K_2[\mathrm{C}]}{1 + K_2[\mathrm{C}]}$$

or

$$\Delta\mu = \frac{[C](\mu_1 - \mu_2)(K_2 - K_1)}{1 + [C](K_1 + K_2) + K_1K_2[C]^2} \tag{1}$$

From this equation it can be derived that the mobility difference will be zero when $K_1 = K_2$ or when $\mu_1 = \mu_2$. Moreover, the mobility difference will also be equal to zero when [C] = 0: thus, when no chiral selector is present in the solution or when [C] is very large. This implies that an optimal concentration of selector exists where the mobility difference, and hence the separation, is maximal.

From the mobility difference model, it could also be derived that the optimum selector concentration is lower at higher equilibrium constants, and hence the optimum selector concentration is compound dependent. Additionally, the mobility difference between two enantiomers will be largest when the migration direction of the analyte is opposite to that of the complex analyte–selector, which implies the usefulness of chiral selectors that carry a charge opposite to that of the analyte. Wren and Rowe also demonstrated the effect of the addition of an organic solvent on the resulting separation [10]. When the chiral selector concentration was above the optimal, the addition of an organic modifier to the buffer results in an increase in the mobility difference, and separation improves. An opposite effect will be observed when the chiral selector concentration is at or below the optimal value.

Next, resolution was explained in terms of differential interactions with a chiral selector and differential migration of the two diastereomeric complexes [11]. The authors consider the changes in the third term, indicated as R3, of the Rs equations:

$$\mathrm{Rs} = \left(\frac{V}{32D}\right)^{1/2}\left(\frac{l}{L}\right)^{1/2}\frac{\Delta\mu_{\mathrm{eph}}}{(\overline{\mu_{\mathrm{eph}}} + \mu_{\mathrm{eo}})^{1/2}} \tag{2}$$

with V the voltage, D the molecular diffusion length, L the total capillary length, l the capillary length to the detector, $\Delta\mu_{\mathrm{eph}}$ the electrophoretic mobility difference between the components, $\overline{\mu_{\mathrm{eph}}}$ the mean electrophoreic mobility, and μ_{eo} the electroosmotic mobility.

In situations where the mobility of both complexes is the same, R3 increases as a function of the cyclodextrin (CD) concentration, reaches a maximum, and then decreases. The magnitude of R3 increases when the procentual difference between the complex formation constants becomes larger. Hence, the CD concentration required to maximize R3 depends on the values of the complex formation constants, with small values needing higher selector concentrations.

When the electrophoretic mobility of the analyte and the EOF are in the same direction, an increase in EOF will always lead to a loss in Rs, which will be the case for cationic analytes. For anionic analytes, the opposite is seen. However, a reduction in EOF will inevitably lead to increased migration times. When the mobility of both complexes is different (e.g., when the enantiomers are oriented in the CD cavity in a slightly different way), and when the enantiomer with the lowest affinity forms the complex with the highest mobility, Rs will be enhanced. Similarly, when the enantiomer with the lowest affinity forms the complex with the lowest mobility, Rs will be lower than when the mobilities of both complexes would be the same.

2.2 Separation Model Proposed by Vigh and Co-workers

Vigh's group has published a series of papers that in addition to the CD concentration, also take into account the effect of pH on the separation [12–16]. They described the effect of the pH and selector concentration for a weak organic acid, and tested their theory by means of analytes such as ibuprofen and fenoprofen [12]. The selectivity of a weak organic acid can be described as follows:

$$\alpha = \frac{\mu_R}{\mu_S} = \frac{\begin{array}{c}\left(1 + (\mu_{RC\text{-}}/\mu_{f\text{-}})K_{RC\text{-}}[C]\right) . \\ \left[1 + K_{SC\text{-}}[C] + ([H_3O^+]/K_H)(1 + K_{HSC}[C])\right]\end{array}}{\begin{array}{c}\left(1 + (\mu_{SC\text{-}}/\mu_{f\text{-}})K_{SC\text{-}}[C]\right) . \\ \left[1 + K_{RC\text{-}}[C] + ([H_3O^+]/K_H(1 + K_{HRC}[C])\right]\end{array}} \tag{3}$$

This equation indicates that the selectivity depends on analyte-specific parameters such as ionic mobilities of the free enantiomers ($\mu_{R\text{-}0}$, $\mu_{S\text{-}0}$) and ionic mobilities of the complexed enantiomers ($\mu_{RC\text{-}0}$, $\mu_{SC\text{-}0}$); the acid dissociation constant of the analyte K_H; the complex formation constants of the ionic enantiomers $K_{RC\text{-}}$ and $K_{SC\text{-}}$; and the complex formation constants of the protonated enantiomers K_{HRC} and K_{HSC}. In addition to these parameters, the pH ($[H_3O^+]$) and the selector concentration ([C]) also play a role.

Hence, selectivity will depend on three different situations that can occur:

1. Only the nonionic forms of the enantiomers interact differently with the CD (desionoselective interactions).
2. Only the ionic forms of the enantiomers interact differently with the CD (ionoselective interactions).
3. Both ionic and nonionic forms of the enantiomers interact differently with the CD (duoselective interactions).

The separation of enantiomers in case (1) is the easiest to achieve. The selectivity will change monotonically as a function of the H_3O^+ and CD concentration; thus, it will increase with decreasing pH and increasing CD concentration. Therefore, it is advised to optimize a separation by decreasing the pH until the desired result is achieved. Any pH lower than the optimal pH will result in an increased separation time (when the separation is performed with a neutral selector and EOF suppression is performed). Reversal of migration order cannot be accomplished in this case.

Enantiomers in case (2) are more complicated to separate. Selectivities below, equal to, or larger than 1 are possible, depending on the relative values of ionic mobilities and equilibrium constants, and hydronium ion and CD concentrations. Reversal of migration order is possible by varying the CD concentration and pH. Chiral selectivities are generally lower than for enantiomers in case (1) because the mobility ratios are also taken into account. Optimization of a separation can still be performed by decreasing the pH until the desired result is obtained. When duoselective interactions occur, separations are most difficult to achieve. A priori, it is unpredictable which BGE will give the best results, but it is advised to start with a low-pH electrolyte.

The outcome of [12] was used further in a later study [13] to discuss the selectivity behavior in cases 1, 2, and 3 for a basic compound. In [14], peak resolution was modeled as a function of CD concentration and pH for separations of case (1) and (3) enantiomers. In [15] and [16], this group developed a resolution model of weak acids and bases, respectively, with HP-β-CD used as a chiral selector.

3 CHIRAL SEPARATIONS IN AQUEOUS CAPILLARY ELECTROPHORESIS

3.1 Cyclodextrins as Chiral Selectors

Undoubtedly, cyclodextrins are the most successful and most frequently used chiral selectors in capillary electrophoresis [17]. They are composed of D(+)-glycopyranose units that are linked through an α(1,4)-glycosidic bonding. The molecule is shaped like a truncated cone. Depending on the number of glycopyranose units, α-cyclodextrin (six units), β-cyclodextrin (seven units), and γ-cyclodextrin (eight units), which are called the *native cyclodextrins*, can be distinguished (Figure 3A). The interior cavity of the cyclodextrin is relatively hydrophobic, which allows hydrophobic interaction between CD and analyte. The exterior is hydrophilic, with the wider rim of the molecule containing secondary hydroxyl groups in the 2- and 3-positions, and the bottom rim composed of primary hydroxyl groups in the 6-position. See Figure 3B for the positions of these hydroxyl groups.

Derivatization of the hydroxyl groups on the outer rim leads to *derivatized cyclodextrins*, which can be either charged or uncharged, depending on the type of groups that are incorporated. The various types of cyclodextrins that are used for chiral separations in CE are discussed further later. Cyclodextrins possess a number of properties that make them desirable chiral selectors in CE. These properties include ultraviolet (UV) transparency, acceptable solubility in commonly used electrolytes of CE (mostly aqueous), low price, stability, and nontoxicity. The mechanism of separation is based on inclusion of the analyte into the relatively hydrophobic cavity, along with additional interactions with the groups that are present on the wider opening of the cyclodextrin. Therefore, the size of the cavity is very important in the chiral recognition mechanism. Forces involved in these interactions include dipole–dipole, van der Waals, and

A

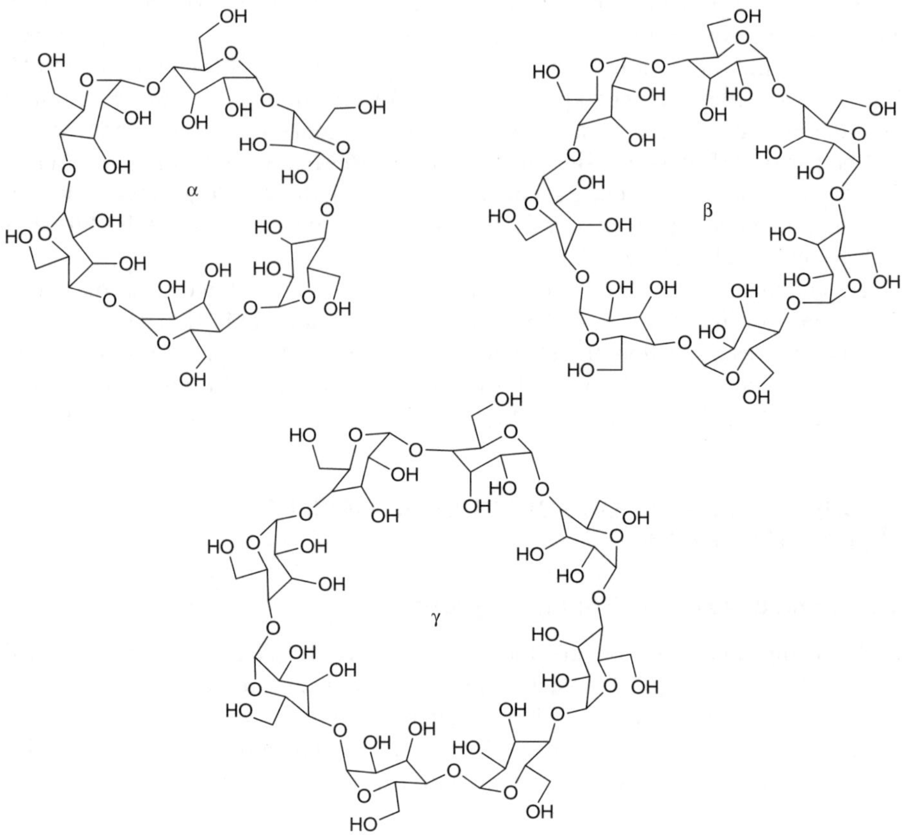

B

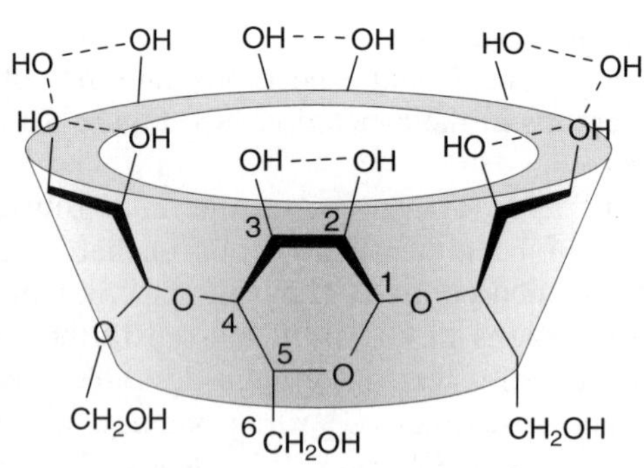

Figure 3 (A) Classical representation of the native cyclodextrins; (B) position of the hydroxyl groups in α-cyclodextrin. (Part A from [18], with permission.)

hydrogen-bonding forces. In the following, more details are given about neutral and charged cyclodextrins, along with the most recent applications, primarily from the last year. A large number of applications (more than 1000 in the past five years) in the enantioseparation field using CE are described. Therefore, we limit ourselves primarily to the most recently published. In Table 1 we present details of selected applications regarding selector type and concentration, electrolyte type, pH, temperature, and applied voltage.

Neutral Cyclodextrins As mentioned earlier, α-, β-, and γ- cyclodextrin (i.e., underivatized cyclodextrins), the native cyclodextrins, are considered to be neutral and have different dimensions due to the number of glycopyranose units. The size of the cavity at the outer rim generally varies around 0.53 nm for α-CD, around 0.65 nm for β-CD, and around 0.85 nm for γ-CD, with a height of about 0.80 nm for all three CDs. The solubility of β-CD in water is the lowest of the three native CDs, which is why most derivatized CD are derivatives of β-CD. To obtain neutral CD derivatives, the hydroxyl groups in the 2-, 3-, and/or 6-position are derivatized with neutral groups. For β-CD derivatives, we can distinguish heptakis-*O*-methyl-β-cyclodextrin (M-β-CD), heptakis(2,6-di-*O*-methyl)-β-cyclodextrin (DM-β-CD), heptakis(2,3,6-tri-*O*-methyl)-β-CD (TM-β-CD), hydroxyethyl-β-cyclodextrin (HE-β-CD), and hydroxypropyl-β-cyclodextrin (HP-β-CD) [19], among other but less used derivatives of α- and γ-CD. Among the most recent applications with neutral CD derivatives, we can find the chiral separation of four major cinchona alkaloids (quinine/quinidine and cinchonine/cinchonidine) using DM-β-CD as a chiral selector [20]. This selector proved to be more efficient than native β-CD in separating the diastereomeric couples. The resolution of the separations increased with increasing selector concentration (range 1 to 7 mM), and the same effect was seen when the EOF was disabled due to coating of the capillary wall. Additionally, peak shapes and reproducibility were also affected positively when disabling the EOF.

Another derivative of β-CD (i.e., TM-β-CD) has been used [21] to enable the separation of ketoprofen enantiomers. The method was validated in terms of selectivity when other drugs are present, and of linearity, limit of detection (LOD), limit of quantitation (LOQ), precision, and accuracy, and could be applied for different matrices, such as plasma, synovial fluid, and urine. This indicated its usefulness in terms of clinical studies. Two chiral derivatizing reagents, (9-anthryl)methoxyacetic acid and (9-anthryl)hydroxyacetic acid, that are used to assign absolute configuration by nuclear magnetic resonance (NMR) analysis, were synthesized and separated by means of capillary electrophoresis using HP-β-CD as a chiral selector [22]. After optimization of CD concentration, pH, applied voltage, and temperature, the method was applied to resolve the enantiomers of (9-anthryl)methoxyacetic acid and (9-anthryl)hydroxyacetic acid with Rs values of 3.27 and 1.92, respectively. Two more pharmaceutically directed applications using HP-β-CD as a chiral selector were described by Kim's group [23,24]. A method was developed to determine the enantiomeric purity of fluvastatin enantiomers [23], and a method was established to determine the enantiomeric impurity of (*S*)-etodolac [i.e., (*R*)-etodolac], down to a level of 0.2% [24].

TABLE 1 Separated Compounds with Cyclodextrins as Chiral Selector

Compounds	Electrolyte	Selector/ Concentration	Applied Voltage, Temperature	SE/LE[a]	Ref.
Amisulpiride	10 mM citrate pH 3.5	S-β-CD, 0.22 % (w/v)	15 kV, 20°C	SE	[33]
Cetirizine	5 mM borate pH 8.7	S-β-CD, 1 % (w/v)	−10 kV, 25°C	LE	[34]
2′-Hydroxyflavanone, 3′-hydroxyflavanone, 4′-hydroxyflavanone	50 mM phosphate pH 3.0	S-β-CD, 2 % (w/v) with or without 2 mM SDS and 2 mM γ-CD or 1.5 mM β-CD	−21 kV, 25°C	LE	[35]
Norepinephrine, epinephrine, isoproterenol, octopamine, synephrine, norephedrine	50–200 mM phosphate pH 3.0	2% (w/v) S-β-CD	−20 kV, 25°C	LE	[36]
Mirtapazine, demethylmirtapazine, 8-OH-mirtapazine	50 mM phosphate pH 2.5	0.55 % (w/v) CM-β-CD	25 kV, 25°C	LE	[39]
Betaxolol, sotalol, metoprolol, bisoprolol, bevantolol, timolol	50 mM acetate pH 4.75	10 mM CM-β-CD	12 kV, 20°C	LE	[40]
Matairesinol	50 mM phosphate pH 5.0	5 mM CM-β-CD	25 kV, 25°C	LE	[41]
Carteolol, atenolol, sotalol, metoprolol, esmolol	20 mM phosphate pH 5.5/MeOH/EtOH:[95:5:5 (v/v/v)]	1.5 % (w/v) CM-β-CD	20 kV, 20°C	LE	[42]
Brompheniramine	25 mM glycine, 100 mM acetate pH 3.1, 0.1% (w/v) hydroxyethylcellulose	5 mg/mL CE-β-CD	NI[b]	LE	[44]
Linezolid	50 mM borate pH 9.0	27.5 mM HDAS-β-CD	15 kV, 15°C	LE	[53]
	50 mM borate pH 8.0	18.75 mM HDAS-β-CD	−15 kV, 15°C	LE	[53]
Promethazine, ethopropazine, trimeprazine, thioridazine	100 mM citrate pH 3.0	0.75% (w/v) heptakis(2,3-dihydroxy-6-sulfato)-β-CD with either 1.5 mM HP-β-CD, 1 mM β-CD, or 17 mM γ-CD	−20 kV, 25°C	LE	[54]

Ecstasy, methadone	15 mM ammoniumformate pH 2.5	0.08% HS-γ-CD	30 kV, 25°C	LE	[58]
Naproxen, ofloxacin, warfarin, ibuprofen, indoprofen, ketoprofen, flurbiprofen, D,L-3-phenyllactic acid, mandelic acid, abscisic acid, tropic acid	50 mM phosphate pH 5	1–9 mM 6-*O*-(2-OH-3-trimethylammoniopropyl)-β-CD	15 kV, 25°C	LE	[59]
8-*O*-sulfated analytes	25 mM phosphate pH 2.54	2–20 mM PEMEDA-β-CD[c]	15 kV, 25°C or 9/15 kV, 25°C	LE	[60]
2-OH-2-phenyllactic acid, 2-cylopentyl-2-phenyllactic acid, 2-cyclohexyl-2-phenyllactic acid, 2-phenylpropionic acid, ibuprofen, flurbiprofen, fenoprofen, ketopropfen, naproxen, carprofen, 2-(2,4-dichlorophenoxy)propionic acid, 2-phenyl-3-methylvaleric acid, mephobarbital, indapamide	25 mM ethanolamine/methanesulfonic acid pH 9.3	1–10 mM PEMEDA-β-CD[c]	12/15/18 kV, 25°C	LE	[60]
11 nonionic analytes	25 mM phosphate pH 2.54	1–10 mM PEMEDA-β-CD[c]	9/10/15 kV, 25°C	LE	[60]
Rotigotine, two related chiral impurities	100 mM phosphate pH 2.5	2% (w/v) S-β-CD + 2% (w/v) methyl-β-CD	−20 kV, 20°C	LE	[66]
Citalopram, citadiol	20 mM phosphate pH 2.5	0.5 mg/mL β-CD + 22 mg/mL S-β-CD	−20 kV, 28°C	LE	[67]

[a]SE, short-end injection method; LE, long-end injection method.
[b]NI, not indicated.
[c]PEMEDA-β-CD, mono-6-deoxy-6-*N*, *N*, *N′*, *N′*, *N′*-pentamethylethylenediaminocyclomaltoheptaose.

Other applications use neutral derivatives of α- or γ-CD as a selector. HP-γ-CD was used to develop a chiral method to assay primaquine in pharmaceutical tablets [25]. The method provided acceptable repeatability for peak areas and migration times, with RSD values ranging between 1.05 and 3.30%, along with good recoveries in the range 96.8 to 104.9%. Chiral recognition of HP-α-CD and HP-β-CD for tryptophan enantiomers has been studied [26]. It was seen that the α-CD derivative was able to separate the enantiomers, whereas β-CD was not. This phenomenon was studied by means of NMR and circular dichroism measurements, which indicated that inclusion on both enantiomers occurred for HP-β-CD, while only D-tryptophan was inserted into the cavity of HP-α-CD, which explained the difference in separation results.

Although less used, some applications with native cyclodextrins can be found in the recent literature, which indicates their usefulness in some situations. Native γ-CD was used for the separation of five basic drugs: alprenolol, propranolol, pindolol, isoxsuprine, and ritodrine [27]. Only the first four could be partially separated using this chiral selector. The addition of alcohols such as ethanol, methanol, isopropanol, and 2-methyl-2-propanol increased, decreased, or did not affected the interaction between the selector and the analyte. Moreover, good correlation between the hydrophobicity of the alcohols and the Rs value, and the bulkiness of the alcohols and the Rs was demonstrated. Using optimal concentrations of 5% 2-methyl-2-propanol and 35 mM γ-CD, propranolol and pindolol could be baseline-separated.

El-Maali's group [28,29] recently published two papers using β-CD as a chiral selector. (+)-Catechin and (−)-epicatechin were baseline-separated and quantified in aqueous and plasma samples [28]. The method used a special high-sensitivity cell from Agilent after the separation path, which allowed more sensitive detection and hence lower LOD and LOQ values. The method was applied to a plasma sample after ingestion of green tea, and 2 h after ingestion, 17.4% of catechin and 1.8% of epicatechin, with respect to their concentration in the tea, were found unchanged in plasma. The same enantiomers, along with ascorbic acid, were separated and quantified in urine in the next study by this group [29]. The method developed was validated in terms of selectivity, accuracy, precision, linearity, LOD, LOQ, and robustness. Then it was applied to determine cathechin, epicatechin, and ascorbic acid in commercial drinks such as green tea, black tea, and coffee, and in urine.

More specialized papers can also be found among the applications with neutral CDs. For example, the effect of achiral *p*-sulfonatocalixarenes on a chiral CE separation using HP-β-CD as a chiral selector was investigated [30]. Here, complete filling techniques and partial filling techniques were used. With the complete filling technique, the total length of the capillary is filled with the chiral selector solution prior to sample introductions, while during analysis both ends of the capillary are connected to BGE without a selector. With the partial filling technique, the selector containing BGE is introduced into the capillary as a plug, shorter than the effective length of the capillary, prior to application of the analyte, while during the run both ends of the capillary are connected to vials

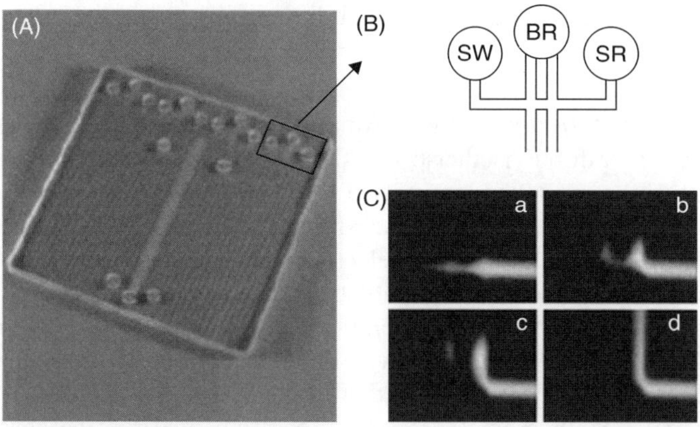

Figure 4 Four-channel chip used by Goa et al. [31]. (A) Photograph of the complete chip. (B) Enlarged view of the sample injection section (SW, sample waste; BR, buffer reservoir; SR, sample reservoir). (C) Fluorescence image of double cross-sample injection. (From [31], with permission.)

with BGE. It was observed that HP-β-CD alone could not initiate the separation of propranolol or brompheniramine using the complete filling technique [30]. Separation could be achieved only when HP-β-CD was loaded using the complete filling technique, followed by a mixture of calixarenes and the CD loaded using the partial filling technique before sample introduction.

Another application demonstrates the use of a four-channel electrophoresis chip (Figure 4) for a rapid screening of chiral selectors [31]. Eight neutral selectors were screened by means of a four-channel chip for their usefulness in the enantioseparation of FITC-labeled norfenefrine and baclofen. Screening of the selectors in two electrophoresis processes could be completed in less than 2 min. For norfenefrine, three selectors were suitable for the separation: two types of DM-β-CD with different substitution degrees, and HP-β-CD. One of the DM-β-CD species allowed the compound to be baseline separated. For baclofen, partial separation with DM-β-CD was obtained, while HP-α-CD and the other DM-β-CD were able to baseline-separate this compound. The work described above also demonstrates that the substitution degree on a CD might affect the enantioseparation. Use of this experimental setup could present a useful and fast alternative for screening purposes in industrial applications.

Charged Cyclodextrins When the hydroxyl groups of a cyclodextrin are derivatized with either positively or negatively charged groups, the net charge of the CD will be the one of its substituents. A distinction is thus made between anionic and cationic derivatives. Charged CDs have the advantage that the separation is often better than that of native CDs, due to an increase in mobility difference when the analyte and CD are oppositely charged [19]. Another advantage is the self-mobility of these CDs, which allows their use in electrokinetic chromatography [32] because their mobility differs from the mobility of the electrolyte.

Anionic Cyclodextrin Derivatives Among the anionic CDs we find sulfated-β-CD (S-β-CD), sulfobutyl ether-β-CD (SBE-β-CD), and sulfoethyl ether-β-CD (SEE-β-CD) as the most frequently used [19]. These CD derivatives are charged throughout the complete pH range. Recent applications with sulfated-β-CD have been described [33–36]. An enantioselective method was developed for the analysis of amisulpiride enantiomers in tablets [33]. After investigation of the effects of CD type and concentration, temperature, buffer concentration, and pH, baseline separation of the enantiomers was obtained within 10 min. The resulting method was validated in terms of linearity, precision, and accuracy. Chou et al. [34] developed a method for the enantioseparation of cetirizine enantiomers in less than 7 min. After method optimization in terms of pH, selector concentration, and buffer concentrations, the method was used successfully on pharmaceutical tablets containing levocetirizine and plasma samples, using cefazolin as an internal standard. In the tablets, levocetirizine was assayed within the range 95 to 105% of the labeled amount. In plasma, both enantiomers were measured. Drugs such as loratadine, diphenhydramine, homochlorcyclizine, and buclizine did not interfere with the separation.

Three hydroxyflavanones, 2′-hydroxyflavanone, 3′-hydroxyflavanone, and 4′-hydroxyflavanone, were separated within one run [35]. Upon the addition of sodium dodecyl sulfate (SDS), the separation between the 4′-hydroxyflavanone enantiomers was lost, but peak shapes of the other four enantiomers were better. In a next step, β-CD or γ-CD were added to the S-β-CD system, and the dual CD system again allowed separation of the 4′ hydroxyflavanone enantiomers. The initial separation, using only S-β-CD in the electrolyte, seemed, however, better for these enantiomers.

The same group investigated the separation of six cathecholamines using four different separation modes [36]: (1) low concentration of S-β-CD in a high-concentration phosphate buffer at low pH in normal polarity, (2) high concentration of S-β-CD at low pH in reversed polarity, (3) moderate concentration of S-β-CD at neutral pH in reversed polarity, and (4) low to moderate concentration of heptakis(2,3-dihydroxy-6-sulfato)-β-CD in a high-concentration phosphate buffer at low pH in normal polarity. From these four systems, the one using a high concentration of S-β-CD in reversed polarity mode gave the best separations. Furthermore, the importance of a high concentration of electrolyte with such a system was demonstrated.

Carboxymethyl-β-cyclodextrin (CM-β-CD) is also a frequently used anionic CD, but it is not negatively charged throughout the complete pH range because it possesses weakly acidic groups. Only at pH values above 4.5 it is fully negatively charged [37,38]. Carboxyethyl-β-cyclodextrin (CE-β-CD) is another anionic analog that is sometimes used. Recent applications using CM-β-CD [39–43] and CE-β-CD [44] have been reported. Malagueño de Santana et al. [39] developed an enantioselective method to determine mirtapazine and metabolites in urine after enzymatic hydrolysis. Influences of CD type (HP-β-CD, CM-β-CD, or S-β-CD), CD concentration, buffer composition temperature, and applied voltage were investigated along with the sample preparation, where the hydrolysis and the extraction were optimized. CM-β-CD was found to give the best separation. The method was validated and applied to urine samples. It was found that the active compound was metabolized primarily to 8-hydroxy analogs.

CM-β-CD was found to give better results than HP-β-CD for the enantioseparation of six beta-blocking agents [40]. After optimization of pH, CD concentration, buffer composition and concentration, capillary length, temperature, and applied voltage, the method could be applied for five of these β-blocking agents in tablets and for one of them in eyedrops. In [41], CM-β-CD was used for the first time to separate enantiomers of matairesinol, a plant lignan that displays phytooestrogen activity [41]. The method developed was able to separate enantiomers better and faster than the HPLC method developed earlier and allowed their quantification.

A field-amplified stacking method was used to increase detection and quantification limits of the developed methods [42,43]. In both papers, CM-β-CD was used as a chiral selector for the enantioseparation of six β-blockers [42] and trihexyphenidyl [43], respectively. The sensitivity of both methods increased dramatically: between 5- and 25-fold [42] to about 500-fold [43]. Both methods could be applied further for the quantification of β-blockers or trihexyphenidyl in human serum with satisfying results.

An application that uses CE-β-CD as chiral selector was presented by Marák et al. [44]. Here, an isotachophoresis (ITP)-CE method was developed to determine the enantiomeric impurity of the model analyte brompheniramine in a pharmaceutical sample. The ITP step before the CE separation made it possible to obtain a lower LOD value of enantiomeric impurity, higher selectivity, higher sample loading capacity, and better reliability. These improvements made it possible to determine the enantiomeric impurity of levobrompheniramine easily in samples containing a 1000-fold excess of dexbrompheniramine.

Because sulfated CDs are often mixtures of isomers with different degrees and positions of substitution, large batch-to-batch variabilities may occur. This can lead to an overall reduction of separation selectivity, a decrease in separation efficiency, excludes fundamental studies of interactions because a mixture of isomers is used, and can compromise the repeatability of a difficult separation [45]. Therefore, a series of single-isomer chiral resolving reagents was introduced in the late 1990s by the Vigh's group, including heptakis (2,3-diacetyl-6-sulfato)-β-CD [45], hepta-6-sulfato-β-CD [or heptakis(2,3-dihy droxy-6-sulfato)-β-CD] [46], and heptakis(2,3-dimethyl-6-sulfato)-β-CD [47]. More recently, other types of single-isomeric CDs were synthesized, such as heptakis(2-*O*-methyl-6-*O*-sulfo)cyclomaltoheptaose [48], hexakis(2,3-di-*O*-methyl-6-*O*-sulfo)-α-cyclodextrin [49], hexakis(6-*O*-sulfo)-α-cyclodextrin [50], octa(6-*O*-sulfo)-γ-cyclodextrin [51], and octakis(2,3-diacetyl-6-sulfato)-γ-cyclodextrin [52].

Recent applications using single-isomeric CD have been reported [53,54]. In [53], a method was developed for the enantiomeric impurity determination of linezolid. The chiral selector used was heptakis(2,3-diacetyl-6-sulfato)-β-CD (HDAS-β-CD), and the authors employed the charged resolving agent migration model (CHARM), a set of guidelines proposed by Williams and Vigh [55] which helps with the selection of optimum separation conditions when charged CDs are used as chiral selectors. After method optimization and validation (linearity, sensitivity, precision, accuracy), determination of the impurity down to a 0.3% level was possible when the impurity eluted in the tail of the major compound (Rs = 3.7). Reversal of enantiomer sequence was achieved by changing from an uncoated to a neutral capillary, which reversed the EOF. Under these conditions, an Rs value of 7 was achieved between the enantiomers. This method was not validated, however.

Heptakis(2,3-dihydroxy-6-sulfato)-β-CD was used as a chiral selector to investigate the enantioseparation of five phenothiazines within one run [54]. Additionally, dual CD systems, where this selector or a mixture of S-β-CD isomers is combined with neutral CDs, such as HP-β-CD, β-CD, and γ-CD, were also investigated. It was observed that the enantioselectivity of the majority of compounds was improved using a combination of heptakis(2,3-dihydroxy-6-sulfato)-β-CD and HP-β-CD. Moreover, the single isomeric CD gave better results in dual systems than the isomeric *S*-β-CD mixture. A reversal of enantiomer migration order was also observed for two compounds, ethopropazine and thioridazine, when a combination of heptakis(2,3-dihydroxy-6-sulfato)-β-CD and γ-CD was used or at varying γ-CD concentrations. This can be assigned to the opposite effects of both CDs on the enantiomers'mobility. Finally, all but one enantiomer could be baseline separated.

Beckman Coulter developed a set of highly sulfated CD (HS-CD) which are mixtures of species with similar sulfation degrees. Characterization of these HS-CDs showed that the average sulfate contents were 11, 12, and 13 for HS-α-CD, HS-β-CD, and HS-γ-CD, respectively [56]. These CDs are known for their generic enantioselectivity; that is, they are able to separate the majority of a set of structurally different compounds and for their excellent batch-to-batch variability [57]. Schappler et al. [58] developed a chiral CE method combined with electrospray ionization mass spectrometry (ESI-MS) to determine ecstasy and methadone in plasma samples using HS-γ-CD as chiral selector. They investigated two types of sample pretreatment steps and observed that protein precipitation allowed quantification in the ppm range, while with liquid–liquid extraction (LLE) one could go down to the ppb level. Both approaches were extensively validated, and the LLE-CE-ESI-MS method was finally applied to real toxicological samples.

Cationic Cyclodextrin Derivatives When screening the literature for applications with cationic CD derivatives, one notices immediately that this type of CD is not commonly used in the chiral separation field, although they allow the separation of negatively charged species. Most cationic CD derivatives can be recognized at the amino group present in their structure, which is responsible for the positive charge.

Some recent applications with positively charged CD derivatives have been reported [59–61]. In [59], 6-*O*-(2-OH-3-trimethylammoninopropyl)-β-CD was synthesized and applied for the separation of 11 acidic analytes. Effects of CD concentration and pH were investigated and the results were compared with those using a commercially available CD, a hydroxypropyltrimethylammonium salt of β-CD. The newly synthesized selector seemed to be more effective. Therefore, the authors concluded that the degree of substitution can play an important role in the enantiorecognition.

In analogy with the anionic single isomer CD, Vigh's group [60] synthesized mono-6-deoxy-6-*N*,*N*,*N*′,*N*′,*N*′-pentamethylethylenediaminocyclomaltoheptaose, a permanently dicationic derivative of β-CD. The CD proved to be very useful for the analysis of negatively charged analytes (i.e., weak acids at high pH and O-sulfated analytes at low pH). Some nonionic compounds could be separated, but the success rate was lower than for the negatively charged analytes. Basic analytes could not be separated at all with this type of CD, indicating that there was no complexation between this selector and cationic analytes.

Tang and Ng have provided an overview of work that has been performed in the group of Ng et al. [61]. This group has synthesized four classes of cationic derivatives of α-, β- and γ-CD and have investigated extensively the effects of pH, selector structure, concentration, temperature, and organic solvents using large sets of dansyl amino acids and hydroxyl acids. They concluded that the use of selectively substituted CDs presented an opportunity to perform mechanistic studies and to develop validated chiral assays. Because of the systematic approach they used in the synthesis, the effects of the CD substituents on the separation outcome could be investigated thoroughly, providing a strong basis for future research regarding cationic CD-based chiral separations.

Dual Cyclodextrin Systems Charged CDs have the ability to reach an increased chiral recognition for oppositely charged species. However, sometimes this CD provides either little enantioselectivity or too high enantioselectivity with no migration as a consequence. The latter can be caused by a too strongly formed analyte-CD complex with a high mobility opposite the EOF, resulting in zero migration. In these cases, opportunities can be found in adding another CD to the electrolyte, to achieve faster elution. The second CD can either provide enantioselectivity itself, compete with the other CD to enable detection of a resolved compound, or provide a differential migration rate [62]. Several possibilities exist to combine two CDs. Either a combination of two neutral CDs can be used [63] or a mixture of two charged CDs [64]. However, the most common approach is use of a combination of a charged and a neutral CD [65].

Recent applications with dual CD systems involve the separation of rotigotine and related impurities [66] and the enantiomeric impurity determination of escitalopram [67]. In [66], three types of CD were investigated: β-CD, methyl-β-CD, and S-β-CD for the enantioseparation of rotigotine from its enantiomeric impurity

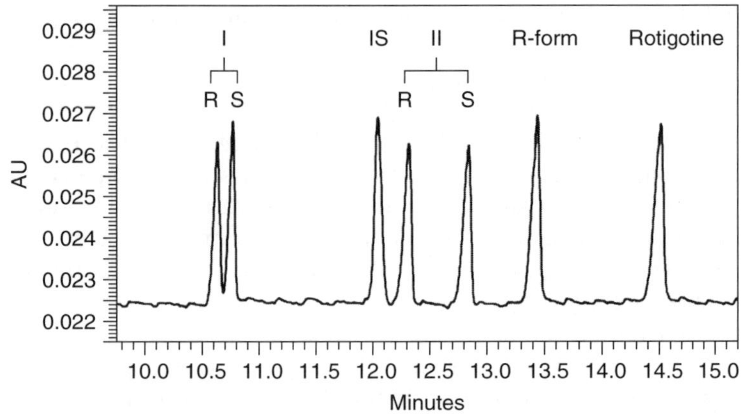

Figure 5 Electrophorogram of the simultaneous separation of rotigotine from its (*R*)-enantiomer and two of its related chiral impurities (I and II). (From [66], with permission.)

[the (*R*)-form] and two other related chiral impurities. S-β-CD provided enough separation power to resolve the enantiomers of the drug, while enantioselectivity for the impurities was given by methyl-β-CD. Therefore, a dual system based on these two CDs was used and enabled separation of the three chiral compounds. Figure 5 displays the resulting separation.

A combination of S-β-CD and β-CD allowed separation of citadiol and escitalopram from their respective enantiomeric impurities [67]. The neutral CD appeared to have a positive effect on the fronting peaks of both racemic mixtures. A face-centered central composite design was hence used to optimize the separation conditions. The factors varied in the design were S-β-CD concentration, buffer concentration, temperature, and applied voltage. The resulting method was validated and applied to the determination of (*R*)-citalopram and citadiol enantiomers in escitalopram bulk samples and tablets (which contained either escitalopram or the racemic mixture) using salicylic acid as an internal standard. In the bulk sample, about 2.3% (*R*)-citalopram and 0.1% (*S*)-citadiol could be detected. For the tablets containing only escitalopram, between 0.37 and 2% of the (*R*)-enantiomer was found, along with 0.45 to 0.72% (*S*)-citadiol. (*R*)-citadiol was not found in these tablets. In tablets containing racemic citalopram, 0.07% and 0.03% (*S*)- and (*R*)-citadiol were found, respectively. Other recent applications using dual CD systems were discussed above [35,54].

Screening Approaches Using Cyclodextrin-Based Capillary Electrophoresis When a broad or generic enantioselectivity is obtained with a certain type of chiral selector (i.e., these selectors show enantioselectivity toward a large number of compounds with different chemical and structural properties), one can consider including these selectors in a screening methodology or a separation strategy. The aim of such methodology or strategy is then to reduce the method development time when a chiral separation method is needed for quantification, enantiomeric impurity determination, or another purpose. The pharmaceutical industry uses such approaches primarily to develop methods that are able to separate potential enantiomeric drug molecules. The more generic CE-based approaches published after 2000 are highlighted below.

Perrin et al. [68] observed a very high chiral recognition capability for highly sulfated cyclodextrins (HS-CD): HS-α-CD, HS-β-CD, or HS-γ-CD in 5% (w/v) concentrations allowed the separation of 26 out of 27 analytes, of which 22 had an Rs value above 2. All experiments were conducted using the short-end injection method. Based on these experiments, a separation strategy was defined by Matthijs et al. [69]. It consists of a screening step and some potential optimization steps, depending on the outcome of the screening. The strategy also refers to the use of dual-cyclodextrin systems when no successful separation is achieved with one CD. This part of the strategy was developed in a later stage [70]. When dual systems are used, long-end injections are recommended in the screening step. The use of short-end injections is proposed when $Rs > 1.5$ is obtained after the screening. When a partial separation is obtained, the strategy indicates that the concentration of neutral CD should be altered, which can result as well in an increased as in a decreased separation, because the outcome seems compound dependent and unpredictable. In a test set of 25 compounds, 89% of the basic compounds were better separated using a dual-CD system.

Another separation strategy using HS-CD was presented by Chapman and Chen [71]. For the screening, they recommend three experiments with a starting concentration of 5% of HS-α-CD, HS-β-CD, or HS-γ-CD in a 25 mM phosphate buffer at pH 2.5. Using this procedure they were able to separate 156 out of 160 drugs with $Rs > 1.0$, indicating the generic applicability of the proposed screening. Optimization steps were also described: In case of a partial resolution, the CD content is varied. Capillary temperature and length, as well as the field strength, can be adapted to further fine-tune the separation. The authors also recommend testing a lower CD concentration when no separation is obtained.

Jimidar et al. [72] developed a very extended separation strategy. Different routes are followed depending on the compound's acidic, basic, or neutral character. Here experimental designs were used throughout the strategy to vary the type of selector (neutral types such as α-CD, HP-β-CD, DM-β-CD, TM-β-CD, γ-CD, and HP-γ-CD, and charged CDs such as HS-α-CD, HS-β-CD, HS-γ-CD, S-β-CD, SBE-β-CD, and CM-β-CD), the concentration of selector, the concentration of buffer, and the amount of organic modifier. The applicability of this strategy was demonstrated by the fact that 10 test substances were baseline separated. The optimization step of this strategy was clarified [73]; for a particular compound, separation conditions using a dual-CD system consisting of a neutral CD and a charged CD were optimized. This was done again using a Box–Behnken experimental design where the concentrations of two CDs, the buffer concentration, and the applied voltage were optimized.

Zhou et al. [74] developed a strategic approach for pharmaceutical basic compounds using sulfated cyclodextrins as chiral selectors. In the first instance, various kinds of selectors from different suppliers (i.e., three brands of randomly sulfated β-CD, randomly sulfated α-CD, single isomer sulfated γ-CD, single isomer sulfated β-CD, HDMA-β-CD, and HDAS-β-CD) are screened for their separation performances. To further optimize the separation, it is recommended that CD concentration, pH, organic modifier content (ACN/MeOH), buffer concentration and type, capillary temperature and length, and applied voltage be optimized sequentially. The approach was tested on 17 compounds, and baseline resolution was achieved for all of them.

El Deeb et al. [75] used a screening procedure where an appropriate selector was chosen based on experiments with 22 cyclodextrins: either the three native CDs or their derivatives (anionic and cationic). After this step, important parameters, such as the concentration of chiral selector, buffer pH, and percentage of organic modifier, were systematically optimized to obtain the final method. The results indicated that the majority of the basic drugs were separated by a CD that carried opposite charges. For acids, only neutral and cationic CDs were tested for their performance. However, only the cationic CD could separate this type of enantiomer successfully.

3.2 Crown Ethers as Chiral Selectors

The most frequently used chiral crown ether in capillary electrophoretic separations is 18-crown-6-tetracarboxylic acid ($18C6H_4$). This compound was synthesized for the first time in 1980 [76] and its structure is displayed in Figure 6. Like CDs, crown ethers form inclusion complexes with the analytes, due to the fact that their polyether ring also forms a cavity. However, to interact with these chiral selectors, the structure of the analyte also plays an important role [77–80]. These selectors are only useful for the separation of small molecules, and are, in particular, used for the separation of amino acids. For example, the presence of a primary amino group in the structure seems to be crucial for enantiorecognition. Bulky substituents on the chiral carbon atoms also favor enantioseparation, while a completely substituted carbon atom (in the α-position) adjacent to the amino group seems to be detrimental. On the other hand, analytes in which the amino groups are adjacent to a chiral center or up to a distance of three to four C atoms are well resolved. Finally, dipeptides containing branched groups were better resolved than those with straight-chain alkyl groups [81].

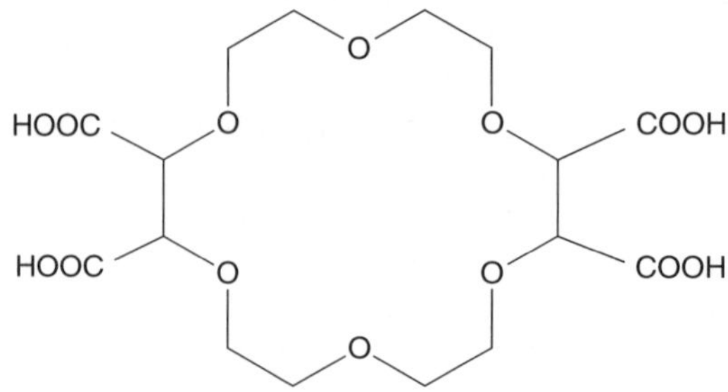

Figure 6 Chemical structure of 18-crown-6-tetracarboxylic acid.

Recent studies [82,83] also pointed out the conformational changes that occur when (+)-18C6H$_4$ interacts with an amino acid [82] or with positively charged species such as glycine$^+$, H_3O^+, $H_5O_2{}^+$, $NH_4{}^+$, and $2CH_3$ $NH_3{}^+$ [83]. When interacting with amino acids, the planar/symmetric ring conformation of the selector changes into a convex/asymmetric one. The amino group of an amino acid forms three hydrogen bonds and five electrostatic interactions with eight oxygen atoms of the carboxyl groups and the polyether ring, while the carboxyl group did not interact directly with the chiral crown ether. Next, it was suggested that the conformational changes of the ring are freely changeable within an allowable range, and the degree of variation depends on the molecular shape of and the interaction mode with the cationic guest.

Recent applications with 18C6H$_4$ have been described [84–86]. In [84], 14 amino acids were tested by means of this selector with a concentration varying between 4 and 15 mM. Nine of them (i.e., glutamine, isoleucine, methionine, phenylglycine, phenylalanine, serine, tyrosine, valine, and threonine) could be separated. Regarding detection, indirect photometric detection was used with chrysoidine as the absorbing compound in the background electrolyte.

Zhou et al. [85] studied the separation of 18 pharmaceutical compounds by means of 18C6H$_4$ as chiral selector. It was seen that this selector was able to separate 11 out of 18 compounds. All of them were primary amines, except for one tertiary amine, which is quite exceptional. Additionally, randomly sulfated β-CD and γ-CD were tested as selector. The sulfated CDs were able to separate more compounds and the resolutions obtained were overall better than with the crown ether.

The separation of di-, tri-, and tetrapeptides was investigated by Gong et al. [86] using contactless conductivity detection to reveal the compounds. Besides the crown ether 18C6H$_4$, a dual system with the crown ether and DM-β-CD gave baseline separations for compounds that could separated partially be only using the crown ether alone. Six dipeptides and three tripeptides were baseline separated using the dual-selector system. Additionally, the diastereomers of two tetrapeptides could be separated using histidine as an additive in the BGE, while 18C6H$_4$ or DM-β-CD could not induce the separation. Linear calibration curves could be constructed for two compounds, and relatively low LOD values were achieved with this type of detection, allowing impurity determinations down to 1%. Finally, this type of detection was also found useful when the separation was performed on a chip electrophoresis system (see below).

The same group [87] used contactless conductivity detection to study the separation of 1-phenylethylamine and 1-cyclohexylethylamine using a dual-selector system consisting of 18C6H$_4$ and DM-β-CD. The crown ether alone was not able to resolve the enantiomers of both compounds, whereas a combination with a CD allowed baseline separation with Rs values above 2.3. This demonstrates that for the crown ether, as for CDs, use of a dual system can be useful to improve or even induce the separation.

Three consecutive papers [88–90] covered the use of self-synthesized crown ethers and tested their potential as chiral selector in CE. Two diaza crown ether derivatives were synthesized and tested by means of five amino acid derivatives: tryptophan, tryptophan butyl ester, tryptophan ethyl ester, tyrosine methyl ester, and tryptophan butyl ester [88]. With the synthesized crown ethers alone, no separation could be induced for any of the analytes. Therefore, they were used in combination with eight CDs, and the new crown ethers induced, or enhanced, the separation compared with systems containing either CD alone, CD with 18-crown-6, or CD with Kryptofix 22 (Figure 7). In only a few cases, a decrease in separation performance was observed when the two new crown ethers were used.

Next, an experimental design approach was used to study the influence of the concentration of the racemic diaza crown ether, the concentration of either M-β-CD or DM-β-CD, and the buffer concentration on the separation of tryptophan methyl ester and tryptophan ethyl ester [89]. In general, the newly synthesized crown ether had a positive influence on the selectivity, but migration times also increased using the dual systems. Baseline separation was not achieved in every case, but the optimal conditions predicted from the design were always better than the starting conditions (i.e., the center point of the design). The best type of CD depended on the type of compound.

In a third study [90], three newly synthesized crown ethers were tested along with three commercially available crown others (18C6, $18C6H_4$, and Kryptofix 22) for separation of the positional isomers of aminophenol, aminobenzoic acid, and aminocresol. The newly synthesized crown ethers did not present an added value for this type of separation. Either the selectors were unable to separate the isomers, performed worse than the commercially available crown ethers, or were not needed to obtain a separation. The studies above indicate that the newly

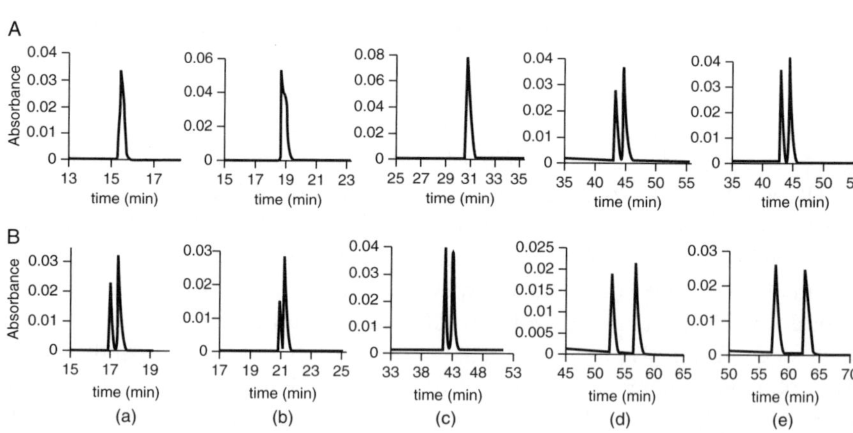

Figure 7 Electropherograms of (A) tryptophan methyl ester and (B) tryptophan ethyl ester using (a) HP-β-CD, (b) HP-β-CD + 18-crown-6, (c) HP-β-CD + tetraoxadiaza-crown ether (Kryptofix 22), (d) HP-β-CD + N-[(2R)-2-(1,4,10,13-tetraoxa-7,16-diazacyclooctadecan-7-yl)propanoyl]glycine, (e) HP-β-CD + racemic *N*-[2-(1,4,10,13-tetraoxa-7,16-diazacyclooctadecan-7-yl)propanoyl]glycine. (From [88], with permission.)

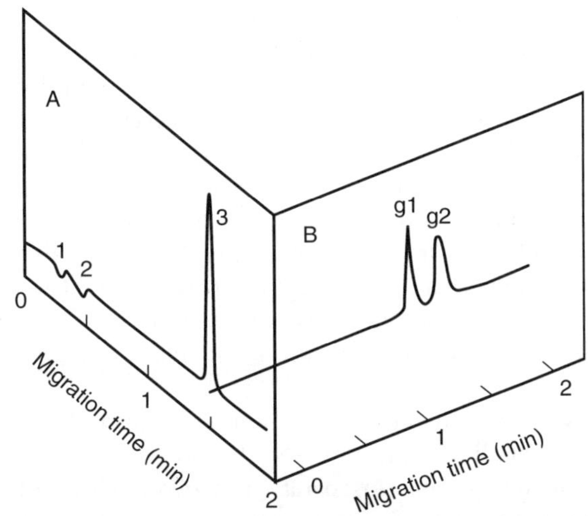

Figure 8 Separation output of an online sample cleanup microchip electrophoresis system for the enantioseparation of gemifloxacin using 18C6H$_4$ as a chiral selector: (A) removal of the metal ions in the first separation channel; (B) separation of gemifloxacin in the second separation channel. Peaks: 1, K^+; 2, Na^+; 3, gemifloxacin racemate; g_1 and g_2: gemifloxacin enantiomers. (From [92], with permission.)

synthesized crown ethers can be useful in the field of chiral separations, but for the separation of positional isomers, they are less suitable.

Also for crown ethers, some specialized applications, such as enantioseparations on microchips, can be found. Gong and Hauser [91] examined several underivatized amines in CE using dual-selector systems consisting of DM-β-CD and 18C6H$_4$ and chips with contactless conductivity detection. For one compound, *trans*-cyclohexane-1,2-diamine, the separation could be transferred to the chip-based system. However, a high resolution (>6) was needed to enable successful transfer from classical CE to the chip. Cho et al. [92] designed a complete sample cleanup and separating device to analyze urinary solutions that contain gemifloxacin. In a first separation channel, the metal ions were separated from the rest of the sample to avoid competition between the analyte and the metal ions for binding with 18C6H$_4$. This separation was followed using indirect fluorescence detection with quinine as a background chromophore. In a second separation channel, the chiral selector was present, which enabled the separation of both gemifloxacin enantiomers. The output of the separation is given in Figure 8.

3.3 Macrocyclic Antibiotics as Chiral Selectors

Macrocyclic antibiotics have been introduced by Armstrong's group as chiral selectors in several techniques, such as liquid chromatography [93] and capillary electrophoresis [94]. Although the macrocyclic antibiotics group contains many compounds with diverse structures, the most investigated, useful, and known chiral selectors are undoubtedly vancomycin, teicoplanin, and ristocetin A (Figure 9).

They are characterized by the presence of multiple chiral centers and have higher-order structures in space. For example, vancomycin has a characteristic basket shape with two side chains. Interactions that are involved in enantioselectivity with these selectors are $\pi-\pi$ interactions, dipole–dipole interactions, hydrogen bonding, and inclusion phenomena, besides the electrostatic interactions that are possible due to their ionizability [95]. A drawback that can be mentioned for these selectors is that they absorb strongly at wavelengths below 250 nm, which limits their use [95]. Also, the stability of solutions can be somewhat troublesome, ristocetin A being the most stable, followed by teicoplanin and vancomycin.

Applications with vancomycin, teicoplanin and a teicoplanin derivative have been described by Fanali's group [96–98]. Fanali et al. [96] used vancomycin as a chiral selector for the enantioseparation of two newly synthesized arylpropionic acids. Three types of CDs (6A-monomethylamino-β-CD, M-β-CD, and TM-β-CD) were also tested as a chiral selector, but vancomycin gave superior results in terms of resolution and analysis time for both compounds. A partial filling technique was used in combination with a polyacrylamide-coated capillary to reduce antibiotic adsorption to the capillary wall, avoid the presence of the antibiotic at the detector, and disable the EOF.

The separation of two non-UV absorbing compounds, aspartic and glutamic acid, was facilitated by Bednar et al. [97] using vancomycin and teicoplanin as chiral selectors. Again, the partial filling technique was used for the abovementioned reasons, and indirect UV detection was performed by means of a sorbic acid/histidine electrolyte. Vancomycin exhibited the best enantioselectivity for both compounds and that method was used successfully on teeth dentine and beer samples. A teicoplanin derivative, MDL 63 246 (Hepta Tyr), was employed by Fanali et al. [98] for the chiral separation of 7α-hydroxy acids. Parameters such as buffer pH, selector concentration, organic modifier type and concentration, and capillary temperature were systematically optimized. Finally, all compounds could be baseline separated. These experiments were conducted using the partial filling countercurrent method, in which the capillary is filled partially with chiral selector solution, followed by a sample injection and applying the electrical field such that the analytes are detected as negatively charged species at the anode while the selector moves in the opposite direction to the cathodic inlet. The latter part of this technique is also referred to as the countercurrent method. Another option to using the countercurrent method is to fill the capillary completely with selector solution prior to injection. Ristocetin A was used in combination with the countercurrent method and a coated capillary to investigate its potential as a chiral selector for the separation of 12 analytes, including nonsteroidal anti-inflammatory drugs, dansylated amino acids, and dinitrophenyl derivatives of amino acids, among others [99]. Using this selector with the countercurrent method and a complete filling of the capillary implies that first a high signal from the selector is detected because it absorbs at wavelengths below 250 nm.

A

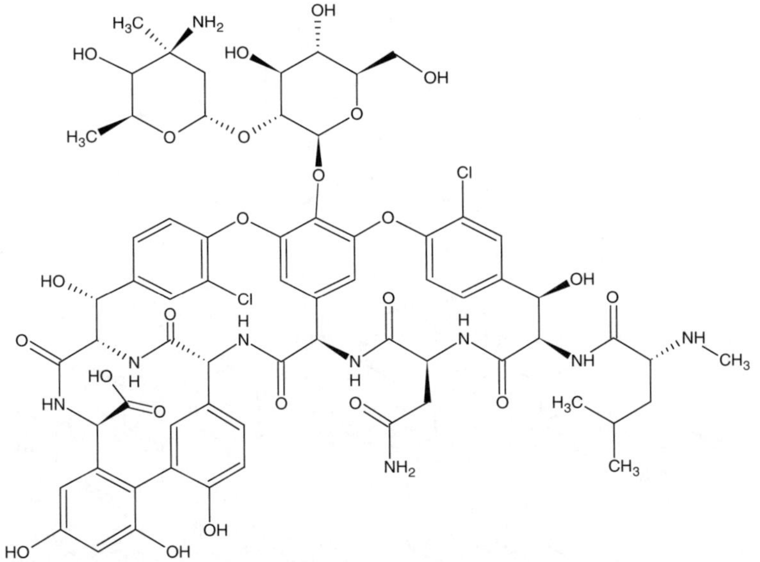

B

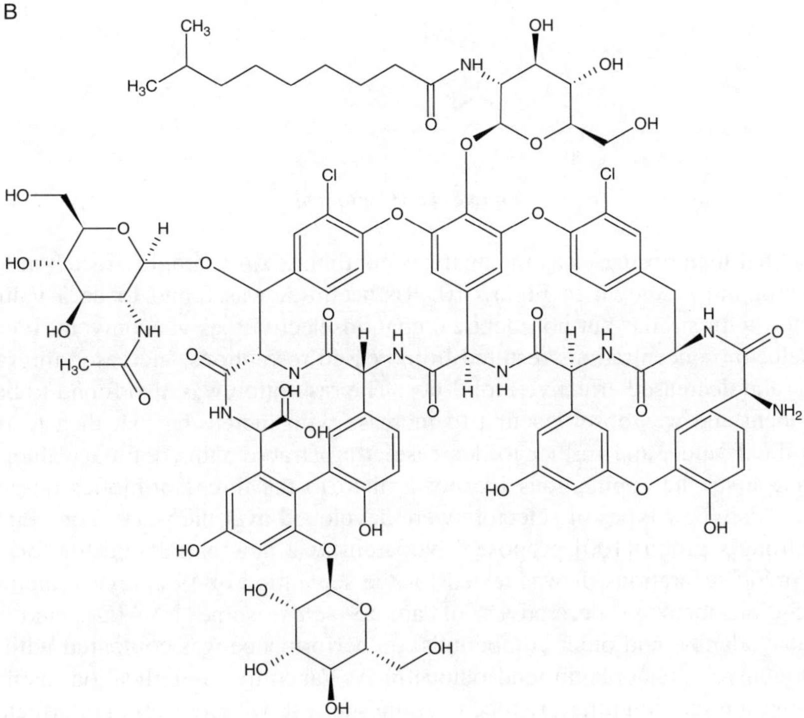

Figure 9 Structures of (A) vancomycin, (B) teicoplanin, and (C) ristocetin A.

C

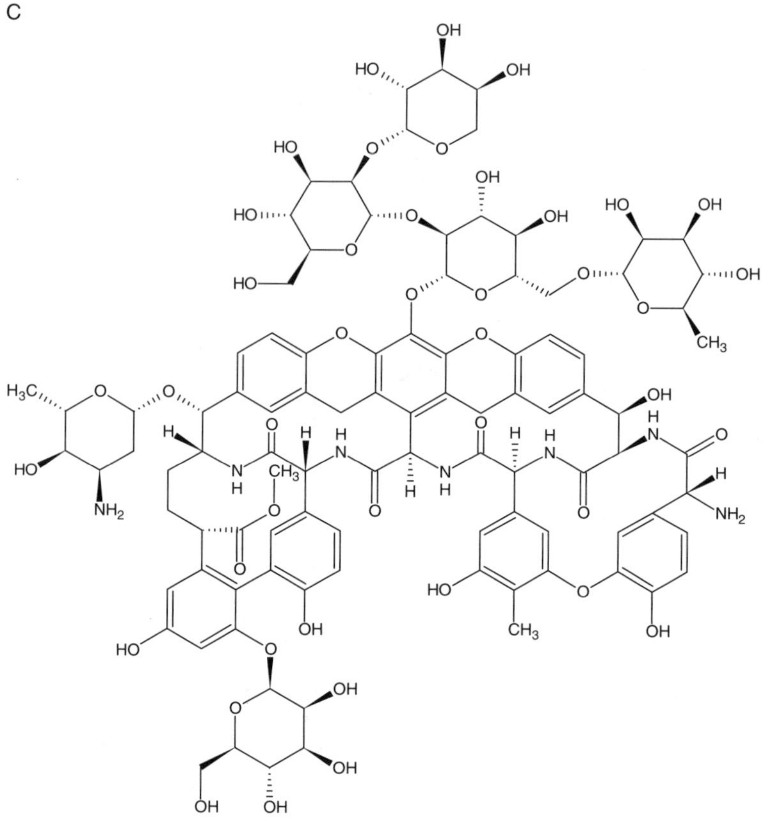

Figure 9 *(Continued)*

The signal then decreases as the analysis continues. An example of such an electropherogram is shown in Figure 10. Ristocetin A was found to be a valuable selector, with similar but not identical enantioselectivity as vancomycin. Increasing selector concentrations resulted in increased resolutions, increased migration times, and decreased effective mobilities. The resolution was also found to be pH dependent and was observed first to increase with increasing pH, then to reach an optimal value, and further to decrease at a pH above this optimal value.

Of course, the applications are not limited to the three antibiotics described above. Also, new types of selectors were developed over the years. For example, Armstrong's group [100] proposed avoparcin as a new chiral selector for electrophoretic separations. It was tested for the separation of 59 analytes, including *N*-3,5-dinitrobenzoyl derivatives of amino acids, some NSAIDs, and dansylphenylalanine and other compounds. Its performance was compared with that of vancomycin, teicoplanin, and ristocetin A. Vancomycin application displayed the longest migration times, ristocetin A the shortest, with avoparcin intermediate. Overall, all compounds would be at least partially separated with this selector. A drawback, also seen with the other macrocyclic antibiotics, is its adsorption

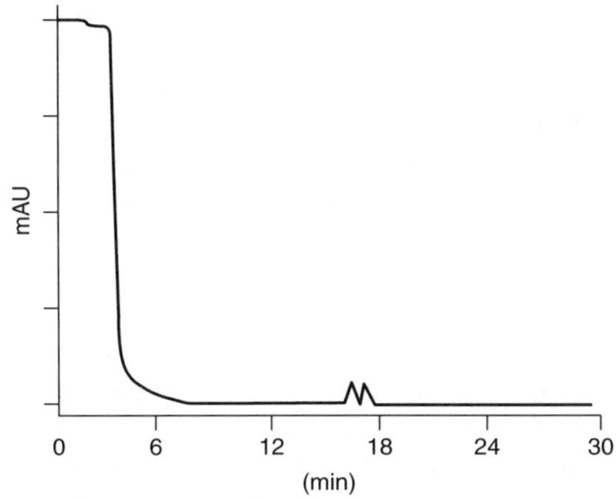

Figure 10 Separation of dansylvaline using the countercurrent method with complete filling of the capillary before injection. Conditions: 0.1 M phosphate buffer pH 6, 1 mM ristocetin A, −8 kV, detection at 230 nm. (From [99], with permission.)

onto the capillary wall and its UV adsorption at wavelengths below 250 nm. However, the latter is reduced by the low concentration that is needed for a separation and by working at detection wavelengths above 250 nm.

The macrocyclic antibiotic A35512B was used by Risley et al. [101] for the separation of 13 dansyl amino acids. The advantages of this selector are its water solubility, making it suitable for CE separations. Second, this selector also does not adsorb much at 254 nm, which allows detection of the enantiomers at this wavelength. However, at lower wavelengths its high absorbance does not allow adequate analysis. Buffer pH, selector concentration, and organic modifier composition were optimized. A concentration of 2.5 mM selector appeared to be sufficient to allow separations. With optimization of the pH, 11 compounds could be baseline separated. For the remaining two, the addition of 2-methoxyethanol was necessary to allow baseline separation.

The new chiral selectors in the macrocyclics antibiotics category most recently described are balhimycin and its analog bromobalhimycin [102]. To enable separations, a polycationic coating that disables the adsorption of the antibiotic was adsorbed to the capillary wall. Additionally, the EOF was reversed by this procedure, which enables a co-EOF separation of negatively charged species, resulting in shorter analysis times. This coating technique was combined with the partial filling method to enable detection at wavelengths below 250 nm. The 16 racemates examined (Dns-norvaline, Dns-phenylalanine, Dns-serine, Dns-leucine, Dns-norleucine, Dns-glutamic acid, Dns-valine, Dns-methionine, Dns-α-amino-*n*-butyric acid, Dns-threonine, Dns-aspartic acid, ketoprofen, 2-(2,4,5-trichlorphenoxy)-propionic acid, pirprofen, tiaprofenic acid, and flurbiprofen) could all be partially or baseline separated with both selectors, except for dansylaspartic acid, which was only resolved with balhimycin. When comparing the results obtained using these two selectors with the vancomycin results, the highest resolution was seen with balhimycin as a selector for all substances, except for tiaprofenic acid, which was best resolved with vancomycin.

Overall, it can be stated that these macrocyclic antibiotics show high potential as chiral selectors. However, their appearance in chiral CE applications tends to decrease in recent years. This is probably due to the fact that the drawbacks (i.e., UV absorption and capillary wall adsorption) are considered more problematic than the advantage of having a broad enantioselectivity.

3.4 Proteins as Chiral Selectors

When one considers the enantioselective processes in a living system, which include interactions of drugs with receptors that are composed of sugars and proteins, it seems evident that proteins can also display enantioselectivity toward chiral entities. Hence, proteins seem to be valuable candidates to be used as chiral selectors. In capillary electrophoresis, human serum albumin (HSA), bovine serum albumin (BSA), α1-glycoprotein, avidin, and riboflavin can be used, among others. Recent applications using human serum albumin were described by Medina-Hernández's group [103–105]. They used affinity electrokinetic chromatography combined with the partial filling technique to induce the separations. With this variant of CE, the capillary is partially filled with the protein solution while the rest of the capillary (with the detection window) contains electrolyte, including the vials over which the separation occurs. Experimental conditions are optimized such that chiral substances can migrate through the protein plug, be separated, and afterward, detected outside the selector plug. An advantage of this approach is that the protein solution is not electrolyzed at the inlet and outlet vials during the analysis because it is introduced hydrodynamically before the separation and not present under the capillary ends when an electrical field is applied. This allows us to use the same protein solution several times and reduces the costs of the analyses.

The separation of 28 basic drugs, including 13β-blockers, 4 phenothiazines, and 11 antihistaminics was investigated by Martínez-Gómez et al. [103]. The effects of structural and physicochemical properties of the compounds, and the percentage of binding between selector and analyte, were brought in relation to the enantioselectivity obtained using chemometric techniques such as discriminant partial least squares. Based on these results, a decision tree was proposed which predicted the capability of HSA to separate a certain substance based on its structural and physicochemical properties. Most influential factors were, not unexpectedly, found to be hydrophobicity, polarity, and molar volume. In a later study [104], the results from earlier work, among them the previous study, were used to develop a method for the enantiomeric purity control of promethazine and trimeprazine. Optimized separation conditions gave resolutions of 2.01 and 1.74, respectively, between the enantiomers. These conditions were applied to assay the racemate of both substances in commercial formulations such as syrup and drops. Concentrations ranged between 95 and 107%, which fell within tolerance limits for these formulations. The authors claim some advantages of this approach, such as simplicity, high sample throughput, reproducibility, accuracy, and low cost.

Martínez-Gómez et al. [105] used experimental design results to develop models of Rs as a function of pH and the concentration of HSA for five antihistaminics. These models made it possible to predict optimum analysis conditions for each compound: brompheniramine, chlorpheniramine, hydroxyzine, chlorcyclizine, and orphenadrine. Only for the three former analytes were the final conditions applied to commercial pharmaceutical formulations, and amounts of both enantiomers were found between 97 and 102% of the concentrations claimed by the manufacturer.

Gübitz's group [106] compared the performance of BSA and canine serum albumin for the separation of tryptophane derivatives. It was found that despite the difference in binding sites, no significant differences in enantioselectivity were found for the investigated substances. Chiral recognition was also found to be dependent on the functional groups present on the tryptophane derivatives, in addition to the classical dependence of pH, selector, and buffer concentration.

Matsannga et al. [107] synthesized and characterized three types of glycoprotein by mass spectrometry: recombinant chicken α_1-glycoprotein, completely deglycosylated α 1-glycoprotein, and native chicken α_1-glycoprotein. Then they were used as chiral selectors in CE for the separation of eperisone, pindolol, and tolperisone. Native chicken α_1-glycoprotein showed the highest enantioselectivity. When propranolol was added to the buffer as a competitive analyte, no resolution was seen for either of the three compounds, indicating that the protein domain is responsible for enantiorecognition and that it contains enantioselective sites for basic drugs.

Recent applications with avidin, riboflavin, or ovomucoid as a chiral selector are executed primarily with chromatographic techniques such as electrochromatography or liquid chromatography [108–110]. A drawback that can be reported using proteins as chiral selectors is their tendency to adsorb to the capillary wall. This results in increased band broadening and decreased resolution when repeating runs. A solution to this problem is either to wash regularly with sodium hydroxide or to use coated capillaries. Depending on the type of application, one or the other approach is most successful.

3.5 Other Chiral Selectors

This section includes the newer trends in chiral CE, where selectors other than the classical are used. The number of applications is hence still quite low. A less common selector, which can also be used in chiral CE, is penicillin G acylase. It is an industrially important enzyme which is used in the production of antibiotics such as penicilins and cephalosporins [111]. In CE it has been found useful to separate acidic species, but as with all proteins, adsorption of this selector to the capillary wall is often seen. Gotti et al. [112] described an approach in which a capillary coated with pullulan, a high-molecular-mass polysaccharide, was used. Seven acidic racemates—4-bromomandelic acid, β-phenyllactic acid, 2-(4-chlorphenoxy)phenyllactic acid, 2-(4-chlorphenoxy)phenylpropionic acid, flurbiprofen, fenoprofen, and suprofen—could be separated using this chiral selector.

Ionic liquids (ILs) are nonmolecular solvents composed of ionic species and have a melting point below 100°C. Some interesting properties are associated with these liquids, including negligible vapor pressure, high ionic conductivity, thermal stability, and tunable viscosity, and they are also miscible with aqueous and organic solvents. In chiral separations they can be used in two modes: either a chiral IL is used as a selector or a chiral selector can be dissolved in an achiral IL [113]. Applications using IL in chiral CE have been described [114,115]. Enantioseparations were achieved for some 2-arylpropionic acid derivatives, such as carprofen, suprofen, indoprofen, ketoprofen, and ibuprofen, using TM-β-CD or DM-β-CD combined with ethyl- and phenylcholine of bis(trifluoromethylsulfonyl)imide as IL [114]. The ILs alone were insufficient to obtain a separation, but the combination with CD appeared to give a synergistic effect. When *S*-[3-(chloro-2-hydroxypropyl)trimethylammonium] [bis((trifluoromethyl)sulfonyl)amide] was used as an IL, the chiral IL alone was insufficient to facilitate the separation of several drugs such as naproxen, propranolol, atenolol, ibuprofen, indoprofen, flurbiprofen, and warfarin [115]. The addition of a chiral anion (sodium cholate) and/or a neutral chiral selector was always needed to enable enantiorecognition, which indicated that the chiral cation of the IL provided at least one of the three interaction points needed for recognition.

Aptamers can also be classified among the newer classes of chiral selectors used. They are single-stranded oligonucleotides with binding properties originating from the SELEX methodology (in vitro selecetion experiments). Due to their capability to bind specifically to target molecules with an equal or better affinity than antibodies, their easy regeneration through denaturation and the relatively good stability of the DNA aptamers, they seem good candidates to enable chiral separations. The first paper that reported the use of an antiarginine RNA aptamer for the enantioseparation of arginine enantiomers was from the group of Peyrin [116]. They used a partial filling technique which allowed detection of the compound without any interference from the UV absorbance of the selector. The study revealed that there was a strong heterogeneity in the target binding sites of the aptamer, due to different unfolding states at different temperatures. This suggested that a narrow temperature range was indispensable to obtain good separation efficiencies and peak shapes when the aptamer is to be used for enantioseparations. The same group also used this aptamer to develop a method that was able to detect trace amounts of D-arginine down to a 0.01% level in the presence of L-arginine, making it suitable for enantiomeric impurity analysis [117].

4 CHIRAL SEPARATIONS IN LIGAND-EXCHANGE CAPILLARY ELECTROPHORESIS

When a ligand-exchange system is used in capillary electrophoresis, the separation mechanism is based on the interchanges of a ligand and an analyte in a complex formed with a central metal ion. Therefore, this technique can only be used for analytes that have complex-formation properties (i.e., act as a ligand). Substances that can be used as ligands are amino acids, hydroxy acids, amino alcohols, diamines, and some dipeptides and polypeptides. Enantiorecognition is based on an enantioselective complexation between the enantiomers that have to be separated and the metal–ligand complex to form a metastable ternary diastereomer [118]. Separation is only seen when a stability constant difference is present for the enantiomer–metal–ligand complexes of both stereoisomers [8,119,120].

The first papers regarding ligand-exchange CE were published by Zare's group [121,122], who used copper complexes of histidine or aspartame for the chiral separation of dansylated amino acids. Underivatized amino acids were separated for the first time by Schmid and Gübitz using copper complexes of L-proline and L-4-hydroxyproline [123]. Further work of this group involved the use of *N*-(2-hydroxy-octyl)-L-4-hydroxyproline/Cu(II) for the separation of α-amino acids [124], the separation of β-methyl amino acids using Cu(II) complexes of L-4-hydroxyproline, *N*-(2-hydroxypropyl)-L-4-hydroxyproline, and *N*-(2-hydroxyoctyl)-L-4-hydroxyproline as chiral selectors [125], among other applications.

Recently, Gübitz's group investigated the possibility of using copper, cobalt, nickel, and zinc complexes of three sugar acids—D-gluconic acid, D-saccharic acid, and L-threonic acid—as selectors for the enantioseparation of amino acids and glycyl peptides [126]. Cu(II), the most frequently used metal ion in ligand exchange CE, was also found to lead to the best separations, in combination with either saccharic acid or threonic acid. With gluconic acid, on the other hand, Co(II) was found to give the highest resolutions for amino acids, while a combination of this sugar acid with Ni(II) resulted in only partial separations. With Zn(II), bad results were obtained. Four ligands were studied by Koidl et al. [127] in combination with Cu(II)—L-4-hydroxyproline, L-histidine, *N*-(2-hydroxypropyl)-L-4-hydroxyproline, *N*-(2-hydroxyoctyl)-L-4-hydroxyproline—for the separation of five halogenated amino acids. *N*-(2-hydroxyoctyl)-L-4-hydroxyproline enabled baseline separation of all analytes investigated.

Chen's group investigated the applicability of Zn(II) complexes in two papers [128,129]. A combination of this metal ion with L- or D-lysine was used as a chiral selector for the enantioseparation of three aromatic amino acids: tryptophane, phenylalanine, and tyrosine [128]. Resolutions up to 7 were obtained. Besides the fact that common factors such as pH, concentration of buffer, metal ion type, ligand concentration with respect to the metal ion concentration, and buffer type were optimized, the authors described the necessity of using an enantiopure ligand: either L- or D-lysine. Reversal of elution order was observed when changing from one to the other enantiomer, whereas no separation was seen when the racemate was used as ligand. Separation of the same three aromatic amino acids was investigated using several ligands, including basic (L-arginine, L-lysine), acidic (L-glutamic acid) and neutral ones (L-proline, L-alanine, L-valine, L-leucine) [129]. The basic ligands performed better than the acidic or neutral ligands, with the best results obtained with L-lysine and L-arginine. However, the performance of a ligand also depended on the pH. The final method developed used L-arginine at pH 8.2 and enabled the quantification of all six enantiomers. This method could be used for the determination of these amino acids in rice-brewed suspensions.

CDs can also be used as ligands in ligand-exchange CE. For example, in one study by Vecchio's group [130], five pure β-CD derivatives were synthesized and used as chiral selectors with Cu(II) as metal ion. All structures contained nitrogen atoms, which were responsible for the electron-pair donor properties of the CDs. The CDs that bear a charge and posses ligand-exchange properties toward metal ions were found to be very good selectors.

Applications using Cu(II) as a central ion with other ligands have been described [131,132]. In L-Prolinamide was used as a ligand for the separation of 12 dansylated amino acids [131]. Depending on the pH, the prolinamide complex with copper was present in various forms. At pH 6, the highest enantioselectivity was obtained. After optimization of copper, ligand, and buffer concentrations, baseline separation of all amino acids was enabled. L-Ornithine was used as ligand for the separation of underivatized and dansyl amino acids, a dipeptide, ephedrine, and pseudoephedrine [132]. The chiral recognition was again found to depend on pH and complex concentration. Positively charged ornithine was adsorbed to the capillary wall, which resulted in decreased peak efficiencies. Therefore, a propane diamine–coated capillary was used, reducing the adsorption and leading to improved separation efficiencies. All analytes investigated were either partially or baseline separated.

5 CHIRAL SEPARATIONS IN NONAQUEOUS CAPILLARY ELECTROPHORESIS

Nonaqueous capillary electrophoresis (NACE) is the latest mode of CE to be developed. From the late 1990s, its application began to rise after a long period of development, starting in the early 1990s [133]. In NACE, mostly methanol, acetonitrile, and their mixtures are employed as organic solvents, but applications with *N*, *N*-dimethylformamide or *N*-methylformamide are also described. Frequently used buffering systems are composed of acetic acid or ammonium acetate. Advantages that are associated with the technique are its ability to analyze analytes that are insoluble in water and analytes that have the same or close electrophoretic mobilities in aqueous environments. Additionally, the use of organic solvents allows the use of interactions for separations that either cannot take place or that are too weak in an aqueous environment, specifically ionic interactions, dipole interactions, or hydrogen bondings. Other advantages are an increased stability of some analytes in NA environments and easy coupling to mass spectrometric detection, for example.

When screening the literature for applications, one observes that more applications in NACE are described than, for example, applications using proteins or macrocylic antibiotics in an aqueous environment. The potential of this mode thus seems fully discovered, and its application will probably continue to rise. Some selected applications from the last five years are discussed below. Table 2 provides an overview of the analytes separated using NACE.

5.1 Applications with Cyclodextrins as Chiral Selectors

The use of a single-isomer CD, heptakis(2,3-diacetyl-6-*O*-sulfo) cyclomaltoheptaose (HDAS-β-CD), in NACE was studied by Vigh's group [134]. They found that the selection of the counter ions (i.e., tetrabutylammonium or sodium salt of the CD) was of importance for the enantioselectivity observed in NACE. A method was developed by Olsson et al. [135] for the simultaneous enantioseparations of omeprazole and its metabolite 5-OH-omeprazole using heptakis-(2,3-di-*O*-methyl-6-*O*-sulfo)-β-CD (HDMS- β-CD) as a chiral selector. Methanol was used in combination with ammonium acetate as BGE. A low temperature (16°C) was used to avoid current breakdown due to a Joule heating effect. The four enantiomers could be baseline separated and the method was validated in terms of precision, linearity, LOD, and LOQ.

Four neutral and water-insoluble pesticides, pyraclofos, profenofos, prothiofos, and sulprofos were analyzed in NACE using sodium cholate with either SDS or γ-CD as chiral selectors [136]. For pyraclofos, SDS had to be added to cooperate with the selector sodium cholate and the separation could only be facilitated using high amounts of SDS and sodium cholate. For profenofos, prothiofos, and sulprofos, only a combination of sodium cholate and γ-CD could enable the enantioseparation. It was, however, found that the addition of a small amount of water in the BGE was necessary to dissolve the cyclodextrin. The final methods were used successfully on spiked soil samples.

Mol et al. described the coupling of NACE with mass spectrometric detection for both chiral and nonchiral applications [137–139]. A setup was presented that used anionic CDs HDAS-β-CD and HDMS-β-CD, as chiral selectors [137]. The experiments were conducted in such a manner that the CD migrated toward the capillary inlet during analysis to reduce the risk of ionization suppression because of CD being present in the ion source. The sodium counterions of the CD still migrated toward the source but did not cause any significant suppression. Using the setup developed, no enantioselectivity was observed with either of the CDs. The addition of (1*S*)-(+)-10-camphorsulfonate, an ion-pairing reagent, allowed the separation of mebeverine and five related compounds within one run. Using this setup, LODs in the sub-μg/mL range were obtained. The same mode of operation, making the CD migrate toward the capillary inlet during analysis, was used [138] for the separation of test mixtures of both chiral and nonchiral acidic substances in combination with positively charged single-isomer CDs: 6-monodeoxy-6-mono(3-hydroxy)propylamino-β-CD or 6-monodeoxy-6-mono(2-hydroxy)propylamino-β-CD. Again, a LOD value below 1μg/mL could be obtained and the ionization suppression observed could be assigned to the chloride counterions of the CD. Five NSAIDs could be enantioseparated using this approach, with Rs values between 0.9 and 2.2. A CD-based

TABLE 2 Compounds Separated Using NACE

Compounds (arranged by reference number)	Background Electrolyte	Selector	Applied Voltage, Temperature	Refs.
Mepenzolate bromide, homatotropine HBr, halostachin, isoproterenol-HCl	50 mM methanesulfonic acid–21 mM TEA–ACN[a]	2 mM sodium salt of HDAS-β-CD	15 kV, 20°C	[134]
Omeprazole, 5-OH-omeprazole	1 M formic acid–30 mM ammonium acetate–MeOH	30 mM HDMS-β-CD	25 kV, 16°C	[135]
Pyraclofos	MeOH/ACN [4:1 (v/v)]	100 mM SDS and 50 mM sodium cholate	30 kV, 25°C	[136]
Profenofos	MeOH/H_2O/ACN [5:4:1 (v/v/v)]	20 mM γ-CD and 50 mM sodium cholate	30 kV, 25°C	[136]
Prothiofos	MeOH/H_2O/ACN [5:4:1 (v/v/v)]	20 mM γ-CD and 75 mM sodium cholate	30 kV, 25°C	[136]
Sulprofos	MeOH/H_2O/ACN [5:4:1 (v/v/v)]	10 mM γ-CD and 50 mM sodium cholate	30 kV, 25°C	[136]
Mebeverine and five related compounds	0.75 M formic acid, 30 mM potassium camphor sulfonate in MeOH	30 mM HDMS-β-CD	30 kV, NI[b]	[137]
Ibuprofen, fenoprofen, flurbiprofen, ketoprofen, indoprofen	20 mM ammonium acetate in MeOH	20 mM 6-monodeoxy-6-mono(3-OH) propylamino-β-CD	−30 kV, NI[b]	[138]
23 N-protected amino acids	100 mM octanoic acid, 12.5 mM ammonia in EtOH/MeOH [60:40 (v/v)]	10 mM *tert*-butyl carbamoylated quinine	−25 kV, 15°C	[140]

Acebutolol, atenolol, betaxolol, celiprolol, metoprolol, pindolol, propranolol, bupivacaine mepivacaine, prilocaine, salbutamol, terbutaline, trimipramine, verapamil	1 M formic acid and 40 mM potassium camphor sulfonate or 40 mM NH_4Cl in MeOH	10 mM HDMS-β-CD	25 kV, 20°C	[141]
Salbutamol, terbutaline, propranolol, mepivacaine, bupivacaine, prilocaine	0.75 M formic acid and 10–33 mM potassium camphor sulfonate in MeOH	30 mM HDMS-β-CD	25 kV, 15°C	[142]
Bupivacaine, mepivacaine, propranolol	Methanolic solution containing 10 mM ammonium formate acidified with 0.75 M formic acid	30 mM HDMS-β-CD	25 kV, 15°C	[143]
Prilocaine	Methanolic solution of 40 mM potassium camphor sulfonate acidified with 0.75 M formic acid	30 mM HDMS-β-CD	25 kV, 15°C	[143]
Atenolol, celiprolol	Methanolic solution of 40 mM potassium camphor sulfonate acidified with 0.75 M formic acid	10 mM HDMS-β-CD	25 kV, 15°C	[143]
Timolol	30 mM potassium camphor sulfonate with 0.75 M formic acid in MeOH	30 mM HDMS-β-CD	25 kV, 15°C	[142,144–146]

(*Continued ovrleaf*)

TABLE 2 (*Continued*)

Compounds (arranged by reference number)	Background Electrolyte	Selector	Applied Voltage, Temperature	Refs.
Ibuprofen, indoprofen	40 mM ammonium acetate in MeOH	5 mM 6-monodeoxy-6-mono(2-hydroxy)propylamino-β-CD	25 kV, 15°C	[147]
Tiaprofenic acid, suprofen, flurbiprofen	40 mM ammonium acetate in MeOH	10 mM 6-monodeoxy-6-mono(3-hydroxy)propylamino-β-CD	25 kV, 15°C	[147]
Fenoprofen, ketoprofen	20 mM ammonium acetate in MeOH	20 mM 6-monodeoxy-6-mono(3-hydroxy)propylamino-β-CD	25 kV, 15°C	[147]
Flurbiprofen	40 mM ammonium acetate in MeOH	10 mM 6-monodeoxy-6-mono(3-hydroxy)propylamino-β-CD	−25 kV, 15°C	[148]
	40 mM ammonium acetate and 20 mM ammonium camphor sulfonate in MeOH	20 mM 6-monodeoxy-6-mono(2-hydroxy)propylamino-β-CD	−25 kV, 15°C	[148]
Atenolol, bambuterol, ephedrine, isoprenaline, pindolol, propranolol, pseudoephedrine, sotalol, terbutaline	40 mM KOH in MeOH/EtOH [2:3 (v/v)]	100 mM diisopropylideneketogulonic acid	30 kV, 25°C	[150]
Atenolol, bambuterol, isoprenaline, pindolol, propranolol, sotalol, terbutaline, timolol	40 mM KOH in MeOH/EtOH [2:3 (v/v)]	100 mM ketopinic acid	30 kV, 25°C	[150]

[a]TEA, triethylamine.
[b]NI, not indicated.

method was used for the separation of several nonchiral drugs, including an artificial test mixture and a one-year-old amiodarone sample [139]. Briefly, it was mentioned that HDMS-β-CD showed some enantioselectivity for atropine, the only chiral drug in the set of compounds investigated.

Crommen's group [140–149] investigated the possibilities of NACE for chiral separation extensively in recent years. Their studies are often characterized by the use of experimental designs to examine simultaneous influences or to optimize several factors. Moreover, the methods developed are validated, and robustness and interlaboratory studies are also described for certain applications. Mostly, their applications consider the use of CD derivatives as chiral selectors, but Piette et al. [140] described the use of a weak anion exchanger, *tert*-butyl carbamoylated quinine. Several N-protected amino acids were selected as test compounds using a countercurrent method. The selectivities obtained by NACE were compared with those obtained by HPLC using a chiral stationary phase with immobilized *tert*-butyl carbamoylated quinine. A good correlation was found for the two techniques, indicating that enantioselectivity for HPLC can be predicted from CE in certain conditions; thus, CE can be used as a preliminary screening tool.

The separation of basic compounds was also investigated in detail. Servais et al. [141], discovered the synergistic effects of ion-paring reagents in combination with an anionic CD derivative on the enantioresolution of basic compounds. The addition of (+)-*S*-camphorsulfonate or alkanesulfonates combined with the CD appeared to have a positive effect on the separation, while use of the CD alone resulted in reduced or lost resolutions. The nature of the CD, the ion-pairing reagent, and their concentrations were found to have a great influence, and the best results were obtained using HDMS-β-CD and camphorsulfonate in a methanolic BGE. When this work was continued [142], an experimental design approach was used to study the influence of the concentrations of camphorsulfonate and HDMS-β-CD on the separation. Compounds showing a high affinity for the CD benefit from the presence of camphorsulfonate, because it acts as a competitor. For others having a lower affinity, the benefit is present only when the optimal CD concentration is reached. For compounds with very low affinities, the combination of CD and ion-pairing reagent always had a negative effect. Separation of timolol enantiomers could be achieved with high resolution in only three experiments based on the prior knowledge derived from the experimental designs.

Servais et al. then studied the influence of the nature of the electrolyte, that is, the cationic component (sodium, potassium, or ammonia) and the anionic compound (chloride, formate, methanesulfonate, and camphorsulfonate) of the BGE, in greater detail [143]. It was found that two electrolytes were interesting in developing chiral separations using NACE. The first is a methanolic solution containing 10 mM ammonium formate and 30 mM HDMS-β-CD acidified with 0.75 M formic acid. This system enabled the separation of compounds that had a very low affinity for the CD. A second system consists of a methanolic solution of 40 mM potassium camphor sulfonate acidified with 0.75 M formic acid and containing varying concentrations of HDMS-β-CD, which is applicable on compounds having either a high affinity for the CD, or a low affinity but where the CD is above its optimal concentration.

Based on the conditions found by Servais et al. when separating timolol enantiomers [142], three papers reported on continued work on this compound. First, the method was validated [144]. Next, a robustness study [145] of this method and an interlaboratory study [146] were undertaken to evaluate the method that enabled the enantiomeric impurity determination of *R*-timolol in *S*-timolol samples. Also, for the enantioseparation of acidic pharmaceuticals, this group conducted additional investigations. Some NSAIDs could be separated using positively charged CDs as chiral selectors [147]. Two types of CD were considered, 6-monodeoxy-6-mono(3-hydroxy)propylamino-β-CD and 6-monodeoxy-6-mono(2-hydroxy)propylamino-β-CD, combined with ammonium acetate in a methanolic BGE. A maximum resolution was seen for all substances with 20 mM 6-monodeoxy-6-mono(3-hydroxy)propylamino-β-CD, but shorter migration times were observed with the other. Three groups of analytes could be distinguished, each having different optimal analysis conditions (for more details, see Table 2). The conditions determined [147] were used further to develop a method to determine flurbiprofen enantiomers in human plasma using flufenamic acid as an internal standard [148]. To clean up the samples, a solid-phase extraction procedure was optimized. The washing steps of the capillary between and after runs appeared to be very important to avoid positively charged CD adsorption onto the capillary wall. The method was validated in terms of selectivity, trueness, precision, accuracy, and linearity.

Another method was developed for the enantiomeric impurity determination of (*R*)-flurbiprofen [149]. The method with 6-monodeoxy-6-mono(3-hydroxy)propylamino-β-CD was not able to induce a separation in samples spiked with a low concentration of (*R*)-flurbiprofen. Therefore, 6-monodeoxy-6-mono(2-hydroxy)propylamino-β-CD was tested, and a baseline separation of both enantiomers was observed. Ammonium camphor sulfonate was found to be necessary to reduce the CD adsorption onto the capillary wall. Using standard addition, 0.1% of the impurity could be determined using the optimized and validated method.

5.2 Other Chiral Selectors

Two selectors were used by Pettersson's group for the separation of 10 basic compounds [150]. One selector, diisopropylideneketogluconic acid, had been developed previously [151], and the other, ketopinic acid, was used for the first time. Both selectors are chiral ion-pair selectors and were compared in terms of performance and enantiomer migration order. Ketopinic acid was able to separate seven out of 10 analytes: atenolol, bambuterol, isoprenaline, pindolol, propranolol, sotalol, terbutaline, and timolol. A reversal of migration order, and generally a lower resolution, were, however, obtained with this selector compared to diisopropylideneketogulonic acid. For timolol, a method with ketopinic acid was developed and the enantiomeric impurity could be detected down to a 0.2% level using a transient isotachophoresis preconcentration technique. For ephedrine, the same method was used but with diisopropylideneketogulonic acid as selector. This method was able to detect the impurity down to 0.033% without any preconcentration.

6 CHIRAL SEPARATIONS IN MICELLAR ELECTROKINETIC CHROMATOGRAPHY

Separation of neutral achiral species is not possible in CE because separation is based on the differences in electrophoretic mobilities of the analytes under the influence of an electrical field. The electrophoretic mobility itself depends on the charge/mass ratio of a substance, but as neutral analytes do not have a charge, their electrophoretic mobility is zero. This means that they all elute together with the EOF. To achieve the separation of these neutral compounds, micellar electrokinetic chromatography (MEKC) was introduced by Terabe et al. in 1984 [152]. With this technique, an ionic surfactant above its critical micellar concentration is used as a pseudostationary phase in the BGE. The surfactant forms micelles when it is above its critical micellar concentration, and the electrophoretic mobility of these micelles differs from that of the surrounding BGE. A very popular, negatively charged surfactant in MEKC is SDS. Analytes, neutral and others, will have different interactions with these micelles; that is, they can be adsorbed on the surface by dipole or electrostatic interactions, they can behave as a cosurfactant by participation in the formation of the micelle, or they can be incorporated into the core of the micelle [8]. This will finally result in different mobilities for the neutral analytes, and hence their separation.

For enantiomer separations, MEKC is also a commonly used technique. Two modes of separation can be adopted: the use chiral surfactants or the use of CD-modified MEKC [153]. When chiral surfactants are used, two approaches are possible: One either uses a chiral ionic surfactant as pseudostationary phase, or one uses a neutral chiral surfactant combined with an achiral ionic surfactant. Usually, neutral chiral surfactants are used in combination with SDS to achieve enantiorecognition. Examples of natural chiral surfactants that are commonly used include bile salts such as sodium cholate, sodium deoxychlote, sodium taurochloate, and sodium taurodeoxycholate, as well as saponins and digitonin. Synthetic chiral surfactants such as *N*-alkanoyl-L-amino acids, *N*-dodecoxycarbonyl amino acids, alkylglucoside surfactants, and chiral high-molecular-mass surfactants such as poly(sodium-*N*-undecylenyl-L-valinate) are also used to facilitate enantioseparations in MEKC.

In CD-modified MEKC, native and derivatized CD can be used as chiral selectors, combined with achiral ionic micelles, which form the pseudostationary phase. Many chiral separations have been performed using MEKC. However, their discussion, if considered relevant, would require a separate chapter. A similar remark could be made concerning capillary electrochromatography (CEC), another technique derived from CE, but where instead of a pseudostationary phase, a real stationary phase is used in the capillary.

7 CONCLUDING REMARKS

When looking at the type of selector that is most frequently used in chiral CE, it is very clear that cyclodextrins are the first choice of selectors. Numerous applications over the past years indicate their potential and popularity. The development of anionic derivatives has boosted their popularity. Most applications with this type of selector also describe method validation to a certain extent, which indicates that quite repeatable results can be obtained that allow assaying enantiomers. Some derivatives, such as the highly sulfated CDs, show generic enantioselectivity toward a large number of structurally diverse compounds. This means that they are suitable for defining screening approaches or separation strategies, which is very interesting in an industrial environment (e.g., in drug development and quality control). This can explain the 18% market share of applications described from the pharmaceutical industry and the continuous growth predicted in this field, whereas applications in academic and governmental environments are predicted to stagnate [154]. In NACE applications, also practically only this type of selector is considered. Besides the number of applications in aqueous environments in recent years, the NACE mode provides the second-highest number. Here also, extended-method validation characterizes the quantitative application of the methods developed.

For crown ethers, macrocyclic antibiotics, and proteins, we observed that fewer applications have been described in recent years. This can probably be explained by the disadvantages that characterize these selectors. With crown ethers, only small molecules bearing an amino group, such as amino acids, can be separated, although occasionally the separation of a small drug molecule was reported. The same applies for ligand-exchange CE, where the analytes must have free electron pairs and where applications are also limited primarily to amino acids. For macrocyclic antibiotics, the number of applications reported has been reduced significantly in recent years. This can be assigned to their limited enantioselectivity in CE and the fact that they absorb UV light at wavelengths below 250 nm and tend to adsorb onto the capillary wall, which imposes the use of countercurrent techniques and partial filling techniques to perform separations. Adsorption onto the capillary wall and limited enantioselectivity might also be the reasons why proteins are not used as frequently. Some newer chiral selectors, such as penicillin G acylase, chiral ionic liquids, and aptamers have also found their way into CE enantioseparations, but still need to prove their broad applicability.

To summarize, it can be expected that the growth of CE in the chiral separation field will continue in both aqueous and nonaqueous environments. Further, it can be expected that the popularity of cyclodextrins as chiral selectors will remain or even increase, with an emphasis on the use of single isomeric derivatives to avoid between-batch variability.

Acknowledgment

D. Mangelings is a postdoctoral fellow of the Research Foundation Flanders (FWO).

REFERENCES

1. I. Ali, V. K. Gupta, H. Y. Aboul-Enein. Chiral resolution of racemic environmental pollutants by capillary electrophoresis. *Crit. Rev. Anal. Chem*., 38:132–146, 2008.
2. G. Gübitz, M. G. Schmid. Chiral separations by capillary electromigration techniques. *J. Chromatogr. A*, 1204:140–156, 2008.
3. B. Chankvetadze. Enantioseparations by using capillary electrophoretic techniques: the story of 20 and a few more years. *J. Chromatogr. A*, 1168:45–70, 2007.
4. A. Rizzi. Fundamental aspects of chiral separations by capillary electrophoresis. *Electrophoresis*, 22:3079–3106, 2001.
5. B. Chankvetadze, G. Blaschke. Enantioseparations in capillary electromigration techniques: recent developments and future trends. *J. Chromatogr. A*, 906:309–363, 2001.
6. S. Fanali. Enantioselective determination by capillary electrophoresis with cyclodextrins as chiral selectors. *J. Chromatogr. A*, 875:89–122, 2000.
7. S. Fanali, Z. Aturki, C. Desiderio. Enantioresolution of pharmaceutical compounds by capillary electrophoresis: use of cyclodextrins and antibiotics. *Enantiomer*, 4:229–241, 1999.
8. B. Chankvetadze. *Capillary Electrophoresis in Chiral Analysis*. Wiley, Chichester, UK, 1997.
9. S. A. C. Wren, R. C. Rowe. Theoretical aspects of chiral separation in capillary electrophoresis. I. Initial evaluation of a model. *J. Chromatogr*., 603:235–241, 1992.
10. S. A. C. Wren, R. C. Rowe. Theoretical aspects of chiral separation in capillary electrophoresis: II. The role of the organic solvent. *J. Chromatogr*., 609:363–367, 1992.
11. S. A. C. Wren, R. C. Rowe, R. S. Payne. A theoretical approach to chiral capillary electrophoresis with some practical implications. *Electrophoresis*, 15:774–778, 1994.
12. Y. Y. Rawjee, D. U. Staerk, G. Vigh. Capillary electrophoretic chiral separations with cyclodextrin additives: I. Acids: chiral selectivity as a function of pH and the concentration of β-cyclodextrin for fenoprofen and ibuprofen. *J. Chromatogr*., 635:291–306, 1993.
13. Y. Y. Rawjee, R. L. Williams, G. Vigh. Capillary electrophoretic chiral separations using β-cyclodextrin as resolving agent. II. Bases: chiral selectivity as a function of pH and the concentration of β-cyclodextrin. *J. Chromatogr*., 652:233–245, 1993.
14. Y. Y. Rawjee, R. L. Williams, G. Vigh. Capillary electrophoretic chiral separations using cyclodextrin additives. III. Peak resolution surfaces for ibuprofen and homatotropine as a function of pH and the concentration of β-cyclodextrin and ibuprofen. *J. Chromatogr. A*, 680:599–607, 1994.
15. Y. Y. Rawjee, R. L. Williams, L. A. Buckingham, G. Vigh. Effects of pH and hydroxypropyl-β-cyclodextrin concentration on peak resolution in the capillary electrophoretic separation of weak bases. *J. Chromatogr. A*, 688:273–282, 1994.
16. Y. Y. Rawjee, G. Vigh. A peak resolution model for the capillary electrophoretic separation of weak acids with hydroxypropyl-β-cyclodextrin containing background electrolytes. *Anal. Chem*., 66:619–627, 1994.
17. G. K. E. Scriba. Cyclodextrins in capillary electrophoresis enantioseparations: recent development and applications. *J. Sep. Sci*., 31:1991–2011, 2008.

18. M. E. Davis, M. E. Brewster. Cyclodextrin-based pharmaceutics: past, present and future. *Nat. Rev. Drug Discov.*, 3:1023–1035, 2004.

19. G. Gübitz, M. G. Schmid. Recent progress in chiral separation principles in capillary electrophoresis. *Electrophoresis*, 21:4112–4135, 2000.

20. D. Tsimachidis, P. Cesla, T. Hájek, G. Theodoridis, P. Jandera. Capillary electrophoretic chiral separation of *Cinchona* alkaloids using a cyclodextrin selector. *J. Sep. Sci.*, 31:1130–1136, 2008.

21. F. K. Glówka, M. KaraYniewicz-Lada. CE determination of ketoprofen enantiomers in clinical samples of plasma, synovial fluid and urine. *Chromatographia*, 67: S97–S105, 2008.

22. B.-L. Li, Z.-G. Shang, L.-L. Du, W. Wang. Chiral resolutions of (9-anthryl)methoxyacetic acid and (9-anthryl)hydroxyacetic acid by capillary electrophoresis. *Chirality*, 20:35–39, 2008.

23. T. Q. Trung, P. T. Dung, N. N. Hoan, D. J. Kim, J. H. Lee, K. H. Kim. Chiral separation of fluvastatin enantiomers by capillary electrophoresis. *Arch. Pharm. Res.*, 31:1066–1072, 2008.

24. P. T. Dung, M. Y. Ko, H. J. Choi, K. S. Sin, K. H. Kim. Determination of enantiomeric impurity of etodolac by capillary electrophoresis using (2-hydroxypropyl-beta-cyclodextrin. *Arch. Pharm. Res.*, 31:1218–1223, 2008.

25. A. A. Elbashir, B. Saad, A. S. M. Ali, M. I. Saleh. Development of a capillary electrophoresis method for the enantioselective estimation of primaquine in pharmaceutical formulations. *J. AOAC Int.*, 91:536–541, 2008.

26. L. F. B. Malta, Y. Cordeiro, L. W. Tinoco, C. C. Campos, M. E. Medeiros, O. A. C. Antunes. Recognition mechanism of D- and L-tryptophan enantiomers using 2-hydroxypropyl-α- or β-cyclodextrins as chiral selector. *Tetrahedron: Asymmetry*, 29:1182–1188, 2008.

27. N. L. Deñola, N. S. Quiming, A. P. Catabay, Y. Saito, K. Jinno. Effects of alcohols on CE enantioseparation of basic drugs with native γ-CD as chiral selector. *J. Sep. Sci.*, 31:2701–2706, 2008.

28. D. A. El-Hady, N. A. El-Maali. Determination of catechin isomers in human plasma subsequent to green tea ingestion using chiral capillary electrophoresis with a high-sensitivity cell. *Talanta*, 76:138–145, 2008.

29. D. A. El-Hady, R. Gotti, N. A. El-Maali. Selective determination of catechin, epicatechin and ascorbic acid in human urine using chiral capillary electrophoresis. *J. Sep. Sci.*, 31:2252–2259, 2008.

30. H. Hashem, M. Kinzig, T. Jira. The effect of achiral calixarenes on chiral separation of propranolol-HCl and brompheniramine maleate in capillary electrophoresis using cyclodextrin as chiral selector. *Pharmazie*, 63:256–262, 2008.

31. Y. Goa, Y. Luo, J. Qin, B. Lin. A multichannel electrophoresis chip platform for rapid chiral selector screening. *Electrophoresis*, 29:1918–1923, 2008.

32. S. Terabe, H. Ozaki, K. Otsuka, T. Nado. Electrokinetic chromatography with 2-*O*-carboxymethyl-β-cyclodextrin as a moving "stationary" phase. *J. Chromatogr.*, 332:211–217, 1985.

33. A. Musenga, R. Mandrioli, E. Morganti, S. Fanali, M. A. Raggi. Enantioselective analysis of amisulpiride in pharmaceutical formulations by means of capillary electrophoresis. *J. Pharm. Biomed. Anal.*, 46:966–970, 2008.

34. Y.-W. Chou, W.-S. Huang, C.-C. Ko, S.-H. Chen. Enantioseparation of cetirizine by sulfated -β-CD-mediated capillary electrophoresis. *J. Sep. Sci*., 31:845–852, 2008.

35. C. H. Lin, W.-R. Fang, C.-M. Kuo, W.-Y. Chang, Y.-C. Liu, W.-Y. Lin, J.-C. Wu, C.-E. Lin. Chiral separation of hydroxyflavanones in cyclodextrin-modified capillary zone electrophoresis using sulfated cyclodextrins as chiral selectors. *J. Chromatogr. A*, 1188:301–307, 2008.

36. C. H. Lin, H.-T. Cheng, I.-J. Fang, Y.-C. Liu, C.-M. Kuo, W.-Y. Lin, J.-C. Wu, C.-E. Lin. Strategies for enantioseparations of catecholamines and structurally related compounds by capillary zone electrophoresis using sulfated β-cyclodextrins as chiral selectors. *Electrophoresis*, 27:3443–3451, 2006.

37. C. Perrin, M. G. Vargas, Y. Vander Heyden, M. Maftouh, D. L. Massart. Fast development of separation methods for the chiral analysis of amino acid derivatives using capillary electrophoresis and experimental designs. *J. Chromatogr. A*, 883:249–265, 2000.

38. T. Schmitt, H. Engelhardt. Charged and uncharged cyclodextrins as chiral selectors in capillary electrophoresis. *Chromatographia*, 37:475–481, 1993.

39. F. J. Malagueño de Santana, V. L. Lanchote, P. S. Bonato. Capillary electrophoretic chiral determination of mirtapazine and its main metabolites in human urine after enzymatic hydrolysis. *Electrophoresis*, 29:3924–3932, 2008.

40. H. Zhang, H. Shao, Y. A. Z. Zhang. Optimized conditions of enantioseparation of β-blockers by CZE using carboxymethyl-β-cyclodextrin as chiral selector. *Chromatographia*, 68:653–658, 2008.

41. U. Müller, Y. Mrestani, R. Neubert, B. Dräger. Chiral separation of the plant lignan matairesinol by capillary electrophoresis. *Electrophoresis*, 29:3582–3587, 2008.

42. L. Huang, J.-M. Lin, L. Yu, L. Xu, G. Chen. Field amplified on-line stacking for simultaneous enantioseparation and determination of some β-blockers using capillary electrophoresis. *Electrophoresis*, 29:3588–3594, 2008.

43. H. Li, P. H. Wang, C. Li, H. Wang, H. S. Zhang. Stereoselective determination of trihexyphenidyl using carboxymethyl-beta-cyclodextrin by capillary electrophoresis with field-amplified sample stacking. *Microchem. J*., 89:34–41, 2008.

44. J. Marák, P. Mikuš, K. Maráková, D. Kaniansky, I. Valášková, E. Havránek. Potentialities of ITP-CZE method with diode array detection for enantiomeric impurity control of dexbrompheniramine in pharmaceuticals. *J. Pharm. Biomed. Anal*., 46:870–876, 2008.

45. J. B. Vincent, A. D. Sokolowski, T. V. Nguyen, G. Vigh. A family of single-isomer chiral resolving agents for capillary electrophoresis: 1. Heptakis(2,3-diacetyl-6-sulfato)-β-cyclodextrin. *Anal. Chem*., 69:4226–4233, 1997.

46. J. B. Vincent, D. M. Kirby, T. V. Nguyen, G. Vigh. A family of single-isomer chiral resolving agents for capillary electrophoresis: 2. Hepta-6-sulfato-β-cyclodextrin. *Anal. Chem*., 69:4419–4428, 1997.

47. H. Cai, T. V. Nguyen, G. Vigh. A family of single-isomer chiral resolving agents for capillary electrophoresis: 3. Heptakis(2,3-dimethyl-6-sulfato)-β-cyclodextrin. *Anal. Chem*., 70:580–589, 1998.

48. M. B. Busby, G. Vigh. Synthesis of a single-isomer sulfated beta-cyclodextrin carrying nonidentical substituents at all of the C2, C3, and C6 positions and its use for the electrophoretic separation of enantiomers in acidic aqueous and methanolic background electrolytes: 2. Heptakis(2-*O*-methyl-6-*O*-sulfo)cyclomaltoheptaose. *Electrophoresis*, 26:3849–3860, 2005.

49. S. L. Li, G. Vigh. Single-isomer sulfated alpha-cyclodextrins for capillary electrophoresis: hexakis(2,3-di-*O*-methyl-6-*O*-sulfo)-alpha-cyclodextrin, synthesis, analytical characterization, and initial screening tests. *Electrophoresis*, 25:2657–2670, 2004.
50. S. L. Li, G. Vigh. Single-isomer sulfated alpha-cyclodextrins for capillary electrophoresis: 2. Hexakis(6-*O*-sulfo)-alpha-cyclodextrin: synthesis, analytical characterization, and initial screening tests. *Electrophoresis*, 25:1201–1210, 2004.
51. W. H. Zhu, G. Vigh. A family of single-isomer, sulfated gamma-cyclodextrin chiral resolving agents for capillary electrophoresis: octa(6-*O*-sulfo)-gamma-cyclodextrin. *Electrophoresis*, 24:130–138, 2003.
52. W. H. Zhu, G. Vigh. A family of single-isomer, sulfated gamma-cyclodextrin chiral resolving agents for capillary electrophoresis: 1. Octakis(2,3-diacetyl-6-sulfato)-gamma-cyclodextrin. *Anal. Chem*., 72:310–317, 2000.
53. K. Michalska, G. Pajchel, S. Tyski. Determination of enantiomeric impurity of linezolid by capillary electrophoresis using heptakis(2,3-diacetyl-6-sulfato)-β-CD. *J. Chromatogr. A*, 1180:179–186, 2008.
54. W.-S. Liao, C.-H. Lin, C.-Y. Chen, C.-M. Kuo, Y.-C. Liu, J.-C. Wu, C.-E. Lin. Enantioseparation of phenothiazines in CD-modified CZE using single isomer sulfated CD as a chiral selector. *Electrophoresis*, 28:3922–3929, 2007.
55. B. A. Williams, G. Vigh. Dry look at the CHARM (charged resolving agent migration) model of enantiomer separations by capillary electrophoresis. *J. Chromatogr. A*, 777:295–309, 1997.
56. F.-T. A. Chen, G. Shen, R. A. Evangelista. Characterisation of highly sulfated cyclodextrins. *J. Chromatogr. A*, 924:523–532, 2001.
57. C. Perrin, H. Fabre, M. Maftouh, D. L. Massart, Y. Vander Heyden. Robustness testing of chiral separations by capillary electrophoresis using highly-sulfated cyclodextrins. *J. Chromatogr. A*, 1007:165–177, 2003.
58. J. Schappler, D. Guillarme, J. Prat, J.-L. Veuthey, S. Rudaz. Validation of chiral capillary electrophoresis–electrospray ionization–mass spectrometry methods for ecstasy and methadone in plasma. *Electrophoresis*, 29:2193–2202, 2008.
59. X. Lin, M. Zhoa, X. Qi, C. Zhu, A. Hao. Capillary zone electrophoretic chiral discrimination using 6-*O*-(2-hydroxy-3-trimethylammoniopropyl)-β-cyclodextrin as chiral selector. *Electrophoresis*, 27:872–879, 2006.
60. K. Nzeadibe, G. Vigh. Synthesis of mono-6-deoxy-6-*N*, *N*, *N'*, *N'*, *N'*-pentam ethylethylenediamino-cyclomaltoheptaose, a single isomer, monosubstituted, permanently dicationic β-CD and its use for enantiomer separation by CE.
61. W. Tang, S. C. Ng. Monosubstituted positively charged cyclodextrins: synthesis and applications in chiral separation. *J. Sep. Sci*., 31:3246–3256, 2008.
62. I. S. Lurie. Separation selectivity in chiral and achiral capillary electrophoresis with mixed cyclodextrins. *J. Chromatogr. A*, 792:297–307, 1997.
63. T. Nhujak, C. Sastravaha, C. Palanuvej, A. Petsom. Chiral separation in capillary electrophoresis using dual neutral cyclodextrins: theoretical models of electrophoretic mobility difference and separation selectivity. *Electrophoresis*, 26:3814–3823, 2005.
64. A. M. Abushoffa, M. Fillet, A. C. Servais, Ph. Hubert, J. Crommen. Enhancement of selectivity and resolution in the enantioseparation of uncharged compounds using mixtures of oppositely charged cyclodextrins in capillary electrophoresis. *Electrophoresis*, 24:343–350, 2003.

65. M. Fillet, Ph. Hubert, J. Crommen. Enantiomeric separations of drugs using mixtures of charged and neutral cyclodextrins. *J. Chromatogr. A*, 875:123–134, 2000.

66. B.-L. Chu, B. Guo, H. Zuo, Z. Wang, J.-M. Lin. Simultaneous enantioseparation of antioparkinsonian medication rotogotine and related chiral impurities by capillary zone electrophoresis using dual cyclodextrin system. *J. Pharm. Biomed. Anal.*, 46:854–859, 2008.

67. B. Sungthong, P. Jác, G. K. E. Scriba. Development and validation of a capillary electrophoresis method for the simultaneous determination of impurities of escitalopram including the *R*-enantiomer. *J. Pharm. Biomed. Anal.*, 46:959–965, 2008.

68. C. Perrin, Y. Vander Heyden, M. Maftouh, D. L. Massart. Rapid screening for chiral separations by short-end injection capillary electrophoresis using highly sulfated cyclodextrins as chiral selectors. *Electrophoresis*, 22:3203–3215, 2001.

69. N. Matthijs, C. Perrin, M. Maftouh, D. L. Massart, Y. Vander Heyden. Knowledge-based system for method development of chiral separations with capillary electrophoresis using highly-sulphated cyclodextrins. *J. Pharm. Biomed. Anal.*, 27:515–529, 2002.

70. N. Matthijs, S. Van Hemelryck, M. Maftouh, D. L. Massart, Y. Vander Heyden. Electrophoretic separation strategy for chiral pharmaceuticals using highly-sulfated and neutral cyclodextrins based dual selector systems. *Anal. Chim. Acta*, 525:247–263, 2004.

71. J. Chapman, F. A. Chen. Implementing a generic methods development strategy for enantiomer analysis. *LCGC North Am.*, 19:427–431, 2001.

72. M. I. Jimidar, W. Van Ael, P. Van Nyen, M. Peeters, D. Redlich, M. De Smet. A screening strategy for the development of enantiomeric separation methods in capillary electrophoresis. *Electrophoresis*, 25:2772–2785, 2004.

73. M. I. Jimidar, T. Vennekens, W. Van Ael, D. Redlich, M. De Smet. Optimization and validation of an enantioselective method for a chiral drug with eight stereo-isomers in capillary electrophoresis. *Electrophoresis*, 25:2876–2884, 2004.

74. L. Zhou, R. Thompson, S. Song, D. Ellison, J. M. Wyvratt. A strategic approach to the development of capillary electrophoresis methods for pharmaceutical basic compounds using sulfated cyclodextrins. *J. Pharm. Biomed. Anal.*, 27:541–553, 2002.

75. S. El Deeb, P. Hasemann, H. Wätzig. Strategies in method development to quantify enantiomeric impurities using CE. *Electrophoresis*, 29:3552–3562, 2008.

76. J.-P. Behr, J.-M. Girodeau, R. C. Hayward, J.-M. Lehn, J.-P. Sauvage. Molecular receptors: functionalized and chiral macrocyclic polyethers derived from tartaric acid. *Helv. Chim. Acta*, 63:2096–2111, 1980.

77. R. Kuhn, F. Stoecklin, F. Erni. Chiral separations by host–guest complexation with cyclodextrin and crown-ether in capillary zone electrophoresis. *Chromatographia*, 33:32–36, 1992.

78. R. Kuhn, S. Hoffstetterkuhn. Chiral separation by capillary electrophoresis. *Chromatographia*, 34:505–512, 1992.

79. R. Kuhn, F. Erni, T. Bereuter, J. Hausler. Chiral recognition and enantiomeric resolution based on host–guest complexation with crown ethers in capillary zone electrophoresis. *Anal. Chem.*, 64:2815–2820, 1992.

80. R. Kuhn, J. Wagner, Y. Walbroehl, T. Bereuter. Potential and limitations of an optically-active crown-ether for chiral separation in capillary zone electrophoresis. *Electrophoresis*, 15:828–834, 1994.
81. M. G. Schmid, G. Gübitz. Capillary zone electrophoretic separation of the enantiomers of dipeptides based on host–guest complexation with a chiral crown ether. *J. Chromatogr. A*, 709:81–88, 1995.
82. H. Nagata, H. Nishi, M. Kamigauchi, T. Ishida. Structural scaffold of 18-crown-6 tetracarboxylic acid for optical resolution of chiral amino acid: x-ray crystal analyses and energy calculations of complexes of D- and L-isomers of tyrosine, isoleucine, methionine and phenylglycine. *Org. Biomol. Chem.*, 2:3470–3475, 2004.
83. H. Nagata, H. Nishi, M. Kamigauchi, T. Ishida. Guest-dependent conformation of 18-crown-6 tetracarboxylic acid: relation to chiral separation of amino acids. *Chirality*, 20:820–827, 2008.
84. Y. Kuwuhara, H. Nagata, H. Nishi, Y. Tanaka, K. Kakehi. Detection and separation of free amino acid enantiomers by capillary electrophoresis with a chiral crown ether and indirect photometric detection. *Chromatographia*, 62:505–510, 2005.
85. L. Zhou, Z. Lin, R. A. Reamer, B. Mao, Z. Ge. Stereoisomeric separation of pharmaceutical compounds using CE with a chiral crown ether. *Electrophoresis*, 28:2658–2666, 2007.
86. X. Y. Gong, D. Dobrunz, M. Kümin, M. Wiesner, J. D. Revell, H. Wennemers, P. C. Hauser. Separating stereoisomers of di-, tri-, and tetrapeptides using capillary electrophoresis with contactless conductivity detection. *J. Sep. Sci.*, 31:565–573, 2008.
87. X. Y. Gong, P. C. Hauser. Enantiomeric separation of 1-phenylethylamine and 1-cyclohexylethylamine in capillary electrophoresis with contactless conductivity detection. *J. Chromatogr. A*, 1094:196–199, 2005.
88. T. Iványi, K. Pál, I. Lázár, D. L. Massart, Y. Vander Heyden. Application of tetraoxadiaza-crown ether derivatives as chiral selector modifiers in capillary electrophoresis. *J. Chromatogr. A*, 1028:325–332, 2004.
89. J. Elek, D. Mangelings, T. Iványi, I. Lázár, Y. Vander Heyden. Enantioselective capillary electrophoretic separation of tryptophane- and tyrosine-methyl esters in a dual system with a tetra-oxadiaza-crown-ether derivative and a cyclodextrin. *J. Pharm. Biomed. Anal.*, 38:601–608, 2005.
90. P. Forlay-Frick, D. Mangelings, T. Iványi, I. Lázár, K. Héberger, Y. Vander Heyden. Newly synthesized tetraoxa-diaza crown ether derivatives versus commercialized crown ethers in the separation of positional isomers with capillary electrophoresis. *J. Pharm. Biomed. Anal.*, 41:1164–1170, 2006.
91. X. Y. Gong, P. C. Hauser. Enantiomeric separation of underivatized small amines in conventional and on-chip capillary electrophoresis with contactless conductivity detection. *Electrophoresis*, 27:4375–4382, 2006.
92. S. I. Cho, J. Shim, M.-S. Kim, Y.-K. Kim, D. S. Chung. On-line sample cleanup and chiral separation of gemifloxacin in a urinary solution using chiral crown ether as a chiral selector in microchip electrophoresis. *J. Chromatogr. A*, 1055:241–245, 2004.
93. D. W. Armstrong, Y. B. Tang, S. S. Chen, Y. W. Zhou, C. Bagwill, J. R. Chen. Macrocyclic antibiotics as a new class of chiral selectors for liquid-chromatography. *Anal. Chem.*, 66:1473–1484, 1994.

94. D. W. Armstrong, K. L. Rundlett, J. R. Chen. Evaluation of the macrocyclic antibiotic vancomycin as a chiral selector for capillary electrophoresis. *Chirality*, 6:496–509, 1994.

95. C. Desiderio, S. Fanali. Chiral analysis by capillary electrophoresis using antibiotics as chiral selector. *J. Chromatogr. A*, 807:37–56, 1998.

96. S. Fanali, C. Cartoni, C. Desiderio. Chiral separation of newly synthesized arylpropionic acids by capillary electrophoresis using cyclodextrins or a glycopeptide antibiotic as chiral selector. *Chromatographia*, 54:87–92, 2001.

97. P. Bednar, Z. Aturki, Z. Stransky, S. Fanali. Chiral analysis of non-UV absorbing compounds by capillary electrophoresis using macrocyclic antibiotics: 1. Separation of aspartic and glutamic acid. *Electrophoresis*, 22:2129–2135, 2001.

98. S. Fanali, Z. Aturki, C. Desiderio, P. G. Righetti. Use of MDL 63 246 (Hepta Tyr) antibiotic in capillary electrophoresis: II. Chiral resolution of α-hydroxy acids. *J. Chromatogr. A*, 838:223–235, 1999.

99. T. M. Oswald, T. J. Ward. Enantioseparations with the macrocylic antibiotic ristocetin A using a countercurrent process in CE. *Chirality*, 11:663–668, 1999.

100. K. H. Ekborg-Ott, G. A. Zientara, J. M. Schneiderheinze, K. Gahm, D. W. Armstrong. Avoparcin, a new macrocylic antibiotic chiral run buffer additive for capillary electrophoresis. *Electrophoresis*, 20:2438–2457, 1999.

101. D. R. Risley, L. Telli-Seifert, Q. J. McKenzie. Enantiomeric separations of dansyl amino acids using the macrocylic antibiotic A35512B as a chiral selector in capillary electrophoresis. *Electrophoresis*, 20:2749–2753, 1999.

102. Z. Jiang, J. Kang, D. Bischoff, B. Bister, R. D. Süssmuth, V. Schurig. Evaluation of balhimycin as a chiral selector for enantioresolution by capillary electrophoresis. *Electrophoresis*, 25:2687–2692, 2004.

103. M. A. Martínez-Gómez, R. M. Villanueva-Cama nas, S. Sagrado, M. J. Medina-Hernández. Potential of human serum albumin as chiral selector of basic drugs in affinity electrokinetic chromatography-partial filling technique. *Electrophoresis*, 27:4364–4374, 2007.

104. M. A. Martínez-Gómez, S. Sagrado, R. M. Villanueva-Cama nas, M. J. Medina-Hernández. Enantioseparation of phenothiazines by affinity electrokinetic chromatography using human serum albumin as chiral selector: application to enantiomeric quality control in pharmaceutical formulations. *Anal. Chim. Acta*, 582:223–228, 2007.

105. M. A. Martínez-Gómez, S. Sagrado, R. M. Villanueva-Cama nas, M. J. Medina-Hernández. Enantiomeric quality control of anitihistamines in pharmaceuticals by affinity electrokinetic chromatography with human serum albumin as chiral selector. *Anal. Chim. Acta*, 592:202–209, 2007.

106. H. Hödl, J. Koidl, M. G. Schmid, G. Gübitz. Chiral resolution of tryptophan derivatives by CE using canine serum albumin and bovine serum albumin as chiral selectors. *Electrophoresis*, 27:4755–4762, 2006.

107. H. Matsanuga, Y. Sadakane, J. Haginaka. Separation of basic drugs using chicken α 1 glycoprotein: insight into chiral recognition mechanism. *Electrophoresis*, 24:2442–2447, 2003.

108. L. L. Zhou, B. Mao, Z. H. Ge. Comparative study of immobilized alpha 1 acid glycoprotein and ovomucoid protein stationary phases for the enantiomeric separation of pharmaceutical compounds. *J. Pharm. Biomed. Anal.*, 46:898–906, 2008.

109. J. Olsson, L. G. Blomberg. Enantioseparation of omeprazole and its metabolite 5-hydroxyomeprazole using open tubular capillary electrochromatography with immobilized avidin as chiral selector. *J. Chromatogr. B*, 875:329–332, 2008.
110. M. C. Millot. Separation of drug enantiomers by liquid chromatography and capillary electrophoresis, using immobilized proteins as chiral selectors. *J. Chromatogr. B*, 797:131–159, 2003.
111. G. Massolini, C. Temporini, E. Calleri. Pennicillin G acylase as chiral selector in LC and CE: exploring the origins of enantioselectivity. *J. Chromatogr. B*, 875:20–29, 2008.
112. R. Gotti, E. Calleri, G. Massolini, S. Furlanetto, V. Cavrini. Penicillin G acylase as chiral selector in CE using a pullulan-coated capillary. *Electrophoresis*, 27:4746–4754, 2006.
113. S. M. Mwongela, N. Siminialayi, K. A. Fletcher, I. M. Warner. A comparison of ionic liquids to molecular organic solvents as additives for chiral separations in micellar electrokinetic chromatography. *J. Sep. Sci*., 30:1334–1342, 2007.
114. Y. François, A. Varenne, E. Juillerat, D. Villemin, P. Gareil. Evaluation of chiral ionic liquids as additives to cyclodextrins for enantiomeric separations by capillary electrophoresis. *J. Chromatogr. A*, 1155:134–141, 2007.
115. C. D. Tran, I. Mejac. Chiral ionic liquids for enantioseparation of pharmaceutical products by capillary electrophoresis. *J. Chromatogr. A*, 1204:204–209, 2008.
116. J. Ruta, C. Ravelet, C. Grosset, J. Fize, A. Ravel, A. Vilet, E. Peyrin. Enantiomeric separation using and L-RNA aptamer as chiral additive in partial-filling capillary electrophoresis. *Anal. Chem*., 78:3032–3039, 2006.
117. J. Ruta, C. Ravelet, I. Buasanne, J.-L. Décout, E. Peyrin. Aptamer-based enantioselective competitive binding assay for the trace enantiomer detection. *Anal. Chem*., 79:4716–4719, 2007.
118. V. A. Davakov. Chiral selectors with chemating properties in liquid chromatography: fundamental reflections and selective review of recent developments. *J. Chromatogr. A*, 666:55–76, 1994.
119. R. Kuhn, S. Hoffstetter-Kuhn. Chiral separation by capillary electrophoresis. *Chromatographia*, 34:505–512, 1992.
120. J. Snopek, I. Jelínek, E. Smolková-Keulemansová. Chiral separation by analytical electromigration methods. *J. Chromatogr*., 609:1–17, 1992.
121. E. Gassmann, J. E. Kuo, R. N. Zare. Electrokinetic separation of chiral compounds. *Science*, 230:813–815, 1985.
122. P. Gozel, E. Gassman, H. Michelsen, R. N. Zare. Electrokinetic resolution of amino-acid enantiomers with copper(II) aspartame support electrolyte. *Anal. Chem*., 59:44–49, 1987.
123. M. G. Schmid, G. Gübitz. Direct enantiomer separation of underivatized amino acids by capillary zone electrophoresis based on ligand exchange. *Enantiomer*, 1:23–27, 1996.
124. A. Vegvari, M. G. Schmid, F. Kilar, G. Gübitz. Chiral separation of alpha-amino acids by ligand-exchange capillary electrophoresis using *N*-(2-hydroxy-octyl)-L-4-hydroxyproline as a selector. *Electrophoresis*, 19:2109–2112, 1998.
125. N. Grobuschek, M. G. Schmid, C. Tuscher, M. Ivanova, G. Gübitz. Chiral separation of beta-methyl-amino acids by ligand exchange using capillary electrophoresis and HPLC. *J. Pharm. Biomed. Anal*., 27:599–605, 2002.

126. H. Hödl, M. G. Schmid, G. Gübitz. Chiral separation of amino acids and glycyl dipeptides by chiral ligand-exchange capillary electrophoresis comparing Cu(II), Co(II), Ni(II) and Zn(II) complexes of three different sugar acids. *J. Chromatogr. A*, 1204:210–218, 2008.
127. J. Koidl, H. Hödl, M. G. Schmid, S. Pantcheva, T. Pajpanova, G. Gübitz. Chiral separation of halogenated amino acids by ligand-exchange capillary electrophoresis. *Electrophoresis*, 26:3878–3883, 2005.
128. L. Qi, Y. Han, M. Zuo, Y. Chen. Chiral CE of aromatic acids by ligand-exchange with zinc(II)-L-lysine complex. *Electrophoresis*, 28:2629–2634, 2007.
129. L. Qi, M. Liu, Z. Guo, M. Xie, C. Qiu, Y. Chen. Assay of aromatic amino acid enantiomers in rice-brewed suspensions by chiral ligand-exchange CE. *Electrophoresis*, 28:4150–4155, 2007.
130. V. Cucinotta, A. Giuffrida, G. Macarone, M. Messina, G. Vecchio. Ligand exchange capillary electrophoresis by cyclodextrin derivatives, a powerful tool for enantiomeric separations. *Electrophoresis*, 27:1471–1480, 2006.
131. Z. Chen, K. Uchiyama, T. Hobo. Chiral resolution of dansyl amino acids by ligand exchange capillary electrophoresis using Cu(II)-L-prolinamides as chiral selector. *Anal. Chim. Acta*, 523:1–7, 2004.
132. Z.-X. Zheng, Y. Wei, J.-M. Lin. Chiral separation based on ligand-exchange capillary electrophoresis using a copper(II)-L-ornithine ternary complex as chiral selector. *Electrophoresis*, 25:1007–1012, 2004.
133. M.-L. Riekkola. Recent advances in nonaqueous capillary electrophoresis. *Electrophoresis*, 23:3865–3883, 2002.
134. S. Sanchez-Vindas, G. Vigh. Non-aqueous capillary electrophoretic enantiomer separations using the tetrabutylammonium salt of heptakis(2,3-diacetyl-6-*O*-sulfo)cyclomaltoheptaose, a single-isomer sulfated β-cyclodextrin highly soluble in organic solvents. *J. Chromatogr. A*, 1068:151–158, 2005.
135. J. Olsson, F. Stegander, N. Marlin, H. Wan, L. G. Blomberg. Enantiomeric separation of omeprazole and its metabolite 5-hydroxyomeprazole using non-aqueous capillary electrophoresis. *J. Chromatogr. A*, 1129:291–295, 2006.
136. L. Huang, J. Lin, L. Xu, G. Chen. Nonaqueous and aqueous-organic media for the enantiomeric separations of neutral organophosphorus pesticides by CE. *Electrophoresis*, 28:2758–2764, 2007.
137. R. Mol, A.-C. Servais, M. Fillet, J. Crommen, G. J. de Jong, G. W. Somsen. Non-aqueous electrokinetic chromatography–electrospray ionization mass spectrometry using anionic cyclodextrins. *J. Chromatogr. A*, 1159:51–57, 2007.
138. R. Mol, G. J. de Jong, G. W. Somsen. Coupling of non-aqueous electrokinetic chromatography using cationic cyclodextrins with electrospray ionization mass spectrometry. *Rapid Commun. Mass Spectrom.*, 22:790–796, 2008.
139. R. Mol, G. J. de Jong, G. W. Somsen. Cyclodextrin-based nonaqueous electrokinetic chromaography with UV and mass spectrometric detection: application to the impurity profiling of amiodarone. *Electrophoresis*, 29:3575–3581, 2008.
140. V. Piette, M. Lämmerhofer, W. Lindner, J. Crommen. Enantiomer separation of N-protected amino acids by non-aqueous capillary electrophoresis and high-performance liquid chromatography with tert.-butyl carbamoylated quinine in either the background electrolyte or the stationary phase. *J. Chromatogr. A*, 987:421–427, 2003.

141. A. C. Servais, M. Fillet, A. M. Abushoffa, P. Hubert, J. Crommen. Synergistic effects of ion-pairing in the enantiomeric separation of basic compounds with cyclodextrin derivatives in nonaqueous capillary electrophoresis. *Electrophoresis*, 24:363–369, 2003.
142. A. C. Servais, M. Fillet, P. Chiap, W. Dewe, P. Hubert, J. Crommen. Enantiomeric separation of basic compounds usin heptakis(2,3-di-*O*-methyl-6-*O*-sulfo)-beta-cyclodextrin in combination with potassium camphorsulfonate in nonaqueous capillary electrophoresis: optimization by means of an experimental design. *Electrophoresis*, 25:2701–2710, 2004.
143. A. C. Servais, M. Fillet, P. Chiap, W. Dewe, P. Hubert, J. Crommen. Influence of the nature of the electrolyte on the chiral separation of basic compounds in nonaqueous capillary electrophoresis using heptakis(2,3-di-*O*-methyl-6-*O*-sulfo)-beta-cyclodextrin. *J. Chromatogr. A*, 1068:143–150, 2005.
144. R. D. Marini, A.-C. Servais, E. Rozet, P. Chiap, B. Boulanger, S. Rudaz, J. Crommen, P. Hubert, M. Fillet. Nonaqueous capillary electrophoresis method for the enantiomeric purity determination of *S*-timolol using heptakis(2,3-di-*O*-methyl-6-*O*-sulfo)-beta-cyclodextrin: validation using the accuracy profile strategy and estimation of uncertainty. *J. Chromatogr. A*, 1120:102–111, 2006.
145. R. D. Marini, E. Rozet, Y. Vander Heyden, E. Ziemons, B. Boulanger, A. Bouklouze, A.-C. Servais, M. Fillet, J. Crommen, P. Hubert. Robustness testing of a chiral NACE method for *R*-timolol determination in *S*-timolol maleate and uncertainty assessment from quantitative data. *J. Pharm. Biomed. Anal*., 44:640–651, 2007.
146. R. D. Marini, C. Groom, F. R. Doucet, J. Hawari, Y. Bitar, U. Holzgrabe, R. Gotti, J. Schappler, S. Rudaz, R. Mol, G. W. Somsen, G. J. De Jong, P. T. T. Ha, J. Zhang, A. Van Schepdael, J. Hoogmartens, W. Briône, A. Ceccato, B. Boulanger, D. Mangelings, Y. Vander Heyden, W. Van Ael, I. Jimidar, M. Pedrini, A.-C. Servais, M. Fillet, J. Crommen, E. Rozet, Ph. Hubert. Interlaboratory study of a NACE method for the determination of *R*-timolol content in *S*-timolol maleate: assessment of uncertainty. *Electrophoresis*, 27:2386–2399, 2006.
147. I. Fradi, A.-C. Servais, M. Pedrini, P. Chiap, R. Ivanyi, J. Crommen, M. Fillet. Enantiomeric separation of acidic compounds using single-isomer amino cyclodextrin derivatives in nonaqueous capillary electrophoresis. *Electrophoresis*, 27:3434–3442, 2006.
148. A. Rousseau, M. Pedrini, P. Chiap, R. Ivanyi, J. Crommen, M. Fillet, A.-C. Servais. Determination of flurbiprofen enantiomers using a single-isomer amino cyclodextrin derivative in nonaqueous capillary electrophoresis. *Electrophoresis*, 29:3641–3648, 2008.
149. A. Rousseau, P. Chiap, R. Ivanyi, J. Crommen, M. Fillet, A.-C. Servais. Validation of a nonaqueous capillary electrophoretic method for the enantiomeric impurity determination of *R*-flurbiprofen using a single-isomer amino cyclodextrin derivative. *J. Chromatogr. A*, 1204:219–225, 2008.
150. Y. Hedeland, J. Lehtinen, C. Pettersson. Ketopinic acid and diisopropylideneketogulonic acid as chiral ion-pair selectors in capillary electrophoresis: enantiomeric impurity analysis of *S*-timolol and 1*R*,2*S*-ephedrine. *J. Chromatogr. A*, 1141:287–294, 2007.
151. Y. Carlsson, M. Hedeland, U. Bondesson, C. Pettersson. Non-aqueous capillary electrophoretic separation of enantiomeric amines with (-)-2,3:4,6-di-*O*-isopropylidene-2-keto-L-gulonic acid as chiral counter ion. *J. Chromatogr. A*, 922:303–311, 2001.

152. S. Terabe, K. Otsuka, K. Ichikawa, A. Tsuchiya, T. Nado. Electrokinetic separations with micellar solutions and open-tubular capillaries. *Anal. Chem.*, 56:111–113, 1984.

153. K. Otsuka, S. Terabe. Enantiomer separation of drugs by micellar electrokinetic chromatography using chiral surfactants. *J. Chromatogr. A*, 875:163–178, 2000.

154. G. Cudiamat. Market profile: capillary electrophoresis. *Column*, 4: 18, 2008.

CHAPTER 12

High-Throughput Screening and Method Development Strategies to Separate Chiral Drug Compounds in HPLC, SFC, and CE

HASRET ATES, DEBBY MANGELINGS, and YVAN VANDER HEYDEN

Department of Analytical Chemistry and Pharmaceutical Technology, Vrije Universiteit Brussel–VUB, Brussels, Belgium

1 INTRODUCTION

Chirality is a molecular phenomenon that has a significant impact in a pharmaceutical context. A *chiral molecule* is defined as a compound that is not superimposable on the molecule that forms its mirror image [1]. This can be caused by the presence of a chiral center, a chiral plane, a chiral axis, or a helical structure in the molecule [1,2]. Such a chiral compound and its mirror image form a pair of enantiomers. Enantiomers have the same chemical and physical properties when they are put in an achiral environment. When the enantiomers are put in a chiral environment, such as the human body, they behave as two different compounds. In pharmaceutical applications this can lead to very hazardous consequences. When a chiral drug molecule is administered as a racemate, it is possible that only one enantiomer is therapeutically active. This enantiomer is called the *eutomer*. The other enantiomer, the *distomer*, can react with the same target, showing less effect, or it may not react at all. But it is also possible that the therapeutically less active or inactive enantiomer can cause effects opposite to those of the eutomer, or its effect can even be toxic. In fact, it can be stated that the eutomer and distomer differ in their pharmacological, pharmacodynamic, pharmacokinetic and toxicological properties.

Chiral Separation Methods for Pharmaceutical and Biotechnological Products, Edited by Satinder Ahuja

To counter these problems, more and more single-enantiomer drugs are being marketed. However, before marketing these drugs, research is needed on both the racemate and the individual enantiomers. Even when enantiopure synthesis is performed, chiral analysis is required since the distomer has to be detected and quantified as an impurity of the eutomer. Consequently, several regulatory authorities, such as the U.S. Food and Drug Administration (FDA) and the European Medicines Evaluation Agency (EMA), imposed guidelines for chiral active substances [3–5]. Furthermore, pharmaceutical companies can exploit enantioseparation to prolong the patent of an already marketed drug by replacing the racemate with the eutomer [6], or they can develop a single enantiomer for another therapeutic indication than that of the racemate [7].

To obtain the individual enantiomers, either for preparative or analytical purposes, rapid separation strategies are needed since time is of crucial importance in drug research and development and in early drug development many separations are required [8,9]. Different techniques can be used to separate enantiomers. Electromigration-based techniques such as capillary electrophoresis (CE) [10–15] and capillary electrochromatography (CEC) [16–22] have already been used in chiral drug analysis, but they are not suited for preparative purposes. Another frequently used technique in the field of chiral separations is high-performance liquid chromatography (HPLC) [23]. Because of its convenience, efficiency, reproducibility, broad applicability, and the occurrence of several subtypes, such as normal-phase liquid chromatography (NPLC), reversed-phase liquid chromatography (RPLC), and polar organic solvent chromatography (POSC), it even became a standard technique for chiral analysis. Some types (e.g., NPLC and POSC) are used in both an analytical- and a preparative-scale context [24]. Supercritical fluid chromatography (SFC) is another technique that is becoming prominent in chiral drug analysis. In some companies it is even considered a first choice for the separation of new compounds because of the high success rate and the often higher-throughput possibilities than those of HPLC [25]. Thin-layer chromatography, gas chromatography, and countercurrent chromatography are also used in the field of enantioseparations. An overview is given in Figure 1.

In this chapter, strategies for high-throughput screening combined with a minimal method optimization are discussed in more detail. These strategies have been developed in house. Strategies in HPLC (normal-phase, reversed-phase, and polar organic solvent chromatography), SFC and CE are considered. All chromatographic strategies are developed on polysaccharide-based chiral selectors, some of which are shown in Figure 2, while the CE strategy is based on the use of highly sulfated cyclodextrins added to the background electrolyte.

Each strategy consists of a screening phase followed by an optimization. In the screening step, the enantioselectivity of a limited number of chromatographic systems (i.e., combinations of stationary and mobile phases) for a given compound is evaluated. On the other hand, the optimization steps will try to improve the initial result, again in a limited number of experiments (Figure 3, p. 387). The strategies are defined to be generic (i.e., applicable on many compounds) and to be high-throughput (i.e., only few experiments are spent on a given compound

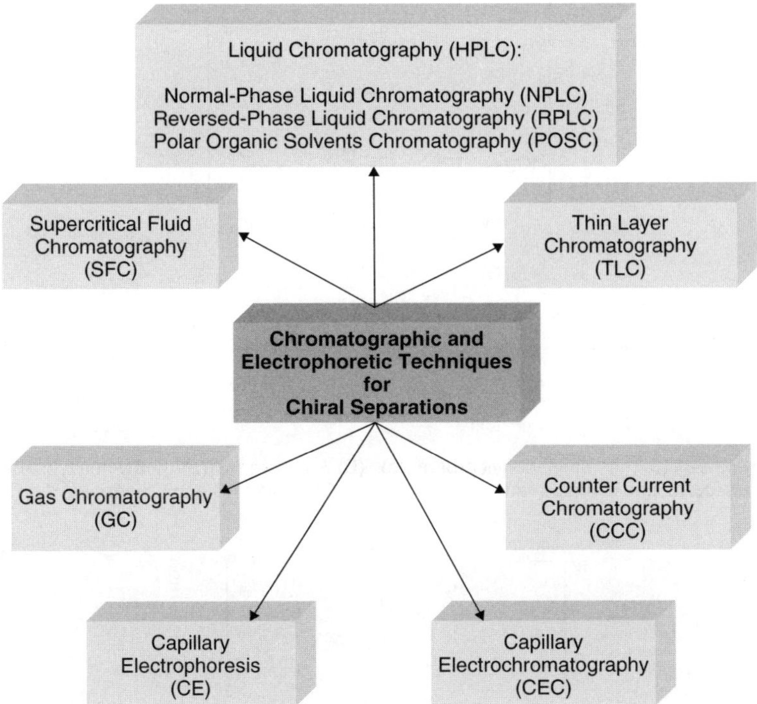

Figure 1 Overview of techniques to separate chiral compounds.

in trying to separate it). This also means that for a particular compound, not all possible efforts are made in trying to optimize its separation.

For each technique, the strategy is shown and discussed. Occasionally, it might be followed by additional discussion related to different aspects of the technique or to recent developments. It should also be clear that we do not try to give an overview of all existing strategies within a given technique, although some may be dealt with in additional discussion. The conditions are chosen based on the knowledge found in the literature and obtained by our own research, as explained in additional discussion. Before using these strategies it is important to note that the purpose of defining a separation strategy is to minimize the trial-and-error approach in analysis and thus to save time while trying to separate chiral compounds. The strategies are not intended to elucidate or study chiral separation mechanisms.

2 SCREENING STRATEGIES IN HIGH-PERFORMANCE LIQUID CHROMATOGRAPHY

As mentioned before, HPLC plays a dominant role in the analysis of chiral compounds. This technique is simple and practical and therefore is well known by

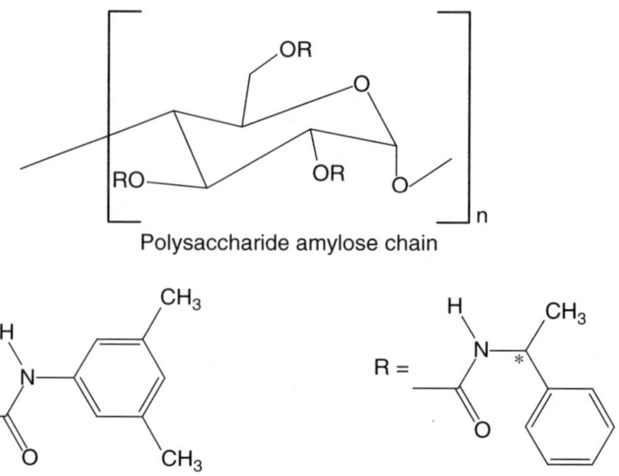

(A) Amylose *tris*(3,5-dimethylphenylcarbamate) Chiralpak AD and Chiralpak AD-RH

(B) Amylose *tris*[(*S*)-methylbenzylcarbamate] Chiralpak AS and Chiralpak AS-RH

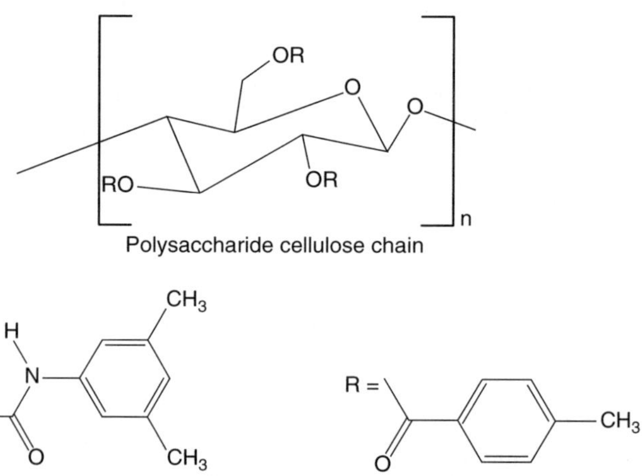

(C) Cellulose *tris*(3,5-dimethylphenylcarbamate) Chiralcel OD and Chiralcel OD-RH

(D) Cellulose *tris*(4-methylbenzoate) Chiralcel OJ and Chiralcel OJ-R

Figure 2 Some polysaccharide-based selectors used as chiral stationary phases.

most analysts. It can be used to analyze a very broad spectrum of analytes having different physicochemical properties, and all analyses can be automated, which is an important advantage in high-throughput analysis [26]. In chiral analysis, this technique can be exploited in many different ways. Chiral selectors can be used as mobile-phase additives to create the chiral environment necessary for enantioseparation. The stationary phases used in combination with these chiral mobile phases are achiral. On the other hand, chiral selectors can also be coated or immobilized onto a silica support, resulting in a chiral stationary phase (CSP).

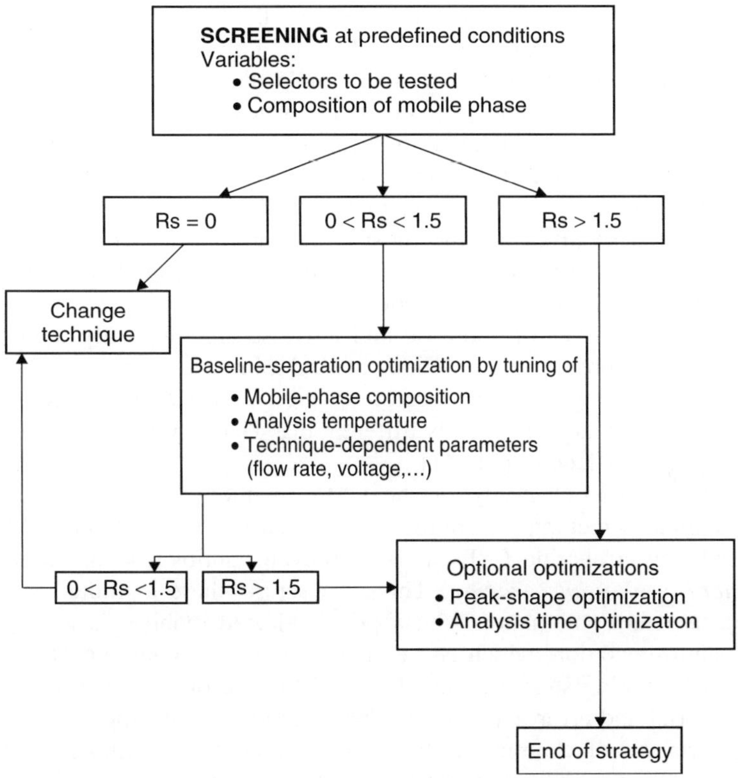

Figure 3 General scheme of a separation strategy.

When a CSP is used, the mobile phase will be achiral, consisting of the regular chromatographic solvents without a chiral additive. When a chiral selector is used to create a chiral mobile or stationary phase, the term *direct enantioseparation* is used [27], referring to a transient interaction of the chiral selector with the enantiomer forming labile reversible diastereomeric complexes. The complexes formed will be distributed differently to the stationary phase, resulting in a separation. However, using chiral mobile-phase additives (CMPAs) has an important drawback in preparative chromatography: The selectors have to be removed from the solvent after the chromatographic separation. The high consumption of CMPAs can be considered to be another drawback. These problems do not occur when CSPs are used. Here the enantiomers will be separated due to the formation of diastereomeric complexes with the selectors present on the CSP. The enantiomer that forms the strongest solute–selector complex will be retained longer on the CSP [26,28]. The term *indirect enantioseparation* is used when the enantiomers are chemically derivatized to diastereomers, which are separated in an achiral environment (e.g., under RPLC conditions with a C_{18} stationary phase). Indirect enantioseparation is not discussed further in this chapter.

Different types of chiral selectors are used on the CSP in HPLC. These selectors are divided into different classes [29–31]. Polymeric selectors, including polysaccharides, synthetic polymers (e.g., molecular imprinted polymers, helical polyacrylates), and proteins, can be distinguished from macrocyclic selectors, with cyclodextrins, glycopeptide antibiotics, and crown ethers as the most important representatives, and from low-molecular-weight selectors such as Pirkle-type columns and ion and ligand exchangers [16,27,29]. Among these selectors, the polysaccharide derivatives are the most widely used [32]. The use of derivatives instead of the native polysaccharides cellulose and amylose is because of the lower enantioselectivity of the latter. In particular, benzoate and phenylcarbamate ester derivatives of cellulose and amylose are frequently used [32,33]. These show enantiorecognition abilities toward a broad spectrum of chiral compounds, due to the higher-order structure of the chiral polysaccharide derivative polymers [32,33]. Many chemically different classes of chiral compounds can be resolved by these selectors [34–41].

The large enantioselectivity of the CSPs mentioned can be exploited for high-throughput screening programs. In many cases even different CSPs (e.g., derivatized polysaccharide CSPs and macrocyclic antibiotics) are used jointly in a *generic separation strategy*. This is a setup where a limited number of CSPs are used in combination with a priori selected mobile phases to quickly obtain enantioseparation, which is the goal of high-throughput analyses. Applying generic screening strategies should avoid the use of more time-and-money-consuming trial-and-error approaches. These strategies are supposed to separate a broad spectrum of chemically different compounds in a limited number of experiments. Thus, newly synthesized chiral compounds can also be subjected to a given strategy to get separated. An important aspect for many screening strategies in HPLC is the possibility of upscaling. The CSPs used are commercialized in columns of different dimensions, which make them, in combination with throughput-improving techniques, ideal for preparative-scale enantioseparations [8]. The mentioned above reasons indicate that using CSPs is by far the most privileged approach in chiral liquid chromatography.

2.1 Normal-Phase Liquid Chromatography

Strategy When liquid chromatography is selected as a technique, a choice of mode should be made. In fact, choosing a mode depends primarily on the properties of the analyte, the goal of the separation, and the CSP that is going to be used. If the analytical separation should be scaled up to the preparative level, it is recommended that solvents be used that are easily removable after separation (i.e., that are volatile). Therefore, working in normal-phase mode might be preferable over reversed phase, although the latter is ecologically and economically more favorable. The choice of mobile phase should be considered very carefully, since the success of separation depends on the combination of the mobile and stationary phases. Different combinations will affect selectivity in an important way: It may appear that unresolved compounds on a given chromatographic system suddenly get resolved after modification of the mobile phase. As a consequence, it remains a challenge to develop chiral separations, since they still cannot be predicted based on the structural properties of the chiral compound [26].

A first important factor in chiral analysis is the type of selector. Therefore, different polysaccharide-derived CSPs are involved in the strategy. A second important factor is the environment in which the solute will reside (i.e., the mobile-phase composition). Several organic modifiers (e.g., ethanol, methanol), compatible with the stationary phase, are proposed. In our NPLC strategy [42], during screening, both factors are combined in a 2^2 full factorial design for basic compounds, whereas a 3×2 experimental design is used for acidic, bifunctional, or neutral compounds (see Figure 4). In the 2^2 design, both selector and organic modifier type are examined at two levels: Chiralpak AD and Chiralcel OD or amylose tris(3,5-dimethylphenylcarbamate) and cellulose tris(3,5-dimethylphenylcarbamate) respectively, for the selector, and 2-propanol and ethanol for the organic modifier type. According to this design, the basic compounds are screened on both columns using the mobile phases *n*-hexane/2-propanol/diethylamine (DEA) [90 : 10 : 0.1(v/v/v)] and *n*-hexane/ethanol/DEA [90 : 10 : 0.1(v/v/v)]. The flow rate in all four experiments is 1.0 mL/min and the temperature is 20°C.

The acidic, bifunctional, and neutral compounds are screened using a 3×2 design (Figure 4). The factors are again the type of selector, in this case Chiralpak AD, Chiralcel OJ [cellulose tris(4-methylbenzoate)] and Chiralcel OD, and the type of organic modifier (levels: 2-propanol or ethanol). Since the number of selectors is three, this design requires six experiments. The compounds are screened on each CSP with hexane/organic modifier/trifluoroacetic acid (TFA) (90 : 10 : 0.1) (v/v/v). Other analysis conditions are identical to those for the basic compounds [42].

Depending on the best result obtained, different optimization steps are included in the strategy. When baseline optimization is achieved, which means that the peaks eluted with a resolution (Rs) >1.5, optimizations for peak shape or retention factor (i.e., runtime) can be performed optionally. These optimizations are rather fine tunings and thus not really required. For peak-shape optimization, the concentrations of DEA or TFA are reduced and examined at three levels: 0.025, 0.050, and 0.075% (v/v), while the analysis temperature is examined at two levels: 15°C and 25°C. These factors are investigated in a 3×2 design. The retention factor is optimized (minimized), first by trying to increase the percentage of organic modifier. If still better results are required, higher temperatures (25°C and 40°C) and flow rates are tested in a 2^2 full factorial design. Flow rates that have to be used are not specified but should be tested on two levels, at values above 0.5 mL/min, taking the (pressure) limitations of the chromatographic system into account [42].

When only a partial separation ($0 < Rs < 1.5$) is obtained, a baseline-separation optimization is prescribed to enhance the separation following different pathways, depending on the retention factor k' of the enantiomers. In different steps, the following factors are adapted: The percentage organic

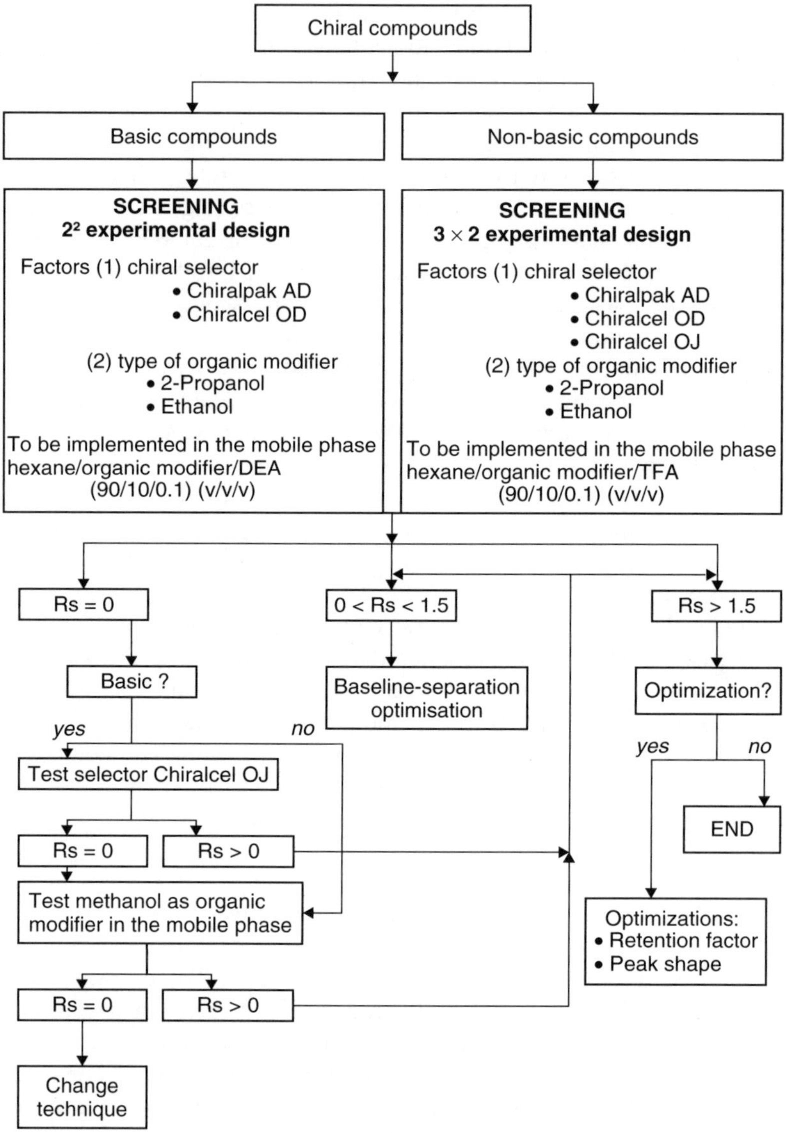

Figure 4 Separation strategy in NPLC. (Adapted from [42].)

modifier is increased, while the concentration of additive is decreased. The flow rate and the analysis temperature are also both decreased.

After screening, a third possibility is that no separation at all is observed (Rs = 0). In this case, the use of a third selector (Chiralcel OJ) is recommended for basic compounds. Then for all compounds, the strategy proposes as a last step use of methanol as an organic modifier in the mobile phase (Figure 4).

When enantioselectivity is not obtained, the use of another technique is recommended. In cases where the compound shows any enantioselectivity after using either Chiralcel OJ and/or methanol, the strategy refers to the baseline-separation optimization step.

Additional Discussion In the literature, different high-throughput-screening approaches in NPLC with polysaccharide-based CSPs are described. Both isocratic [43] and gradient elution proposals [44] exist in order to get an idea of the chiral separation possibilities for a compound in the shortest time possible [42–45]. Some strategies are only developed for groups of molecules (e.g., phenylethanolamine derivatives [38], allylic alcohols [46], or amino acids [47]). Other strategies are developed using only one specific CSP type (e.g., cellulose-based polysaccharide CSP) [48]. However, in many cases the same successive steps are observed.

A general screening strategy in NPLC mode starts very often with the use of a mobile phase based on *n*-hexane (can be replaced by the less toxic *iso*-hexane) or *n*-heptane since these proved to give the best results on polysaccharide derivative CSPs, regardless of whether the selectors are coated or immobilized on the silica support [37,38,42–52]. Very often, alcohols such as methanol, ethanol, or 2-propanol are added to the hydrocarbon solvent. The reason for using such mixtures is to increase the polarity of the mobile phase, resulting in changed interactions between selector and solute, hopefully ending up in a better separation [26]. Depending on the type, structure, and concentration of the alcohol used, the selector may show small structural changes with a nonnegligible effect on the chiral selectivity [53].

Besides the organic modifiers (polarity), there is another major factor influencing selectivity in chiral analysis: organic acids and organic bases used as additives in the mobile phase. It has been demonstrated that these compounds are essential for the separation of acidic and basic drugs [54,55]. Trifluoroacetic acid is a frequently used acid. Acetic acid and alkylsulfonic acids, such as ethanesulfonic acid, can also be dissolved in the mobile phase. Triethylamine and diethylamine are the common basic additives, while butylamine and dimethylethylamine are possible alternatives [45,55]. The influence of these additives, generally used in concentrations of 0.1% (v/v) in the mobile phase, is versatile. The enantiomers should be in a neutral form in order to interact with the polysaccharide selector. Acidic additives protonate the residual silanol groups of the silica support of the CSP, to prevent peak tailing of basic compounds. They also convert acidic solutes into their neutral form, causing a reduction in retention and suppressing tailing as shown by Tang [54].

For basic additives the same goals of minimizing retention times and peak tailing and improving enantioresolution are believed to be obtained due to ion-pair formation with unprotonated silanol groups on the CSP and with acidic groups on the chiral analytes [54,55]. It can be stated that in general acidic additives are used for the separation of acid compounds, while basic additives are used for basic compounds, although trifluoroacetic acid is also used as an additive when amino-containing basic compounds are analyzed [38]. Even combinations of both types in one mobile phase are common, as we discuss later. The choice of additive is not so important for compounds with bifunctional (amphoteric) or neutral properties.

In several screening strategies, method development or screening conditions are defined according to an experimental design described earlier. Experimental designs are predefined experimental setups in which certain variables, which in the actual context are considered important, are examined at a given number of levels in a fixed number of experiments [56]. In the experimental setups used in chiral chromatography, the sequence of selectors and organic modifiers to be tested is often prescribed. In fact, the systems with the broadest enantioselectivity are evaluated first.

By changing the CSPs or the type of organic modifiers, inversion of elution order may appear. The change of elution order is of importance when developing analyses that can be scaled up for preparative purposes and when developing analyses that are used to assay the distomer impurity, because it is important to be able to choose which enantiomer elutes first [57,58]. Besides isocratic elution, there are also rapid NPLC screening strategies in gradient elution mode. De la Puente et al. [44] describe an automated screening method for high-throughput chiral analyses. The mobile phase is composed of the same solvents as in isocratic mode: The basic solvent is a hydrocarbon such as *n*-hexane or *n*-heptane, with an additive, while the organic modifiers are short-chain alcohols (ethanol, 2-propanol). A gradient range from 20 to 70% (v/v) alcohol in a time range of 20 min was investigated [44].

2.2 Reversed-Phase Liquid Chromatography

Separating chiral compounds in NPLC can cause solubility problems for polar substances. Aqueous solvents may offer a solution for the analysis of compounds that could not be examined in NPLC. Another advantage of the reversed-phase mode, besides enlarging the spectrum of analyzable compounds, is the use of less toxic solvents. Aqueous buffers are less expensive, and this mode facilitates analysis in biological matrices, since sample preparation from plasma and serum does not need to end in apolar mobile-phase solvents [59].

Strategy Matthijs et al. [42] defined a screening strategy for chiral compounds in RPLC (Figure 5). This strategy can be applied either by executing the complete screening experimental design or by screening the test compounds sequentially at given conditions. The design is a 3 × 2 design. The first factor is the chiral selector examined at three levels (Chiralcel OD-RH, Chiralcel OJ-R, and Chiralpak AD-RH), while the second factor is the buffer pH, examined at two levels (pH 2.0 and 9.0). In practice, this means that test compounds are screened on three CSPs with two mobile phases. The first mobile phase is a mixture of 100 mM KPF_6 in 50 mM H_3PO_4 at pH 2.0 and acetonitrile in the ratio 60 : 40 (v/v). The second consists of a borate buffer (20 mM H_3BO_3) at

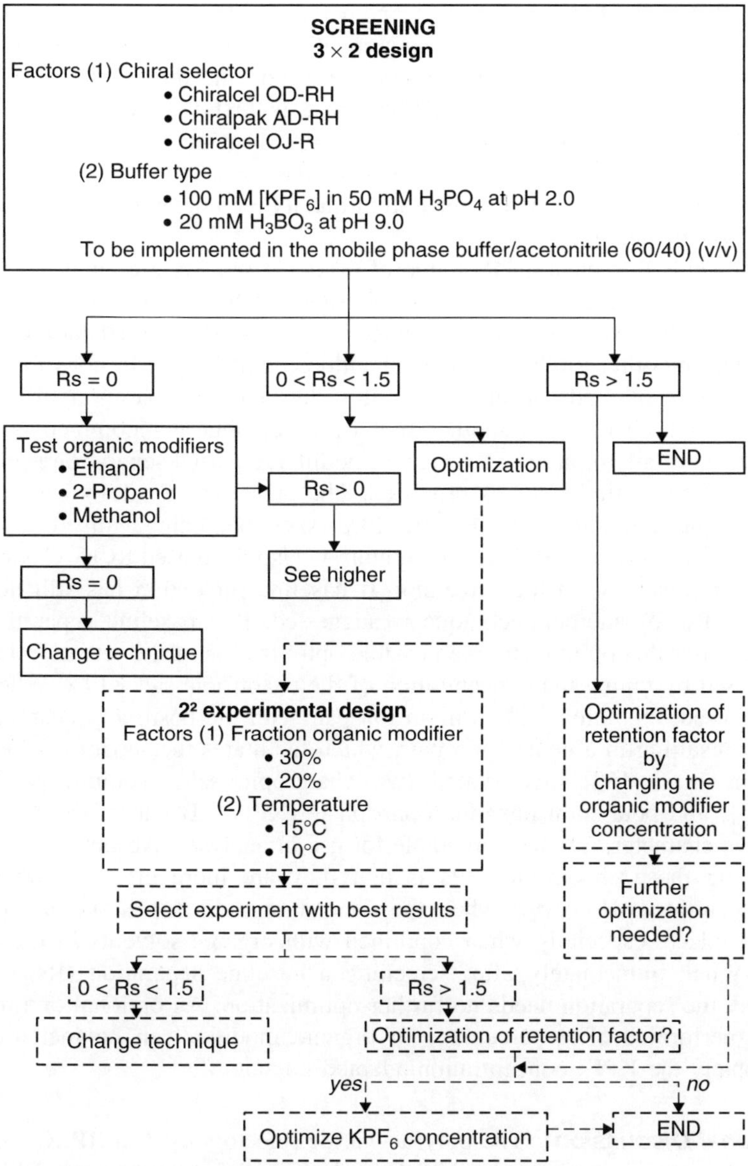

Figure 5 Separation strategy in RPLC. (Adapted from [42].)

pH 9.0, again mixed with acetonitrile in a 60 : 40 (v/v) ratio. Although execution of all conditions gives the most complete results, it is rather time consuming. The proposed alternative is, as mentioned above, a sequential screening of the CSPs in a given sequence: Chiralcel OD-RH > Chiralpak AD-RH > Chiralcel OJ-R, first with the acidic mobile phase. This approach is frequently performed because of the complementarity of the CSPs: If one CSP does not show enantioselectivity, it is possible that another does. The sequence is defined by the authors according to the decreasing enantioselectivity of the systems.

When no separation is achieved after screening the three columns with the acidic mobile phase, the basic mobile phase is first used on Chiralpak AD-RH. Earlier results showed that usually no improvement in enantioselectivity is observed with a basic mobile phase on the other two CSPs [60]. Based on the resolution, which is used as the separation-quality parameter, the screening again ends in three categories.

Compounds that have an Rs value of 0 after screening are analyzed with an adapted mobile phase. The acetonitrile is replaced by another organic modifier according to the proposed sequence ethanol > 2-propanol > methanol. Changing the organic modifier might give new enantiorecognition possibilities toward the analytes, as mentioned earlier. If after this step, enantioseparation still has not been achieved, changing to another analytical technique is recommended. However, when initially or after changing the modifier, a partial separation is obtained ($0 < Rs < 1.5$), optimization is proposed. The prescribed 2^2 design reduces the fraction organic modifier to 30% and 20% (v/v) (beginning with 40% (v/v) in the original mobile phase). The temperature is also decreased to 15°C and 10°C, whereas it was 20°C in the screening. If baseline separation has still not been achieved, use of another technique is suggested. If a baseline separation was achieved after this optimization, a possible optimization of the retention factor k' is proposed by tuning the concentration of the chaotropic salt KPF_6. When such a salt is used, the anion will form an ion pair with the positively charged basic analyte, resulting in a neutral ion pair, which facilitates the separation [43,60].

Perrin et al. [60] investigated two chaotropic salts, sodium perchlorate ($NaClO_4$) and potassium hexafluorophosphate (KPF_6). Because $NaClO_4$ has an explosive character, it is less favorable for use in analyses. An important remark concerning these chaotropic salts is that they are more effective when they are concentrated. However, when the concentration is too high, precipitation can take place, especially when combined with organic solvents in the mobile phase. When immediately after screening a baseline separation ($Rs \geq 1.5$) is obtained, the separation needs no further optimization. An optional optimization can be performed: The percentage of organic modifier can be increased and occasionally the KPF_6 concentration adapted.

Additional Discussion The polysaccharide selectors used in RPLC mode are exactly the same as those used in NPLC mode. The difference between RPLC-and NPLC-polysaccharide CSPs is not revealed by the manufacturer but is probably related to the used silica, the coating material, and the coating mechanism. In general, the mobile phase consists of an aqueous buffer and organic solvents, (e.g., acetonitrile, methanol, ethanol) [42,59,60]. The RPLC strategy described above no longer uses organic acids and bases as mobile-phase additives (as in NPLC) to keep the chiral solutes in their nonionized form. Neutrality of the compound is important when using polysaccharide derivatives as selector. When the analyte is charged, it will not interact with the selector; thus, no enantioseparation will take place. To prevent ionization of acidic and basic compounds, acidic and basic mobile phases, respectively, are used. It should be mentioned that polysaccharide-derived CSPs have a workable pH range between 2.0 and 9.0; thus, work with acidic or basic mobile phases must be in this range.

In generic screening strategies, conditions must be found to screen chiral compounds within a small number of experiments. In fact, a "universal" mobile phase is preferably used. In such a mobile phase in RPLC, the aqueous part (buffer) should, in theory, be both acidic and basic, for acidic and basic compounds, respectively. Since this is impossible, an acidic buffer at pH 2.0 is prepared. Under these conditions, ionization of acidic compounds will be suppressed. To analyze basic compounds with the same mobile phase, the addition of a chaotropic salt is proposed [42,59–61]. The combination of an acidic buffer and a chaotropic salt in the buffer seems a good compromise for a multifunctional mobile phase. The possibility exists that some basic compounds still are not resolved under these conditions. In these cases it is opportune to screen with a basic buffer at pH 9.0, preferably a borate buffer [59,60]. However, many basic compounds need an even higher pH than 9.0 to be uncharged. Although in the screening stage the role of additives such as DEA and TFA is replaced by the pH-controlled buffer, the use of these organic additives might reappear in optimization steps.

2.3 Polar Organic Solvent Chromatography

Using polysaccharide derivatives as CSPs has become very common in chromatographic analysis. The classical normal- and reversed-phase modes make it possible to separate a wide spectrum of chiral compounds. However, a number of compounds might still not be resolved. Using only polar organic solvents, without aqueous buffers as part of the mobile phase, can offer a possible solution. Polar organic solvent chromatography is not only advantageous in the case of solubility problems, but also offers different chiral recognition mechanisms than those of NPLC and RPLC [44]. In addition, these polar organic solvents are, as the apolar normal-phase solvents, more easily removed from the analytes after separation than are the salty buffers in RPLC, which is an important advantage from a preparative point of view [62].

Separation Strategy The first step of the polar organic solvent strategy (Figure 6) [63] involves the sequential screening on four polysaccharide CSPs in a given sequence: Chiralpak AD-RH > Chiralcel OD-RH > Chiralpak AS-RH > Chiralcel OJ-R. These CSPs are packed with cellulose- and amylose-derived selectors as described earlier (the selector for Chiralpak AS-RH is amylose tris[(*S*)-α-methylbenzylcarbamate]). Screening on these CSPs happens with two mobile phases. The first mobile phase consists of acetonitrile/DEA/TFA [100 : 0.1 : 0.1 (v/v/v)], while the second is composed of methanol/DEA/TFA [100 :0.1 : 0.1 (v/v/v)]. This first screening step has again three possible outcomes, depending on the obtained resolution. No enantioseparation (Rs = 0) after the first screening leads initially to the same action as when a partial

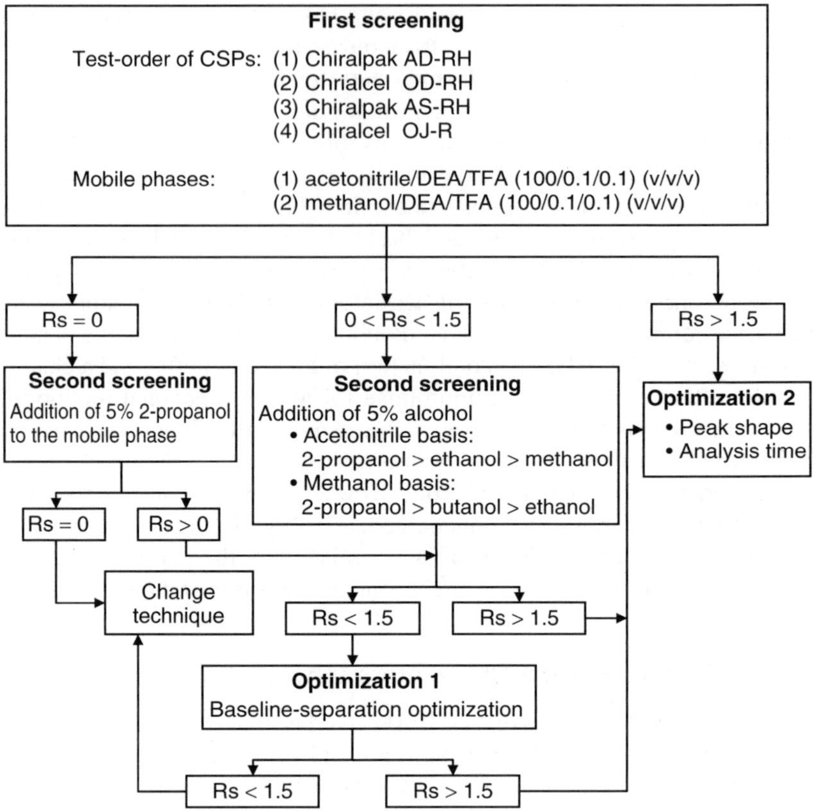

Figure 6 Separation strategy in POSC. (Adapted from [63].)

separation (0 < Rs < 1.5) is obtained; in both cases, the composition of the mobile phase is changed. The organic solvent is decreased to 95% (v/v) while 2-propanol is added in a concentration of 5% (v/v). A second screening is thus executed. The mobile phases are acetonitrile/2-propanol/DEA/TFA [95 : 5 : 0.1 : 0.1 (v/v/v/v)] and methanol/2-propanol/DEA/TFA [95 : 5 : 0.1 : 0.1 (v/v/v/v)]. If enantioseparation is not achieved, changing the technique is recommended.

Similar mobile-phase changes are recommended when, initially, after the first screening, partial separation is obtained (0 < Rs < 1.5). In this case only one mobile phase (i.e., based on acetonitrile or methanol) is examined, depending on the best result obtained from the first screening. Not only 2-propanol, but also two other short-chained alcohols, are potentially tested. In fact, the second screening again becomes a sequential screening on the best CSP, either with an acetonitrile-based mobile phase and 5% (v/v) of 2-propanol, ethanol, or methanol, respectively, or with a methanol-based mobile phase with 5% (v/v) 2-propanol, butanol, or ethanol. Each mobile phase also contains the additives DEA/TFA in the concentrations indicated earlier. The simultaneous use of both additives is preferred for different reasons. The mobile phase is again multifunctional and thus made as generic as possible. But besides this, the combination of DEA and TFA is supposed to minimize the memory effect of the additives [64]. This phenomenon is discussed in more detail later.

When after the second screening, $0 < Rs < 1.5$ is still obtained, a baseline-separation optimization is proposed. This optimization is done following a 3×2 design. Varying factors are the organic modifier concentration and the analysis temperature. The mobile- and stationary-phase combination of the second screening that gave the highest Rs (which is still < 1.5) is selected. The organic modifier fraction used previously is increased at three levels. For acetonitrile-based mobile phases the fraction will become 10, 15, and 20% (v/v), while in methanol-based mobile phases the levels are 15, 50, and 85% (v/v). Simultaneously, the analysis temperature is decreased. Where the nominal value was 20°C the optimization proposes analysis at 15°C and 5°C. In chiral analysis, temperature changes may influence enantioselectivity and thus the separation, due to differences in thermodynamic parameters that determine the selector–solute interactions [65]. If these experiments still result in $Rs < 1.5$, it is suggested that another technique be tried.

When at any moment sufficient separation is obtained ($Rs > 1.5$), the development can be stopped unless additional optimization is desired. A possible optimization that is proposed is to improve the peak shape. In this optimization step a 2^3 full factorial design examines three factors at two levels in eight experiments. The factors that have an influence on peak shape are the types and concentrations of organic additives. Alternative compounds for TFA and DEA are tested. Acetic acid and formic acid replace TFA, while isopropylamine and ethanolamine are the alternatives for DEA. Each additive is tested at concentration levels of 0.025 or 0.050% (v/v). Thus, three factors (acetic additive, basic additive, and concentration) are tested at two levels. Besides peak-shape optimization, there is also an option for retention time optimization by increasing the column temperature and the flow rate taking into consideration the limitations of the chromatographic equipment used.

Additional Discussion The POSC strategy uses the same CSPs as the RPLC strategy. Although the mobile phases are somewhat less polar than the aqueous mobile phases in reversed-phase mode, the same CSPs can be used in both modes. This is an example of multimodality: the application of different modes on the same column. Török et al. [66] investigated the possibility of using normal-phase CSP in polar organic mode and confirmed the possibility without observing a loss of selectivity (α) or resolution (Rs). There are also reports about the use of normal-phase columns in reversed-phase mode [59]. The applicability of the CSPs in diverse modes is probably because they are very similar [67].

The aforementioned phenomenon is becoming more common; several new polysaccharide-based CSPs are marketed on which normal-phase, reversed-phase, and polar organic solvents chromatography can be applied. Sepapak-2 [cellulose tris(3-chloro-4-methylphenylcarbamate)], Sepapak-3 [amylose tris (5-chloro-2-methylphenylcarbamate)], Chiralpak IA [immobilized amylose tris(3,5-dimethylphenylcarbamate)], Chiralpak IB [immobilized cellulose tris (3,5-dimethylphenylcarbamate)], and Chiralpak IC [immobilized cellulose tris (3,5-dichlorophenylcarbamate)] are such phases.

The use of one column in different modes of analysis should be done with care, because memory effects can occur, especially when organic mobile-phase additives are used to enhance separation quality. They are, for instance, used in polar organic solvent chromatography and normal-phase liquid chromatography: no aqueous buffers are applied in these mobile phases; thus, the ionization of the analytes is tuned by using organic acids and bases. Studies on mainly polysaccharide-derived CSPs have been made to investigate the effect of such acidic and basic additives on the chromatographic results [55,68–70]. During these experiments a *memory effect* was noted: a phenomenon where these mobile-phase additives leave a trace on the chiral selector of the CSP. The consequences of this contamination are that the separation results are affected. Stringham et al. [69] did some experiments on polysaccharide-based stationary phases where the memory effect of DEA was clearly shown.

An important remark in this context can be made about the type of mobile phase used. When it concerns apolar-type eluents (e.g., *n*-hexane based), the memory effect of DEA on the column lasts much longer than it does with polar organic eluents. Effective washing methods are derived from these findings: To minimize the memory effect of DEA, the column used should be washed with organic solvents and preferably with 2-propanol, which removes the basic additive memory effect almost completely [69]. The same suggestions are made by Matthijs et al. [64]. Both references also confirm the findings of Ye et al. [70] concerning the benefits of rinsing with an acidic additive to remove the basic. However, when rinsing with the acid TFA, a new problem occurs since this additive also has to be removed. Although the effect of this additive seems to be even more persistent, thus difficult to remove, a washing step with pure acetonitrile should give minimal memory effects [65]. The best solution for avoiding memory-effect problems is to dedicate one polysaccharide CSP for each additive.

In screening strategies, the goal is to standardize the conditions such that many compounds can be analyzed at the same generic conditions. Therefore, acidic and basic mobile-phase additives are used together, which even results in a more advantageous situation, since this combination gives fewer memory-effect problems compared to mobile phases with only one additive [65,69,70].

3 STRATEGY IN SUPERCRITICAL FLUID CHROMATOGRAPHY

3.1 Introduction

In chiral analysis where the emphasis switches more and more to "fast" enantioseparations, different analytical techniques for high-throughput are evaluated continually. For a long time, the success of the HPLC modes overshadowed the search for other techniques. Since SFC made its revival in chiral analysis, it became a strong player and quickly secured an important place next to the usual separation techniques. When mentioning SFC, in fact, both supercritical and subcritical fluid chromatography are implied. In supercritical chromatography the mobile phase is at a temperature and pressure above the critical point in the phase diagram of the mobile-phase substances. A supercritical fluid combines the solvating power and density of a liquid with the viscosity and compressibility of a gas [71]. In subcritical chromatography analysis the mobile phase is at a temperature and a pressure just below the critical point, but since the physical properties remain similar, a distinction is rarely made between them [72].

The supercritical fluid used most frequently in pharmaceutical analysis is CO_2 because of its convenient supercritical pressure and temperature properties (73.8 bar and 31.1°C, respectively), but also because of its limited toxic, nonflammable, and environmentally friendly properties [26,73–75]. The presence of apolar CO_2 as a principal compound in SFC mobile phases makes this technique comparable with that of the normal-phase mode. Therefore, practically all chiral analyses on polysaccharide-derived CSP in SFC are executed with normal-phase columns [35,76–86]. When SFC is executed with CSP, the term *packed column SFC* is used because in SFC there is also a possibility of using open columns: narrow capillary tubes in which the stationary phase is coated onto the capillary wall. According to the aforementioned conditions, it seems that SFC is reserved primarily for rather apolar analytes. The polarity of CO_2 can be tuned by increasing the temperature and pressure [26], but this is insufficient to analyze more polar analytes. Luckily, CO_2 has the property of being miscible with polar organic solvents such as acetonitrile and short-chain alcohols (e.g., methanol, ethanol, 2-propanol). Using these modifiers expands the range of SFC for chiral analysis.

The recognition mechanisms for chiral separations are not entirely revealed yet, but it is general knowledge that organic modifiers change the polarity of the mobile phase and with this the selector–enantiomer interactions on polysaccharide-derived CSPs. Changing conditions in chiral separations can also affect the shape of the selector and thus the enantioselectivity. Using different types of modifiers leads to different selectivities, while tuning modifier concentrations affects retention times. When higher concentrations of these solvents are used, the mobile phase might switch in the phase diagram from supercritical to subcritical. However, as mentioned before, the physical properties remain similar; thus, usually no special precautions must be taken. It is also possible to keep the mixed mobile phase in the supercritical state by applying higher operating temperatures [71].

To complete the comparison with NPLC mobile phases, the place of organic bases and acids as additives is also discussed briefly. As mentioned before, these additives are generally used to improve separation quality. The main mechanisms are ion-pair formation and suppression of ionization. Because of the intrinsic acidic character of CO_2, acidic additives seem unnecessary for SFC. Nevertheless, some acidic additives, such as ethanesulfonic acid, have been examined [83]. Depending on the type of modifier, the analyte, and the additive used, a changed enantioselectivity may occur. This is a very general statement which, in fact, indicates the necessity of more research in the field of these additives in SFC. The same is true for basic additives [81]. Therefore, it is difficult to define which additive should be used in given circumstances. However, it is clear that they might have positive effects on separation.

Although chiral SFC is still a relatively young technique and not everything is well understood or examined, it offers a number of advantages over standard analytical NPLC. Higher efficiencies are obtained due to lower viscosities of solvents and high diffusion coefficients of solutes. In SFC, solvent consumption is low; CO_2 replaces the toxic and flammable solvents, and thus SFC can be labeled as a "green" technique. Rapid column equilibration, high flow rates, short analysis times, and high sample throughput speed up a screening procedure. Higher resolutions result in increased enantiomeric purity of collected fractions. It is thus worthwhile trying to implement SFC in both analytical and preparative chiral separation programs. The main drawback that should be considered is the cost and complexity of instrumentation when preparative-scale enantioselective SFC is wanted.

3.2 Strategy

Because of its advantages and complementarity to other techniques, high-throughput screening strategies are defined in SFC. Similarly to Matthijs et al. [43,64] for NPLC, RPLC and POSC, Maftouh et al. [25] proposed a SFC strategy (see Figure 7). The CSPs to be screened are ranked as Chiralpak AD > Chiralcel OD > Chiralcel OJ > Chiralpak AS. The first modifier to be used in combination with CO_2 is methanol, and the second is 2-propanol. The fraction of the modifiers is 10 and 20% (v/v), respectively. Of course, as for all strategies, there are exceptions: compounds that show selectivity only with other modifiers (e.g., ethanol). But again, the most favorable modifiers are included to define a generic strategy. Acetonitrile is not included in the strategy because it showed hardly any benefit. In the strategy proposed, a strong basic additive (isopropylamine) is used in the separation of basic compounds, and a strong acidic additive (TFA) is used for acidic compounds. For both types, 0.5% (v/v) was chosen as concentration.

As in the NPLC strategy, the analytes are divided into two groups: the first group contains basic, bifunctional, and neutral compounds. These are screened with the mobile phases CO_2/methanol/isopropylamine [90 : 10 : 0.5 (v/v/v)] and CO_2/2-propanol/isopropylamine [80 : 20 : 0.5 (v/v/v)]. Bifunctional compounds could also be analyzed with acidic additives in the mobile phase, while neutral compounds do not need any additive. These compounds were arbitrarily considered together with the basic compounds, whereas in the NPLC strategy they were considered with the acidic compounds. The second group contains the acidic compounds, screened with mobile phases CO_2/methanol/TFA [90 : 10 : 0.5 (v/v/v)] and CO_2/2-propanol/TFA [80 : 20 : 0.5 (v/v/v)]. The complete screening (i.e., an eight-experiment system) might be performed, but again the analyst may choose to stop the screening from the moment that a satisfying resolution is

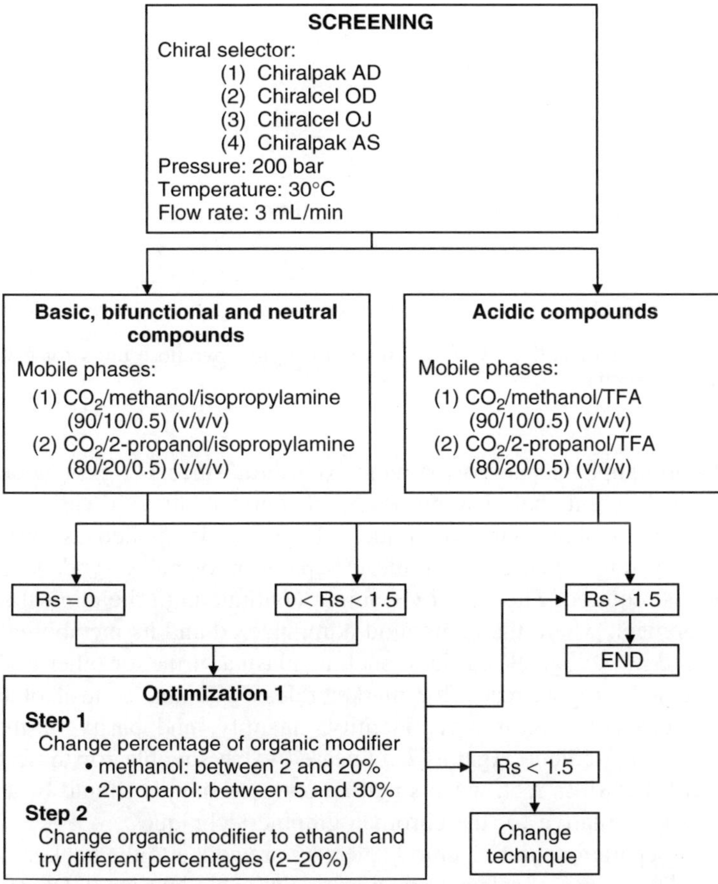

Figure 7 Separation strategy in SFC. (Adapted from [25].)

obtained. If optimization is required after screening, the first step is changing the organic modifier concentration. For methanol both a decrease to 2% (v/v) and an increase to 20% (v/v) are prescribed. For 2-propanol the concentrations tested range between 5 and 30% (v/v). In case these modifications do not give improved results, the modifier should be changed to ethanol with concentrations varying from 2 to 20% (v/v) [25]. The analysis temperature was set at 30°C, the flow rate at 3 mL/min, and the detection wavelength at 220 nm. The results obtained after application of this strategy confirm the complementarity of the various modes (apolar NPLC–aqueous RPLC–polar organic solvents–supercritical fluids) [64].

4 STRATEGY IN CAPILLARY ELECTROPHORESIS

A final technique discussed in this chapter is capillary electrophoresis. Although CE is not applicable for preparative-scale enantioseparations, the development

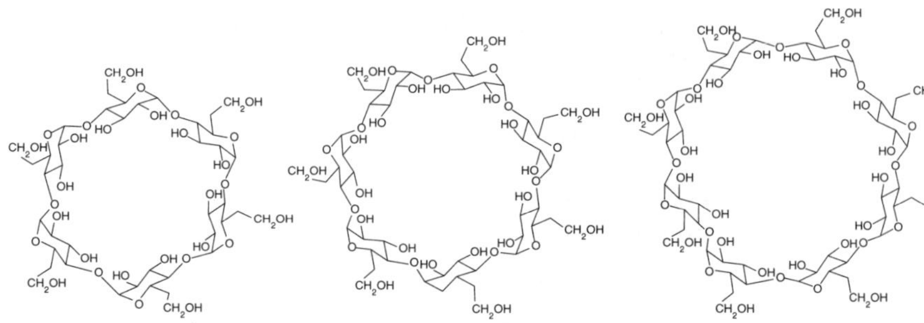

Figure 8 Structure of native cyclodextrins with 6 glucopyranose units for α-CD, 7 for β-CD, and 8 for γ-CD.

of a high-throughput separation strategy is useful since CE is a widely used technique for different analytical purposes. In pharmaceutical research, chiral CE (besides the chromatographic techniques of course) is applied as early as the drug-discovery stage, where enantiomeric separation of many synthesized chiral compounds is required. The use of CE might continue until the clinical stages of drug development, where the compound administered and its metabolites should be determined in biological matrices, such as plasma, urine, or other body fluids [12,14]. When a drug is ready for market release, quality control of the drug product is needed to confirm the identity, quantity, and purity of the active pharmaceutical ingredient. Again CE appeared to be suitable for these types of assays [12,14,87,88]. In fact, when small amounts of sample should be analyzed, CE can be an alternative for the chromatographic techniques.

For direct separations in CE, chiral selectors are added to the background analyte (BGE). The choice of selectors is diverse, but cyclodextrins (CDs) (Figure 8) are by far the most popular [11,15,30,89–93]. Regardless of whether or not they are derivatized, the popularity of CDs can be attributed to their high aqueous solubility (except for native β-CD), their stability in aqueous media, their commercial availability, and their relatively low cost. In addition, these selectors do not show UV absorption, which eliminates a possible drawback of decreased sensitivity [15]. For chiral purposes, the ability of enantiorecognition is obviously most important. CDs do show good selectivity for a wide range of compounds [91].

4.1 Strategy

In CE strategy (Figure 9) [92], only one type of chiral selector is used: highly sulfated and multiply charged anionic cyclodextrins (HS-CD). Three different derivatives, HS-α-CD, HS-β-CD, and HS-γ-CD, are implemented in the strategy. A sequential screening of the compounds with the three selectors is recommended. As starting conditions for the screening, a 5% (w/v) CD concentration at pH 2.5 is tested, starting with the γ derivative, followed by the β and ending with the α derivative. The BGE is a 50-mM phosphate buffer (prepared

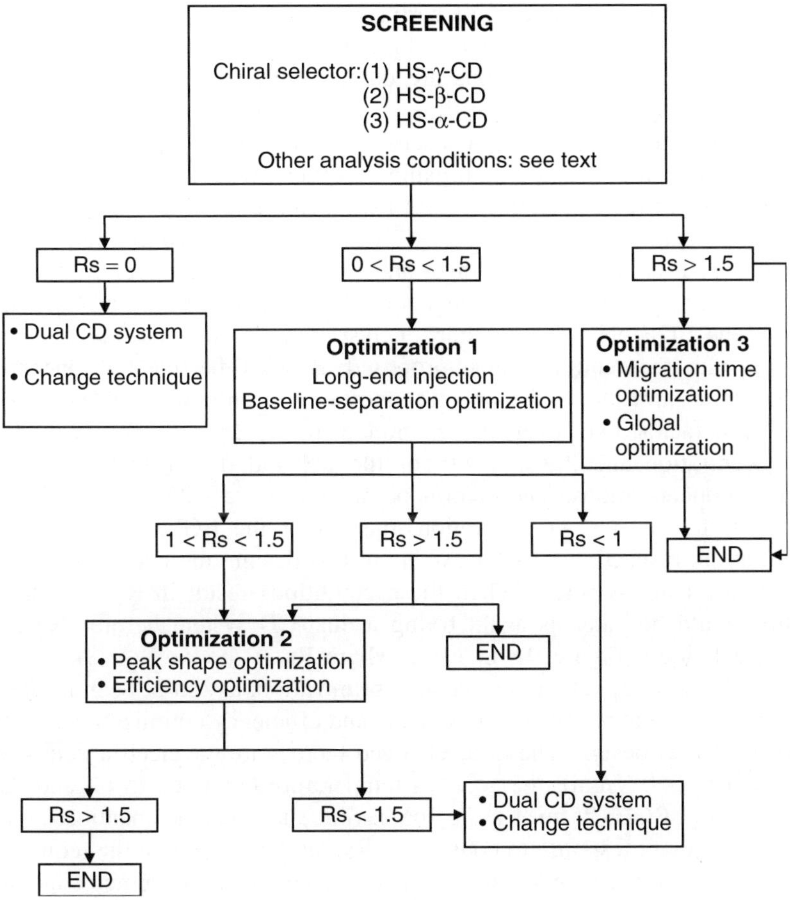

Figure 9 Separation strategy in CE. (Adapted from [92] and [94].)

from 85% (w/w) orthophosphoric acid) adjusted to the prescribed pH with triethanolamine (TEA). Besides adjusting the pH, TEA also seems to reduce the adsorption of basic analytes to the capillary wall, which results in a reduced peak tailing. The analysis temperature is 20°C, the electric field strength 300 V/cm (reversed polarity) with an anodic detection at a wavelength of 200 or 220 nm. All experiments are performed as short-end injection analyses. After screening, three options are again considered, depending on the obtained resolution. First, when no separation occurred (Rs = 0), a dual-selector system was recommended. This implies a combination of a highly sulfated CD (α, β, or γ) with a neutral CD. The neutral CDs used are heptakis (2,3,6-tri-*O*-methyl)-β-CD (TMCD), heptakis(2,6-di-*O*-methyl)-β-CD (DMCD), and hydroxypropyl-β-CD (HPCD). This dual-system testing is, in fact, an extension of the strategy [94] and is discussed later.

When the screening resulted in a partial separation ($0 < \text{Rs} < 1.5$), a first optimization is proposed. Optimization 1 (Figure 10) is, in fact, a baseline-separation optimization where the conditions are defined differently for compounds with either a long or a short migration time (t_M). Compounds are considered to have a long migration time when t_M exceeds 15 min. For this group, the optimization does not include the use of organic modifier (MeOH) since its behavior is unpredictable, which can result in even longer migration times [92]. The factors that are changed are merged in a 2×3 design. The CD concentration and the pH of the BGE are evaluated at three and two levels, respectively; 2.5, 5.0, and 10% (w/v) are the concentration levels of the selector, and 2.5 and 4.0 are the levels of the BGE pH. When a short analysis time is obtained ($t_M < 15$ min) a 2^3 design plus the central point is proposed. Besides the selector concentration and the pH, a third factor, the organic modifier concentration, will be tested.

For each factor, two levels are examined: being 2.5 and 10% (w/v) for the CD concentration and 2.5 and 4.0 for the pH and 0 and 15% (v/v) for the methanol concentrations. The central point values are 6.25% (w/v), 3.25, and 7.5% (v/v) for the CD concentration, the pH of the BGE, and the methanol concentration, respectively. After executing this design, the conditions that gave the best result are selected. When these conditions result in $\text{Rs} < 1$, the next step that could be taken is again trying a dual-CD system before changing to another technique (Figure 9). In cases where $\text{Rs} > 1.5$ or $1 < \text{Rs} < 1.5$ are obtained, the strategy refers to a second optimization, as can be seen in Figure 9. Optimization 2 (Figure 10) is a peak shape and efficiency optimization prescribed by means of a 2^3 design. The three changed factors are the electric field strength (250 V/cm to 350 V/cm), the analysis temperature (15°C to 25°C) and the ion strength of the BGE (25 mM to 75 mM). Since it concerns enantioseparations, the most important response to consider is Rs, but depending on the requirements of the analyses, other responses, such as plate number or asymmetry factor, can also be taken into account. When $\text{Rs} > 1.5$ is reached after optimization 2 or even after optimization 1, the method development may end. When baseline separation is not reached, a dual-cyclodextrin system is again proposed as an optional possibility before switching to another technique. However, when at any moment $\text{Rs} > 1.5$ is obtained, either development can stop or optimization 3 (Figures 9 and 11) can be executed. This final optimization step is divided into a migration time optimization and a global optimization. The migration time optimization is again meant for analyses taking too long (more than 15 min). A 2^3 design examines the temperature (25 and 30°C), electric field strength (350 and 400 V/cm), and CD concentration [7.5 and 10% (w/v)]; the increased factor levels should decrease the migration times. In addition to the migration time, efficiency is also a response that should be evaluated. For global optimization a 2^{4-1} design is described. Besides the three factors examined in optimization 2, a fourth, the CD concentration, is also evaluated. The values of the first three factors (electric field strength, analysis temperature, and ion strength) are the same as in optimization 2. The CD concentration is examined at two levels in this global optimization: at 2.5 and at 10% (w/v). In this optimization step, the

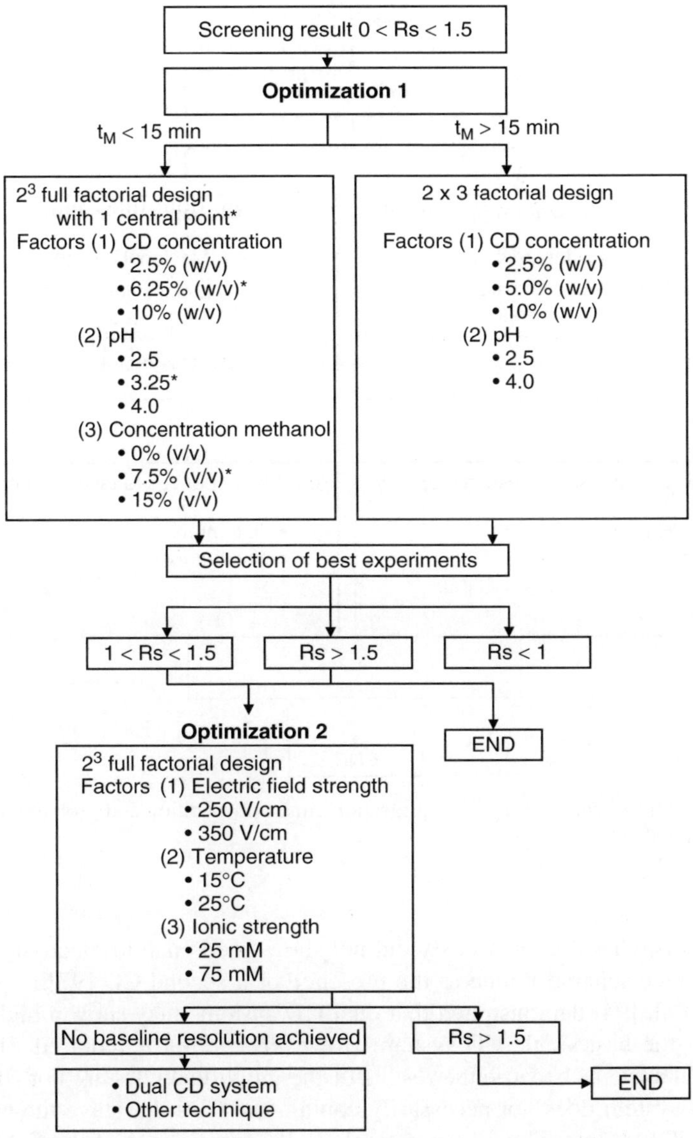

Figure 10 Optimizations 1 and 2: baseline-separation optimization and peak shape optimization. (Adapted from [92].)

responses evaluated in addition to migration time and resolution are asymmetry factor and plate number.

As mentioned before, this CE strategy is extended with a dual-selector system approach. This offers a possibility to expand the selectivity. With dual-selector systems, a combination of two types of CDs in the running buffer is meant: neutral–anionic, neutral–cationic, or cationic–anionic CDs [13,30,91–96]. In

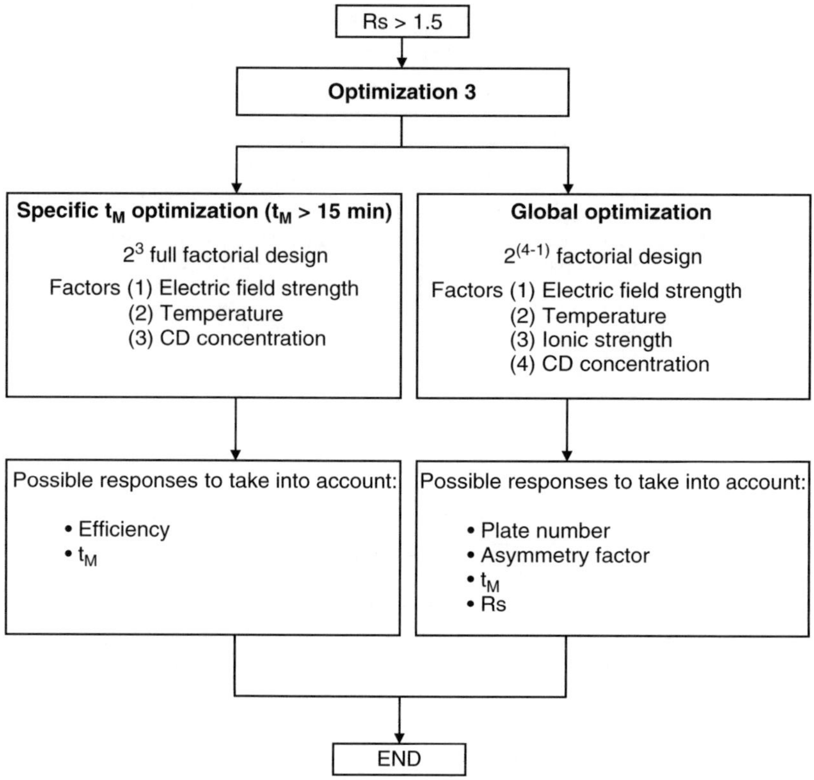

Figure 11 Optimization 3: specific migration-time optimization and global optimization. (Adapted from [92].)

such systems, CDs that previously did not show good enantioselectivity can suddenly enhance separation due to the presence of a second CD [97].
Matthijs et al. [94] demonstrated that dual-CD systems may show a higher selectivity than the best single-CD system. However, caution is required when composing dual systems because the results of the individual CDs are not additive. A good dual system does not necessarily contain the CDs that gave the best result in single-CD systems. This is due not only to the interactions of the CDs with the solute, but also to factors such as the viscosity of the BGE and the directions and sizes that the mobilities themselves play in enhancing the separation [98–100]. Matthijs et al. [94] defined conditions with dual-CD systems (Figure 12) potentially to be applied when no enantioseparation is obtained after sequential screening with the three HS-CDs used initially or when no baseline separation was achieved after optimization 1. Such a dual system can also be tested if a higher resolution is still wanted when a baseline separation (Rs > 1.5) has already been achieved after completing the strategy. In all three cases, the same conditions are executed (Figure 12). The CDs that are combined are the anionic HS-CD with one

of the following neutral selectors: heptakis(2,3,6-tri-*O*-methyl)-β-CD (TMCD), heptakis(2,6-di-*O*-methyl)-β-CD (DMCD), or hydroxypropyl-β-CD (HPCD).

The dual system uses different conditions for basic and nonbasic compounds. As mentioned earlier it is not always evident from the results of the single-CD systems which HS-CD should best be combined with which neutral, but starting with HS-γ-CD for basic analytes and HS-β-CD for nonbasic analytes is recommended. The sequence of the tested neutral selectors is the same for both groups: TMCD > DMCD > HPCD. The BGEs used for the analysis of basic compounds are composed of 50 mM phosphoric acid adjusted to a pH 2.5 with TEA and (1) 100 mM TMCD with 5% (m/v) HS-x-CD, (2) 50 mM DMCD with 5% (m/v) HS-x-CD, or (3) 50 mM HPCD with 5% (m/v) HS-x-CD. Since this extended strategy is a sequential approach, the development can stop as soon as a desired result is obtained. When no separation has been obtained after all experiments proposed, a switch to one of the chromatographic techniques is recommended. Whenever a partial separation (Rs between 0 and 1.5) occurs, the optimization is done by changing the concentration of the neutral CD in the

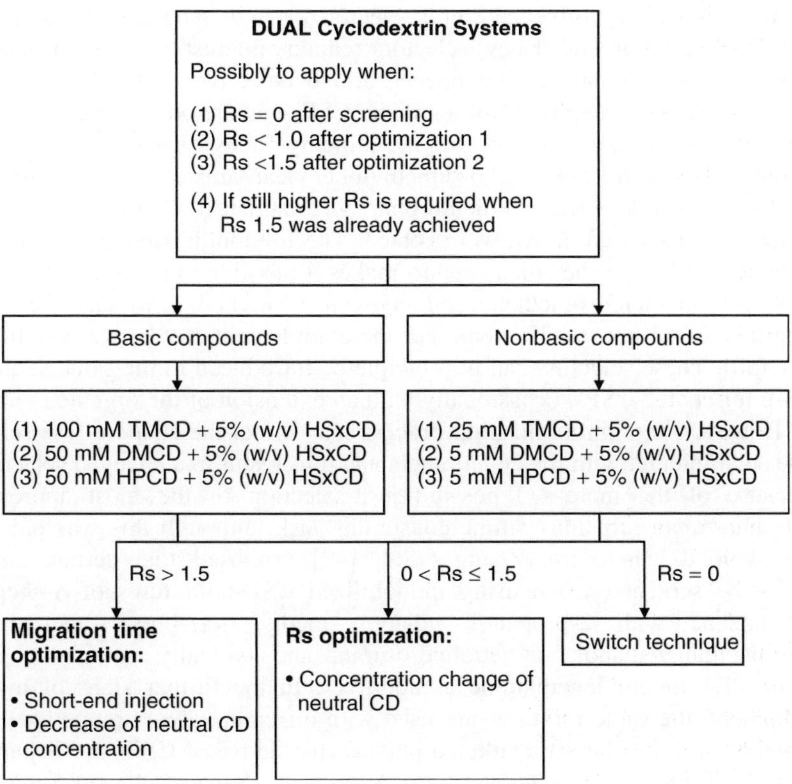

Figure 12 Dual-CD system as an extension of the CE strategy defined. (Adapted from [94].)

buffer. When Rs > 1.5 is obtained, migration time optimization is prescribed by executing a short end injection. When this fails, a decreased concentration of the neutral CD might be a possible solution [94].

5 NEW DEVELOPMENTS

Most experiments in the various strategies explained in this work should be executed sequentially. The reason is that screening or development can then be stopped as soon as a desired result is obtained. However, there is a drawback attached to this way of working: The best conditions for a given compound might be ignored. Therefore, for chromatographic systems, parallel setups have been developed in which several columns can be screened with the same mobile phase. The detection is made by means of a multichannel detector (e.g., UV detector) monitoring each column. Parallel setups where the screening can be stopped automatically according to predefined criteria have also been described [101–103]. In this way, method development time is kept as short as possible.

Besides parallel setups, new chiral selectors are also appearing on the market. Since there is still no universal chiral selector that will separate all chiral compounds, the development of new selectors remains interesting. The introduction of immobilized stationary phases opened new perspectives in the field of chiral separations. Some examples of immobilized CSPs that are on the market are Chiralpak IA [immobilized amylose tris(3,5-dimethylphenylcarbamate)], Chiralpak IB [immobilized cellulose tris(3,5-dimethylphenylcarbamate)], and Chiralpak IC [immobilized cellulose tris(3,5-dichlorophenylcarbamate)]. The selectors used in the strategies discussed above were coated. The immobilization of the polysaccharide derivatives on the silica support makes it possible to use solvents such as tetrahydrofuran, dichloromethane, and methyl *tert*-butyl ether, which were forbidden on the coated phases. This enlarges the complementarity between stationary phases [50]. These selectors can in principle be introduced in the same strategies as those for coated CSPs, occasionally with an extension of the organic modifiers used. However, the immobilization process may affect the shape of the selector cavities that interact with the enantiomers and thus result in a changed selectivity.

Because of the increased possibilities, selection of the most appropriate mobile phase can turn into a time-consuming task, although this was precisely to be avoided. Therefore, Zhang et al. [49] proposed that certain solvent mixtures be screened when using immobilized CSPs: mixtures of *n*−heptane or *n*−hexane with 2-propanol, ethanol, methyl *tert*-butyl ether/ethanol, dichloromethane/ethanol, or tetrahydrofuran, and typically containing 0.1% (v/v) of TFA or ethylenediamine as additives. In the further steps of method development, the same mixtures are used with different solvent ratios [49].

Another group of newly marketed polysaccharide-based CSPs are Sepapak-2 or Lux Cellulose-2, Sepapak-3 or Lux Amylose-2, Sepapak-4, and Sepapak-5 with cellulose tris(3-chloro-4-methylphenylcarbamate), amylose tris(5-chloro-2-methylphenylcarbamate), cellulose tris(4-choloro-3-methylphenylcarbamate), and cellulose tris(3,5-dichlorophenylcarbamate), respectively, as selectors. Some research concerning their place in the existing strategies is already ongoing [104]. It seems that these CSPs are rather easy to implement in the chromatographic strategies: They show a broad selectivity at the predefined analysis conditions and are complementary to some selectors already used in the strategy. In the near future, updated strategies will probably be defined, based primarily on the new selectors.

6 CONCLUSIONS

As difficult chiral recognition mechanisms are not fully understood, this makes the prediction of enantioseparations hardly possible, some generic screening and method development strategies have been developed to avoid time-consuming trial-and-error approaches. These strategies are developed for different techniques. Those in NPLC, RPLC, POSC, super- and subcritical fluid chromatography, and capillary electrophoresis have been discussed in this chapter. When one technique fails to separate certain compounds, it is possible that another technique will succeed in obtaining a baseline resolution. This complementarity between the techniques enlarges the spectrum of chiral compounds that can be separated with one of the strategies defined.

Acknowledgment

Debby Mangelings is a postdoctoral fellow of the Fund for Scientific Research Flanders (FWO).

REFERENCES

1. IUPAC Compendium of Chemical Terminology, Electronic Version. goldbook.iupac.org/C01058.html (accessed Nov. 17, 2008).
2. www.diss.fu-berlin.de/2003/141/chirality.pdf (accessed Nov. 17, 2008).
3. FDA Policy Statement for the Development of New Stereoisomeric Drugs. www.fda.gov/cder/guidance/stereo.htm (accessed Nov. 17, 2008).
4. Investigation of Chiral Active Substances, Directive 75/318/EEC as amended. www.ema.europa.eu/pdfs/human/qwp/3cc29aen.pdf (accessed May 3, 2010).
5. ICH Topic Q6A Specifications: Test Procedures and Acceptance Criteria for New Drug Substances and New Drug Products: Chemical Substances. www.ema.europa.eu/pdfs/human/ich/036796en.pdf (accessed May 3, 2010).
6. I. Agranat, H. Caner, J. Caldwell. Putting chirality to work: the strategy of chiral switches. *Nat. Rev. Drug Discov.*, 1:753–768, 2002.
7. N. R. Srinivas, R. H. Barbhaiya, K.K. Midha. Enantiomeric drug development: issues, considerations, and regulatory requirements. *J. Pharm. Sci.*, 90:1205–1215, 2001.

8. Y. Zhang, D. R. Wu, D. B. Wang-Iverson, A. A. Tymiak. Enantioselective chromatography in drug discovery. *Drug Discov. Today*, 10:571–577, 2005.
9. S. Andersson, S. G. Allenmark. Preparative chiral chromatographic resolution of enantiomers in drug discovery. *J. Biochem. Biophys. Methods*, 54:11–23, 2002.
10. K. Verleysen, P. Sandra. Separation of chiral compounds by capillary electrophoresis. *Electrophoresis*, 19:2798–2833, 1998.
11. A. Rizzi. Fundamental aspects of chiral separations by capillary electrophoresis. *Electrophoresis*, 22:3079–3106, 2001.
12. G. Blaschke, B. Chankvetadze. Enantiomer separation of drugs by capillary electromigration techniques. *J. Chromatogr. A*, 875:3–25, 2000.
13. M. Fillet, Ph. Hubert, J. Crommen. Enantiomeric separation of drugs using mixtures of charged and neutral cyclodextrins. *J. Chromatogr. A*, 875:123–134, 2000.
14. A. Van Eeckhaut, Y. Michotte. Chiral separations by capillary electrophoresis: recent developments and applications. *Electrophoresis*, 27:2880–2895, 2006.
15. S. Fanali, Z. Aturki, C. Desiderio. New strategies for chiral analysis of drugs by capillary electrophoresis. *Forensic Sci. Int.*, 92:137–155, 1998.
16. G. Gübitz, M. G. Schmid. Chiral separation by chromatographic and electromigration techniques: a review. *Biopharm. Drug Dispos.*, 22:291–336, 2001.
17. B. Preinerstorfer, M. Lämmerhofer. Recent accomplishments in the field of enantiomer separation by CEC. *Electrophoresis*, 28:2527–2565, 2007.
18. D. Mangelings, M. Maftouh, Y. Vander Heyden. Capillary electrochromatographic chiral separations with potential for pharmaceutical analysis. *J. Sep. Sci.*, 28:691–709, 2005.
19. D. Mangelings, C. Perrin, D. L. Massart, M. Maftouh, S. Eeltink, W. Th. Kok, P. J. Schoenmakers, Y. Vander Heyden. Optimization of the chlorthalidone chiral separation by capillary electrochromatography using an achiral stationary phase and cyclodextrin in the mobile phase. *Anal. Chim. Acta*, 509:11–19, 2004.
20. D. Mangelings, M. Maftouh, D. L. Massart, Y. Vander Heyden. Comparison of the miniaturised techniques capillary electrochromatography and capillary liquid chromatography for the chiral separation of chlorthalidone. *J. Pharm. Biomed. Anal.*, 35:789–799, 2004.
21. D. Mangelings, N. Hardies, M. Maftouh, C. Suteu, D. L. Massart, Y. Vander Heyden. Enantioseparations of basic and bifunctional pharmaceuticals by capillary electrochromatography using polysaccharide stationary phases. *Electrophoresis*, 24:2567–2576, 2003.
22. D. Mangelings, M. Maftouh, D. L. Massart, Y. Vander Heyden. Enantioseparations by capillary electro-chromatography: differences exhibited by normal- and reversed-phase versions of polysaccharide stationary phases. *Electrophoresis*, 25:2808–2816, 2004.
23. E. Francotte. Isolation and production of optically pure drugs by enantioselective chromatography. In *Chirality in Drug Research*, Vol. 33, E. Francotte, W. Lindner, Eds. Wiley-VCH, Weinheim, Germany, 2006, pp. 155–187.
24. T. E. Beesley, J. T. Lee, A. X. Wang. Method development and optimization of enantiomeric separations using macrocyclic glycopeptide chiral stationary phases. In *Chiral Separation Techniques: A Practical Approach*, G. Subramanian, Ed. Wiley-VCH, Weinheim, Germany, 2001, p. 25.

25. M. Maftouh, C. Granier-Loyaux, E. Chavana, J. Marini, A. Pradines, Y. Vander Heyden, C. Picard. Screening approach for chiral separation of pharmaceuticals: III. Supercritical fluid chromatography for analysis and purification in drug discovery. *J. Chromatogr. A*, 1088:67–81, 2005.
26. N. M. Maier, W. Lindner. Stereoselective chromatographic methods for drug analysis. In *Chirality in Drug Research*, Vol. 33, E. Francotte, W. Lindner, Eds. Wiley-VCH, Weinheim, Germany, 2006, pp. 189–260.
27. M. Lämmerhofer, W. Lindner. Recent developments in liquid chromatographic enantioseparation. In *Handbook of Analytical Separations*, Vol. 1, K. Valkó, Ed. Elsevier, Amsterdam, 2000, pp. 337–437.
28. C. Pettersson, B.-A. Persson. HPLC applications for chiral pharmaceutical analysis. In *Handbook of HPLC*, E. Katz, R. Eksteen, P. Schoenmakers, N. Miller, Eds. Marcel Dekker, New York, 1998, pp. 669–693.
29. E. Francotte, A. Junker-Buchheit. Preparative chromatographic separation of enantiomers. *J. Chromatogr.*, 576:1–45, 1992.
30. Z. Wang, J. Ouyang, W. R. G. Baeyens. Recent developments of enantioseparation techniques for adrenergic drugs using liquid chromatography and capillary electrophoresis: a review. *J. Chromatogr. B*, 862:1–14, 2008.
31. J. Haginaka. Pharmaceutical and biomedical applications of enantioseparations using liquid chromatographic techniques. *J. Pharm. Biomed. Anal.*, 27:357–372, 2002.
32. Y. Okamoto, Y. Kaida. Resolution by high-performance liquid chromatography using polysaccharide carbamates and benzoates as chiral stationary phases. *J. Chromatogr. A*, 666:403–419, 1994.
33. Y. Okamoto, E. Yashima. Insight into the separation of enantiomers by HPLC on polysaccharide derivatives. *Angew. Chem. Int. Ed.*, 37:1020–1043, 1998.
34. R. Cirilli, R. Feretti, E. De Santis, B. Gallinella, L. Zanitti, F. La Torre. High-performance liquid chromatography separation of enantiomers of flavanone and 2′-hydroxychalcone under reversed-phase conditions. *J. Chromatogr. A*, 1190:95–101, 2008.
35. L. Toribio, M. J. del Nozal, J. L. Bernal, C. Alonso, J. J. Jiménez. Enantiomeric separation of several antimycotic azole drugs using supercritical fluid chromatography. *J. Chromatogr. A*, 1144:255–261, 2007.
36. X. Zhang, J. Ouyang, W. R. G. Baeyens, S. Zhai, Y. Yang, G. Huang. Enantiomeric separation of β-blockers by HPLC using (*R*)-1-naphtylglycine and 3,5-dinitrobenzoic acid as chiral stationary phase. *J. Pharm. Biomed. Anal.*, 31:1047–1057, 2003.
37. B. Chankvetadze, C. Yamamoto, Y. Okamoto. Enantioseparation of selected chiral sulfoxides using polysaccharide-type chiral stationary phases and polar organic, polar aqueous-organic and normal-phase eluents. *J. Chromatogr. A*, 922:127–137, 2001.
38. J. Yang, J. Guan, L. Pan, K. Jiang, M. Cheng, F. Li. Enantioseparation and impurity determination of the enantiomers of novel phenylethanolamine derivatives by high performance liquid chromatography on amylose stationary phases. *Anal. Chim. Acta*, 610:263–267, 2008.
39. W. H. Pirkle, K. Z. Gan. Facile and predictable means of separating the enantiomers of 5-arylhydantoins. *J. Chromatogr. A*, 790:65–71, 1997.
40. I. Kartozia, M. Kanyonyo, T. Happaerts, D. M. Lambert, G. K. E. Scriba, B. Chankvetadze. Comparative HPLC enantioseparation of new chiral hydantoin derivatives on three different polysaccharide type chiral stationary phases. *J. Pharm. Biomed. Anal.*, 27:457–465, 2007.

41. O. P. Kleidernigg, C. O. Kappe. Separation of enantiomers of 4-aryldihydropyrimidines by direct enantioselective HPLC: a critical comparison of chiral stationary phases. *Tetrahedron: Asymmetry*, 8:2057–2067, 1997.

42. N. Matthijs, C. Perrin, M. Maftouh, D. L. Massart, Y. Vander Heyden. Definition and system implementation of strategies for method development of chiral separations in normal- or reversed-phase liquid chromatography using polysaccharide-based stationary phases. *J. Chromatogr. A*, 1041:119–133, 2004.

43. C. Perrin, V. A. Vu, N. Matthijs, M. Maftouh, D. L. Massart, Y. Vander Heyden. Screening approach for chiral separation of pharmaceuticals: I. Normal-phase liquid chromatography. *J. Chromatogr. A*, 947:69–83, 2002.

44. M. L. de la Puente, C. T. White, A. Rivera-Sagredo, J. Reilly, K. Burton, G. Harvey. Impact of normal-phase gradient elution in chiral chromatography: a novel, robust, efficient and rapid chiral screening procedure. *J. Chromatogr. A*, 983:101–114, 2003.

45. M. L. de la Puente. Highly sensitive and rapid normal-phase chiral screen using high-performance liquid chromatography–atmospheric pressure ionization tandem mass spectrometry (HPLC/MS). *J. Chromatogr. A*, 1055:55–62, 2004.

46. S. Morante-Zarcero, I. Del Hierro, M. Fajardo, I. Sierra. Development and validation of a chiral HPLC method for rapid screening of allylic alcohol asymmetric epoxidation processes. *Anal. Chim. Acta*, 618:102–109, 2008.

47. Y. K. Ye, B. S. Lord, L. Yin, R. W. Stringham. Enantioseparation of amino acids on polysaccharide-based chiral stationary phases. *J. Chromatogr. A*, 945:147–159, 2002.

48. K. M. Kirkland. Optimization of chiral selectivity on cellulose-based high-performance liquid chromatographic columns using aprotic mobile-phase modifiers. *J. Chromatogr. A*, 718:9–26, 1995.

49. T. Zhang, D. Nguyen, P. Franco. Enantiomer resolution screening strategy using multiple immobilised polysaccharide-based chiral stationary phases. *J. Chromatogr. A*, 1191:214–222, 2008.

50. A. Ghanem, H. Y. Aboul-Enein. On the solvent versatility in immobilized amylose tris(3,5-dimethylphenylcarbamate) chiral stationary phase in high performance liquid chromatography: application to the asymmetric cyclopropanation of olefins. *Anal. Chim. Acta*, 548:26–32, 2005.

51. A. Ghanem, H. Y. Aboul-Enein. Comparison, applications, advantages and limitations of immobilized and coated amylose tris-(3,5-dimethylphenylcarbamate) chiral stationary phases in HPLC. *J. Liq. Chromatogr. Rel. Technol.*, 28:2863–2874, 2005.

52. A. Ghanem, L. Naim. Immobilized versus coated amylose tris(3,5-dimethylphenylcarbamate) chiral stationary phases for the enantioselective separation of cyclopropane derivatives by liquid chromatography. *J. Chromatogr. A*, 1101:171–178, 2006.

53. T. Wang, R. M. Wenslow. Effect of alcohol mobile-phase modifiers on the structure and chiral selectivity of amylose tris(3,5-dimethylphenylcarbamate) chiral stationary phases. *J. Chromatogr. A*, 1015:99–110, 2003.

54. Y. Tang. Significance of mobile phase composition in enantioseparation of chiral drugs by HPLC on a cellulose-based chiral stationary phase. *Chirality*, 8:136–142, 1996.

55. Y. K. Ye, R. W. Stringham. The effect of acidic and basic additives on the enantioseparation of basic drugs using polysaccharide-based chiral stationary phases. *Chirality*, 18:519–530, 2006.

56. Y. Vander Heyden, C. Perrin, D. L. Massart. Optimization strategies for HPLC and CZE. In *Handbook of Analytical Separations: Separation Methods in Drug Synthesis and Purification*, Vol. 1, K. Valkó, Ed. Elsevier, Amsterdam, 2000, pp. 163–212.

57. J. A. Whatley. Rapid method development for the separation of enantiomers by means of chiral column switching. *J. Chromatogr. A*, 697:257–261, 1995.

58. J. G. Ning. Direct chiral separation with Chiralpak®-AD converted to the reversed phase mode. *J. Chromatogr. A*, 805:309–314, 1998.

59. K. Tachibana, A. Ohnishi. Reversed-phase liquid chromatographic separation of enantiomers on polysaccharide type chiral stationary phases. *J. Chromatogr. A*, 906:127–154, 2001.

60. C. Perrin, N. Matthijs, D. Mangelings, C. Granier-Loyaux, M. Maftouh, D. L. Massart, Y. Vander Heyden. Screening approach for chiral separation of pharmaceuticals: II. Reversed-phase liquid chromatography. *J. Chromatogr. A*, 966:119–134, 2002.

61. J. G. Hou, Y. L. Wang, C. X. Li, X. Q. Han, J. Z. Gao, J. W. Kang. Reversed-phase separations on cellulose tris(3,5-dimethylphenylcarbamate) chiral stationary phase. *Chromatographia*, 50:89–95, 1999.

62. B. Chankvetadze, I. Kartozia, C. Yamamoto, Y. Okamoto. Comparative enantioseparation of selected chiral drugs on four different polysaccharide-type chiral stationary phases using polar organic mobile phases. *J. Pharm. Biomed. Anal*., 27:467–478, 2002.

63. N. Matthijs, M. Maftouh, Y. Vander Heyden. Chiral separation strategy in polar organic solvent chromatography and performance comparison with normal-phase liquid and supercritical-fluid chromatography. *J. Sep. Sci*., 29:1353–1362, 2006.

64. N. Matthijs, M. Maftouh, Y. Vander Heyden. Screening approach for chiral separation of pharmaceuticals: IV. Polar organic solvent chromatography. *J. Chromatogr. A*, 1111:48–61, 2006.

65. M. Kazusaki, H. Kawabata, H. Matsukura. Influence of temperature on enantioseparation employing an amylose-derivative stationary phase. *J. Liq. Chromatogr. Rel. Technol*., 23:2937–2946, 2000.

66. G. Török, L. Goetelen, R. Luyckx, P. Van Broeck. Evaluation of the performance of commercially available polysaccharide-based chiral stationary phases after multicycle operation in multimodel elution mode. *J. Pharm. Biomed. Anal*., 39:425–430, 2005.

67. D. Mangelings, M. Maftouh, D. L. Massart, Y. Vander Heyden. Enantioseparations by capillary electrochromatography: differences exhibited by normal- and reversed-phase versions of polysaccharide stationary phases. *Electrophoresis*, 25:2808–2816, 2004.

68. Y. K. Ye, R. Stringham. Effect of mobile phase amine additives on enantioselectivity for phenylalanine analogues. *J. Chromatogr. A*, 927:53–60, 2001.

69. R. W. Stringham, K. G. Lynam, B. S. Lord. Memory effect of diethylamine mobile phase additive on polysaccharide stationary phases. *Chirality*, 16:493–498, 2004.

70. Y. K. Ye, B. Lord, R. W. Stringham. Memory effect of mobile phase additives in chiral separations on chiralpak AD column. *J. Chromatogr. A*, 945:139–146, 2002.

71. P. J. Schoenmakers. Open columns or packed columns for supercritical fluid chromatography: a comparison in supercritical fluid chromatography. In *Supercritical Fluid Chromatography*, R. M. Smith, Ed. The Royal Society of Chemistry, St. Edmundsbury Press, London, 1988, pp. 102–136.
72. K. W. Phinney. Sub- and supercritical fluid chromatography for enantiomer separations. In *Chiral Separation Techniques: A Practical Approach*, G. Subramanian, Ed. Wiley-VCH, Weinheim, Germany, 2001, pp. 299–315.
73. L. S. Daintree, A. Kordikowski, P. York. Separation processes for organic molecules using SFC technologies. *Adv. Drug Deliv. Rev*., 60:351–372, 2008.
74. R. W. Stringham, B. R. Krueger, J. Marshall. Use of elevated flow rates in preparative subcritical fluid chromatography. *J. Chromatogr. A*, 1175:112–116, 2007.
75. L. T. Taylor. Supercritical fluid chromatography for the 21st century. *J. Supercrit. Fluids*, 47:566–573, 2009.
76. A. Medvodivici, P. Sandra, L. Toribio, F. David. Chiral packed column subcritical fluid chromatography on polysaccharide and macrocyclic antibiotic chiral stationary phases. *J. Chromatogr. A*, 785:159–171, 1997.
77. J. A. Blackwell. Manipulation of chiral resolution for isoxazoline-based IIb/IIIa receptor antagonists using various mobile phase additives in subcritical fluid chromatography. *Chirality*, 10:338–342, 1998.
78. M. S. Villeneuve, R. J. Anderegg. Analytical supercritical fluid chromatography using fully automated column and modifier selection valves for the rapid development of chiral separations. *J. Chromatogr. A*, 826:217–225, 1998.
79. K. Yaku, F. Morishita. Separation of drugs by packed-column supercritical fluid chromatography. *J. Biochem. Biophys. Methods*, 43:59–76, 2000.
80. Y. Zhao, G. Woo, S. Thomas, D. Semin, P. Sandra. Rapid method development for chiral separation in drug discovery using sample pooling and supercritical-fluid chromatography-mass spectrometry. *J. Chromatogr. A*, 1003:157–166, 2003.
81. Y. K. Ye, K. G. Lynam, R. W. Stringham. Effect of amine mobile phase additives on chiral subcritical fluid chromatography using polysaccharide stationary phases. *J. Chromatogr. A*, 1041:211–217, 2004.
82. C. White. Integration of supercritical fluid chromatography into drug discovery as a routine support tool: I. Fast chiral screening and purification. *J. Chromatogr. A*, 1074:163–173, 2005.
83. R. W. Stringham. Chiral separation of amines in subcritical fluid chromatography using polysaccharide stationary phases and acidic additives. *J. Chromatogr. A*, 1070:163–170, 2005.
84. Y. Zhao, W. A. Pritts, S. Zhang. Chiral separation of selected proline derivatives using a polysaccharide-type stationary phase by supercritical fluid chromatography and comparison with high-performance liquid chromatography. *J. Chromatogr. A*, 1189:245–253, 2008.
85. W. W. Barnhart, K. H. Gahm, Z. Hua, W. Goetzinger. Supercritical fluid chromatography comparison of the poly(*trans*-1,2-cyclohexanediyl-bisacrylamide) (P-CAP) column with several derivatized polysaccharide-based stationary phases. *J. Chromatogr. B*, 875:217–229, 2008.
86. K. H. Gahm, H. Tan, J. Liu, W. Barnhart, J. Eschelbach, S. Notari, S. Thomas, D. Semin, J. Cheetham. Purification method development for chiral separation in supercritical fluid chromatography with the solubilities in supercritical fluid chromatographic mobile phases. *J. Pharm. Biomed. Anal*., 46:831–838, 2008.

87. M. Fillet, L. Fotsing, J. Bonnard, J. Crommen. Stereoselective determination of *S*-naproxen in tablets by capillary electrophoresis. *J. Pharm. Biomed. Anal.*, 18:799–805, 1998.

88. T. de Boer, R. A. de Zeeuw, G. J. de Jong, K. Ensing. Recent innovations in the use of charged cyclodextrins in capillary electrophoresis for chiral separations in pharmaceutical analysis. *Electrophoresis*, 21:3220–3239, 2000.

89. G. G. Yowell, S. D. Fazio, R. V. Vivilecchia. Enantiomeric separation of *n-tert*-butoxycarbonyl amino acids by capillary electrophoresis using hydroxypropyl-substituted cyclodextrins. *J. Chromatogr. A*, 745:73–79, 1996.

90. M. Blanco, I. Valverde. Choice of chiral selector for enantioseparation by capillary electrophoresis. *Trends Anal. Chem.*, 22:428–439, 2003.

91. Z. Juvancz, R. B. Kendrovics, R. Ivànyi, L. Szente. The role of cyclodextrins in chiral capillary electrophoresis. *Electrophoresis*, 29:1701–1712, 2008.

92. N. Matthijs, C. Perrin, M. Maftouh, D. L. Massart, Y. Vander Heyden. Knowledge-based system for method development of chiral separations with capillary electrophoresis using highly-sulphated cyclodextrins. *J. Pharm. Biomed. Anal.*, 27:515–529, 2002.

93. C. Perrin, Y. Vander Heyden, M. Maftouh, D. L. Massart. Rapid screening for chiral separations by short-end injection capillary electrophoresis using highly sulphated cyclodextrins as chiral selectors. *Electrophoresis*, 22:3203–3215, 2001.

94. N. Matthijs, S. Van Hemelryck, M. Maftouh, D. L. Massart, Y. Vander Heyden. Electrophoretic separation strategy for chiral pharmaceuticals using highly-sulphated and neutral cyclodextrins based dual selector systems. *Anal. Chim. Acta*, 525:247–263, 2004.

95. C. Perrin, M. G. Vargas, Y. Vander Heyden, M. Maftouh, D. L. Massart. Fast development of separation methods for the chiral analysis of amino acid derivatives using capillary electrophoresis and experimental designs. *J. Chromatogr. A*, 883:249–265, 2000.

96. B. Chankvetadze. Separation selectivity in chiral capillary electrophoresis with charged selectors. *J. Chromatogr. A*, 792:269–295, 1997.

97. M. Fillet, L. Fotsing, J. Crommen. Enantioseparation of uncharged compounds by capillary electrophoresis using mixtures of anionic and neutral β-cyclodextrin derivatives. *J. Chromatogr. A*, 817:113–119, 1998.

98. W. Lindner, B. Böhs, V. Seidel. Enantioselective capillary electrophoresis of amino acid derivatives on cyclodextrin evaluation of structure-resolution relationships. *J. Chromatogr. A*, 697:549–560, 1995.

99. A. M. Abushoffa, M. Fillet, Ph. Hubert, J. Crommen. Prediction of selectivity for enantiomeric separations of uncharged compounds by capillary electrophoresis involving dual cyclodextrin systems. *J. Chromatogr. A*, 948:321–329, 2002.

100. M. I. Jimidar, W. Van Ael, P. Van Nyen, M. Peeters, D. Redlich, M. De Smet. A screening strategy for the development of enantiomeric separation methods in capillary electrophoresis. *Electrophoresis*, 25:2772–2785, 2004.

101. E. Francotte. Multi-parallel chiral screening. G.I.T. Lab. J., 3 (2006) 46–48.

102. L. Zeng, R. Xu, D. B. Laskar, D. B. Kassel. Parallel supercritical fluid chromatography/mass spectrometry system for high-throughput enantioselective optimization and separation. *J. Chromatogr. A*, 1169:193–204, 2007.

103. P. Sajonz, W. Schafer, X. Gong, S. Schultz, T. Rosner, C. J. Welch. Multiparallel microfluidic high-performance liquid chromatography for high-throughput normal-phase chiral analysis. *J. Chromatogr. A*, 1145:149–154, 2007.

104. H. Ates, D. Mangelings, Y. Vander Heyden. Chiral separations in polar organic solvent chromatography: updating a screening strategy with new chlorine-containing polysaccharide-based selectors. *J. Chromatogr. B*, 875:57–64, 2008.

CHAPTER 13

Use of Enantioselective Synthesis and Preparative Chiral Chromatography to Access a Challenging Enantiopure Pharmaceutical Candidate from a Mixture of Four Stereoisomers

CHRISTOPHER J. WELCH, DEREK H. HENDERSON, WILLIAM R. LEONARD, MIRLINDA BIBA, MIKE ZACUTO, FRED FLEITZ, AMUDE KASSIM, CHENG-YI CHEN, and PETER SAJONZ

Separation and Purification Center of Excellence, Merck Research Laboratories, Rahway, New Jersey

1 INTRODUCTION

Preparative chiral chromatography is frequently used for the rapid purification of enantiomers during pharmaceutical discovery and development, owing to the simplicity and speed offered by this technique [1,2]. A number of recent developments have been helpful in making preparative chiral chromatography more routine and predictable [3]. Modern chromatographic method development is substantially automated, often making identification of candidate separation conditions routine and straightforward [4]. Small-scale loading studies allow fast evaluation of different candidate separation conditions, enabling rapid optimization of chromatographic productivity (the amount of purified product that can be obtained with a given amount of stationary phase in a given time) [5]. Scale-up from small- to production-scale chromatography is also relatively straightforward and highly predictable [6]. Despite these recent advances, practicing preparative chromatographers frequently encounter challenging separation problems that are not easily solved using conventional approaches that have been worked out for

Chiral Separation Methods for Pharmaceutical and Biotechnological Products, Edited by Satinder Ahuja

general enantioseparations. In this study we describe the use of preparative chiral chromatography for the preparation of a challenging drug candidate with multiple stereocenters, a class of preparative chiral separation that can be quite difficult.

2 EXPERIMENTAL

2.1 Chemicals

Methanol, 2-propanol, and *n*-heptane were obtained from EMD Chemicals (Gibbstown, NJ). Ethanol was ACS/USP grade (Pharmco, Brookfield, CT). Compounds 1 to 4 were obtained Merck Process Research (Merck Research Laboratories, Rahway, NJ).

2.2 Chiral Stationary Phases

The chiral stationary phases (CSPs) used for the column screening study were Chiralpak AD and AS and Chiralcel OD, OJ, and OF (Chiral Technologies, Inc., West Chester, PA). The CSP for the method optimization, loading studies, and final preparative separation was Chiralpak AD (particle size 20 μm).

2.3 Equipment

Chiral SFC Method Development Screening Chiral supercritical fluid chromatographic (SFC) method development screening was carried out using a Berger–Thar analytical supercritical fluid chromatograph (Thar Instruments, Inc., Pittsburgh, PA) fitted with a six-position column selection valve and Agilent model 1100 diode array ultraviolet (UV)-visible detector (Agilent Technologies, Palo Alto, CA). Column screening was carried out using the standard gradient approach described previously [7].

Preparative HPLC Method Development and Loading Studies Preparative method development and loading studies were conducted on an Agilent 1100 system with a G1311A quaternary pump, G2260 Prep ALS autosampler, and G1315B diode array UV-visible detector.

Preparative HPLC Preparative high-performance liquid chromatography (HPLC) separations were carried out using a Varian preparative HPLC system (Varian, Palo Alto, CA) with PrepStar SD-1 binary pumps, 210 injection pump, 320 UV detector, advanced laser polarimeter (PDR Chiral, Lake Park, FL), and a ReSzonator fluid module. Prochrom dynamic axial compressions columns were used with dimensions of 11 × 25 cm or 5 × 25 cm (Novasep, Boothwyn, PA).

2.4 Results and Discussion

Compound **1** in Figure 1 is a recent pharmaceutical development candidate from these laboratories. The compound contains two stereocenters and can exist in four

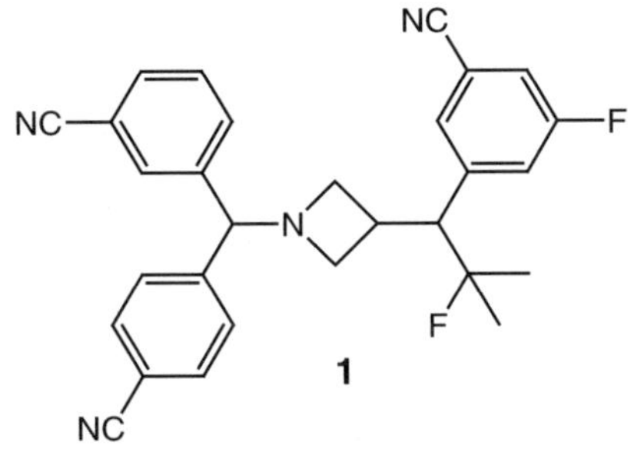

1

Figure 1 Target (S,S)-**1** stereoisomer prepared in this study.

different stereoisomeric configurations, only one of which, the (S,S) stereoisomer, possesses the desired activity.

There are a multitude of possibilities for the use of preparative chiral chromatography in combination with organic synthesis for accessing the desired stereoisomer of **1**. The two stereocenters in the molecule are fairly remote from one another, suggesting that the use of one stereocenter to influence formation of the other may be somewhat problematic, and arguing instead for a synthetic strategy in which two enantiopure fragments are prepared independently, then coupled. The similarity of the two aryl substituents at the amine stereocenter suggests that many of approaches to the synthesis or resolution of the "western" fragment of this molecule will be quite challenging. The presence of the azetidine ring spacer and the quaternary alkyl fluoride in the "eastern" fragment represent additional synthetic challenges for this target molecule.

In cases such as this, where multiple possibilities exist for employing preparative chiral chromatography, we have found that it is often advisable to carry out initial screening on whatever relevant intermediates may be available or easily synthesized, allowing the results to inform the selection of the best path forward [8]. In this study, more than a dozen compounds representing several distinct synthetic routes to **1** were examined by chiral SFC using a previously described standard method development screening approach [9].

We were somewhat surprised to find that the enantiomers of the diarylmethylamine, **2**, were well resolved by chiral SFC (Figure 2A). This molecule is nearly symmetrical, being chiral only by virtue of the different aromatic substitution patterns, and was predicted to be only marginally resolved. The excellent separation of the enantiomers of the diarylmethlyamine, **2**, emphasizes the value of screening everything, including compounds that are not expected to be well resolved. Several analogs and derivatives of amine **2** were also resolved, but the ready resolution of the amine itself confirmed the viability of synthetic routes based on enantiopure **2**.

In contrast, synthetic precursors to the eastern fragment were poorly resolved, one of the better examples being the malononitrile intermediate, **3** (Figure 2B). Fortunately, the four stereoisomers of the target, **1**, are remarkably well resolved

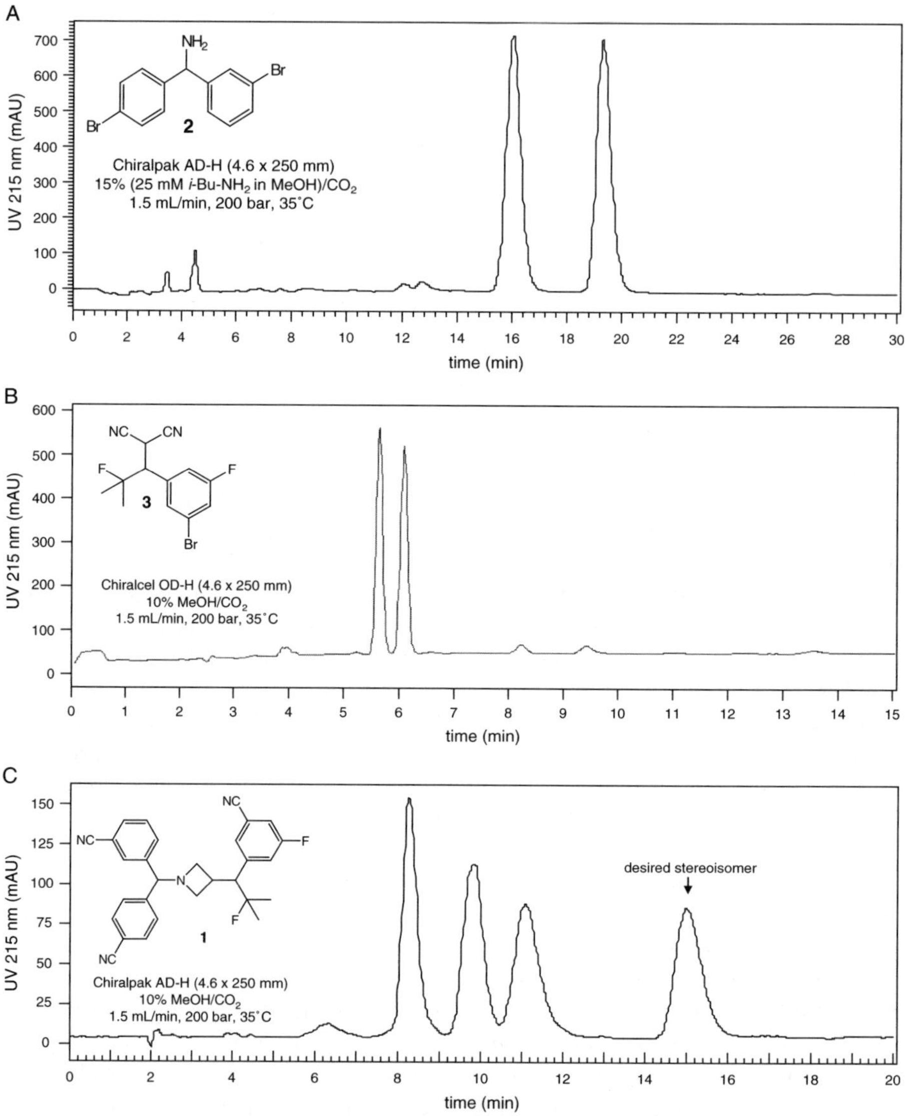

Figure 2 Chromatographic separation of possible intermediates and final product for the target compound, **1**, using chiral SFC. Conditions as noted in the chromatograms.

on the Chiralpak AD-H CSP, with the desired (*S*,*S*)-stereoisomer eluting last, well resolved from the other components (Figure 2C).

Based on a consideration of first principles, the synthetic approach where two enantiopure starting materials are obtained independently, then coupled, would generally be preferred to the "brute force" approach in which two racemates are coupled, then the desired stereoisomer chromatographically purified. The

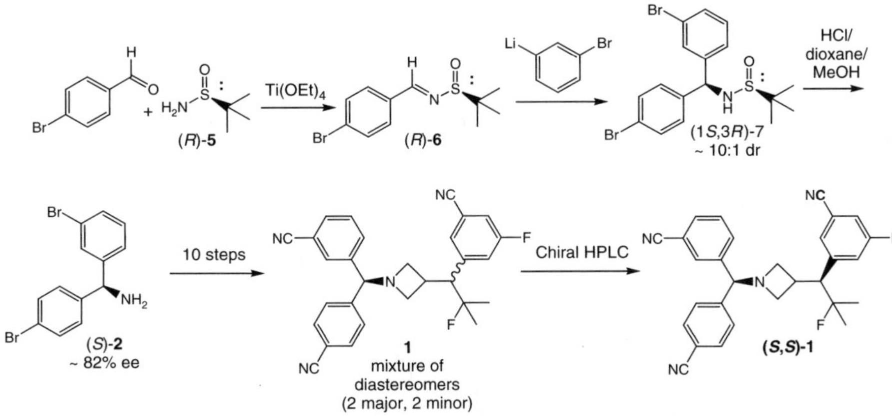

Figure 3 Synthetic route to enantiopure compound 1 utilizing a combination of enantioselective Ellman imine addition chemistry and preparative chiral chromatography.

first approach can in principle afford product in 50% yield, as opposed to the maximum 25% yield (assuming no diastereoselectivity in the coupling reaction) obtainable via the second approach. In the present case, the lack of a good method for accessing the eastern portion of the target compound in enantiopure form forced the consideration of options where the stereoisomers of the final product were resolved.

We next considered whether it would be preferable to resolve the enantiomers of amine **2** first and then use this enantiopure starting material to prepare the product as a mixture of diastereomers, or simply to use racemic amine **2** to prepare and then resolve the resulting 1 : 1 : 1 : 1 mixture of stereoisomers. Neither approach was especially attractive, as both afforded only a maximum 25% yield, squandering valuable starting materials and intermediates. At about this time, investigations into possible approaches to the enantioselective synthesis of amine **2** identified a reasonable approach involving diastereoselective aryl lithium addition to an enantiopure sulfinyl imine, **6**, derived from the enantiopure (*R*)-*t*-butanesulfinamide, **5** (Figure 3). This general synthetic strategy, developed originally by Ellman et al. [10], often affords high levels of enantioselectivity in the formation of the sulfinylamide product, which can then be deprotected to afford the target amine in high enantiopurity. In the present case, formation of the sulfinylamide, **7**, occurred with a maximum 4 : 1 preference for formation of the desired diastereomer, affording amine **2** in about 75% enantiomeric excess (ee) upon deprotection.

Enantioenriched amine **2** was carried forward through a series of 10 synthetic steps to afford product, **1**, as an enriched mixture of stereoisomers. Figure 4 is an analytical chromatogram of the mixture of isomers. Again, the desired isomer elutes last, well resolved from the other isomers, but now constituting about 40% of the total as opposed to the approximate 25% composition obtained previously.

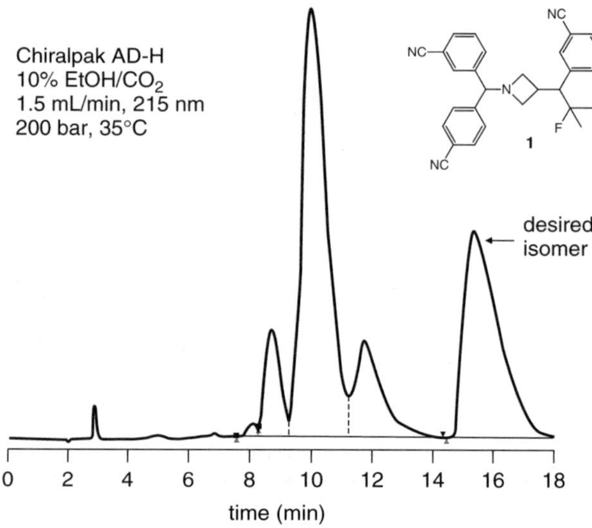

Figure 4 Analytical SFC chromatogram showing an enriched mixture of the product, **1**, prepared using the enantioselective Ellman approach.

We next investigated the use of preparative chromatography to isolate the desired stereoisomer from the enriched mixture of isomers shown in Figure 4. Our typical approach for carrying out separations of developmental compounds would be to scale the existing analytical method first to the gram scale, and then up to the multikilogram scale using SFC [11]. We generally prefer SFC for preparative separations, owing to the advantages in speed and solvent savings relative to the corresponding HPLC approaches. However, at the time this study was conducted, we had not yet added our capabilities to carry out SFC separations on a larger scale, and such purifications were carried out at that time by HPLC. Consequently, the analytical SFC method was translated to HPLC using the larger particle size 20-μm ID version of the Chiralpak AD CSP that would be used for large-scale preparative separation. It is very important for the sake of accurate predictions that stationary phase and all conditions used in modeling studies parallel those that will be used in the large-scale separation. Suitable conditions were identified using either ethanol–heptane or 2-propanol/heptane eluents, as depicted in Figure 5. In both methods, the three undesired isomers coelute early in the chromatogram, with the desired isomer eluting last, in a broad peak. Such broad peaks are typically undesirable in preparative chromatography, as they tend to increase the overall cycle time of the method and also afford product as a large volume of dilute solution. The selectivity of the 2-propanol can be seen to be greater than that of EtOH; however, the broadening seen in the 2-propanol chromatogram is much worse than that observed with EtOH. Consequently, we decided to investigate the performance of both of these methods under preparative load.

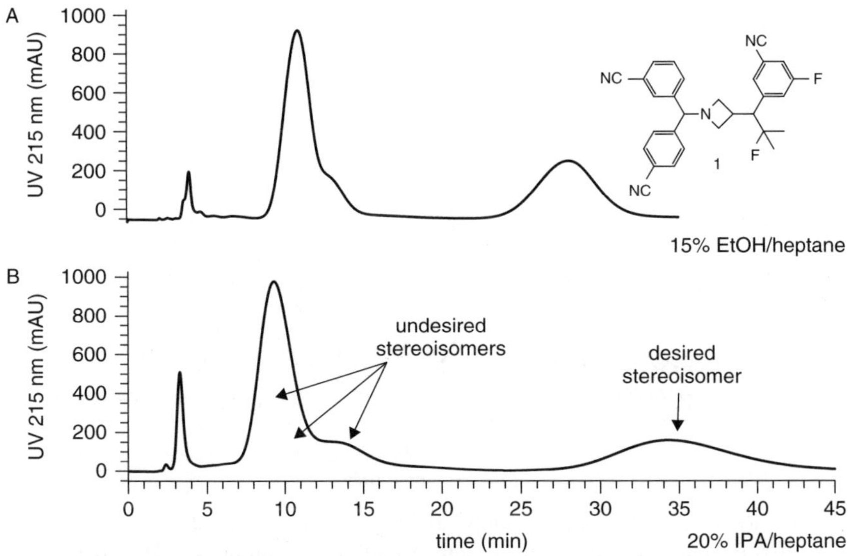

Figure 5 Isocratic chiral HPLC methods for purification of the desired isomer of **1** from a mixture of four stereoisomers. Conditions: Chiralpak AD, 20μm, 250 × 4.6 mm; flow: 1.4 mL/min; 15% EtOH/heptane, 20% IPA/heptane, 25% IPA/heptane.

The chromatographic productivity of the two methods was compared by carrying out loading studies on a small analytical column. Such studies allow one to estimate the amount of purified product that can be obtained with a given amount of stationary phase in a given amount of time. This productivity factor, typically expressed in units of kkd (kilograms product per kilogram stationary phase per 24-hour day), allows one to estimate very accurately the time, solvent, and equipment needed to carry out a larger-scale separation. In the representative loading study shown in Figure 6, considerable distortion of the peaks can be observed at increasing load, especially for the desired, late-eluting stereoisomer. Fractionation and analysis of the product peak showed that considerable contamination by undesired stereoisomers was found at the highest load (24.3 mg), whereas product of high purity was obtained with the intermediate load conditions (12.2 mg). Problems with "tailing" of undesired components into the desired peak of interest are often encountered in situations where the desired component elutes after the undesired: the problem being particularly difficult in some instances. In the present case, loading studies were used to estimate a chromatographic productivity of about 0.08 kkd for both the 15% EtOH/heptane and the 20% IPA/heptane eluent methods, far short of the value of 0.3 kkd desired for a good separation and >1 kkd for an excellent separation.

In addition to the poor chromatographic productivity obtained for these isocratic elution methods, both conditions afford the desired isomer as a very dilute solution, necessitating the evaporation of large volumes of solvent. Wishing to avoid this problem and to increase the productivity of the method,

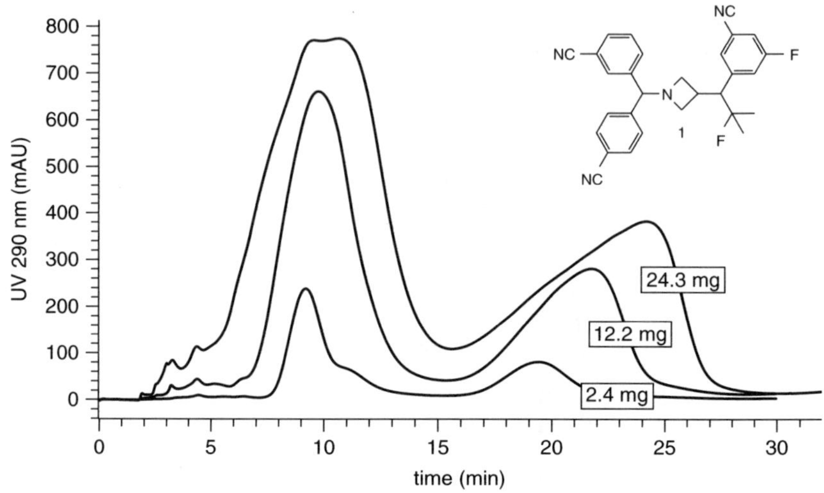

Figure 6 Representative loading study showing the injections of varying amounts of isomer mixture onto an analytical column. Conditions: Chiralpak AD, 4.6 mm ID × 25 cm, 20-μm particle size, 15% EtOH/heptane, 1.4 mL/min.

the use of a step gradient to speed elution of the desired component was investigated. Gradient elution methods are typically of limited value for improving stereoisomeric separations, the time gained in speeding the elution of the second component often being equaled or surpassed by the time required for column reequilibration prior to the subsequent injection. In the present case, the band broadening observed for the target peak is so extreme that implementation of a step gradient significantly improved productivity from 0.08 to 0.13 kkd (Figure 7). Equally important, the volume of solvent that would have to be evaporated to recover the desired stereoisomer coming from the purification of 700 g of the isomer mixture was estimated at only 218 L for the step gradient method as opposed to 1182 L for the isocratic method.

Similar results were obtained upon inserting a step gradient of 65% IPA/heptane into the 20% IPA/heptane method, affording a productivity of 0.12 kkd and an estimated solvent-rich cut volume of 404 L (as opposed to 0.8 kkd and 1058 L for the isocratic method). The step gradient methods are clearly superior to the isocratic methods, with the ethanol eluent showing a small but definite superiority to the IPA eluent. We were, however, somewhat concerned that the ethanol step gradient conditions cross into the "unsafe" region of 15 to 60% ethanol composition, which is outside the recommended operation range for the CSP advocated by the vendor. In fact, isocratic elutions within this region are sometimes carried out with little or no detectable leaching of the adsorbed stationary phase [12]. We were, however, somewhat reluctant to subject the CSP to repeated rapid excursions through the unsafe 15 to 60% EtOH region that the use of the ethanol step gradient method would necessitate. Consequently, the 2-propanol step gradient method was employed for the large-scale separation.

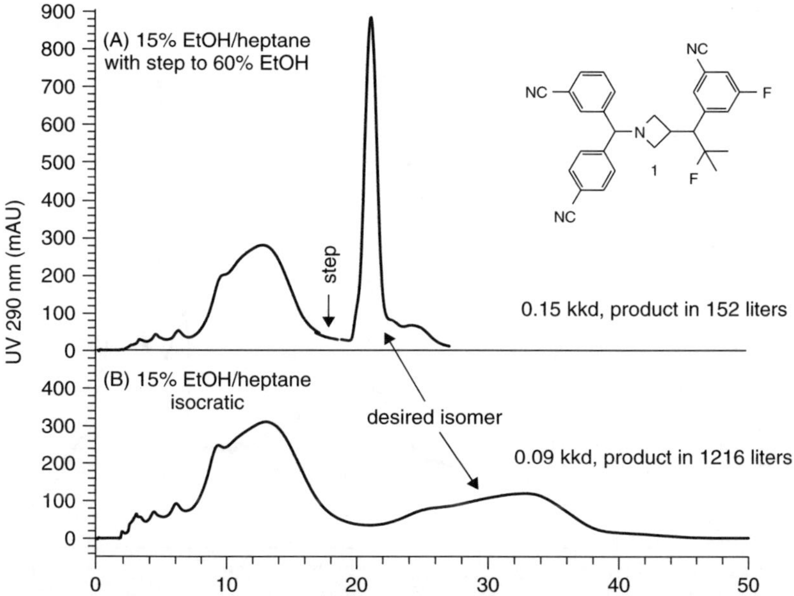

Figure 7 Comparison of isocratic (B) and step gradient (A) methods for preparative purification of the desired stereoisomer of **1**, illustrating the increased productivity and more concentrated fractions obtained with the step gradient method.

A representative chromatogram from the preparative campaign using an 11-cm ID column is depicted in Figure 8. Several interesting points can be made about this separation. First, the UV trace looks absolutely terrible, providing a good illustration of the adage that "prep chromatography begins where the pretty peaks end." The distorted product peak obtained in the chromatogram makes fraction triggering somewhat difficult, however, the use of a polarimetric detector (PDR ALP) affords a second trace with a more uniform peak, which was used as the basis for triggering fraction collection in the preparative campaign. Automated injection and fraction collection afforded a total of 173 g of the desired (1*S*,5*R*)-stereoisomer of product 1 with excellent purity (>99% ee and >99% de), corresponding to a product recovery of 85% based on assayed content of the crude feed mixture. The preparative campaign was completed with 40 h of instrument time, and the actual chromatographic productivity obtained during the campaign was 0.112 kkd, with 387 L of solvent evaporated for product recovery, all in good agreement with the estimated values.

3 CONCLUSIONS

The current study provides an example of the combination of enantioselective synthesis with preparative chiral chromatography in the challenging preparation of an enantiopure single diastereomer of a pharmaceutical intermediate from a

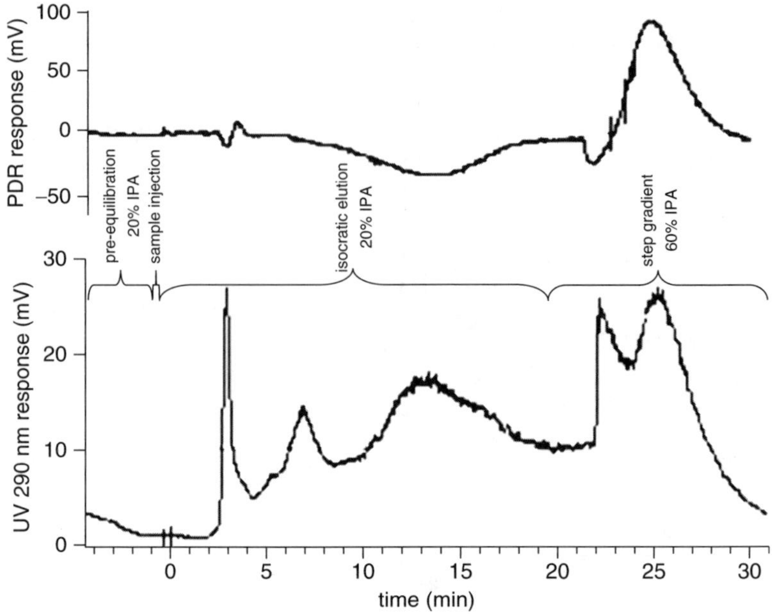

Figure 8 Representative injection from the large-scale preparative HPLC campaign used in the preparation of 173 g of enantiopure compound 1. Conditions: 20% IPA/heptane (step to 65% IPA/heptane at 20 min), 800 mL/min, ca. 45 mL injection at 225 mg/mL, detection by UV 290 nm and PDR advanced laser polarimeter.

mixture of four stereoisomers. A somewhat unusual application of a step gradient increased the productivity of the method while dramatically increasing the concentration of the target molecule in the chromatographic fractions, greatly reducing the volume of solvent needed in product isolation.

REFERENCES

1. C. J. Welch. In *Preparative Enantioselective Chromatography*, G. Cox, Ed. Blackwell, London, 2005, pp. 1–18.
2. C. J. Welch, F. Fleitz, F. Antia, P. Yehl, B. Waters, N. Ikemoto, J. D. Armstrong, D. J. Mathre. *Org. Process Res. Dev*., 8:186–191, 2004.
3. W. Lindner, E. Francotte, Eds. *Chirality in Drug Research.* Wiley, Hoboken, NJ, 2007.
4. P. Sajonz, X. Gong, W. R. Leonard, M. Biba, C. J. Welch. *Chirality*, 18:803–813, 2006.
5. W. R. Leonard, D. W. Henderson, R. A. Miller, G. A. Spencer, O. S. Sudah, M. Biba, C. J. Welch. *Chirality*, 19:693–700, 2007.
6. C. J. Welch, P. Sajonz, G. A. Spencer, W. R. Leonard, W. A. Schafer, F. Bernardoni. *Org. Process Res. Dev*., 12:674–677, 2008.

7. M. Biba, C. J. Welch, J. O. DaSilva, D. J. Mathre. *Am. Pharm. Rev*., 8:68–74, 2005.
8. T. D. Nelson, C. J. Welch, J. D. Rosen, J. H. Smitrovich, M. A. Huffman, J. M. McNamara, D. J. Mathre. *Chirality*, 16:609–613, 2004.
9. C. J. Welch, M. H. Kress, M. Beconi, D. J. Mathre. *Chirality*, 15:143–147, 2003.
10. J. A. Ellman, T. D. Owens, T. P. Tang. *Acc. Chem. Res*., 35:984, 2002.
11. C. J. Welch, W. R. Leonard, J. O. DaSilva, M. Biba, J. Albaneze-Walker, D. W. Henderson, B. Laing, D. J. Mathre. *LCGC*, 16–29, 2005.
12. C. J. Welch, J. Fairchild, P. Sajonz. *Chirality*, 19:607–611, 2007.

CHAPTER 14

A Look into the Future: Chiral Analysis Using Chemical Sensor Technology

GREGORY K. WEBSTER

Global Analytical Research and Development, Abbott Laboratories, Abbott Park, Illinois

WILLIAM J. BUTTNER

International Center for Sensor Science and Engineering, Illinois Institute of Technology, Chicago, Illinois

1 INTRODUCTION

Currently, it is conservatively estimated that nearly 40% of the drugs under development within the pharmaceutical industry are chiral [1]. Often, in the early stages of drug development, the synthetic process yields racemic mixtures that need to be separated from the production perspective and assayed from an analytical perspective. Chiral screening systems based on liquid or supercritical fluid chromatography or capillary electrophoresis systems have been investigated to provide a more systematic approach to chiral method development [2,3]. Testing for enantiomeric purity is neither as straightforward nor as systematic as it is with achiral chromatography. Selection of chromatographic conditions and stationary phases for chiral applications involves a great deal of trial and error, which can be time consuming and costly. Such screens involve hours of continuous operation per sample, use several thousand dollars' worth of chromatographic columns, and may or may not lead to a promising result. Although chiral screening for method development is an improvement in efficiency relative to the less structured trial-and-error scenarios, it still has significant time and material costs. Investment in chiral method development alternatives is justified. One leading alternative is the use of chemical sensors. Sensor technology can potentially bring about improved cost and time efficiencies in the development of chiral purity assays.

Chiral Separation Methods for Pharmaceutical and Biotechnological Products, Edited by Satinder Ahuja

Sensor technology currently explored in research laboratories is available commercially in a variety of designs and principles. Sensors have been developed primarily for analytes that span various industrial and medical needs and at present are of particular interest for military and fuel cell applications. The applicability of sensor technology to applications of pharmaceutical interest is growing. Chemical sensors have been used effectively within analytical chemistry to provide reliable, efficient, low-cost detection devices [4–9]. Use of chemical sensors to discriminate the enantiomers of the drug or drug product matrix directly, or to improve current chromatographic capabilities has the potential to improve impurity identification and speed up analysis in all stages of product development. Sensor techniques implemented for drug substance development potentially carry through to the manufacturing process. The early stages of product development could utilize the high throughput and short analysis times to extract impurity information quickly, thereby enabling rapid method development that will continue into preclinical development and assisting in the timely analysis and release of the bulk drug supply.

Several sensor platforms have been chosen for preliminary investigations of chiral detection. As sensor technology is readily in the form of gas detectors, initial enantiospecific detection investigations began studying chiral analysis in gas-phase applications [10–13]. This work used principally surface acoustic wave sensors (SAW) [14,15]. However, other platforms, such as emissive cadmium selenide [16], TSR [17,18], optical [19–23], field effect [24], immunological [25], and electrochemical [26–29] sensors, have been investigated. In terms of moving chromatographic chiral analysis to routine chemical sensing, the quartz crystal microbalance platform looks to be the most promising [30–40]. Like earlier attempts at developing chiral sensors, initial applications for the quartz crystal microbalance began with gas-phase detection [30–35]. This work used cyclodextrins [30,31,34,35], and molecularly imprinted polymer (MIP) films [32,33]. Chiral selectivity based on sensor technology was compared to traditional gas chromatography [31].

Pharmaceutical applications require analysis in a condensed phase, for the drug itself is seldom volatile. Cao et al. [32] used a coated quartz-crystal microbalance (QCM) sensor and an imprinted polymer film to provide enantiomeric selectivity with dansylphenylalanine and established a detection limit of 5 ppm and a working range from 5 to 500 ppm in a pH 10 solution. Haupt et al. [33] used a similar QCM technique with (*S*)-propanolol in acetonitrile. These studies established that chiral discrimination in liquid media could be performed selectively and with stability. Lu et al. [36,37] began looking at chiral interactions with chiral stationary phases bonded to QCM sensors. Inagaki et al. [38] and Buttner et al. [40] showed that applications of films from chiral stationary phases used in liquid chromatography illustrate chiral recognition using the QCM platform. Reimhult et al. [41] used MIP films for chiral recognition on a QCM platform. The Inagaki and Reimhult groups used a static cell design, while the Buttner group used a flow design to mirror traditional chromatographic applications.

2 QCM SENSORS

A QCM sensor is a mass detector that can measure very small mass changes on a quartz-crystal resonator. The sensitivity of a QCM sensor in a gas-phase experiment is approximately 0.1 mg or lower. Thus, QCM sensors are able to measure the changes in mass at a small proportion of a monolayer. As a sensor, the QCM has the advantages of high sensitivity and real-time monitoring of mass changes on the sensor crystal.

QCM sensors operate using a converse piezoelectric effect. Application of an electric field across a piezoelectric material induces a deformation of the material. Ascentric materials such as quartz produce an electrical potential when mechanical stress is applied to the surface. The magnitude of this potential is proportional to the stress applied. Materials such as quartz have a polar axis due to dipoles associated with the arrangement of atoms in the crystalline lattice. The mechanical stress displaces atoms resulting in a net change in dipole moment. This change in dipole moment generates a charge in the crystal.

QCM sensors employ α-quartz crystals when used as chemical sensors, due to their mechanical and piezoelectric properties. The AT cut crystal is used for its low-frequency drift at room temperature. The quartz crystal is placed between two metal vapor–deposited electrodes. When an alternating electric field is applied, the quartz crystal oscillates. The vibrations of the quartz crystal establish a transverse acoustic wave that propagates across the crystal and reflects back into the crystal at the surface. This standing-wave condition with acoustic wavelength is equal to twice the combined thickness of the crystal and electrodes. QCM sensors typically operate at 5 to 10 MHz for sensor applications.

The QCM sensors was a frequency control element until in 1959, Sauerbrey showed that the frequency shift of a quartz-crystal resonator is directly proportional to the added mass [41]. This breakthrough enabled the sensor platform to measure very small masses. The Sauerbrey equation is written

$$\Delta f = -\frac{2f_0^2 \Delta m}{A(\rho_q \mu_q)^{1/2}} \tag{1}$$

where

Δf = measured frequency shift
f_0 = resonant frequency of the fundamental mode of the crystal
Δm = mass change per unit area (g/cm^2)
A = piezo-electrically active area
ρ_q = density of quartz, 2.648 g/cm^3
μ_q = shear modulus of quartz, 2.947×10^{11} $g/cm \cdot s^2$

Sauerbrey's equation does not hold unless the QCM coating is uniform.

After Kanazawa showed that the change in resonant frequency of a QCM sensor taken from air to a liquid phase is proportional to the square root of the liquid's density–viscosity portent, it was realized that a quartz crystal could be excited to a stable oscillation when it is completely immersed in a liquid [42]. Kanazawa's equation is written

$$\Delta f = -f_u^{3/2}\sqrt{\frac{\rho_L \eta_L}{\pi \rho_q \mu_q}} \tag{2}$$

where

Δf = measured frequency shift
f_u = resonant frequency of the unloaded crystal
ρ_L = density of liquid in contact with the crystal
η_L = viscosity of liquid in contact with the crystal
ρ_q = density of quartz, 2.648 g/cm^3
μ_q = shear modulus of quartz, 2.947×10^{11} g/cm $\cdot$s^2

QCM sensors have been studied for a variety of vapor-phase and liquid-phase analyses, such as pollutants and bacteria. According to the piezoelectric effect, the change of frequency of a QCM sensor corresponds to the negative value of variation of mass. QCM sensors can be applied to both vapor- and condensed-phase analyses. A high degree of selectivity can be engineered into the QCM sensor through the use of specialized coatings. Due to the specific chemical–chemical interaction between the analyte molecules and coating on the surface of a QCM sensors, the performance of the QCM sensor is controlled. Accordingly, application of a chiral-specific coating should provide a sensor platform for selective enantiomeric detection. One challenge, however, with these sensor-based methods, including QCMs, is to achieve sufficient sensitivity for practical analyses.

3 QCM CHIRAL SENSOR APPLICATIONS

The true challenge for using chemical sensors for stereoselectivity applications is resolving the response of the enantiomers from a single theoretical plate separation [35]. Figure 1 illustrates the "one-plate resolution" achieved in the gas phase with chemical sensors. Basically, this has been achieved using molecularly imprinted polymer coatings or chiral chromatographic stationary phases that are coated or bonded to the QCM sensor surface. Bonding the chiral selector should yield more robust sensors [38].

Two primary applications have been sought for chiral chemical sensors. The goal is to use the chemical sensor to supplant the need for chromatographic analysis for enantiomeric purity determinations. Expanding the capability, performance, and applicability of enantiomeric purity determinations through chiral chemical sensor detection sensors would lower the overall cost of analysis as well as potentially lead to process analytical technology (PAT) applications. Use of electronic sensors for direct discrimination of the enantiomers of a drug

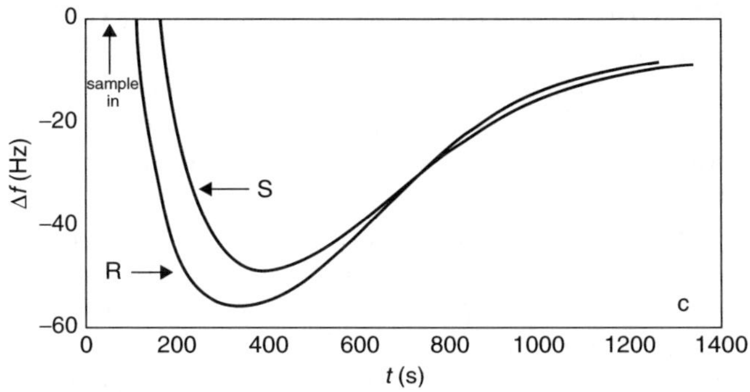

Figure 1 Gas-phase chiral sensor resolution of carvone eantiomers. (From [34].)

substance or drug product matrix, or to improve current chromatographic capabilities, has the potential to improve impurity identification and speed up analysis in all stages of analytical development. Sensor techniques developed for the drug substance are likely to carry through to the drug product Early development compounds will utilize the high-throughput and short analysis times to extract impurity information quickly, enabling rapid method development that will continue into preclinical development, assisting in the timely analysis and release of the bulk drug supply.

Testing for enantiomeric purity is neither straightforward nor systematic as it is with achiral chromatography. Selection of chromatographic conditions and stationary phases for chiral applications still involve a great deal of trial and error. Pharmaceutical laboratories routinely use chiral screening systems to assist in a systematic approach to chiral method development. However, such screens involve several hours of continuous operation per sample, the use of thousands of dollars worth of chromatographic columns and may or may not lead to a promising result. The LC columns employed have a finite lifetime and must be replaced periodically. Investment in chiral method development is worthwhile, for it leads to an approvable analytical data package for the U.S. Food and Drug Administration (FDA) and, subsequently, product approval. Sensor technology can potentially bring improved cost and time efficiencies to the development of such chiral drugs. Preliminary studies in using chiral coatings not only have the goal of direct enantiomeric purity determination, but have also been developed with the goal of stationary-phase screening [36–38].

Future studies with chiral sensors will address the potential issue of nonspecific binding to the chiral surface. Because chiral selectivity comes from surface coatings, nonspecific adsorption on these chiral coatings potentially will interfere with the signal of interest. Current studies have focused on screening-type applications where only the enantiomers of interest are present. This is probably true for initial for PAT applications as well.

4 CHROMATOGRAPHIC STATIONARY-PHASE COATINGS

Chiral stationary-phase coating of QCM sensors for chiral detection have been used in preliminary studies for chiral discriminations. Lu et al. used the (*R*,*R*)- and (*S*,*S*)- forms of the Whelk-O stationary phase to produce complementary responses to the *R* and *S* forms of the analyte, respectively. [36,37,40] Inagaki et al. [38] used CGGA and DNPAP stationary phases. Both groups bonded the stationary phase to the gold electrode surface of the QCM as self-assembled monolayers [43]. The Whelk-O phase was used in a flow cell to mirror its use as a chromatographic phase, while the CGGA and DNPAP phases were used in a static cell (Figure 2). Representative enantioselective response with chemical sensors coated with chromatographic stationary-phase films is shown in Figure 3.

In the Lu et al. investigations, the olefin form of Whelk-O was attached to the gold electrode via a two-step surface-modification process. The first step relied on the formation of a stable Au–S bond. Normally, this would be accomplished by a "traditional approach" to add a terminal thiol group to the coating of interest and then using this thiolated coating to bind to the gold surface. The thiol would form a covalent bond with the Au surface to immobilize the coating. The synthetic pathway to transform the terminal olefin to a thiol is a complex synthesis and would require extensive cleanup steps for purification of the desired product. Instead, an alternative approach was developed. An Au–S–R–SH complex was first formed on the surface of a QCM sensor using 1,4-butanedithiol. A gold–sulfur bond also formed, in principle, free thiol groups at the C_4 position,

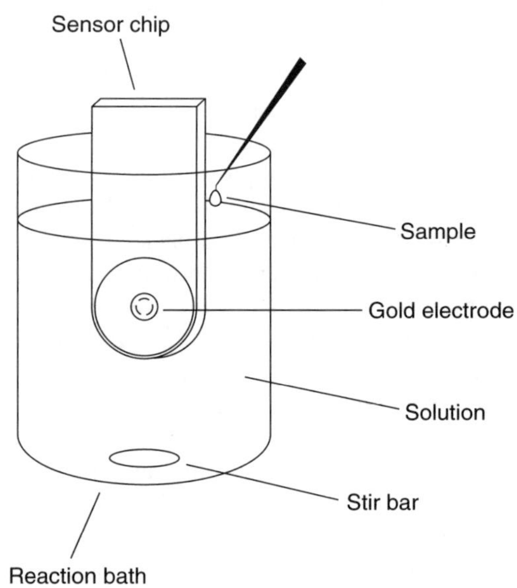

Reaction bath in the QCM instrument.

Figure 2 Liquid-phase chiral sensor placement. (From [37].)

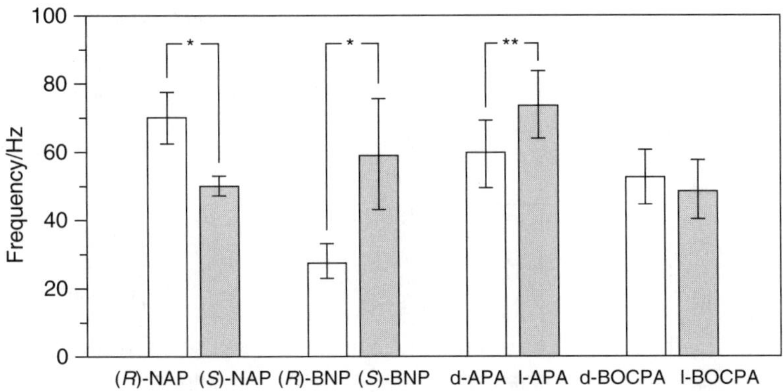

Figure 3 Frequency mean for enantiomers resolved using chromatographic stationary phases. (From [37].)

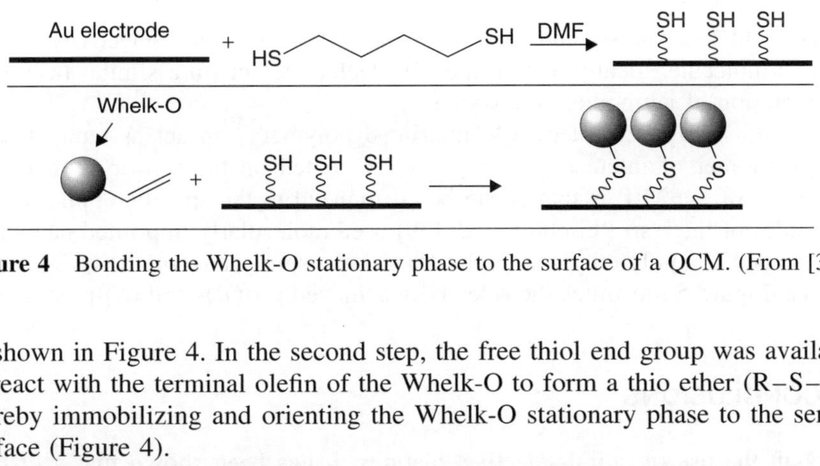

Figure 4 Bonding the Whelk-O stationary phase to the surface of a QCM. (From [39].)

as shown in Figure 4. In the second step, the free thiol end group was available to react with the terminal olefin of the Whelk-O to form a thio ether (R–S–R′), thereby immobilizing and orienting the Whelk-O stationary phase to the sensor surface (Figure 4).

In a similar fashion, Inagaki et al. used amido-coupling reactions to bond to the QCM sensor surface. An Au–S–R–NH_2 complex was first formed on the surface of the QCM using aminoethanedithiol, after which the CGGA or DNPAP selector was reacted with *N*-hyroxysuccinimide and 1-(3-dimethlyaminopropyl)–3-ethylcarbodiimide hydrochloride to activate the carboxyl groups. The stationary phase was then bonded to the QCM.

5 MIP COATINGS

Several investigations have established the use of molecularly imprinted polymers as chiral selectors for chemical sensors [44–48]. A molecular imprinted polymer (MIP) is simply a polymer that was formed in the presence of a molecule of interest. This molecule is removed after the polymer is formed, leaving a matching template of the molecule in the polymer. The MIP formed is commonly used in

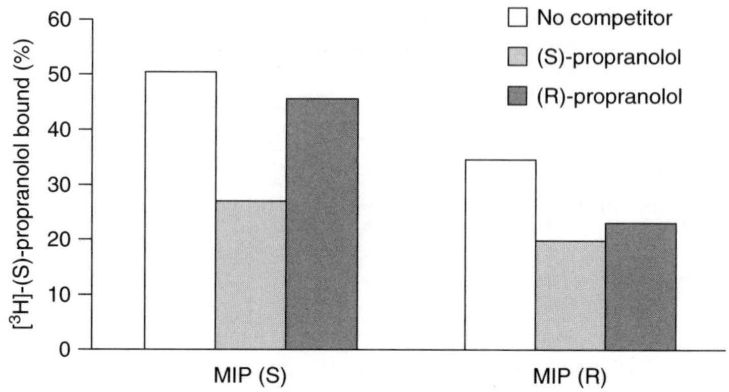

Figure 5 Displacement of labeled (*S*)-propranolol bound to imprinted nanoparticles. (From [38].)

sensors and in catalysis and separation methods to enhance selectivity for the original molecule. Ideally, use of the MIP technique acts in a similar fashion to the operation of antibodies or enzymes.

In chiral analysis, molecularly imprinted polymers can act as templates for the enantiomer of interest and are typically coated on the sensor surface. The advantage of the MIP is that it can be customized to the specific application or molecules of interest. Reimhult et al. [39] used molecularly imprinted nanoparticles trapped in a poly(ethylene terphthalate) layer spin coated on the QCM sensor surface. Figure 5 illustrates the selectivity achieved for (*R*)- and (*S*)-propranolol.

6 CONCLUSIONS

Through the use of chiral-selective coatings, it has been shown that a distinct concentration dependence and stereo-specific response with QCM sensors can be realized. This work has set the groundwork for future developed in using chemical sensor technology for chiral analysis. Chemical sensors will be developed not only for direct enantiomeric purity determinations but also as a platform for rapid stationary-phase screening for chromatographic applications. As many new molecular entities are isolated using preparative chiral chromatography in early stages of development, sensor technology will make significant cost and efficiency contributions to the pharmaceutical industry.

REFERENCES

1. S. Ahuja. Commercial manufacture of chiral pharmaceuticals. In *Chiral Separations: Applications and Technology*, S. Ahuja, Ed. American Chemical Society, Washington, DC, 1996, p. 13.

2. A. Akin, F. J. Antosz, J. L. Ausec, K. F. Greve, R. L. Johnson, L. E. Magnusson, T. Ramstad, S. L. Secreast. An orthogonal approach to chiral method development screening. *Curr. Pharm. Anal.*, 3(1):53–70, 2007.
3. G. Gübitz, M. G. Schmid, Eds. *Chiral Separations: Methods and Protocols*. Methods in Molecular Biology, Vol. 242. Humana Press, New York, 2004.
4. T. Wessa, S. Kuppers, M. Rapp, J. Reibel, U. Stahl. Industrial applications of SAW sensors: on-line control of preparative HPLC. *Fresenius' J. Anal. Chem.*, 363:744–748, 1999.
5. T. Wessa, S. Kuppers, G. Mann, M. Rapp, J. Reibel. On-line monitoring of process HPLC by sensors. *Org. Process Res. Dev.*, 4:102–106, 2000.
6. L. M. Dorozhkin, I. A. Rozanov. Acoustic wave chemical sensors for gases. *J. Anal. Chem.*, 56(5):399–416, 2001.
7. G. K. Webster, D. A. Farrand, M. A. Litchman. A chemical sensor based scheme to evaluate packaging containment effectiveness: 1. Development and optimization. *J. Process Anal. Chem.*, 8(1):11–17, 2003.
8. G. K. Webster, D. A. Farrand, M. A. Litchman. A chemical sensor based scheme to evaluate packaging containment effectiveness: 2. Scale-up and qualification of the inspection scheme. *J. Process Anal. Chem.*, 9(1):30–37, 2004.
9. G. K. Webster, T. Hobson, W. B. Holzheuer, J. R. Stetter, W. J. Buttner. Application of a low cost amperometric sensor for detection of alcohols in process related headspace applications. *J. Process Anal. Chem.*, 10(2):34–43, 2008.
10. M. Trojamwicz, M. Ceislo. Electrochemical and piezoelectric enatioselective sensors and biosensors. *Anal. Lett.*, 38:523–547, 2005.
11. K. Bodenhoffer, A. Hierlemann, J. Seemann, G. Guaglitz, B. Kopperhoefer, W. Gopel. Chiral discrimination using piezoelectric and optical gas sensors. *Nature*, 387:577–580, 1997.
12. U. Weimar, W. Gopel. Chemical imaging: trends in practical multiparameter sensor systems. *Sensors Actuators B*, 52:143–161, 1998.
13. A. Hierlemann, A. J. Ricco, K. Bodenhoffer, W. Gopel. Effective use of molecular recognition in gas sensing: results from acoustic wave and in situ FT-IR measurements. *Anal. Chem.*, 71:3022–3035, 1999.
14. A. Hierlemann, H. Baltes, V. Schurig. Search for extraterrestrial enantioenrichment by using chemical microsensors. *Enantiomer*, 6:129–139, 2001.
15. W. Weng, J. Han, Y. Chen, X. Huang. Progress in chiral sensors. *Huaxue Jinzhan*, 19(11):1820–1825, 2007.
16. D. E. Moore, K. Meeker, A. B. Ellis. Detection of chiral analytes through adduct formation with chiral films onto emissive cadmium selenide substrates. *J. Am. Chem. Soc.*, 118:12997–13001, 1996.
17. K. Bodenhoffer, A. Hierlemann, J. Seemann, G. Guaglitz, B. Christian, B. Kopperhoefer, W. Gopel. Chiral discrimination in the gas phase using different transducers: thickness shear mode resonators and reflectometric interference spectroscopy. *Anal. Chem.*, 69:3058–3068, 1997.
18. K. Bodenhoffer, A. Hierlemann, M. Jurza, V. Schurig, W. Gopel. Chiral discrimination of inhalation anesthetics and methyl propionates by thickness shear mode resonators: new insights into the mechanisms of enantioselectivity by cyclodextrins. *Anal. Chem.*, 69:4017–4031, 1997.

19. T. D. James, K. R. A. Sandanakumara, S. Shinkai. Chiral discrimination of monosaccharides using fluorescent molecular sensor. *Nature*, 374(6520):345–347, 1995.

20. Y. Kubo, S. Maeda, S. Tokita, M. Kubo. Colorimetric chiral recognition by a molecular sensor. *Nature*, 382(6591):522–524, 1996.

21. W. Guo, J. Wang, C. Wang, J. Q. He, X. He, J. P. Cheng. Design, synthesis and enantiomeric recognition of dicyclodipeptide-bearing calyx[4]arenas: a promising family for chiral gas sensor coatings. *Tetrahedron Lett.*, 43:5665–5667, 2002.

22. I. Sato, Y. Ohgo, H. Igarashi, D. Nishiyama, T. Kawasaki, K. Soai. Determination of absolute configurations of amino acids by asymmetric autocatalysis of 2-alkynylpyrimidyl alkanol as a chiral sensor. *J. Organomet. Chem.*, 692(9):1783–1787, 2007.

23. Z.-H. Chen, Y.-B. He, Hu, G. Chen, X.-H. Huang, L. Hu. Synthesis and chiral recognition properties of a novel colorimetric chiral sensor for carboxylic anions. *Aust. J. Chem.*, 61(4):310–315, 2008.

24. L. Torsi, Farinola, M. Gianluca, F. Marinelli, M. C. Tanese, O. H. Omar, L. Valli, F. Babudri, F. Palmisano, P. G. Zambonin, F. Naso. A sensitivity-enhanced field-effect chiral sensor. *Nat. Mater.*, 7(5):412–417, 2008.

25. O. Hofstatter, H. Hofstetter, M. Wilchek, V. Schurig, B. S. Green. Chiral discrimination using an immunosensor. *Nat. Biotechnol.*, 17(4):371–374, 1999.

26. P. S. Bates, R. Kataky, D. Parker. Chiral sensors based on lipophilic cyclodextrins: interrogation of enantioselectivity by combined NMR, structural correlation and electrode response studies. *J. Chem. Soc.*, 4:669–675, 1994.

27. R. Kataky, P. S. Bates, D. Parker. Functionalized. α-cyclodextrins as potentiometric chiral sensors. *Analyst*, 117(8):1313–1317, 1992.

28. P. S. Bates, R. Kataky, D. Parker. A chiral sensor based on a peroctylated α-cyclodextrin. *J. Chem. Soc. Chem. Commun.*, 2:153–155, 1992.

29. S. Jo, Y. Jin, J. Kim, H. Suh. Synthesis of ferrocenyl and diphenyl substituted bispyridino-18-crown-6 ether for chiral recognition. *Bull. Korean Chem. Soc.*, 28(11):2015–2019, 2007.

30. Palaniappan, X. Lia, F. E. H. Tay, J. Li, X. Sua. Cyclodextrin functionalized mesoporous silica films on quartz crystal microbalance for enhanced gas sensing. *Sensors Actuators B*, 119:220–226, 2006.

31. May, I. P. Byfield, M. Linstrom, L. F. Wunsche. Chiral discrimination using a quartz crystal microbalance and comparison with gas chromatographic retention data. *Chirality*, 9:225–232, 1997.

32. L. Cao, X. C. Zhou, S. F. Y. Li. Enantioselective sensor based on microgravimetric quartz crystal microbalance with molecularly imprinted polymer film. *Analyst*, 126:184–188, 2001.

33. K. Haupt, K. Noworyta, W. Kutner. Imprinted polymer-based enantioselective acoustic sensor using a quartz crystal microbalance. *Anal. Commun.*, 36:391–393, 1999.

34. C. Fietzek, T. Hermle, W. Rosenstiel, V. Schurig. Chiral discrimination of limonene by use of β-cyclodextrin-coated quartz crystal-microbalances (QCMs) and data evaluation by artificial neuronal networks. *Fresenius' J. Anal. Chem.*, 371:58–63, 2001.

35. S. C. Ng, T. Sun, S. O. Chan. Durable chiral sensor based on quartz crystal microbalance using self-assembled monolayer of permethlyated β-cyclodextrin. *Macromol. Symp.*, 192:171–118, 2003.

36. C. L. Lu, V. Perez-Luna, J. R. Stetter, W. J. Buttner, G. W. Webster. Evaluation of surface–enantiomer interactions for chiral selective detection in QCM sensors. Presented at the Pittsburgh Conference and Exposition on Analytical Chemistry and Applied Spectroscopy, Chicago, Mar. 2007.

37. W. J. Buttner, V. H. Perez-Luna, C. L. Liu, J. R. Stetter, G. K. Webster, Strategies for a chiral selective chemical sensor. Presented at the International Meeting on Chemical Sensors, Columbus, OH, July 2008.

38. S. Inagaki, J. Z. Min, T. Toyo'oka. Prediction for the separation efficiency of a pair of enantiomers during chiral high-performance liquid chromatography using a quartz crystal microbalance. *Anal. Chem.*, 80:1824–1828, 2008.

39. K. Reimhult, K. Yoshimatsu, K. Risveden, S. Chen, L. Ye, A. Krozer. Characterization of QCM sensor surfaces coated with molecularly imprinted nanoparticles. *Biosens. Bioelectron.*, 23:1908–1914, 2008.

40. W. J. Buttner, C. L. Lu, V. Perez-Luna, J. R. Stetter, G. K. Webster. Preliminary characterization of a commercial chiral stationary phase as a selector for chemical sensor applications using a quartz crystal microbalance. *Sensors Transducers*, 101:1–11, 2009.

41. G. Saurbrey. *Z. Phys.*, 155:206–222, 1959.

42. K. K. Kanazawa, J. Gordon, Frequency of a quartz microbalance in contact with a liquid. *Anal. Chem.*, 57:1770–1771, 1985.

43. T. Nakanishi, N. Banno, M. Matsunaga, T. Asahi, T. Osaka. Enantioselective crystallization on self-assembled monolayers and its application to chiral sensing. *Chem. Sensors*, 20(Suppl. B):862–863, 2004.

44. D. Kriz, M. Kempe, K. Mosbach. Introduction of molecularly imprinted polymers as recognition elements in conductometric chemical sensors. *Sensors Actuators B*, 33:178–181, 1996.

45. Y. Kubo, T. Shin'ya, S. Tokita, M. Kubo. Colorimetric chiral recognition by a molecular sensor. *Nature*, 382:522–524, 1996.

46. K. Haup. Molecularly imprinted polymers in analytical chemistry. *Analyst*, 126:747–756, 2001.

47. M. Zayats, M. Lahav, A. B. Kharitonov, I. Willner. Imprinting of specific molecular recognition sites in inorganic and organic thin layer membranes associated with ion-sensitive field-effect transducers. *Tetrahedron*, 58:815–824, 2002.

48. D. Nopper, O. Lammershop, G. Wulff, G. Gauglitz. Amidine-based molecularly imprinted polymers:new sensitive elements for chiral chemosensors. *Anal. Bioanal. Chem.*, 377(4):608–613, 2003.

CHAPTER 15

Chirality of Biomolecules and Biotechnology Products

SATINDER AHUJA
Ahuja Consulting, Calabash, North Carolina

1 INTRODUCTION

A *biomolecule* may be defined as any organic molecule that is produced by a living organism. Biomolecules can be small or large; it is important to determine a molecule's chirality and any impact it may have on their biological activity [1–10]. This concern stems from the administration of racemic thalidomide, which resulted in a large number of birth defects; these defects were attributed to the fact that the L-form of thalidomide is teratogenic, whereas the D-form is soporofic [11]. This conclusion remains controversial because racemization of thalidomide was overlooked. However, since the fiasco of thalidomide, the chirality of small molecules has been studied and controlled rigorously. Some examples of small biomolecules are hormones, metabolites, vitamins, and natural products. The chirality of adrenalin, a small biomolecule that acts like a hormone, is well known because of the differences in biological activity of its enantiomers. Euler [12] proved that noradrenaline rather than adrenaline was the transmitter released by the postganglionic sympathetic neurons. At that time, the actions of (±)-noradrenaline were compared with the physiological effects produced by stimulation of the sympathetic nerve. Two years later, Tullar meticulously resolved pure diastereoisomers, (−)-noradrenaline-(+)-bitartrate and (+)-noradrenaline-(+)-bitartrate, from the racemate [13]. The potency of the resolved (−)-noradrenaline was similar to that produced by nerve stimulation. These studies emphasize that knowledge of absolute configuration and conformation of the neurotransmitters is necessary for better understanding of the molecular mechanism involved in the activation of transmembrane G-protein–linked receptors.

Chiral Separation Methods for Pharmaceutical and Biotechnological Products, Edited by Satinder Ahuja

TABLE 1 Biological Activities of Stereoisomers of Some Small Molecules

Compound	Activity
Adrenalin (epinephrine)	*l*-isomer is 10 times more active as a vasoconstrictor than the *d*-isomer.
Synephrine	*l*-isomer has 60 times the pressor activity of the *d*-isomer.
Felodipine (metabolites)	metabolism rate of (*R*)- and (*S*)-felodipine is related to a lower K_m value for the (*R*)-isomer.
Amphetamine	*d*-isomer is a potent central nervous system stimulant; the *l*-isomer has little effect.
Propanolol	*S*(−)-isomer has only β-adrenergic blocking activity.
Warfarin	*S*(−)-isomer is five times more potent as an anticoagulant than the *R*(+)-isomer.
Ascorbic acid (vitamin C)	The *d*-isomer is antiascorbutic; the *l*-isomer is not active.

TABLE 2 Leading Biotech Drugs

Product	Biological Basis	Activity/Use	Sales ($ millions)
Epogen	Eythropoietin	Red blood cell growth	1161
Neupogen	Colony-stimulating factor	White blood cell growth	1056
Procrit	Eythopoietin	Red blood cell growth	1000
Humulin	Insulin	Diabetes	936
Intron-A	Interferon α-2b	Anticancer, infections	598
Engerix-B	Hepatitis B	Vaccine	584
Cerezyme	Glucoce-rebrosidase	Genetic deficiency	333
Activase	Tissue plasminogen activator	Heart attack/stroke	261
Humatrope	Somatotropin	Growth deficiencies	260
ReoPro	GPIIb/IIIa antibody	Blood clot prevention	254
Avonex	Interferon β-1a	Multiple sclerosis	240
Protropin/nutropin	Somatrem/somatropin	Growth deficiencies	224

Table 1 shows some small molecules whose enantiomers exhibit different biological activity.

Large biomolecules are exemplified by peptides, oligopeptides, polypeptides, proteins, and nucleic acids. A large number of successful biotechnology products based on proteins (see Table 2 and the Appendix) have been intoduced into our armamentarium of modern medicine to ameliorate human suffering. Proteins are complex organic macromolecules whose structures are coded in an organism's DNA. Each protein has a unique genetically defined amino acid sequence that determines its specific shape and functions (e.g., enzymes, hormones, and immunoglobulins involved in oxygen transport, muscle contractions, or electron transport). It is also well known that proteins are composed of chiral amino acids (see the detailed discussion on proteins in Section 5). Unfortunately, chiral studies are largely ignored for biomacromolecules such as proteins. In this chapter we briefly review the chirality of our biological world and discuss what is being done or not done to monitor biomolecules such as proteins and biotechnology products based on proteins, including their building blocks (i.e., amino acids and peptides).

2 CHIRALITY IN OUR BIOLOGICAL WORLD

Asymmetry or chirality (handedness) is ubiquitous in our world, from small molecules to living organisms, even though it is not always apparent. Pasteur was the first to grasp the importance of chirality in the biological world [14]. He observed that all living species are primordially, in their structure and their external forms, functions of cosmic asymmetry. In 1861, he stated that the difference in properties of corresponding right and left substances when they are subjected to asymmetric forces suggests to us the "mysterious cause" that presides over the asymmetric arrangement of the atoms in natural organic substances. He wondered: Why this asymmetry? Why one asymmetry rather than its inverse? It led him to the conclusion that perhaps asymmetric forces exist at the moment of the elaboration of natural organic products. He further concluded that enantioenrichment forms the only sharply defined boundary between the chemistry of dead and living matter. This led him to advance the idea that our world is dissymetrical. His brilliant thinking was confirmed by the discovery in 1956 that parity is not conserved for the weak interactions governing beta decay, and that matter is inherently chiral [15,16; see Section 3].

Many examples of asymmetry in the animal world were compiled by Neville [17] in 1976 (Figure 1). For example, most snails (*Helix pomatia*) are dextral. The workers at an escargot-processing facility in Dijon, France, use a specially designed chiral knife to extract the snails from their shells; occasionally, they encounter a sinistral shell into which this knife does not fit [18]. The occurrence of such individual shells is approximately 1 in 20,000, which represents an enantiomer-rich excess of 99.99% for the dextral form.

A casual look at the human body suggests that it is bilaterally symmetrical, but there are some common exceptions. For example, there is the very widespread intense predilection for right-handedness—approximately 95% of all people are right-handed. Why a small percentage of individuals are left-handed remains shrouded in mystery. It is reasonable to assume that probability works so that nature produces a small number of other forms.

During Pasteur's lifetime, all of the substances that were found to be optically active in solution, where the optical rotation necessarily had a molecular origin, were natural products or derivatives of natural products. This could have encouraged Pasteur to speculate that living organisms had access to a chiral natural force for the biosynthesis of their optically active products. He believed that the inherent chirality of matter might have played a role in the origin and persistence of biomolecular homochirality in the natural world (i.e., the sole occurrence

Figure 1 *Amphidromus inversus* on Kapas Island, Malaysia. (a) One dextral (top) and five sinistral adults (approximate shell height: 40 mm); (b) interchiral copulation between a dextral (left) and a sinistral (right) individual; (c) a freshly laid clutch of eggs; (d) a sinistral (left) and a dextral (right) adult shell; (e) a large number of apically crushed shells in a rat midden; (f) a recently hatched egg clutch, all sinistral (mean shell diameter: 3 mm). (From http://krauthammerlab.med.yale.edu/imagefinder/Home.) (*See insert for color representation of the figure*.)

of L-amino acids in proteins and of D-sugars in nucleic acids and biologically important polysaccharides) as it relates to the influence of forces of dissymmetry on the evolution of homochirality. Many natural plant products used in ancient times contain active components that are chiral (e.g., alkaloids). As chemical sciences advanced, active substances from plants were isolated. Quinine, strychnine, brucine, veratradine, and emetine were characterized in the early 1820s by Pelletier and Caventou. However, the asymmetric carbon atoms of these alkaloids were discovered much later.

2.1 True and False Chirality

Barron [19] with his decisive work on "true" and "false" chirality shed considerable light on what represented a compelling chiral influence. Most supposedly chiral influences are actually falsely chiral: parity-odd, time-odd, and of the wrong symmetry. Barron recognized that unqualified asymmetric synthesis could be generated only by something with the same symmetry, that is, a parity-odd, time-even influence. The characteristics of true chirality are natural optical rotation, parity-odd, time-even pseudoscalar. This provides the following definition, which allows chirality to be differentiated from other types of dissymmetry: *True chirality* is demonstrated by arrangements that exist in two discrete enantiomeric states that are interconverted by space inversion but not by time reversal united with any proper spatial rotation.

3 BIOMOLECULAR HOMOCHIRALITY

An important characteristic of life is its *homochirality*, because nature uses only one enantiomer and not the other, as exemplified by animals made of proteins based exclusively on L-amino acids, coded by DNA based exclusively on D-deoxyribose. However, a few unnatural D-amino acids and L-sugars do occur with specific roles, as in bacterial cell walls (but not bacterial protein or nucleic acids). D-Amino acids are used in cell walls because the enzymes of predators digest only L-amino acids. Furthermore, the genetic code does not code for these D-amino acids, which are synthesized by enzymatic inversion of the L-form [20].

3.1 Homochirality and the Origin of Life

When we consider what came first in the origin of life, it is important to realize that only the parity-violating energy differences (PVEDs) of ancestral biomolecules is pertinent (the handedness of recent biomolecules is predetermined by a diastereomeric connection with their ancestors). Three major categories that should be considered: nucleic acids, proteins, and clay minerals. The PVEDs of molecules from all three classes of possible ancestral replicators have been calculated by ab initio methods [21]. The natural L-amino acids L-alanine, L-valine, L-serine, and L-aspartate were all found to be more stable than their unnatural L-enantiomers in their solution conformations and also the α-helix and β-sheet conformations, by 10^{-17} kT. Among the natural L-sugars, it was found that D-glyceraldehyde, the parent of the higher sugars, is indeed PVED-stabilized, by about 10^{-17} kT. These ab initio estimations, including the chiral electroweak force, indicate that the L-amino acids and the D-sugars are more stable than the corresponding enantiomers. The small energy difference between these enantiomeric pairs, with Darwinian reaction kinetics in a flow reactor, account for the choice of biomolecular handedness made when life began.

It is apparent from the discussion above that homochirality may be a prerequisite for life. This is because polymerization to provide the essential long-chain homochiral polymers (e.g., all-D nucleic acids or all-L polypeptides) will not go well in racemic solution even when directed by a homochiral template. Joyce and colleagues [22] showed that poly(C)-directed configuration of poly(G) is strongly chiroselective, since only the D-monomers get incorporated into the increasing poly(G) chain, but that the reaction is rigorously repressed by the existence of the L-monomer, and this enantiomeric cross-inhibition effect was established in a study by Orgel [23]. As a result, a nearly homochiral monomer solution appears to be needed for effective polymerization. This suggests that a homochiral prebiotic chemistry could be a precondition for life. If this were the case, the origin of biomolecular chirality is not just a matter of interest but a central problem of the origin of life; a chiral stimulus might be required for initiating life, and the characteristic of that chiral influence will establish which enantiomer is selected.

3.2 Homochirality and Metabolism

The importance of homochirality in biological systems is highlighted by reality in that the "wrong" molecule often has negative effects. If an enzyme integrates even one D-amino acid, it will be the wrong shape and, as a result, will be inactivated. Also, the processes of aging and carcinogenesis are affected by the inclusion of the wrong molecule [21]; the body therefore contains special enzymes called *d*-amino acid oxidases to eradicate amino acids of the wrong hand.

Homochirality is necessary for efficient metabolism, like the universal implementation of right-hand screws in engineering. As early as 1894, Fischer recognized this in his stereochemical lock-and-key premise: A chiral molecule can "sense" the disparity between the enantiomers of another chiral molecule, just as a left hand can feel the difference between right and left gloves and will only fit easily into the left. But molecules do not always act together preferentially with other molecules of the same hand as with hands and gloves—the preference depends on the situation; for example, a right hand prefers to shake another right hand (standing facing each other) but prefers to hold a left hand (standing side by side). Where chirality exists, for efficient biochemistry there must be homochirality. So, is life inevitably chiral? The answer appears to be yes, because only the tiniest and least complicated molecules are achiral. Molecules such as proteins that are large enough and intricate enough to support life are almost always chiral and, therefore, homochiral.

Stereoselectivity in binding can have a significant effect on drug disposition, such as first-pass metabolism, metabolic clearance, renal clearance, and protein and tissue binding [24]. For example, human serum albumin (HSA) is able to bind a great number of various endogenous and exogenous compounds stereoselectively. Various experimental data suggested that the two major drug-binding cavities, sites I and II, do not seem to be the stereoselective binding sites of HSA. Stereoselective binding of HSA under disease conditions such as renal and hepatic diseases was found to be enhanced. In addition, site-to-site displacement of a site II–specific drug by another site II–specific drug was also found to be stereoselective. Endogenous compounds such as long-chain fatty acids and uremic toxins are likely to cause combined direct and cascade effects that contribute to the preferential binding of a particular drug enantiomer. Taking together the findings of other studies, it is highly possible that the stereoselective binding site exists at the interface of the subdomains.

4 BIOTECHNOLOGY PRODUCTS

A number of biotechnology products have been introduced successfully into the market (Table 2). Some basic information on these biotechnology products are included in the Appendix. A review of Table 2 and the Appendix shows that a large number of biotechnology products are based on proteins. For example, human growth hormone (hGH, somatotrophin) is a 22-kDa, 191-amino acid single-chain protein produced by somatroph cells of the anterior pituitary gland. It is the major physiological regulator of growth, and deficiencies in growth hormone levels have long been recognized as the underlying cause of growth disorders (dwarfism). Initial preparations were made by extraction and purification from cadaveric pituitary tissue. Since 1984, hGH has been prepared by recombinant deoxyribose nucleic acid (rDNA) technology. As is usually the case with biologicals, characterization of the drug substance depended on a combination of physicochemical and biological methods, and the hGH molecule became well characterized fairly early in its life as a drug. Indeed, by 1980 the major degradation forms and structural variants of the hGH molecule had been described and reviewed. In the early 1990s a series of collaborative studies on analysis of recombinant hGH (somatropin) established that available bioassays were much less selective that physicochemical methods in detecting and quantifying structural degradation, and 1994 saw an international consensus to replace the bioassays with physicochemical analytical methods for routine batch release of somatropin. During the last decade in most markets, somatropin has, unusually for a protein, been subject to batch release and control dependent entirely on physicochemical analysis, without the routine use of any form of bioassay. During that time there has been a continuous development and refinement of methods, and the identification of a range of structural variants and degradation products of the molecule. Proteins, including structural characterization, are discussed in detail in Section 5.

5 PROTEINS

Proteins are composed of more than 40 amino acids in peptide linkages that fold back on themselves in a particular way. In biochemistry, the term *amino acid* is used for amino acids in which the amino and carboxylate functionalities are attached to the same carbon, plus proline, which is not actually an amino acid. The amino acids are the building blocks of long polymer chains. With 2 to 10 amino acids, such chains are called *peptides*, with 10 to 40 they are called *polypeptides*, and longer chains are known as *proteins*. These protein configurations have many structural and enzymatic roles in organisms. Twenty amino acids are encoded by the standard genetic code, but there are more than 500 natural amino acids. When amino acids other than the set of 20 are observed in proteins, it is usually the result of modification after translation (protein synthesis). Only two amino acids other than the standard 20 are known to be incorporated into proteins during translation, in certain organisms.

The characteristic three-dimensional shape of a protein is imposed on it by the secondary and tertiary structure of the peptide chain. The ability of the protein polypeptide chain to adopt complex three-dimensional conformations comes from double-bond characteristics of the α-carbon (which carries the amino acid side group). This allows the bonds in either direction of the α-carbon to freely rotate. The angles formed at these bonds are designated *phi* (ϕ) and *psi* (ψ). Of all possible combinations of ϕ and ψ, only a small subset can occur physically because of collisions of the peptide oxygen and hydrogen atoms. Of these allowable combinations, some exhibit better stability than others. A polypeptide chain with the given angles can form left- or right-handed helix structures.

5.1 Protein Folding

Proteins are synthesized as linear polymers that develop biological activity as a result of folding into characteristic three-dimensional structures. Stuctural proteomics covers the three-dimensional structures of various proteins. Protein folding is a physical process by which a polypeptide folds into its characteristic and functional three-dimensional structure from a random coil [25]. Each protein exists as an unfolded polypeptide or random coil when translated from a sequence of mRNA to a linear chain of amino acids. This polypeptide lacks any developed three-dimensional structure. However, amino acids interact with each other to produce a well-defined three-dimensional structure, the folded protein, known as the *native state*. The resulting three-dimensional structure is determined by the amino acid sequence [26]. A protein molecule folds spontaneously during or after synthesis. For many proteins, the correct three-dimensional structure is essential to function [27]. Failure to fold into the intended shape usually produces inactive proteins with different properties, including toxic prions. Several neurodegenerative and other diseases are believed to result from the accumulation of misfolded (incorrectly folded) proteins. The characterisitc three-dimensional shape is called *protein conformation*. Protein folding results from a rapid biochemical reaction that begins before a protein has been completely synthesized and proceeds through primary, secondary, and tertiary structures before the final quaternary structure is developed. Although these macromolecules may be regarded as folding themselves, the process also depends on the solvent (water or lipid bilayer), the concentration of salts, the temperature, and the presence of molecular chaperones [28].

5.2 Stuctural Characterization of Proteins

Protein preparation purity, correct folding, monodispersity, and homogeneity are analyzed by mass spectrometry (MS), gel electrophoresis, size-exclusion chromatography, circular dichroism, dynamic light scattering, and nuclear magnetic resonance (NMR) [2]. The most widely used approaches for stuctural characterization of proteins are x-ray crystallography and NMR spectroscopy.

An incisive probe of chirality and of biomolecular structure and behavior can be conducted with Raman optical activity (ROA). The reader may want to review the theory, instrumentation, and applications of ROA, which measures vibrational optical activity by means of a small difference in the intensity of Raman scattering from chiral molecules in right- and left-circularly polarized incident light or, equivalently, a small circularly polarized component in scattered light [29]. Besides providing the absolute configuration of small chiral molecules, the application of ab initio methods to the analysis of experimental ROA spectra offers promise for determination of the three-dimensional structure and conformational distribution in unprecedented detail. The rich ROA spectra of aqueous solutions of biomolecules provide detailed structural information, including, in the case of proteins, the tertiary fold in addition to secondary structure elements. ROA studies of unfolded and partially folded proteins provide new insight into the residual structure in denatured proteins and the aberrant behavior of proteins responsible for misfolding diseases.

Identifying protein targets for structural investgation is a crucial strategic issue. For example, finding a novel protein fold could provide valuable clues to its activity.The protein targets are also tested for their stability, oligomeric state, and level of posttranslational modifications. In earlier texts, bioseparations of proteins were discussed at length [6,8]. It was pointed out that interfacial protein adsorption and inactivation can occur during bioseparations. This can generally be attributed to unfolding of proteins during processing. For further information on this subject, see Section 9.

Peptide and protein arrays are a modern tool of proteomics used to search for particular peptides and proteins and to analyze their activities [2]. Protein and peptide arrays have been developed in analogy with DNA arrays that were developed earlier. Protein and peptide arrays facilitate the use of proteomic methods in diagnostics, replacing complicated chromatographic or electrophoretic analyses. They are used in basic research and by industry for drug discovery purposes. For example, the array components may be designed by in silico proteolytic digestion of proteins, so that conclusions on the activity of the full protein molecules can be drawn without having to obtain proteins by DNA recombination or by isolation from natural sources.

Protein arrays consist of a solid support with immobilized protein molecules. The support may have different characteristics, depending on the type of array. Basically, it may be a glass plate, membrane, beads, or a microtiter plate. There are various methods of preparation and uses of protein arrays. For investigation of protein–protein interactions (see the following section), functional arrays are used that contain thousands of different proteins. They are employed to assess interactions of proteins with other proteins, DNA, or low-molecular-mass ligands in particular; the interactions investigated may encompass those between receptor and ligand (or, specifically, drug), or between the enzyme and its substrate or inhibitor. A dedicated array makes possible complete analysis of the interactions of several versions of one protein, including its domains, mutants, or alternative splicing variant with another ligand of choice.

5.3 Protein Interactions

Proteins exert their roles in living organisms by interacting with other proteins, nucleic acids, lipids, saccharides, and other classes of compounds; therefore, the functions of proteins can be examined by analyzing their interactions. As a first step to understanding the role of an unknown protein, its subcellular localization should be probed, because the two are often strongly correlated.

Structural proteomics proceeds from gene sequence to the three-dimensional structure of a protein encoded by this gene, thereby creating a potential to link pure structural information to the biochemical and biological function of the macromolecule. It offers a systematic planning and high-throughput approach which was found very useful in initiatives such as the human genome project [2]. At times, the structural similarity of a target structure to the functionally characterized protein in the database is so striking that the function of the protein of interest might be identified almost directly. In the absence of clear structural homology on a global level, it may still be possible to recognize specific three-dimensional substructures that hold the essence of protein biochemical function. The examples of such motifs mentioned most often are the helix–turn–helix motif characteristic of DNA-binding proteins and the helix–loop–helix structural domain (EF hand), which is responsible for calcium binding. The motif-based function prediction methods include construction of three-dimensional templates that capture amino acid residues important for the function by superimposing the functionally related structures, followed by assigning the novel protein to the template that is most alike. Another emergent approach is based on the assumption that intermolecular interactions between proteins (or between proteins and other molecules) occur because of the complementarity of their molecular surfaces. The surface of the protein of interest is modeled and analyzed with regard to features such as hydrophobicity, electrostatic potential, or the presence of either exposed protein–protein interaction sites or the clefts on the surface that can be accessed by a solvent or ligand. The most recent and powerful structure–function prediction algorithms combine various approaches and involve the following steps: First search for proteins having the same, or a related, overall fold, then analyze the molecular surfaces, and finally, investigate a number of structural templates.

It is important to emphasize that the assignment of function to a protein based on its structure is a complex process that requires human intervention at many stages since one biochemical function may be associated with a variety of folds, while the tie fold is often found in numerous, clearly distinct protein families that perform various functions. Therefore, to correlate protein structure and its biochemical function, it is always necessary to examine carefully the architecture of the active site of the molecule being investigated.

By blocking the activity of a particular protein by gene-silencing techniques, it is possible to identify the processes in which it plays significant roles. The best time to switch off the functioning of proteins is before they are formed (i.e., at the level of translation). It is well known that each mRNA molecule gives rise to many protein copies. Antisense oligonucleotides (AOs) have the power to block the biosynthesis of a particular protein without influencing the expression of other genes. They are single-stranded oligonucleotides (DNA and RNA), typically 15 to 25 nucleotides, whose sequence is complementary to mRNA that results from gene transcription.

5.4 Chirality of Proteins

Based on information from gene sequences, we can roughly estimate that the human body contains some 60,000 to 70,000 proteins [2]. Taking into consideration posttranslational modifications (over 100 identified) and alternative splicing, we may end up with an approximate total of 700,000 proteins. One cell contains an average of approximately 10,000 proteins, with their quantity and proportions subject to changes in space and time. Isolation and identification of a particular protein, with its search for all modifications, its three-dimensional structure, and the function of such a huge number of molecules is a real challenge for today's scientist. The flow of information within a cell depends on many factors, such as the protein biosynthesis pathways, whose general scheme is presented in Figure 2.

It is logical to assume that all proteins are chiral because they are produced from L-amino acids [5,9]. Proteins can fold to form topological knots; other protein-folding motifs, such as β-sheet and α-helix, are also chiral. For example, many transmembrane proteins are based on a structure containing seven membrane-spanning α-helices, which forms a tubelike supramolecular structure. Such proteins may have an appearance that looks very symmetric; however, they are all chiral because they are composed of L-amino acids.

A variety of cellular components having interesting asymmetric shapes play structural roles in the cell. Many structural proteins have a helical shape. Whenever a tube, strut, rod, or spar is needed at the subcellular level, nature often turns to a helical solution. Muscle proteins actin and myosin, collagen fibrils, microtubules, and prokaryotic flagella are a few common examples.

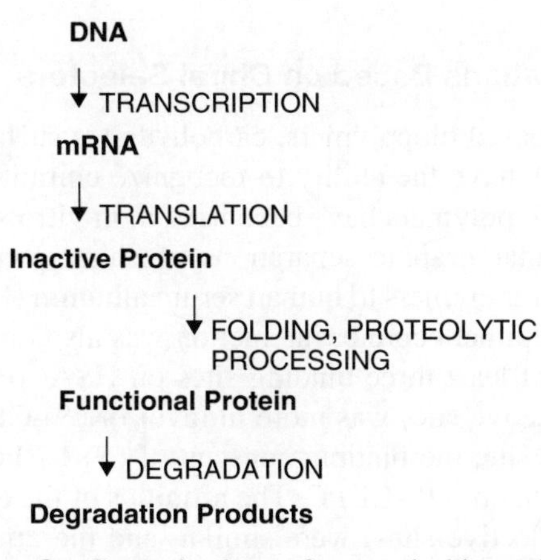

Figure 2 General scheme for protein biosynthesis.

The structures of some rotary proteins have been determined. For example, the catalytic domain of adenosine triphosphate (ATP) synthetase was shown to rotate in a counterclockwise direction when ATP is present [30]. One very remarkable and long-studied protein motor is the bacterial flagellar motor [31]. The motor runs on a proton gradient and is used for propulsion by some prokaryotic bacteria, such as *Escherichia coli*. Proton flow through the motor brings about rotation, which turns the flagellum. Essentially, this motor is a wheel, a chiral tool essential to many of humankind's inventions, but it is not often found in nature.

Chiral selectors in biological systems are frequently biopolymers that develop their chirality from L-amino acid structures and the resulting secondary and tertiary structures. The best-known examples are enzymes, receptors, and carrier proteins: for example, serum albumin. It is well known that binding of drugs to proteins can be stereoselective (see Sections 6 and 8), which can produce different pharmacological and toxicological properties of enantiomers; this emphasizes the need to monitor the structure of proteins closely.

6 CHIRAL SELECTORS

The first account of molecular chiral recognition may be attributed to Pasteur, who in 1858 described how the dextro form of ammonium tartrate is more quickly destroyed by the mold *Penicillium glaucum* than the levo isomer. In biological systems, the separation of enantiomeric compounds is conceptually similar to the chemical process (i.e., it is based on transformation of enantiomers into diastereomers; see the next section for further discussion). However, the crucial step is the creation of transient diastereomeric complexes rather than the formation of diastereomeric compounds,. The mechanism is based on the capability of one chiral molecule (the selector) to interact with the enantiomer of another (the selectand). Molecular chiral recognition or delineation is the consequence of energy differences between the diastereomeric selector–selectand complexes.

6.1 In Vitro Separations Based on Chiral Selectors

Besides amino acid–based biopolymers, carbohydrates such as cellulose or amylose and cyclodextrins have the ability to recognize chirality. Innate and derivatized forms of these biopolymers have been used in in vitro systems for analytical and preparative chromatographic separations of chiral compounds. Binding of carbenicillin (CBPC) for epimers to human serum albumin (HSA) was found to be stereoselective [32]. Epimer–epimer interaction was also observed in the binding to HSA. There were at least three binding sites on HSA for CBPC epimers, one of which (a stereoselective site) was more in favor of (*S*)-CBPC than (*R*)-CBPC. At the stereoselective site, the binding constant of (*S*)-CBPC was approximately fourfold greater than that of (*R*)-CBPC. The affinities of the epimers to other binding sites (nonstereoselective sites) were similar, and the affinity of (*S*)-CBPC of the nonstereoselective sites was much smaller than that for the stereoselective site.

(R)-CBPC and (S)-CBPC appeared to displace each other at all the binding sites (i.e., the binding of the epimers was competitive at the nonstereoselective sites as well as at the stereoselective site). By using site marker ligands, it was revealed that CBPC epimers may bind to site I (warfarin-binding site), but not to site II (diazepam-binding site). A binding model with an assumption of competitive interactions at all the binding sites simulated the binding characteristics of CBPC epimers quite well. Binding of ibuprofen (IB) enantiomers to HSA was studied using a chiral fluorescent derivatizing reagent, which enabled the measurement of IB enantiomers at a concentration as low as 5×10^{-8} M [33]. Scatchard analyses revealed that there were two classes of binding sites for both enantiomers. For the high-affinity site, the number of binding sites was one for both enantiomers, and the binding constant of (R)-IB was 2.3-fold greater than that of (S)-IB. The difference in the affinity at the high affinity site may result in the stereoselective binding of IB enantiomers at therapeutic concentrations. By using site marker ligands, it was confirmed that the high-affinity site of IB enantiomers is site II (diazepam-binding site). Also, significant enantiomer–enantiomer interactions were observed in the binding. The binding data were quantitatively analyzed, and a binding model with an assumption of competitive interactions only at the high-affinity site simulated the binding characteristics of IB enantiomers fairly well.

Proteins are used as natural chiral selectors in chiral stationary phases (CSPs) for separations of a variety of compounds to (5). This is a logical step because these biomacromolecules are responsible for the chiral discrimination of drugs in the body, and it is this enantioselectivity that can bring about different pharmacological and toxicological properties of enantiomers that require the separation of the enantiomers in the first place. This suggests that it would be worthwhile to utilize them in chromatographic systems as CSPs because these large molecules have many more potential points for interaction.. CSPs based on albumin are useful mainly for anionic drugs. α_1-Acid glycoprotein CSP, Enantiopac, offers good enantioresolution for a wide range of both cationic and anionic drugs. The most useful of the newer protein CSPs seemed to be that based on ovomucoid, a glycoprotein found in the whites of chicken eggs. In a comparative study of a commercial CSP based on immobilized ovomucoid protein and a second-generation α_1-acid glycoprotein CSP, there was a proliferation of protein CSP; that is, besides bovine albumin, different types of serum albumin phases have been produced (e.g., human, rat, and rabbit).

Medium-sized glycopeptides (macrocyclic antibiotics) with relative molecular masses in the range 1400 to 2100 have many stereogenic centers, semirigid macrocyclic rings, and diverse functionality that provide many opportunities for hydrogen bonding, dipole stacking, electrostatic interactions, steric repulsion, hydrophobic interaction, and to a lesser extent, n–n interactions. This can be seen from the structure of vancomycin (see Chapter 3). Besides vancomycin, CSPs based on teicoplanin, ristocetin, and avoparcin are available commercially.

7 ENZYMATIC ENANTIOSELECTIVITY

The capability of an enzyme to differentiate between the enantiomers of a chiral substrate can be influenced by kinetic and/or thermodynamic factors. Molecular chiral recognition can take place at the binding step, generating differences in affinities between the substrate and the enzyme. Thus, Michaelis–Menten kinetics can be dissimilar for each enantiomer [34]. The kinetics-based enantioselectivity is exemplified by in vitro oxidation of (R)-felodipine, which occurs more rapidly than that of (S)-felodipine [35]. The difference between the metabolism rate of (R)- and (S)-felodipine is related to a lower K_m value for the (R)-isomer. Accordingly, the enantioselectivity (defined as α) of the oxidation of (R)- and (S)-felodipine can be illustrated in terms of their relative K_m values as

$$\alpha = \frac{K_{mR}}{K_{mS}} \tag{1}$$

where K_{mR} is the K_m value for (R)-felodipine and K_{mS} is the K_m value for (S)-felodipine.

Enzymatic enantioselectivity may also take place at the catalytic stage because of variations between the energies of the activated transition states, ΔG.

7.1 Modeling Enantioselectivity

In 1933, Easson and Stedman [36] described the first three-point contact model between receptor and ligand. In their mechanism, enantioselective receptor binding resulted from the disparate binding of two enantiomers to a common site formed by a "three-point contact" model between receptor and ligand (Figure 3). In this example, enantiomer 1 was more active than enantiomer 2 because 1 was more firmly connected to receptor 3. The binding differential results from the progression of the substituents, BCD, surrounding the chirally substituted carbon atom. They form a triangular facet of the tetrahedral bond array. For 1, the sequence equals the complementary triad of binding sites on 3, B′C′D′, leading to a three-point interaction. Enantiomer 2 has a mirror-image order, DCB, and its interaction with 3 occurs at only two of the three sites on the receptor surface, bringing about a comparatively weaker ligand–receptor interaction.

The three-point interaction model was ignored for 15 years until it was invoked again by Ogston to explain the enzymatic decarboxylation of L-serine to glycine [37]. The first supposition of this model that must be considered is the assignment of three separate binding sites on the enzyme. The catalytic area of an enzyme is best portrayed as a spatial environment or a hollow space, not as a position or single interaction location. Binding sites that situate and limit the substrate in relation to the enzyme do not inevitably sit within the catalytic spot: for example, the hydrophobic binding site of chymotrypsin. If this is the case, Ogston's model includes only two interaction sites that produce binding interactions between the enzyme and the substrate. Because the binding sites are not equal, the bound substrate is accessible to the enzyme in two different spatial configurations. The

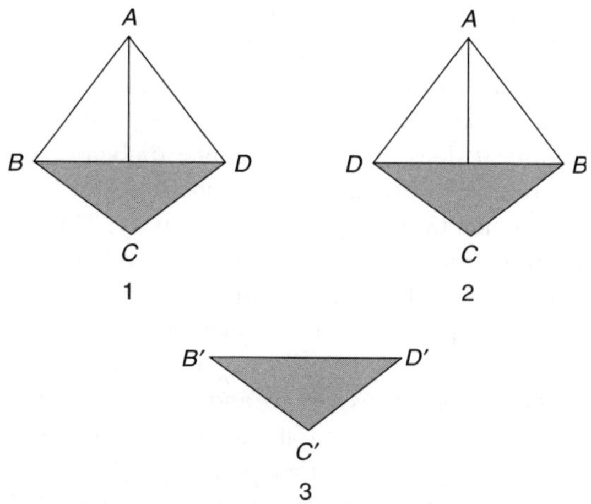

Figure 3 Easson and Stedman's chiral recognition model.

chirality of the enzyme situates the catalytic site in an arrangement such that only one of the configurations can be decarboxylated, leading to the production of just one enantiomeric compound. These observations suggest that enantioselectivity can result from a two-point directional contact and from the chirality of the enzyme. It should be noted that this model is somewhat like the rocking tetrahedron model suggested by Sokolov and Zefirov [38].

8 STEREOSELECTIVE DRUG–RECEPTOR INTERACTIONS

The stereoselective drug–receptor interaction originates from the chirality of the fundamental building blocks of receptors: the L-amino acid of proteins. This chirality is conserved and regenerated through the dominant or exclusive stereochemistry of amino acid and protein synthesis. The ratio of D/L amino acids in proteins is increased by spontaneous racemization, a process facilitated by heat, water, and other life-damaging processes.

Receptors derived from D-amino acids would be expected to have the same physical properties as the L-receptor, but the opposite stereochemistry of ligand recognition. The synthesis of the D-enzyme HIV-1 protease revealed the expected reciprocal chirality to the natural L-enzyme with respect to both substrates and inhibitors [39]. In fact, D-amino acid–containing peptides are relatively common in nature, being found in antibiotic families such as the gramicidins and lantibiotics; in the dermorphin and other peptides from the South American tree *Phyllomedusa sauvagei*, which contains D-alanine; and the Ca^{2+} channel toxins from the funnel-web spider, *Agelenopsis aperta*, which contains a posttranslationally introduced D-serine that is important both to biological activity and channel subtype selectivity.

Although chirality of interaction has long been regarded as a primary characteristic of biological recognition processes, this might not be an invariant characteristic. Insulin C-peptide, proteolytically cleaved from the insulin prohormone, has been assumed to be devoid of biological activity [40]. Although C-peptide lacks insulinlike action, it does prevent or reduce diabetes- and hyperglycemia-induced nerve and vascular dysfunction by a nonchiral process, because both the D-amino acid and retro-enantiomers have equal activity. The nature of the underlying process—receptor, ion channel formation, or other process—remains to be defined, but the underlying lack of chirality is of great interest. This suggests that stereoselectivity of drug action can be derived from both pharmacodynamic and pharmacokinetic processes and can arise from any or all of the processes involved in drug action, from transport to and from the site of action, storage in depots, interaction with binding proteins, interaction with receptors, metabolic events, and terminal transport in excretory pathways (Figure 4).

The chirality of agonist recognition has been defined around the β-OH group of catecholamines, including the selective agonist isoproterenol. With an arrangement of helixes in a circular and counterclockwise manner, current models place transmembrane domains III and V as interacting with the amine and catechol functions, respectively, of the agonist catecholamines. Asn-293 of transmembrane helix VI might be important in defining stereoselectivity because its mutational replacement in the human β_2-receptor by leucine or alanine leads to loss of agonist affinity and stereoselective recognition [41]. This was principally because the loss of affinity of the more active enantiomer and agonists that lack the β-OH group showed no loss, nor did β-antagonists.

Although D-peptides and D-proteins are attractively degradation-resistant species, the exclusivity of the natural L world means that such D species cannot interact with the mirror-image surface of the L-protein or peptide. Thus, a D-peptide might be protease-resistant but be unable to fit a receptor site because

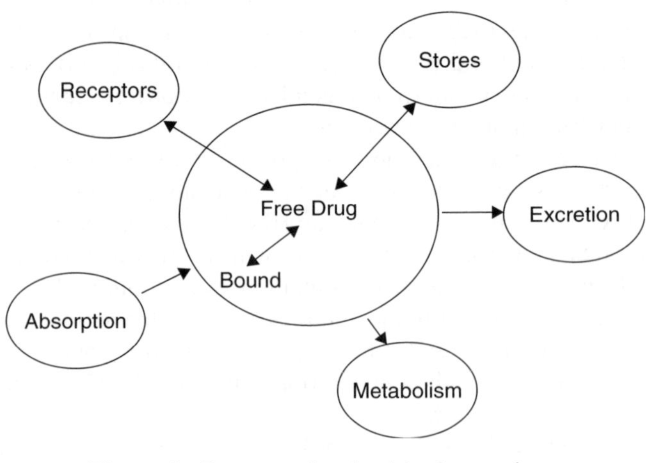

Figure 4 Processes involved in drug actions.

the amino acid side chains will be oriented in the opposite direction to that of the natural L-peptide. If, however, the D-amino acids run in the opposite sequence to that of the L-peptide, the resulting retro-enantiomer will have the same side-chain orientation, albeit with a different backbone, and have a similar topochemical surface [42].

9 PROTEIN MODIFICATIONS

Identification of protein modifications (PTMs) or mutations is extremely difficult. A fairly large number of examples of protein modifications with functional relevance in biological systems may be found in the book by Kraj and Silberring [2]. Peptide maps or peptide tags do not cover the entire protein sequence, and therefore many PTMs are not detected. This fact is particularly important, as PTMs or mutations often influence protein activity and function. PTMs are very common in many proteins, but modification is a statistical process, and such modified molecules may not be detected during batch analysis.

Although the code for the formation of proteins is laid down in the genetic information, protein products may deviate from the programmed sequence because of functionally directed processes and/or environmental influences. On the DNA level, coding errors or mutations may cause changes in the protein sequence [43]. For example, a single amino acid change from hydrophilic glutamic acid to hydrophobic in hemoglobin α is responsible for the formation of sickle-shaped red blood cells. This placement of a nonpolar residue on the outside of hemoglobin causes disadvantageous structural changes in red blood cells but is also correlated with protection against malaria.

9.1 Protein Variants

Protein variants are those with the same amino acid sequence but different folds or different carbohydrate residues. The particular series of amino acids that form a protein is known as that protein's *primary structure*. This sequence is determined by the genetic makeup of the individual. Proteins have several well-classified elements of local structure formed by intermolecular attraction; this forms the *secondary structure* of protein. They are broadly divided in two types, α-helix and β-sheet (or β-pleated sheets). Alpha-helixes are formed by coiling of protein due to attraction between the amine group of one amino acid and the carboxylic acid group of other. The coil contains about 3.6 amino acids per turn, and the alkyl group of amino acid lies outside the plane of the coil. β-Pleated sheets are formed by strong continuous hydrogen bonds over the length of a protein chain. Bonding may be parallel or antiparallel in nature. Structurally, natural silk is formed of β-pleated sheets. Usually, a protein is formed by action of both these structures in variable ratios. Coiling may also be random. As mentioned earlier, the overall three-dimensional structure of a protein is termed its *tertiary structure*. It is formed as the result of various forces, such as hydrogen bonding, disulfide bridges, hydrophobic interactions, hydrophilic interactions, and van der Waals force. When two or more different polypeptide chains cluster to form a protein, a *quaternary structure* is formed. A quarternary structure is a unique attribute of polymeric and heteromeric proteins such as hemoglobin, which consists of two α and two β peptide chains. Other protein variations that include folding and unfolding of proteins are discussed below.

9.2 Folding and Unfolding of Proteins

The first step in comparative protein modeling is fold assignment (or fold recognition), accomplished by searching databases to identify protein structures with sequences similar to the target sequence [44, p. 211]. This is usually done by multiple or pairwise sequence alignment, or if the basic sequence-to-sequence alignment does not recognize structurally characterized homologs of the protein being investigated, by *threading*, which scans the linear amino acid sequence of target protein against databases of known three-dimensional structures. The search returns the set of most compatible protein structures, which may serve as three-dimensional templates in modeling. Next, more detailed sequence–structure alignment is performed to find the most accurate fit between structurally equivalent residues in the target and template sequences. Subsequent model building is based initially on the alignment and structures of the templates, while at a later stage it might require supplementary modeling of amino acid segments that differ substantially from the template (e.g., by ab initio methods). Finally, the quality of the model is tested and refined in iteratively repeated cycles of modeling and evaluation until the structure meets specific geometric and, most important, energetic criteria.

The effectiveness of the comparative modeling approach is derived from the limited number of distinct stable folds that can be adopted by a polypeptide chain. The most abundant folds, such as an eight-stranded α/β barrel of triosephosphate isomerase, may be found in as many as 20 individual protein families.

One criterion that contributes to the selection of targets for structure determination is the novelty criterion: the likelihood of identifying a novel protein fold solving the structure of a member of a large protein family for which no structural data are available. It is estimated that directing efforts at the experimental determination of 10,000 to 20,000 representative protein structures could result in selecting information on all naturally occurring protein folds. Subsequently, databases constructed on that foundation would enable us to build structural models of remaining proteins, minimizing the amount of further experimental work required.

9.3 Coformational or Stereoelectronic Effects

Discussed below are some interesting studies relating to molecular flexibility and biological activity. Also included here are some interesting studies on solution structural studies on insulin, chiral mutagenesis of insulin, chiral inversion mechanisms and their impact on properties of enzymes, and refolding of proteins.

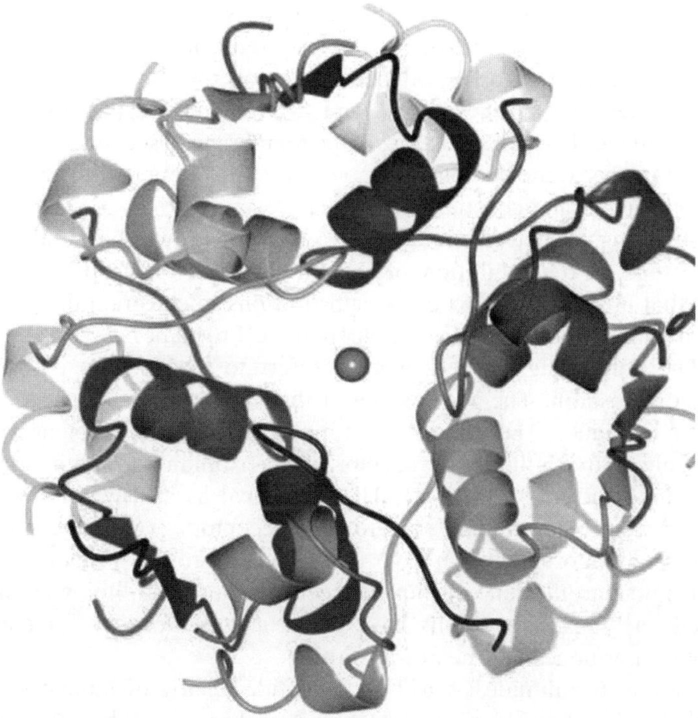

Figure 5 Insulin hexamer. Each of the 12 chains is shown in a different color. (*See insert for color representation of the figure*.)

Insulin is a polypeptide hormone (Figure 5) produced in the beta cells of the islets of Langerhans situated in the pancreas of all vertebrates. It is synthesized in vivo by a single-chain polpeptide precursor, proinsulin, which is converted by proteolysis to insulin after the native conformation has been established. Human insulin was the first commercial health care product produced by recombinant DNA technology.

The effect of molecular flexibility on biological activity have been described for soft (e.g., hGH peptides) and hard molecules (e.g., biscoclaurine-type alkaloids). These molecules have a macrocyclic structure. During molecular mechanics analysis, the minimum essential unit was observed, which affects insulin-involved fatty acid synthesis [45]. The flexibility of the molecular center can affect biological activity through the diversification of structural features when compared to two types of molecules that have a rigid (haloacetylcarbamoyl-2-nitroimidazole analogs: chiral-TXs) or flexible (bis-quaternary ammonium compounds: bis-QACs) molecular center. Center flexibility reflects the conformation occurrence in TXs and bis-QACs. A parameter (solvation-free energy: dGW) that reflects structural hydrophobicity was utilized for the molecular design of the brefeldin A analog. This hydrophobic index was found useful for conformational analysis of chiral-TXs and bis-QACs. In molecular dynamics analysis of cholesterol-dependent cytolysin (e.g., streptolysin O) and cholesterol-independent cytolysin (e.g., intermedilysin), the whole molecules moved like a bow, and different conformations were shown in every moment. In such situations, the membrane-associated 11-mer region in these cytolysins were flexible and could always interact with extramolecular factors (e.g., membrane constitution).

Peptoid analogs selectively inhibit the release of growth hormone in vivo but have no effect on the inhibition of insulin [46]. A novel approach has been presented that utilizes β-methyl chiral substitutions to constrain the peptoid side-chain conformation of a cyclic hexapeptide c[Pro6-Phe7-*d*-Trp8-Lys9-Thr10-Phe11]. The numbering in the sequence refers to the position of the residues in native somatostatin. The Pro residue in this compound is replaced with aryl-alkyl peptoid residues. The studies led to molecules that show potent binding and increased selectivity to the hsst2 receptor (weaker binding to the hsst3 and hsst5 receptors compared to L-363, 301). The biological assays, which include binding to five recombinant human somatostatin receptors, were carried out in two independent laboratories, and in vivo inhibition of growth hormone and insulin provide insight into the relationship between structure and biological activity of somatostatin analogs. The results have important implications for the study of other peptide hormones and neurotransmitters.

The α-aminosuccinimide (Asu11) octapeptide analog of human growth hormone hGH [6–13] (Leu6-Ser-Arg-Leu-Phe-Asu-Asn-Ala13) has been reported to have hypoglycemic activity, while the corresponding peptide with Asp at position 11 is inactive [47]. To determine whether this change in activity is caused by conformational and/or stereoelectronic effects, the incorporation of two different isomeric γ-lactam structures at position 11 has been investigated. One lactam structure (I) is of the type developed by Freidinger and co-workers, while the isomeric γ-lactam structure (II) represents a new type of constrained synthon for use in peptide synthesis. The chiral type II γ-lactam was synthesized via a suitably protected desoxodipeptide prepared in several ways from L-aspartic acid. The solution conformations of the [Asu11]- and the [γ-lactam11]-containing hGH [6–13] peptide analogs were investigated with the aid of two-dimensional NMR (COSY and NOESY) spectroscopy. Conformational similarities were found for these hGH [6–13] peptide analogs. For example, for all peptide analogs studied, weak NOEs were evident between the Phe10 ring protons and protons of the amino acid residues at the C-terminus. Overall, however, the NOESY NMR spectra of the [Asu11]- and the [γ-lactam11]-containing peptides related to hGH [6–13] suggest the presence of an extended structure in solution with a possible weak type II′β-turn at position 11. The extent of conformational constraint introduced into these hGH [6–13] peptide analogs by substitution of the Asu11 residue with either isomeric γ-lactam structure was reflected as differences in their hypoglycemic activity. In particular, the hGH [6–13] peptide analog derived from the new chiral type II γ-lactam exhibits both lower activity in intravenous insulin tolerance tests in vivo and weaker NOEs than the isomeric hGH [6–13] peptide analog derived from the type I γ-lactam structure. The relative change in blood glucose levels from 20 to 90 min for the racemic (*R*, *S*)-form of the type II γ-lactam compared to the control values was approximately half that of the (*S*)-stereoisomer.

The C-peptide of proinsulin is important for the biosynthesis of insulin but has long been considered to be biologically inert. Recent studies in diabetic patients have stimulated a new debate about its possible regulatory role, suggesting that it is a hormonally active peptide. Structural studies were conducted on the C-peptide using two-dimensional NMR spectroscopy, in aqueous solution. NOE patterns and chemical shifts indicate that the ensemble is a nonrandom structure and contains substructures with defined local conformations [48]. These are more clearly visible in 50% H_2O/50% 2,2,2-trifluoroethanol. The N-terminal region (residues 2 to 5) forms a type I β-turn, whereas the C-terminal region (residues 27 to 31) presents the best-defined structure of the whole molecule, including a type III′ β-turn. The C-terminal pentapeptide (EGSLQ) has been suggested to be responsible for chiral interactions with an uncharacterized receptor, probably G-protein coupled. The three central regions of the molecule (residues 9–12, 15–18, and 22–25) show tendencies to form β-bends. The proposed structure described above for the C-terminal pentapeptide is consistent with the previously postulated CA knuckle, believed to represent the active site of the C-peptide of human proinsulin.

Insulin contains a β-turn (residues B20 to B23) interposed between two receptor-binding elements, the central α-helix of the B chain (B9 to B19) and its C-terminal β-strand (B24 to B28). The turn contains conserved glycines at B20 and B23. Although insulin exhibits marked conformational variability among crystal forms, these glycines consistently maintain positive ϕ dihedral angles within a classic type I β-turn. Because the Ramachandran conformations of GlyB20 and GlyB23 are ordinarily forbidden to L-amino acids, turn architecture may contribute to structure or function. Chiral mutagenesis was employed for comparison of corresponding D- and L-Ala substitutions to investigate this turn [49]. Control substitutions are introduced at GluB21, a neighboring residue exhibiting a conventional (negative) ϕ angle. The D- and L-Ala substitutions at B23 are associated with a marked stereospecific difference in activity. Whereas the D-AlaB23 analog retains native activity, the L analog exhibits a 20-fold decrease in receptor binding. By contrast, D- and L-AlaB20 analogs each exhibit high activity. Stereospecific differences between the thermodynamic stabilities of the analogs are nonetheless more pronounced at B20 [$\Delta\Delta G(u)$ 2.0 kcal/mol] than at B23 [$\Delta\Delta G(u)$ 0.7 kcal/mol]. Control substitutions at B21 are well tolerated without significant stereospecificity. Chiral mutagenesis thus defines the complementary contributions of these conserved glycines to protein stability (GlyB20) or receptor recognition (GlyB23).

The chiral inversion mechanism and the properties of involved enzymes were investigated using rat liver subcellular fractions [50]. (±)-4-(4-Cyanoanilino)-5, 6-dihydro-7-hydroxy-7*H*-cyclopenta*d*-pyrimidine is a selective and reversible MAO-A inhibitor. The (*S*)-enantiomer of RS-8359 has been demonstrated to be inverted to the (*R*)-enantiomer after oral administration to rats. The 7-hydroxy function of RS-8359 was oxidized at least by the two different enzymes. The cytosolic enzyme oxidized enantiospecifically the (*S*)-enantiomer with NADP as a cofactor. On the other hand, the microsomal enzyme catalyzed more preferentially the oxidation of the (*S*)-enantiomer than the (*R*)-enantiomer with NAD as a cofactor. As to product enantioselectivity of reduction of the 7-keto derivative, it was found that only the alcohol bearing the (*R*)-configuration was formed by the cytosolic enzyme with NADPH and the microsomal enzyme with NADH at an almost equal rate. The reduction rate was much larger than the oxidation rate of the 7-hydroxy group. The results suggest that chiral inversion might occur via an enantioselectivity of two consecutive opposing reactions, oxidation and reduction of the keto-alcohol group. In this case, the direction of chiral inversion from the (*S*)-enantiomer to the (*R*)-enantiomer is governed by the enantiospecific reduction of intermediate 7-keto group to the alcohol with (*R*)-configuration. The enzyme responsible for the enantiospecific reduction of the 7-keto group was purified from rat liver cytosolic fractions and identified as 3α-hydroxysteroid dehydrogenase (3α-HSD) by way of a database search of peptide mass data obtained by nano-LC/MS/MS.

RhNTA protein is a new thrombolytic agent that has potential medicinal and commercial value [51]. Protein refolding is a bottleneck for large-scale production of valuable proteins expressed as inclusion bodies in *Escherichia coli*. The denatured rhNTA protein was refolded by an improved size-exclusion chromatography refolding process achieved by combining an increasing arginine gradient and a decreasing urea gradient (two gradients) with a size-exclusion chromatography refolding system. The refolding of denatured rhNTA protein showed that this method could significantly increase the activity recovery of protein at high protein concentrations. The activity recovery of 37% was obtained from the initial rhNTA protein concentration up to 20 mg/mL. After refolding by two gradient size-exclusion chromatography refolding processes, the refolded rhNTA was purified by ion exchange and affinity chromatography. The purified rhNTA protein showed one band in SDS-PAGE, and the specific activity of purified rhNTA protein was 110,000 IU/mg.

10 CONCLUSIONS

It is well known that binding of drugs to proteins can be stereoselective, which can result in different pharmacologic and toxicologic properties of enantiomers; this stresses the need to monitor the structure of proteins closely as well. A number of protein-based drugs produced by biotechnology are chiral; however, these macromolecules are not monitored thoroughly to assure that all of their components have the desired chiral structure. This may stem from the fact that it would be difficult and time consuming to carry out such analyses. Furthermore, the observed biological activity of these large protein molecules requires that they have certain finite structural components to assure primary, secondary, and tertiary structures to produce the final quaternary structure. Detailed physicochemical testing, coupled with studies of biological acivities, need to be done to assure that denaturation of proteins has not occurred during production. Also, it is necessary to assure that they are refolded properly in case any unfolding were to occur during processing. Failure to fold into the intended shape usually produces inactive proteins with different properties, including toxic prions. Several neurodegenerative and other diseases are believed to result from the accumulation of misfolded (incorrectly folded) proteins. ROA studies of unfolded and partially folded proteins provide new insight into the residual structure in denatured proteins and the aberrant behavior of proteins responsible for misfolding diseases.

REFERENCES

1. S. Ahuja. Overview of preparatory separation science and biotechnology. Presented at the AAPS Biotechnology Conference, Toronto, Ontario, Canada, June 22–25, 2008.
2. A. Kraj, J. Silberring. *Proteomics*. Wiley, Hoboken, NJ, 2008.
3. S. Ahuja. Are chiral separations of drugs really easy? Presented at the American Chemical Society Meeting, Boston, Aug. 19–23, 2007.
4. S. Ahuja. *LCGC*, 25:2, Nov. 2007.
5. W. J. Lough, I. W. Wainer, Eds. *Chirality in Natural and Applied Science*. CRC Press, Boca Raton, FL, 2000.
6. S. Ahuja. *Handbook of Bioseparations*. Academic, San Diego, CA, 2000.
7. S. Ahuja. *Chiral Separations by Chromatography*. Oxford University Press, New York, 2000.
8. A. Sadana. In *Bioseparation of Proteins*, S. Ahuja, Series Ed. Academic Press, San Diego, CA, 1998.
9. S. Ahuja. *Chiral Separations: Applications and Technology*. American Chemical Society, Washington, DC, 1997.
10. S. Ahuja. *Chiral Separations by Liquid Chromatography*. American Chemical Society, Washington, DC, 1991.
11. G. Blaschke, H. P. Kraft, K. Fickentcher, F. Koehler. *Arzneim. Forsch*., 29:1640, 1979.
12. V. U. S. Euler. *Acta Physiol. Scand*., 12:73, 1946.
13. B. F. Tullar. *J. Am. Chem. Soc*., 70:2067, 1948.
14. R. Dubos. *Louis Pasteur: Free Lance of Science*. Charles Scribner, New York, 1950.
15. T. D. Lee, C. N Yang. *Phys. Rev*., 104:254, 1956.
16. H. Latal. In *Chirality: From Weak Bosons to Alpha-Helix*, R. Janoschek, Ed. Springer-Verlag, Berlin, 1991, pp. 1–17.
17. A. C. Neville. *Animal Asymmetry*. Studies in Biology, Vol. 57. Edward Arnold, London, 1976.
18. C. Welch. See p. 285 in ref. 5.
19. L. D. Barron. See p. 53 in ref. 5.
20. G. Spach, A. Brack. In *Structure, Dynamics, and Evolution of Biological Macromolecules*, C. Helen, Ed. Reidel, Dordrecht, The Netherlands, 1983, p. 383.

21. A. J. MacDermott. See p. 23 in ref. 5.
22. G. F. Joyce, G. M. Visser, C. A. A. van Boeckel, J. van Boom, L. E. Orgel, J. van Westresen. *Nature*, 310:602, 1984.
23. L. E. Orgel. *Orig. Life Evol. Biosphere*, 26:261, 1996.
24. V. T. G. Chuang, M. Otagiri. *Chirality*, 18:159, 2006.
25. B. Alberts, A. Johnson, J. Lewis, M. Raff, K. Roberts, P. Walters, The shape and structure of proteins. In *Molecular Biology of the Cell*, 4th ed. Garland Science, New York, 2002.
26. C. Anfinsen. The formation and stabilization of protein structure. *Biochem. J.*, 128(4): 737–749, 1972.
27. J. M. Berg, J. L. Tymoczko, L. Stryer; Web content by N. D. Clarke. Protein structure and function. In *Biochemistry*. W.H. Freeman, San Francisco, 2002.
28. B. van den Berg, R. Wain, C. M. Dobson, R. J. Ellis. Macromolecular crowding perturbs protein refolding kinetics: implications for folding inside the cell. *EMBO J.*, 19(15): 3870–3875, Aug. 2000.
29. L. D. Barron, F. Zhu, L. Hecht. *Vibrat. Spectrosc.*, 42:15, Oct. 2006.
30. M. Brennan, *C & EN*, p. 27, Mar. 31, 1997.
31. R. M. Macnab, S. Aizawa. *Annu. Rev. Biophys. Bioeng.*, 13:51, 1984.
32. T. Itoh, K. Nakashima, Y. Tsuda, H. Yamada. *Chirality*, 8:201–206, 1996.
33. T. Itoh. *Chirality*, 9:643, 1997.
34. B. Testa, J. M. Mayer. *Prog. Drug Res.*, 32:249, 1988.
35. V. G. Eriksson, J. Lundahl, M. Baarnhiel, C. G. Regardh. *Drug Metab. Dispos.*, 19:889, 1991.
36. E. H. Easson, E. Stedman. Studies on the relationship between chemical constitution and physiological action: molecular dissymmetry and physiological activity. *Biochem. J.*, 27:1257, 1933.
37. A. G. Ogston. *Nature*, 29:963, 1948.
38. V. I. Sokolov, N. S. Zefirov. *Dokl. Akad. Nauk SSSR*, 319:1382, 1991.
39. R. C. deL Milton, S. C. F. Milton, S. B. H. Kent. *Science*, 256:1445, 1992.
40. D. J. Triggle. See p. 109 in ref. 5.
41. K. Weiland, H. M. Zuurmond, C. Krasel, A. P. Ijzerman, M. J. Lohse. *Proc. Natl. Acad. Sci. USA*, 93:9276, 1996.
42. M. Chorev, M. Goodman. *Trends Biotechnol.*, 13:438, 1995.
43. S. Konig. See p. 153 in ref. 2.
44. M. Rzychon. See p. 199 in ref. 2.
45. K. Ohkura. *J. Biol. Chem.*, 281:2236, 2006.
46. T. A. Tran, R. H. Mattern, M. Afargan, O. Amitay, O. Ziv, B. A. Morgan, J. E. Taylor, D. Hoyer, M. Goodman. *Int. J. Peptide Protein Res.*, 44:568, 1994.
47. N. J. Ede, I. D. Rae, M. T. Hearn. *J. Med. Chem.*, 35:3537, 1992.
48. C. E. Munte, L. Vilela, H. R. Kalbitzer, R. C. Garratt. *Biol. Pharm. Bull.*, 30:1025, 2007.
49. S. H. Nakagawa, Q. X. Hua, S. Q. Hu, W. Jia, S. Wang, P. G. Katsoyannis, M. A. Weiss. *FEBS J.*, 272:4284, 2005.
50. K. Itoh, K. Hoshino, A. Endo, T. Asakawa, K. Yamakami, C. Noji, T. Kosaka, Y. Tanaka. *Chirality*, 18:698, 2006.
51. X. Fan, D. S. Xu, B. Lu, J. Xia, D. Wei. *Biomed. Chromatogr.*, 23:257, 2009.

APPENDIX OF BIOTECHNOLOGY PRODUCTS

Activase (Alteplase) is a tissue plasminogen activator produced by recombinant DNA technology. It is a sterile, purified glycoprotein of 527 amino acids, and it is synthesized using complementary DNA (cDNA) for natural human tissue-type plasminogen activator obtained from a human melanoma cell line. The manufacturing process involves the secretion of the enzyme alteplase into the culture medium by an established mammalian cell line (Chinese hamster ovary cells) into which the cDNA for alteplase has been genetically inserted. Fermentation is carried out in a nutrient medium containing the antibiotic gentamicin, 100 mg/L. However, the presence of the antibiotic is not detectable in the final product.

Avonex (interferon β-1a) is a 166-amino acid glycoprotein with a predicted molecular weight of approximately 22,500 Da. It is produced by recombinant DNA technology using genetically engineered Chinese hamster ovary cells into which the human interferon β gene has been introduced. The amino acid sequence is identical to that of natural human interferon β. Using the World Health Organization (WHO) natural interferon β standard, Second International Standard for Interferon, Human Fibroblast (Gb-23-902-531), Avonex has a specific activity of approximately 200 million international units (IU) of antiviral activity per milligram of interferon β-1a, determined specifically by an in vitro cytopathic effect bioassay using lung carcinoma cells (A549) and encephalomyocarditis virus (ECM). Thirty micrograms of Avonex contains approximately 6 million IU of antiviral activity using this method. The activity against other standards is not known.

Cerezyme (imiglucerase for injection) is an analog of the human enzyme β-glucocerebrosidase, produced by recombinant DNA technology. β-Glucocerebrosidase (β-D-glucosyl-*N*-acylsphingosine glucohydrolase) is a lysosomal glycoprotein enzyme that catalyzes the hydrolysis of the glycolipid glucocerebroside to glucose and ceramide. It is produced by recombinant DNA technology using mammalian cell culture (Chinese hamster ovary). Purified imiglucerase is a monomeric glycoprotein of 497 amino acids, containing four N-linked glycosylation sites. Imiglucerase differs from placental glucocerebrosidase by one amino acid at position 495, where histidine is substituted for arginine. The oligosaccharide chains at the glycosylation sites have been modified to terminate in mannose sugars. The modified carbohydrate structures on imiglucerase are somewhat different from those on placental glucocerebrosidase. These mannose-terminated oligosaccharide chains of imiglucerase are specifically recognized by endocytic carbohydrate receptors on macrophages, the cells that accumulate lipid in Gaucher disease.

Epogen (erythopoietin) is a glycoprotein that stimulates red blood cell production. It is produced in the kidney and stimulates the division and differentiation of committed erythroid progenitors in the bone marrow. It is a 165-amino acid glycoprotein manufactured by recombinant DNA technology and has the same biological effects as those of endogenous erythropoietin. It has a molecular weight of 30,400 Da and is produced by mammalian cells into which the human erythropoietin gene has been introduced. The product contains the identical amino acid sequence of isolated natural erythropoietin.

Humulin is synthesized in a special non-disease-producing laboratory strain of E.coli bacteria that has been genetically altered to produce human insulin. Humulin N [human insulin (rDNA origin) isophane suspension] is a crystalline suspension of human insulin with protamine and zinc, providing an intermediate-acting insulin with a slower onset of action and a longer duration of activity (up to 24 hours) than that of regular human insulin. The time course of action of any insulin may vary considerably in different people or at different times in the same person.

Intron-A (interferon α-2b) is a recombinant interferon product. It has been classified as an α interferon and is a water-soluble protein with a molecular weight of 19,271 Da, produced by recombinant DNA techniques. It is obtained from the bacterial fermentation of a strain of *E. coli* bearing a genetically engineered plasmid containing an interferon α-2b gene from human leukocytes. The fermentation is carried out in a defined nutrient medium containing the antibiotic tetracycline hydrochloride at a concentration of 5 to 10 mg/L; the presence of this antibiotic is not detectable in the final product. The specific activity of recombinant interferon α-2b is approximately 2.6×10^8 IU/mg protein as measured by HPLC assay.

Neupogen (filgrastim) is a human granulocyte colony-stimulating factor (G-CSF), produced by recombinant DNA technology. Neupogen is the Amgen Inc. trade name for filgrastim, which has been selected as the name for recombinant methionyl human granulocyte colony-stimulating factor. It is a 175-amino acid protein manufactured by recombinant DNA technology. It is produced by *E. coli* bacteria into which has been inserted the human granulocyte colony-stimulating factor gene; it has a molecular weight of 18,800 Da. The protein has an amino acid sequence that is identical to the natural sequence predicted from human DNA sequence analysis, except for the addition of an N-terminal methionine necessary for expression in *E. coli*. The product is nonglycosylated and thus differs from G-CSF isolated from a human cell.

Nutropin [somatropin (rDNA origin)] is a human growth hormone (hGH) produced by recombinant DNA technology. Nutropin has 191-amino acid residues and a molecular weight of 22,125 Da. The amino acid sequence of the product is identical to that of pituitary-derived human growth hormone. The protein is synthesized by a specific laboratory strain of *E. coli* as a precursor consisting of the rhGH molecule preceded by the secretion signal from an *E. coli* protein. This precursor is directed to the plasma membrane of the cell. The signal sequence is removed and the native protein is secreted into the periplasm so that the protein is folded appropriately as it is synthesized. Nutropin's biological potency is determined using a cell proliferation bioassay.

Procrit (epoetin α) is a 165-amino acid glycoprotein manufactured by recombinant DNA technology and has the same biological effects as endogenous erythropoietin. It has a molecular weight of 30,400 Da and is produced by mammalian cells into which the human erythropoietin gene has been introduced. The product contains the identical amino acid sequence of isolated natural erythropoietin.

Proleukin is a highly purified human recombinant interleukin-2 product, with a molecular weight of approximately 15,300 Da. The chemical name is des-alanyl-1, serine-125 human interleukin-2. A lymphokine, it is produced by recombinant DNA technology using a genetically engineered *E. coli* strain containing an analog of the human interleukin-2 gene. Genetic engineering techniques were used to modify the human IL-2 gene, and the resulting expression clone encodes a modified human interleukin-2. This recombinant form differs from native interleukin-2 in the following ways: (1) It is not glycosylated because it is derived from *E. coli*; (2) the molecule has no N-terminal alanine; the codon for this amino acid was deleted during the genetic engineering procedure; (3) the molecule has serine substituted for cysteine at amino acid position 125 (this was accomplished by site-specific manipulation during the genetic engineering procedure); and (4) the aggregation state of proleukin is likely to be different from that of native interleukin-2.

Protropin (somatrem) is a polypeptide hormone produced by recombinant DNA technology. Protropin has 192-amino acid residues and a molecular weight of about 22,000 Da. The product contains the identical sequence of 191 amino acids constituting pituitary-derived human growth hormone plus an additional amino acid, methionine, on the N-terminus of the molecule. Protropin is synthesized in a special laboratory strain of *E. coli* bacteria which has been modified by the addition of the gene for human growth hormone production. Its biological potency is determined by measuring the increase in body weight induced in hypophysectomized rats.

Pulmozyme (dornase α) is recombinant human deoxyribonuclease I (rhDNase), an enzyme that selectively cleaves DNA. Pulmozyme is administered by inhalation of an aerosol mist produced by a compressed air–driven nebulizer system. Launched in 1994, Pulmozyme was the first new therapeutic drug for the management of cystic fibrosis in over 30 years. Daily administration of Pulmozyme in conjunction with standard therapies is indicated in the management of cystic fibrosis patients to improve pulmonary function. This drug has also been shown to reduce the risk of respiratory tract infections requiring parenteral antibiotics. The proposed mechanism of action in vitro indicates that Pulmozyme cleaves extracellular DNA in the mucus of cystic fibrosis patients, reducing the adhesiveness and viscoelasticity of the mucus. Dornase α activity can be affected by excipients such as benzalkonium chloride and disodium edetate.

INDEX

The "n" after a page number refers to an entry that appears in a footnote on that page.

Chiral Separation Methods for Pharmaceutical and Biotechnological Products, Edited by Satinder Ahuja